Excitable Cells in Tissue Culture

Excitable Cells in Tissue Culture

Edited by

PHILLIP G. NELSON
National Institutes of Health
Bethesda, Maryland

and

MELVYN LIEBERMAN
Duke University Medical Center
Durham, North Carolina

PLENUM PRESS • NEW YORK AND LONDON

Library of Congress Cataloging in Publication Data

Main entry under title:

Excitable cells in tissue culture.

Includes index.

1. Neurons. 2. Muscle cells. 3. Tissue culture. 4. Electrophysiology. I. Nelson, Phillip Gillard, 1931- II. Lieberman, Melvyn. [DNLM: 1. Tissue culture—Collected works. 2. Neurons—Physiology—Collected works. 3. Muscles—Physiology—Collected works. 4. Electrophysiology—Collected works. WS530 E96]

QH585.E95 599.08'76 80-8106

ISBN 0-306-40516-4

A Division of Plenum Publishing Corporation
227 West 17th Street, New York, N.Y. 10011

Printed in the United States of America

Contributors

JAMES F. AITON • Department of Physiology, Duke University Medical Center, Durham, North Carolina 27710. *Present address:* Department of Physiology, University of St. Andrews, St. Andrews, Scotland

JEFFERY L. BARKER • Laboratory of Neurophysiology, National Institute of Neurological and Communicative Disorders and Stroke, National Institutes of Health, Bethesda, Maryland 20205

RICHARD P. BUNGE • Department of Anatomy and Neurobiology, Washington University School of Medicine, St. Louis, Missouri 63130

HAROLD BURTON • Department of Anatomy and Neurobiology, Washington University School of Medicine, St. Louis, Missouri 63130

WILLIAM A. CATTERALL • Department of Pharmacology, University of Washington, Seattle, Washington 98195

THEODORE R. COLBURN • Research Services Branch, National Institute of Mental Health, National Institutes of Health, Bethesda, Maryland 20205

LISA EBIHARA • Department of Physiology, Duke University Medical Center, Durham, North Carolina 27710

C. RUSSELL HORRES • Department of Physiology, Duke University Medical Center, Durham, North Carolina 27710

EDWARD A. JOHNSON • Department of Physiology, Duke University Medical Center, Durham, North Carolina 27710

YOSHIAKI KIDOKORO • The Salk Institute, San Diego, California 92138

YOSEF KIMHI • Department of Neurobiology, The Weizmann Institute of Science, Rehovot, Israel

GLENN A. LANGER • Departments of Medicine and Physiology and the American Heart Association, Greater Los Angeles Affiliate Cardiovascular Research Laboratories, University of California at Los Angeles, Center for the Health Sciences, Los Angeles, California 90024

HAROLD LECAR • Laboratory of Biophysics, Intramural Research Program, National Institute of Neurological and Communicative Disorders and Stroke, National Institutes of Health, Bethesda, Maryland 20205

MELVYN LIEBERMAN • Department of Physiology, Duke University Medical Center, Durham, North Carolina 27710

ROBERT L. MACDONALD • Department of Neurology, University of Michigan, Ann Arbor, Michigan 48109

ELAINE A. NEALE • Laboratory of Developmental Neurobiology, National Institute of Child Health and Human Development, National Institutes of Health, Bethesda, Maryland 20205

PHILLIP G. NELSON • Laboratory of Developmental Neurobiology, National Institute of Child Health and Human Development, National Institutes of Health, Bethesda, Maryland 20205

ROBERT D. PURVES • Department of Anatomy and Embryology, University College London, London WC1E 6BT, England

FREDERICK SACHS • Department of Pharmacology and Therapeutics, State University of New York, Buffalo, New York 14214

NORIKAZU SHIGETO • Department of Physiology, Duke University Medical Center, Durham, North Carolina 27710. *Present address:* Division of Cardiology, Hiroshima National Chest Hospital, Hiroshima, Japan

BRUCE M. SMITH • Research Services Branch, National Institute of Mental Health, National Institutes of Health, Bethesda, Maryland 20205

THOMAS G. SMITH, Jr. • Laboratory of Neurophysiology, National Institute of Neurological and Communicative Disorders and Stroke, National Institutes of Health, Bethesda, Maryland 20205

ILAN SPECTOR • Laboratory of Biochemical Genetics, National Heart, Lung, and Blood Institute, National Institutes of Health, Bethesda, Maryland 20205

Preface

The tissue culture approach to the study of membrane properties of excitable cells has progressed beyond the technical problems of culture methodology. Recent developments have fostered substantive contributions in research concerned with the physiology, pharmacology, and biophysics of cell membranes in tissue culture. The scope of this volume is related to the application of tissue culture methodology to developmental processes and cellular mechanisms of electrical and chemical excitability. The major emphasis will be on the body of new biological information made available by the analytic possibilities inherent in the tissue culture systems.

Naturally occurring preparations of excitable cells are frequently of sufficient morphological complexity to compromise the analysis of the data obtained from them. Some of the limitations associated with dissected preparations have to do with the direct visualization of and access to the cell(s) in question and maintenance of steady-state conditions for prolonged periods of time. Since preparations in tissue culture can circumvent these problems, it is feasible to analyze the properties of identifiable cells, grown either singly or in prescribed geometries, as well as to follow the development of cellular interactions. A crucial consideration in the use of cultured preparations is that they must faithfully capture the phenomenon of interest to the investigator. This and other potential limitations on the methodology are of necessary concern in the present volume.

The twelve chapters of this book are divided into four sections: The first (Chapters 1–3) deals with peripheral and central neuronal systems, the second (Chapters 4 and 5) discusses certain biophysical techniques as applied to tissue culture systems, the third (Chapters 6–8) concerns the analysis of clonal systems, and the fourth (Chapters 9–12) has four chapters dealing with skeletal, smooth, and cardiac muscle.

The control of the cellular environment during differentiation and development has been exploited in peripheral synaptic systems with particular success. Regulation of the biochemical nature of sympathetic neurons (their cholinergic or catecholaminergic determination) has been shown to be dependent on the interaction among certain diffusible molecules produced by nonneuronal cells and electrical activity in the neurons. In both the sympathetic neurons and neuromuscular systems, the morphological and physiological events that occur during synapse formation have been achieved with a high degree of resolution. Molecular and membrane processes involved in the synthesis and distribution of acetylcholine receptors have been extensively analyzed in nerve and muscle cultures. Some of these areas of research are not dealt with in detail in the present volume; recent reviews are available (Fambrough, *Physiol. Rev.* **59,** 1979; Fischbach *et al., Pharmacol. Rev.* **30,** 1978).

The tissue culture technique has made accessible a variety of central neurons for biophysical and morphological study. Central synaptic mechanisms can be analyzed in some detail, and morphophysiological correlations clearly established. The relationship between synaptic responses and the responses elicited by the application of putative transmitter agents can be examined by a variety of techniques. Voltage clamp techniques relevant to tissue cultured preparations and the recent striking advances in noise and single-channel analysis of voltage sensitive and chemically activated ionic mechanisms are described. The prospect of applying these techniques to a broad range of hitherto inaccessible cells is particularly exciting.

Clonal systems have been utilized with two rather different approaches over the last ten years. Genetic studies have been directed at either the common or mutually exclusive expression and regulation of different neurobiologically relevant properties such as transmitter-related enzymes or receptors. Clonal preparations have also been the subject of biophysical analyses of such variables as the voltage-sensitive sodium and/or calcium channels.

Skeletal, smooth, and cardiac muscle cells in culture have provided suitable model systems for applying electrophysiological and radiotracer techniques to study their membrane properties. Microelectrodes have been used to monitor the development of electrogenicity in skeletal muscle cell lines as well as primary cultures of skeletal myoblasts. The effect of innervation on skeletal muscle action potentials and their sensitivity to acetylcholine and tetrodotoxin have also been explored. Although cultured smooth muscle cells from a variety of tissues are electrically excitable, the results are somewhat preliminary because of the difficulties in recording from these cells. Iontophoretic application of neurotransmitters and drugs have been more successful in preparations of smooth muscle cells, presumably because of the favorable tissue geometry. Cardiac cells have proven to be readily amenable to ^{45}Ca and ^{42}K tracer studies when grown either directly on the surface of a scintillation disk or around nylon monofilament. The role of the sarcolemma–glycocalyx complex in the regulation of calcium exchange in cultured cardiac cells is being pursued using a new membrane preparation from monolayer cultures. Linear and spherical preparations of heart cells have been successfully used in studies of cable and membrane properties, respectively.

The present volume is a heterogeneous compilation of recent results and directions in the field of tissue culture as applied to excitable membrane systems. The relatively minor emphasis that has been placed on an exhaustive review of the literature is intended to reflect the present extremely active and result-oriented state of tissue-culture-based research. A broad variety of problems ranging from the determinants of neuronal differentiation to the molecular structure of the sodium channel are under intense study. The results to date are clearly substantial and point to many new areas that have yet to be explored. We hope these chapters capture some of the excitement and optimism in the field.

ACKNOWLEDGMENTS. It is a pleasure to acknowledge the skillful preparation of the Index by Patricia Spivey and the editorial assistance of Rita Lohse.

Phillip G. Nelson
Melvyn Lieberman

Contents

Chapter 1

The Expression of Cholinergic and Adrenergic Properties by Autonomic Neurons in Tissue Culture

Harold Burton and Richard P. Bunge

Chapter 4

Voltage Clamp Techniques Applied to Cultured Skeletal Muscle and Spinal Neurons

Thomas G. Smith, Jr., Jeffery L. Barker, Bruce M. Smith, and Theodore R. Colburn

Chapter 5

Membrane Noise Analysis

Harold Lecar and Frederick Sachs

Chapter 8

Studies of Voltage-Sensitive Sodium Channels in Cultured Cells Using Ion-Flux and Ligand-Binding Methods

William A. Catterall

Chapter 9

Electrophysiological Properties of Developing Skeletal Muscle Cells in Culture

Yoshiaki Kidokoro

The Expression of Cholinergic and Adrenergic Properties by Autonomic Neurons in Tissue Culture

HAROLD BURTON and RICHARD P. BUNGE

1. INTRODUCTION

Both circumstance and paradox dictate that this chapter should center primarily on the autonomic adrenergic (sympathetic) neuron in tissue culture. The particular circumstance is the availability of a well-defined trophic factor, nerve growth factor (NGF), for this neuron, allowing its long-term culture isolated from all other tissues; no comparable factor has been characterized for neurons from other portions of the autonomic nervous system. The paradox is that adrenergic neurons in tissue culture have provided much new information on the development of cholinergic mechanisms: when taken for culture from the perinatal animal, the adrenergic neuron has the unexpected property of developing the ability to synthesize and release acetylcholine and to establish cholinergic interactions with adjacent neurons of its own type. Because of the intense interest in exploring the nature and mechanism of this transmitter shift, a substantial portion of recent work on the adrenergic neuron in culture is centered on these problems. The sympathetic neuron may also provide adrenergic innervation to a variety of target tissues in culture and thus has been useful for the study of both adrenergic and cholinergic mechanisms.

Because a well-defined trophic factor for parasympathetic neurons is not yet available, the long-term culture of pure populations of this neuron type in the dissociated state has presented substantial difficulties. As we will discuss in

HAROLD BURTON and RICHARD P. BUNGE • Department of Anatomy and Neurobiology, Washington University School of Medicine, St. Louis, Missouri 63130.

Section 3, however, some success has been achieved when parasympathetic neurons are combined with target tissues (or with medium conditioned by these tissues); they are thus capable of expressing their cholinergic properties in a manner available for detailed physiological analysis. Tissue culture work with the third major part of the autonomic nervous system, the enteric plexus, has been recently revived and will also be discussed.

2. SOME HISTORICAL ASPECTS OF THE CULTURE OF AUTONOMIC NEURONS

The study of autonomic nerve tissue has played an important role in the development of culture techniques. Several years after Harrison's (1910) seminal report established the basic techniques of nerve tissue culture, Lewis and Lewis (1912) reported impressive axonal outgrowth from explants of chick enteric plexus neurons. Szantroch (1933) and Levi and Delorenzi (1934) used cultures of chick autonomic ganglia in studies of nerve fiber growth and interrelationships in culture. These studies confirmed the active forward movement of the growth cone and established its activity in pinocytosis. Unfortunately, these early studies became enmeshed with the question of whether nerve fibers broke apart and reanastomosed, a question the limited optical resolution available at that time was not able to resolve. Relatively long-term cultures of autonomic ganglia were first obtained in the 1940s, and in 1947 Murray and Stout successfully cultured fragments of adult human autonomic ganglia obtained during surgical sympathectomy. Most importantly, Levi-Montalcini *et al.* (1954) utilized sympathetic ganglia (as well as spinal ganglia) in the first demonstration *in vitro* of the ability of a factor from mouse sarcoma to engender massive neurite outgrowth from ganglion explants in culture and thus established the diffusible nature of the factor now known as NGF. Later experiments by Levi-Montalcini and Angeletti (1963) established the dependence of dissociated neurons of chick autonomic spinal ganglia on the presence of NGF. The first physiological work on peripheral neurons in culture was, however, to center on sensory rather than autonomic neurons (see review by Crain, 1976).

It was not until relatively recently that the culture of sympathetic neurons after enzymatic or mechanical dissociation (allowing subsequent intracellular physiological recordings) was extensively undertaken. Of particular importance was the development of methods for culturing neurons after dissociation of the neonatal rat superior cervical ganglion (Bray, 1970). After dissociation and a period of growth and interaction in culture, these neurons permit stable microelectrode penetration and detailed physiological study. Because these neurons come from rat, physiological results of work in culture can be directly compared to an extensive literature on the physiology of this ganglion *in vivo*

(Nishi, 1974). It is, therefore, possible to provide during the course of this review a detailed comparison of the basic properties of the rat adrenergic neuron *in vivo* and *in vitro.*

3. THE PARASYMPATHETIC NEURON IN CULTURE

Successful culture of parasympathetic ganglia has been reported recently by several laboratories. Coughlin (1975) has grown the mouse submandibular ganglion with the anlage of the submandibular gland. The neurons only produced a substantial neuritic network when grown together with gland tissue. This observation indicated the necessity of trophic support from target tissues for the successful culture of this neuronal species. No physiological studies were undertaken. Nishi and Berg (1977) have demonstrated that skeletal muscle myotubes provide substantial trophic support for dissociated ciliary ganglion neurons, which survive only a few days in culture without added target tissue. Neurons destined to be lost in the course of normal development were "rescued" from this fate if grown with skeletal muscle prior to the period of naturally occurring cell death (known to occur in this ganglion between 8 and 14 days *in ovo*). More recently, Nishi and Berg (1979) have demonstrated that medium conditioned with heart or skeletal muscle (or in the presence of embryo extract) permits long-term survival and the development of choline acetyltransferase in cultures of chick ciliary ganglion neurons grown without supporting cells.

Hooisma *et al.* (1975) successfully maintained the chick ciliary ganglion in culture with embryonic chick striated leg muscle. Intracellular recordings from the muscle cells established the presence of excitatory junctional potentials which were never recorded in myotubes in the absence of a ciliary ganglion. These potentials were completely blocked by D-tubocurarine at a concentration of 1 μg/ml but not by atropine, indicating that avian ciliary neurons establish nicotinic cholinergic junctions with somatic striated muscle. It should be pointed out that *in vivo* a portion of these chick ciliary parasympathetic neurons (unlike those in the mammal) normally innervate fast striated muscle of the ciliary body and sphincter iridis.

Studying explant cultures of greater maturity (up to 3 weeks *in vivo*), Betz (1976) noted considerable spontaneous excitatory activity and raised the question of whether this might be generated by cholinergic interaction among neurons. In fact, physiological evidence of synaptic interaction between parasympathetic neurons in culture has not been demonstrated (in contrast to mammalian adrenergic neurons, as discussed in Section 8). This is somewhat surprising because chick ciliary ganglion neurons are known to be sensitive to acetylcholine, at least directly after excision (Betz, 1976). Chick ciliary ganglion neurons in whole ganglia *in vivo* are not known to form synapses among themselves when their input is removed, but profiles of synaptic nature have been observed in dissociated cultures of the avian ciliary ganglion; in these same cultures, no evidence for physiological interaction could be obtained (Tuttle *et al.*, 1978). In certain species of amphibian, parasympathetic neurons normally form synapses among themselves (Roper, 1976a,b); this is also known to occur after denervation (Sargent and Dennis, 1977).

It is now also known from transplantation experiments in the embryo that cells destined to become ciliary ganglion neurons are capable of altering their course of differentiation to express adrenergic characteristics (LeDouarin *et al.*, 1978). In contrast, no instances have been reported where neurons of the ciliary ganglion in tissue culture have expressed other than cholinergic properties.

4. NEURONS OF THE ENTERIC PLEXUS IN CULTURE

The culture of enteric plexus neurons from the chick gut was first undertaken over 65 years ago with notable success (Lewis and Lewis, 1912). The paucity of subsequent attempts may derive from the diffuse anatomical disposition of the neuronal cell bodies and from the general recognition that this complex plexus is not representative of autonomic ganglia in general. In fact, these aspects may be advantages rather than disadvantages for *in vitro* studies.

Several recent reports indicate the possible uses of enteric plexus cultures. Dreyfus and her colleagues (1977a,b) have utilized organotypic cultures of fragments of intestine wall to demonstrate the uptake and synthesis of serotonin by neurons intrinsic to the gut wall. This approach eliminates the possibility of uptake mechanisms deriving from axons extrinsic to the plexus.

More recently, Jessen *et al.* (1978) have described methods of culturing Auerbach plexus neurons from guinea pig colon in cultures that are completely free of smooth muscle components. After enzymatic treatment, small aggregates of cells are obtained that are complexly interconnected by neuritic strands. Neuritic outgrowth from the explanted tissue is largely confined to a cellular substrate that is established as glialike cells proliferate from the explant margins. The plexus provides a preparation in which neurons are well dispersed, permitting intracellular recording under direct visualization. Most impaled cells demonstrated action potentials with small and short-lived afterhyperpolarizations. A small number of cells showed action potentials followed by intense prolonged afterhyperpolarization. Neurons with these two basic physiological characteristics have been demonstrated in Auerbach's plexus *in situ*. The pharmacology of these synaptic interactions has not yet been reported.

Jessen *et al.* (1978) cite literature indicating the remarkable complexity expressed by the variety of neurons present in Auerbach's plexus. On the basis of anatomical studies, nine different types of nerve cell bodies have been defined, as have eight morphologically different axon terminals; on the basis of extracellular recordings, six neuronal types have been noted. Pharmacological studies indicate cholinergic, adrenergic, and "purinergic" nerve components; recent studies suggest that several peptides (enkephalin, vasoactive intestinal peptide, somatostatin, substance P, and bradykinin) may act as neurotransmitters or neuromodulators in Auerbach's plexus. Our knowledge concerning how these agents may be utilized in effecting the integrative activity of the gut is at present rudimentary; in attacking this problem tissue culture approaches would seem particularly advantageous.

5. THE SYMPATHETIC NEURON IN CULTURE: BASIC PROPERTIES

Many properties of sympathetic ganglion neurons are expressed *in vitro* despite the altered environment in which these cells are maintained. We will primarily examine the expression of some of these characteristics by the principal neuron from rat superior cervical ganglia (SCGN) in dissociated cell cultures with comparison to adrenergic neurons cultured from other sources. The small intensely fluorescent cells will receive little attention because they apparently survive well only in explant cultures where physiological analysis has not been systematically undertaken.

5.1. Morphology

The growth of the SCGN in culture has been described in detail by a number of investigators (Bray, 1973; M. Bunge, 1973; R. Bunge *et al.*, 1974; Chun and Patterson, 1977a–c; Johnson *et al.*, 1980a,b; among others). In brief, individual or small groups of neurons settle onto a collagen substrate and usually begin to send out processes within 24 hr. These fibers ramify over the culture dish to form an elaborate and interlaced random network (Fig. 1). As the neurons mature in culture, the general morphology of the neuron is similar to that of the principal sympathetic ganglion neuron *in vivo* (Eränkö, 1972; Matthews, 1974). The large cell body has a flattened round or fusiform shape, contains a prominent nucleus, and has one or more organelle-filled large proximal processes (Fig. 2). Studies by Wakshull *et al.* (1979a,b) provide the following observations. The somal diameters of cells in cultures prepared from embryonic or perinatal sympathetic ganglia are smaller (average 35.3 $\pm$ 0.88 μm after 3–4 weeks *in vitro*) than neurons *in vivo* of comparable total age. In contrast, the somal diameters of SCGN in cultures prepared from postnatal or adult rats are indistinguishable from those of *in vivo* cells (average 48.9 ± 2.9 μm). SCGN grown in culture from postnatal tissues also regenerate a greater number of large proximal processes (4.2 ± 0.21 processes vs. 2.3 ± 0.16), so that the overall form of these cells is more comparable to sympathetic neurons in the animal (Fig. 3). However, a considerable degree of variability occurs in the complexity of the distribution of processes elaborated by all ages of SCGN in culture. In addition, changes in the substrate over which these cells grow alter the spatial alignment of processes. Despite the problems caused by the physical constraints of the culture dish and the absence of a "normal" preganglionic connection, the SCGN in dissociated cell cultures present a morphological appearance similar to their *in vivo* counterparts and neurons in general.

Given the complexity of the networks formed *in vitro* and the randomness with which processes from a single cell are distributed, it has been difficult in the living culture to determine which of the neuritic processes of neurons in culture are dendrites and which are axons. Previous attempts to unravel this problem by using electrophysiological criteria on a variety of neuronal types grown in culture have not been successful because both large and small processes have been found to fire all-or-none action potentials (Hild and Tasaki, 1962; Okun, 1972; Varon and Raiborn, 1971; R. Bunge *et al.,* 1974). The recent application of intracellular labeling with horseradish peroxidase in culture (Neale *et al.,* 1978; Wakshull *et al.,* 1979a,b) has permitted visualization of more of the morphology of individual neurons, including processes of <1 μm that course with many other similar processes in neuritic bundles across the culture dish. The thick, tapering proximal processes of SCGN in culture are probably dendrites since they can be stained with thionin (Fig. 2) and have been shown to contain ribosomes and other cytoplasmic organelles in electron micrographs (Landis, 1977; Rees *et al.,* 1976). Synapses are also common on these processes (Landis, 1977; Rees *et al.,* 1976). The greatest number and most extensive processes are thin and nontapering. These may reach several millimeters in length as they ramify and branch repeatedly; they contact and interact with other processes and neurons in the culture dish. Some of these thin processes are smooth, whereas others develop varicosities (Fig. 4) along their entire length. These varicosities reveal formaldehyde-induced fluorescence, and electron micrographs show that they contain clusters of small dense-cored vesicles (Chamley *et al.,* 1972; Landis, 1977; Johnson *et al.,* 1980a,b). It is likely that the varicose processes are axonal, since these join cells that are synaptically coupled (Wakshull *et al.,* 1979a) and resemble the ground plexus of adrenergic endings in the animal (Burnstock and Costa, 1975; Jacobowitz, 1974; Matthews, 1974). The function of the nonvaricose thin processes is unclear. They contain few synapticlike vesicles in electron micrographs, few cytoplasmic organelles are present, and they do not receive synapses. Defining all of the thin processes as axons seems inappropriate because this would imply that a cell in culture issues dozens of axons from all over its cell body and proximal dendrites. Some of these smooth processes may be axons that will

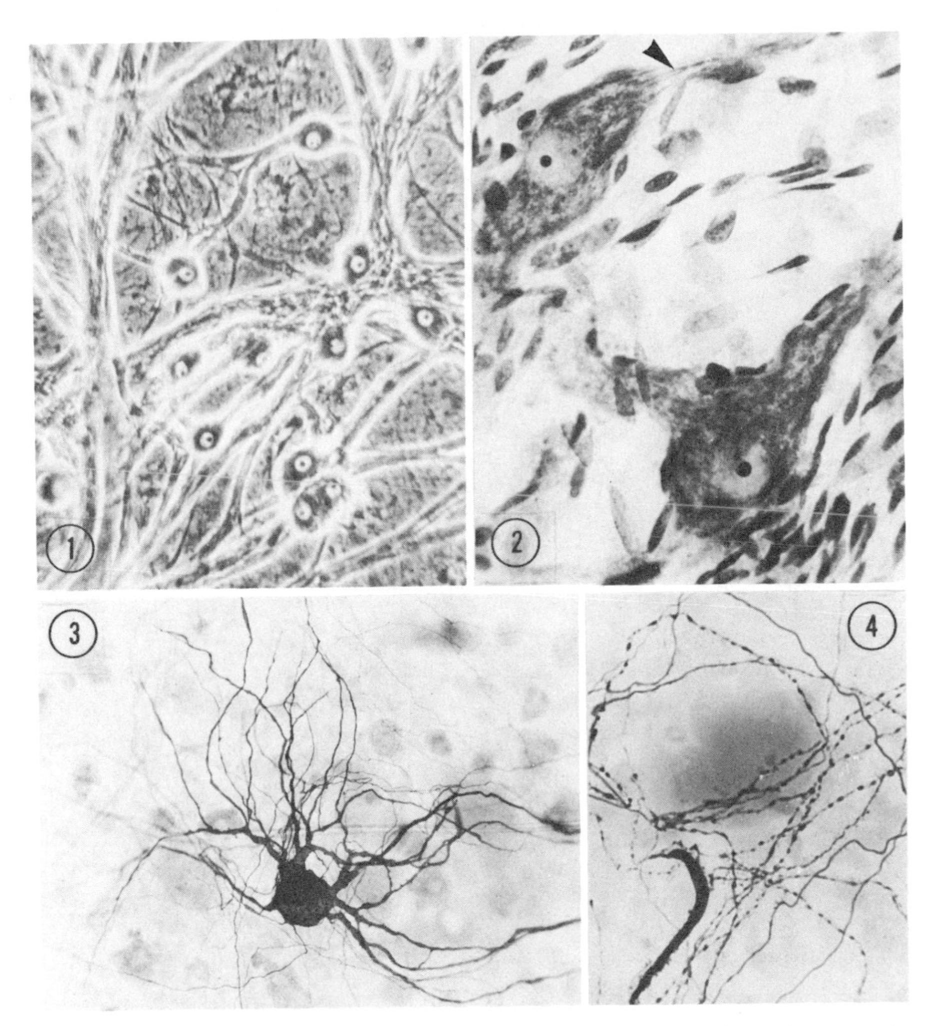

FIGURE 1. Dissociated neuron from the neonatal rat superior cervical ganglion grown in culture in the presence of supporting cells. A rich network of neurites form in culture between the neurons. The neurons are similar in size and type. Phase micrograph of living culture, 7 weeks in culture. 200 ×, reproduced at 82%.

FIGURE 2. A thionin-stained preparation of cultured neurons similar to those illustrated in Fig. 1. Note that proximal processes contain Nissl bodies, often in a linear array as at the arrowhead. Light micrograph of neurons after 6 weeks in culture. 750 ×, reproduced at 82%.

FIGURE 3. Photomicrograph of adult SCGN injected intracellularly with horseradish peroxidase. Note the two types of processes: large and tapering and thin, nontapering. Age *in vivo* 70 days; age *in vitro* 27 days. 170 ×, reproduced at 82%. (From Wakshull *et al.*, 1979a.)

FIGURE 4. Photomicrograph showing distribution of neuronal processes from an HRP-labeled SCGN over another neuron. Note the large number of varicosities on the processes that pass over the unlabeled neuron. Culture prepared from 28-day-old pup and grown *in vitro* for 34 days. 444 ×, reproduced at 82%. (From Wakshull *et al.*, 1979a.)

develop varicosities after making synaptic contact with some target (Chamley *et al.*, 1973). Some of them may be abortive sprouts that are destined to be resorbed.

5.2. Biophysical Membrane Properties

5.2.1. Passive Properties

A close correlation exists between the passive electrical properties (input resistance, membrane capacitance, and time constant) of sympathetic neurons grown *in vitro* (Ko, 1975; R. Bunge *et al.*, 1974; O'Lague *et al.*, 1978a) and *in vivo* (see Nishi, 1974).

The resting and action potential amplitudes recorded from SCGN in culture are also similar in size to those reported from the animal, especially if only the results from the most stable penetrations are considered (Perri *et al.*, 1970; R. Bunge *et al.*, 1974; O'Lague *et al.*, 1978a). The resting potentials of dissociated SCGN observed by various investigators have ranged between 20 and 90 mV, and average values have been between 40 and 65 mV. The values of 70–75 mV probably reflect the true undisturbed resting potential (O'Lague *et al.*, 1978a). The action potential amplitudes reported vary widely, but values greater than 75 mV may be most representative.

5.2.2. Active Properties

The action potentials of mammalian sympathetic neurons have been shown to be dependent on Na^+ conductance changes, since the responses can be blocked by tetrodotoxin (TTX) *in vivo* (Tashiro and Nishi, 1972; Hirst and Spence, 1973) and *in vitro* (Ko *et al.*, 1976a; O'Lague *et al.*, 1978a). However, in the presence of TTX and TEA, additional depolarization of 10–15 mV over the threshold to elicit an action potential will evoke another all-or-none regenerative response whose amplitude can be directly related to the external Ca^{2+} concentration; this second Ca^{2+} response can be blocked by Co^{2+} in adult rat and rabbit superior cervical ganglion neurons *in vivo* (Tashiro and Nishi, 1972; Yarowsky and McAfee, 1977; Horn and McAfee, 1978), in neurons within guinea pig's Auerbach plexus (Hirst and Spence, 1973), in bullfrog sympathetic neurons (Koketsu and Nishi, 1969), and *in vitro* in dissociated SCGN grown from newborn rats (O'Lague *et al.*, 1978a). The presence of a calcium action potential can sometimes be suspected in normal medium because of the addition of a secondary repolarization phase that appears as a slope change or hump during the falling phase of the action potential (see Fig. 4A2 in Bunge *et al.*, 1974) and because of a prolonged or late afterhyperpolarization (LAH) lasting $>$ 300 msec. This LAH was previously shown to result from a long-lasting increase in K^+ conductance in frog sympathetic ganglion cells (Blackman *et al.*, 1963); it has recently been demonstrated that this increase in potassium permeability is brought about by the Ca^{2+} action potential (Yarowsky and McAfee, 1977; Horn and McAfee, 1978; O'Lague *et al.*, 1978a). A calcium action potential has also been observed in cultures from chick (Dichter and Fischbach, 1977) and fetal mouse (Matsuda *et al.*, 1978) sensory ganglion neurons (DRG). Three classes of action potentials were noted in comparable age explant cultures of mouse DRG: (1) calcium-dependent action potentials that were TTX-resistant in Na-free medium; (2) action potentials that were insensitive to TTX but failed without sodium; and (3) TTX-sensitive, Na-dependent action potentials. Matsuda *et al.* (1978) suggest that these different classes of response may reflect stages in the development of Na^+ channels in immature neurons and that Ca^{2+} spikes may, therefore, only be seen in developing neurons. However, Dichter and Fischbach (1977) argue, by citing previous work on amphibian sensory neurons, that all sensory neurons can generate Ca^{2+} action potentials. Studies on adult sympathetic (Tashiro and Nishi, 1972; Horn and McAfee, 1977; Yarowsky and McAfee, 1978) and parasympathetic (Hirst and Spence, 1973) neurons imply that all neural crest derivative neurons have the capability for regenerative calcium currents, even in the adult. A study of the parameters affecting Ca^{2+} currents as a function of the actual age of a neuron remains to be done (O'Lague *et al.*, 1978a).

6. INPUT TO SYMPATHETIC NEURONS IN CULTURE

6.1. Mechanisms of Neurotransmission *in Vivo*

The pharmacology of autonomic ganglion neurotransmission in the animal is complex, and summarizing the work on this subject is beyond the scope of this chapter (see Libet, 1975; Nishi, 1974; Phillis, 1970). Of immediate interest is the correlation between the responses of sympathetic cells in culture and the multilevel response of these neurons to preganglionic stimulation in the intact ganglion. Previous studies (reviewed in Nishi, 1974) have demonstrated a short latency, fast rising excitatory response following preganglionic stimulation; this EPSP is evoked by the release of acetylcholine (ACh) from preganglionic terminals onto nicotinic, cholinergic receptors. In addition, a slow, long-lasting hyperpolarization of 1- to 4-sec duration and additional slow and late depolarizations, which may last tens of seconds, have been reported (Nishi, 1974). The hyperpolarization phase corresponds with a positive potential recorded from the postganglionic trunk; it has been associated with a reduction in impulse conduction through the ganglion (Nishi, 1974) and has, therefore, been called a slow IPSP. The slow IPSP is presumably due to the preganglionic activation of muscarinic cholinergic receptors on small, intensely fluorescent cells (Matthews and Raisman, 1969; Williams and Palay, 1969; Dun and Karczmar, 1978) that then release catecholamines (dopamine in rats) onto receptor sites on the principal ganglionic neuron. The slow depolarizations have been attributed to binding of released ACh with muscarinic receptors on the principal neurons. Several analyses of the processes underlying the generation of the slow IPSP have suggested that the catecholamine interactions release cyclic nucleotides within the principal neurons and these, acting as second stage neurotransmitters, produce hyperpolarization (Greengard, 1978). Tests for the existence and properties of the various membrane receptors proposed to be responsible for these complex responses in sympathetic ganglia have been indirect, have frequently involved nonphysiological manipulations of the ganglia (such as greater than normal tetanic stimulation of preganglionic fibers), and have often required circulating pharmacological blocking agents (such as nicotine), through the ganglion. The necessity of using these procedures has led to some skepticism regarding the interpretation of ganglionic transmission (Phillis, 1970; Volle, 1969) and has raised some questions regarding the distinctions between conductance changes associated with synaptic transmission and with alterations in the metabolism of the ganglion cells (Weight, 1971).

Neurons grown in culture have provided an opportunity to test, directly and unequivocally, for the existence of various receptors on the surface of the principal sympathetic neuron. Iontophoretic electrodes can be guided directly to the exposed neuronal membrane; the concentrations of the pharmacological agents in the medium surrounding the neurons can be closely monitored; and possibly drug-induced changes in the health of the neurons can be observed during the course of an experiment.

6.2. Acetylcholine Responses of Sympathetic Neurons in Culture

In each of the iontophoretic studies on embryonic (Obata, 1974, 1977; Ko *et al.*, 1976a; O'Lague *et al.*, 1978c; Wakshull *et al.*, 1979a) and adult (Wakshull *et al.*, 1979a) SCGN grown in culture, only acetylcholine consistently evoked a brief depolarizing response. ACh appropriately resulted in decreased membrane resistance that was brought about by conductance changes insensitive to TTX. Prolonged exposure to ACh produced specific desensitization which was equivalent to similar phenomena reported at other cholinergic junctions (Ko *et al.*, 1976a; Dennis *et al.*, 1971; Katz and Thesleff, 1957). Hexamethonium and mecamylamine, nicotinic ganglionic cholinergic antagonists, more effectively blocked the ACh responses than did curare (Ko *et al.*, 1976a; Wakshull *et al.*, 1979b). Low concentrations of atropine (2.8×10^{-6} M) had no effect, but 10^{-4} M reversibly blocked the ACh responses (Ko *et al.*, 1976a). Negative findings have been obtained with α-bungarotoxin (Ko *et al.*, 1976a; Nurse and O'Lague, 1975). Recent data have shown that α-bungarotoxin also fails to block iontophoretically evoked ACh or synaptic potentials in chick superior cervical ganglia *in vivo* (Carbonetto *et al.*, 1977, 1978) or in rat superior cervical ganglia *in vivo* (Brown and Fumagalli, 1977). These studies with α-bungarotoxin suggest that the membrane receptor sites for ACh are distinct from those for α-bungarotoxin (Carbonetto *et al.*, 1978). Thus, it has been established that specific ACh receptors survive or are reestablished on SCGN after the denervation and dissociation procedures accompanying their preparation for culture.

Recent evidence (Carbonetto and Fambrough, 1978) has shown that the distribution of ACh receptors on these neurons, when expressed as a percentage of surface membrane, is the same in somal and neurite outgrowth zones for explant cultures. However, areas of differing sensitivity over the surface of the cell soma and proximal processes have been observed in both embryonic and postnatal SCGN cultures (Ko, 1975; O'Lague *et al.*, 1978c; Wakshull, 1978). The overall sensitivity to ACh was lower, and the time course of the potentials was slower than values reported previously for skeletal muscle (Kuffler and Yoshikami, 1975) and frog parasympathetic neurons (Dennis *et al.*, 1971). No evidence of supersensitivity has been reported (Ko *et al.*, 1976a; Obata, 1974; O'Lague *et al.*, 1978c; Wakshull *et al.*, 1979a).

6.3. Spinal Cord Projections to Sympathetic Neurons in Culture

The presence of ACh receptors on neurons grown in culture was verified by demonstrating that cholinergic transmission formed between explants of fetal rat spinal cord grown with dissociated sympathetic neurons (R. Bunge *et al.*, 1974; Ko *et al.*, 1976a). Spontaneous and evoked excitatory postsynaptic potentials have been recorded that have similar parameters to the fast EPSP recorded from mammalian sympathetic ganglia (Ko *et al.*, 1976a; Nishi, 1974). The EPSPs excited by stimulating the spinal cord explants fluctuated in amplitude and occasionally failed with constant current stimulation; they were blocked by high Mg^{2+}/Ca^{2+} ratios, hexamethonium or mecamylamine, and only high concentrations of atropine. These synaptic responses were also unaffected by α- or β-adrenergic blocking agents (Ko *et al.*, 1976a). These experiments indicated that cholinergic, nicotinic "preganglionic" connections from spinal cord may reform *de novo* in culture, and that ACh was probably released in discrete packets (Ko, 1975).

6.4. Catecholamine Sensitivity of Sympathetic Neurons in Culture

No slow potentials were recorded from SCGN that responded to electrical stimulation of co-cultured spinal cord explants, even following several attempts to demonstrate their presence with repetitive stimulation (Ko *et al.*, 1976a). One possible explanation for this discrepancy with the work on sympathetic neurons *in vivo* (Nishi, 1974) was the absence of the small, intensely fluorescent (SIF) cell in the dissociated SCGN cultures. SIF cells have been seen in explant culture (Chamley *et al.,* 1972; Olson and Bunge, 1973), but no recordings have been made in this type of culture. However, a variety of pharmacological tests have been performed to ascertain whether receptors for catecholamines exist on the membranes of dissociated SCGN in culture (Obata, 1974; Ko *et al.,* 1976a,b; O'Lague *et al.,* 1978c; Wakshull *et al.,* 1979a). In all cases, including embryonic and adult SCGN, no physiologically measurable changes in the resting potentials or action potentials have been detected. No slow or fast time course conductance changes have been obtained with bath, pressure, or iontophoretic application of norepinephrine or dopamine (Wakshull *et al.,* 1979a). O'Lague *et al.* (1978c) have, however, reported that catecholamines at concentrations of 30–500 μM will reduce the strength of cholinergic transmission between SCGN. These effects were never reversed and, consequently, the basis of these changes remains ambiguous. Dun and Nishi (1974), Christ and Nishi (1971a,b), and Dun and Karczmar (1977) have demonstrated similar reductions in the mean quantal content of preganglionic transmitter release in the animal. Whether the effects in culture are also due to catecholamine actions on the presynaptic terminal remains to be resolved (O'Lague *et al.,* 1978c).

This discrepancy between the existence of slow potentials in recordings from sympathetic neurons *in vivo* and *in vitro* may occur because SCGN in culture are somehow changed by the culture environment. Perhaps without the presence of the SIF cells in dissociated cell cultures, the SCGN lose their catecholamine receptors or lose some component of the complex membrane interactions involving cyclic nucleotides that have been postulated to be involved with activation of these receptors (Greengard, 1978; Hashiguchi *et al.,* 1978; Kobayashi *et al.,* 1978). The hypothesis suggesting receptor loss would require a specific trophism between the SIF cells or the catecholamines released by these cells and the membrane receptors of the principal neurons. If this were the case, one might expect evidence of a slow IPSP in explant cultures (which contain SIF cells) that are innervated from some cholinergic "preganglionic" source or in dissociated SCGN that were grown with catecholamines in the bathing medium.

6.5. The Specificity of Innervation to Sympathetic Neurons in Culture

An important issue in the analysis of synaptic connections formed by SCGN in culture is the relationship between the specificity of connections formed in the animal and in culture. In the animal, SCGN receive an overlapping but restricted series of connections from neighboring thoracic spinal cord segments (Njå and Purves, 1977a,b; Yip *et al.,* 1978). Each neuron is primarily innervated by preganglionic fibers from one segment but receives additional inputs from adjacent segments. The specification of preganglionic connections seems to depend upon some label that the ganglion cells have, although this provides a relatively weak influence on preganglionic connections since foreign nerves (e.g., vagal) will innervate the ganglion cells (Purves, 1976). It is also known that a certain degree of trial and error characterizes the development of connections to the submandibular ganglion (Lichtman, 1977). The factor responsible for the specification of preganglionic innervation is not the post-ganglionic target tissue (Lichtman and Purves, 1978; Purves and Lichtman, 1978; Thompson *et al.,* 1978), although the anatomic position of the postganglionic neuron and its peripheral synapses are important for the maintenance of (Njå and Purves, 1977a) and for some selectivity in the distribution (Purves *et al.,* 1978) of synaptic input from the preganglionic neuron (Njå and Purves, 1977b). This latter trophism may be related to NGF levels since NGF injections can substitute for the postganglionic neurons to prevent the loss of preganglionic synapses (Purves and Njå, 1976).

The relatively weak influence that SCGN exert on the pattern of preganglionic connections in the animal is consistent with the ability of these neurons to accept diverse innervation in culture. Detailed studies of the specification of connections from potentially different preganglionic sources and sympathetic cells in culture have not been attempted. However, using electron microscopic analysis of synaptic profiles, Olson and Bunge (1973) observed that the number of synapses found in organotypic superior cervical ganglia (SCG) cultures was greatest when these explants were co-cultured with explants from thoracic spinal cord as compared with SCG explants grown alone or with explants from the cerebral cortex. These results suggest that some specificity of synapse formation may still be exerted in culture by the SCGN since they did not receive innervation from any of the varied population of neurons within the cerebral cortex explants. A diverse source of "preganglionic" neurons is still probable in these cultures since foreign (vagal) and native (thoracic preganglionic) cholinergic neurons may maintain functional connections *in vivo* on the superior cervical ganglia of guinea pigs (Purves, 1976). In addition, dissociated SCGN will accept synapses from other SCGN (see Section 8.1.) while still receiving synapses from explants of spinal cord grown in the same culture (Fig. 5) (R. Bunge *et al.,* 1974; Ko *et al.,* 1976b). These results indicate that postganglionic autonomic neurons may accept innervations from a limited variety of sources. Persistence of these synapses may then be the determining issue in specifying the connections (Lichtman, 1977). The factors associated with this maintenance are unknown, although in culture a prerequisite for the long-term retention of a synapse may be the ability of the innervating neuron to release acetylcholine.

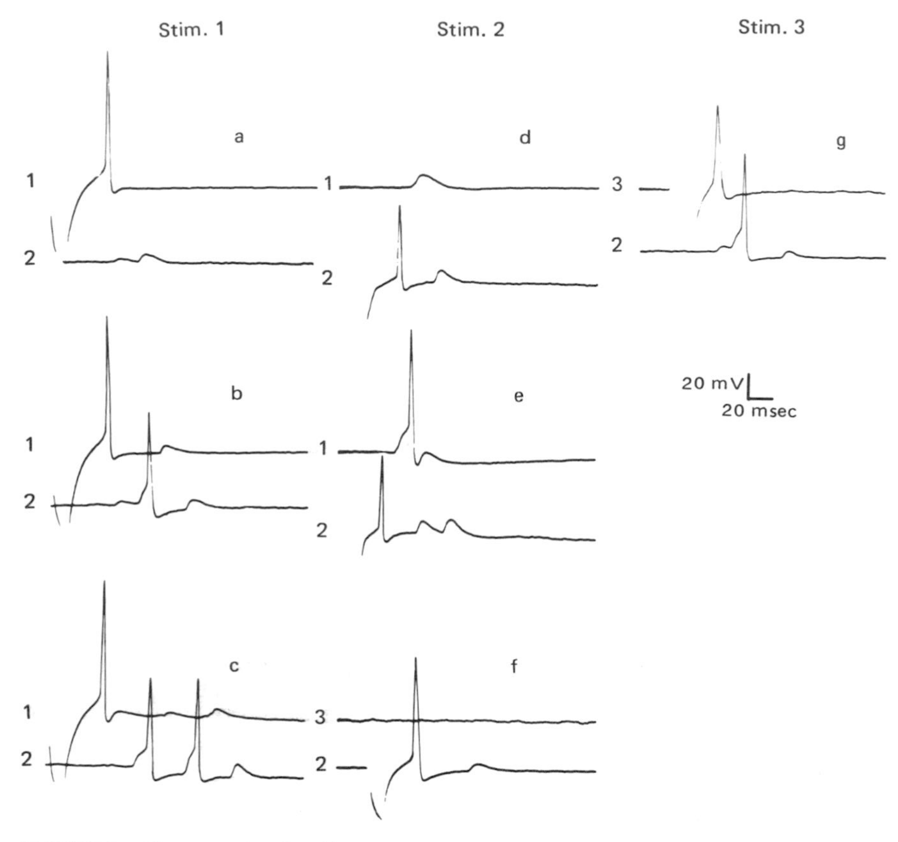

FIGURE 5. "An example of multisynaptic networks between SCGN grown with spinal cord. Stimulation of cell 1 evokes EPSPs or APs in cell 2 (a–c). Stimulation of cell 2 evokes reciprocal synaptic response in cell 1 (d,e). There is also a recurrent EPSP following the AP in cell 2 (b–e). Since complex interactions exist between these two cells, multiple responses may occur. For example, in b stimulation of cell 1 evokes a synaptically mediated AP in cell 2 which, in turn, causes an EPSP back onto cell 1 due to the reciprocal connection, and an EPSP in cell 2 via a recurrent synaptic pathway. If this recurrent EPSP is also suprathreshold and fires a second AP in cell 2 (c), then a second EPSP occurs in cell 1 via reciprocal connections, and an additional EPSP appears in cell 2 due to the recurrent synaptic innervation. Similarly in e stimulation of cell 2 evokes a recurrent EPSP in cell 2 and a suprathreshold synaptic response in cell 1 which, in turn, causes an additional EPSP in cell 2 via reciprocal connections. After withdrawing one microelectrode from cell 1 and penetrating cell 3 (f,g) an additional contact is seen with cell 2. Stimulation of cell 2 now causes only a recurrent EPSP in cell 2 and no response in cell 3 (f). Stimulation of cell 3, however, evokes a synaptically mediated AP plus a recurrent EPSP in cell 2 (g). Cell 2 thus receives convergent inputs from at least three sources: cells 1, 2, and 3." (From Ko *et al.*, 1976b.)

7. INNERVATION OF VISCERAL TARGET TISSUES GROWN WITH SYMPATHETIC NEURONS

7.1. General Aspects of the Morphology of Visceral Innervation *in Vivo*

Several recent reviews (Burnstock and Bell, 1974; Furness and Costa, 1974; Uehara *et al.*, 1976) provide the following data. The typical sympathetic neuron possesses a long, thin (<1 μm in diameter) axon that becomes distinctively varicose as it reaches its peripheral innervation field. The linear array of varicosities occupying the terminal branches of the axon form the "autonomic ground plexus." These varicosities (as studied, for example, in the vas deferens) were found to be 1–2 μm dilations of axons 0.1–0.5 μm in diameter; there were about 250–300 dilations per millimeter over a total of several centimeters of terminal axon field. The varicosities are filled with mitochondria and vesicles of various sizes and sometimes exhibit small areas of density below the plasmalemma, suggesting possible sites of neurotransmitter release. Whereas the preterminal axon is invariably ensheathed by Schwann cells, the most distal regions may lose this ensheathment and thus directly expose the varicosity to the extracellular space.

The relationship of the varicosity to the target tissue varies substantially. Nerve–smooth muscle apposition is known to be closest (15–30 nm) in the vas deferens, nictitating membrane, and sphincter pupillae (Uehara *et al.*, 1976). Merrillees (1968) concluded, after correlating ultrastructural and physiological studies, that transmitter released from varicosities further than about 100 nm from muscle cells may not have a significant effect on these cells. In certain organs (e.g., vas deferens), nerve endings may invaginate smooth muscle cells, ending deep within them; in contrast, adrenergic nerves terminate in most blood vessels at the adventitial–medial border, approaching only within 50–80 nm of the underlying smooth muscle component. Burnstock and Bell (1974) have emphasized that many individual cells of visceral targets are linked by low-resistance junctions, and, therefore, a widespread effect can be obtained without innervation of individual effector cells.

Unlike the somatic neuromuscular junction, the visceral junction may exhibit interruption of the intervening basal lamina in areas of close neuromuscular apposition. Postjunctional specializations in relation to the target cell axolemma are inconsistent features; there are sometimes present aggregations of vesicles, some increased density of the postjunction membrane, or subsurface cisternae (Merrillees, 1968). Frequently, varicosities occur substantial distances from cells of target tissue which, in turn, show no regions of "synaptic" specialization to mark receptor sites. Consequently, no degree of certainty exists regarding where the nerve-to-target signal is being passed.

7.2. Sympathetic Innervation *in Vitro*

It has been known for some time that sympathetic neurons in culture generate axons which, for a substantial part of their length, may exhibit periodic varicosities (Chamley *et al.,* 1972; Silberstein *et al.,* 1971; Wakshull *et al.,* 1979a; Fig. 4). This property is expressed by sympathetic neurons grown either alone or together with target tissues (cf. Chamley *et al.,* 1973; Silberstein *et al.,* 1971; Wakshull *et al.,* 1979a,b). These are presumably sites of neurotransmitter release *in vitro* as they are thought to be *in vivo.*

7.2.1. Cardiac Muscle

The first demonstration of functional sympathetic innervation of peripheral target tissue in culture was provided in a brief report by Crain (1968), who described an increased rate of contraction in heart explants following stimulation of the fibers from co-cultured sympathetic ganglion. Masurovsky and Benitez (1967) demonstrated close apposition between the sympathetic fibers and the heart rudiment in this type of culture. Vesicle-containing varicosities were observed within 15 nm of the myocardial cells. No specialized pre- or postsynaptic membrane specializations were seen. Mark *et al.* (1973) observed that cultures containing myocardial cells with apposed nerve fibers appeared to maintain a more highly differentiated state than those without innervation. Subsequently, Purves *et al.* (1974) demonstrated that stimulation of sympathetic chain ganglia co-cultured with heart caused an acetylcholine-modulated inhibition of spontaneous cardiac beat.

In elegant experiments designed to determine the nature of the transmitter released from sympathetic neurons taken from newborn rat, Furshpan and colleagues (1976) cultured single newborn rat superior cervical ganglion neurons on islands of heart muscle cells (Fig. 6). As will be presented in greater detail below (see Section 8.2), the culture conditions used are now known to promote the development of cholinergic properties in young SCG neurons. Some neurons were found to inhibit cardiac muscle (by acetylcholine release), others to excite (by catecholamine release), and several neurons were found which first inhibited, and then excited, heart muscle (Fig. 6). Those neurons that slowed the beat of the underlying myocytes often excited themselves at nicotinic, cholinergic synapses. The cytochemistry of the synaptic endings in cultures of this type has been studied by Landis (1976). Neurons previously characterized by electrophysiological recordings were fixed with $KMnO_4$ to mark the dense core characteristic of synaptic vesicles in adrenergic neurons.

The synaptic endings of neurons secreting catecholamines contained many small granular vesicles, whereas neurons secreting acetylcholine contained none. The endings of "dual function" neurons secreting both acetylcholine and catecholamines contained occasional small granular vesicles. Proximity of nerve ending to heart muscle was observed, but no synaptic specializations were seen. Interestingly, both putatively cholinergic and adrenergic neurons made morphologically distinct endings on themselves, but only cholinergic "autapses" were detected electrophysiologically. Thus, the catecholaminergic endings were physiologically silent.

7.2.2. Smooth Muscle

Although it has been known for more than a decade that spinal neurons cultured with striated muscle develop functional synapses in culture (for review, see Shimada and Fischman, 1973; Crain, 1976), the first physiological demonstration of innervation to smooth muscle in culture was not reported until 1974. Several earlier reports had demonstrated morphological association between nerve and smooth muscle in culture. (1) Silberstein *et al.* (1971) presented light microscopic evidence for the reinnervation of rat iris in organ culture by added sympathetic ganglia. A dense plexus of highly fluorescent, varicose fibers formed over the iris muscle. (2) Mark *et al.* (1973) studied the interactions between nerve fibers from guinea pig and rat sympathetic ganglia and single muscle cells from rat vas deferens and heart. Long-lasting (3 days) associations were formed in these instances in contrast to the temporary (several hours) contacts established between nerve fibers and fibroblasts. (3) Chamley *et al.* (1973) expanded this work to seek evidence for influences of visceral target tissues on sympathetic axon growth and survival. Smooth muscle explants were observed to provide attraction for nerve fibers, while single muscle cells did not. They also observed long-lasting association of nerve with muscle cells but not with fibroblasts, as well as an accelerated rate of growth over muscle cell surface. Palpation of smooth muscle cells by exploring growth cones increased the rate of spontaneous contraction in these cells; this increase was greater when several nerve fibers were involved. They also provided evidence that after substantial axon–muscle cell interaction, subsequently arriving nerve fibers were excluded from innervating already contacted muscle cells (see also Chamley *et al.*, 1974).

The report of Purves *et al.* (1974) demonstrated physiologically effective nerve–smooth muscle combinations maintained for 3–14 days *in vitro*. Iris sphincter muscle grown with rat sympathetic chain ganglia showed contractile responses after delivery of extracellular repetitive stimulation to the surface of the nerve explant. Similar responses were observed upon stimulation of guinea pig ciliary ganglia grown with either taenia coli muscle or with explants from

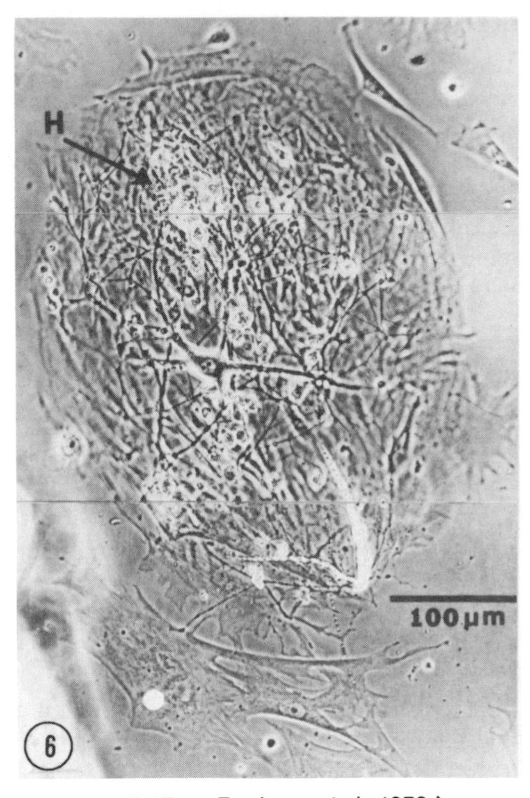

FIGURE 6. "A microculture containing a dual-function solitary neuron after 13 days in culture. Arrow at H shows cluster of myocytes. All records were from this neuron. a–c, neuronal impulse and autopic EPSPs before (a), during (b), and after (c) perfusing with hexamethonium (0.5 mM). d, A train of neuronal impulses (deflection of lower trace) produced inhibition and then excitation of spontaneous myocyte activity (upper trace). e, inhibition was blocked by atropine (0.1 μM). In d and e, hexamethonium (0.5 mM) was present. f, the effect of atropine (0.1 μM; no hexamethonium in f–i at higher sweep speed. g, block of excitation by propranolol (0.6 μM; atropine still present). h, about 45 min after removal of propranolol (atropine still present) with excitation restored. i, the dual effect restored by perfusion with drug-free fluid. Scales are (for *y* and *x* axes, respectively): a–c, 50 mV and 30 msec; d and e, 100 mV and 12.5 sec; f–i, 100 mV and 5 sec." (From Furshpan *et al.*, 1976.)

the vas deferens. These contractions were sometimes accompanied by an increase in the amplitude of spontaneous rhythmic movements. Effective junctions between sympathetic nerves and vas deferens cells were found as early as 1–2 days after contact. The cholinesterase inhibitor neostigmine (5×10^{-6} to 10^{-10} g/ml) potentiated, and hyoscine (10^{-7} g/ml) blocked the muscle responses of sympathetic ganglia–iris and ciliary ganglion–taenia co-cultures. Hyoscine also abolished muscle responses in ciliary ganglion–vas deferens cultures.

The subsequent observations by Hill and colleagues in 1976 on the innervation of iris musculature in culture are of particular interest in light of the subsequent detailed discussion concerning plasticity of transmitter production in the newborn rat sympathetic neuron. Both sphincter and dilator portions of the newborn rat iris were grown with superior cervical or lumbar paravertebral sympathetic ganglia. Based on both morphological and physiological studies, the two muscles received distinctly different innervation patterns *in vitro*. Var-

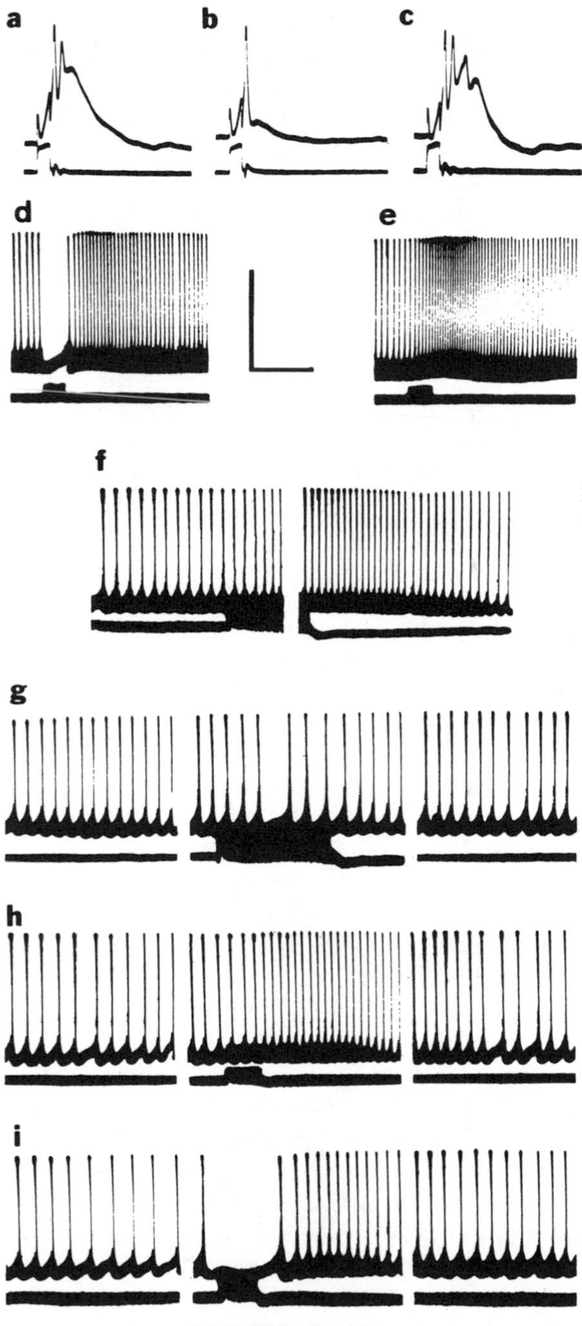

FIGURE 6. (*cont.*)

icose fibers, believed to be adrenergic on the basis of specific fluorescence, were consistently associated with the dilator muscle; these were not shown to be functionally active. Junctions with the sphincter were judged to be cholinergic on the basis of hyoscine-sensitive contractions evoked by nerve stimulation and sensitivity to iontophoretically applied acetylcholine.

In commenting on the dominant expression of cholinergic function by neurons in these sympathetic ganglia, these authors make the logical assumption that the cholinergic fibers derive from the small subpopulation of cholinergic neurons thought to be present in most sympathetic ganglia. In light of subsequent work, as is discussed in detail below (Section 8.1), the possibility that this cholinergic function was expressed by adrenergic neurons that had added or acquired the ability to synthesize acetylcholine must also be considered. It seems worth noting that in all but one instance dealing with heart cell innervation (Furshpan *et al.,* 1976), physiological studies have revealed cholinergic (rather than adrenergic) mechanisms at work in mixed cultures of autonomic neurons and their targets.

7.2.3. Adipose Tissue

Ko *et al.* (1976b) cultured neonatal rat SCGN together with fat cells taken from the intrascapular fat pad. Electron microscopic study of these cultures showed nerve processes within 40 nm of the fat cell plasma membrane. As is seen *in vivo,* ensheathing Schwann cells were absent in this region, but a thin basal lamina remained interposed. Some of the vesicles in the profiles contained dense cores after $KMnO_4$ fixation, but the physiological properties of this junction have not been tested. This "adrenergic" target did not prevent the development of cholinergic interactions between SCGN.

8. SYNAPTIC INTERACTIONS BETWEEN SYMPATHETIC NEURONS IN CULTURE

8.1. Synaptic Interactions between Sympathetic Neurons

Morphological studies have established that many synaptic contacts develop between SCGN grown in dissociated cell cultures (R. Bunge *et al.,* 1974; Claude, 1973; Rees and Bunge, 1974). Several observations suggest that these kinds of connections in culture might be expected on the basis of the growth and development of sympathetic neurons in the animal. As discussed above in Section 6.5, under certain experimental conditions, sympathetic cells can receive and form a variety of connections, and many of these without a great deal of specificity. Consequently, novel connections between SCGN in culture might be expected. There is also evidence of normally occurring intrinsic connections between principal ganglion cells in some mammalian sympathetic ganglia (Jacobowitz, 1970; Raisman *et al.,* 1974; Elfvin, 1971; Grillo, 1966; Dail and Evan, 1978; Joó *et al.,* 1971; Tamarind and Quilliam, 1971; Yokota and Yamauchi, 1974; Matthews, 1974). The synaptic profiles *in vivo* contain small dense core vesicles and are, therefore, considered to be distinct from synapses formed by SIF cells. The synaptic profiles present in SCGN cultures prepared from embryonic or neonatal tissues and grown *in vitro* for less than 3 weeks also contain small dense core vesicles and, thus, resemble the intrinsic connections seen *in vivo* (R. Bunge *et al.,* 1974; Claude, 1973; Rees and Bunge, 1974; Johnson *et al.,* 1976).

The synapses between SCGN in culture have been studied intensively during the past several years because, unexpectedly, it was found that the signaling between the cells involves the release of acetylcholine (Ko *et al.,* 1976b; Johnson *et al.,* 1976; O'Lague *et al.,* 1974, 1976, 1978b,c; Wakshull *et al.,* 1979b). In addition, cholinergic synapses in culture have been recorded between SCGN and skeletal muscle (Nurse and O'Lague, 1975; O'Lague *et al.,* 1976; Wakshull *et al.,* 1979b) and cardiac myocytes (Fig. 6; Furshpan *et al.,* 1976). The network of connections between SCGN can be of any form including autaptic, reciprocal, and multisynaptic (Fig. 5; Ko *et al.,* 1976b; Johnson *et al.,* 1976; O'Lague *et al.,* 1974, 1976, 1978b; Wakshull *et al.,* 1979a). The cholinergic nature of the neuron–neuron synapses was demonstrated by reversibly blocking the responses with nicotinic antagonists such as mecamylamine, hexamethonium, or curare (Ko *et al.,* 1976b; O'Lague *et al.,* 1974, 1976, 1978c). Atropine causes blockade only at high concentrations (Ko *et al.,* 1976b; O'Lague *et al.,* 1978c). The neuromuscular junctions are preferentially blocked by curare or α-bungarotoxin (O'Lague *et al.,* 1976; Wakshull *et al.,* 1979b), but the neuronal synapses are unaffected by α-bungarotoxin (Ko *et al.,* 1976b; O'Lague *et al.,* 1976, 1978b). Adrenergic receptor antagonists have had no consistent effects on the interneuronal synapses (Ko *et al.,* 1976a,b; O'Lague *et al.,* 1978c; Wakshull *et al.,* 1979b). An additional indicator that these synapses are cholinergic has been the demonstration of high sensitivity ACh responses that arise from restricted spots and that mimic the appearance of the excitatory postsynaptic potentials (Ko *et al.,* 1976b; O'Lague *et al.,* 1974, 1978c).

The presence of cholinergic synapses in cultures of nominally adrenergic neurons could be explained in several ways. One possibility is the selective survival of separate populations of cholinergic and adrenergic neurons *in vitro.* The cholinergic neurons in culture would derive from the small number of cholinergic sympathetic cells thought to exist in the animal (Aiken and Reit, 1969; Sjöquist, 1963; Yamauchi *et al.,* 1973). These cells would then be responsible for all of the cholinergic synapses recorded in culture. Alternatively, a group of undifferentiated cells may exist in embryonic and newborn ganglia from which the cultures are prepared. This group of cells may then proliferate and/ or develop *de novo* within a culture environment that forces all of these cells to develop cholinergic properties. All differentiated neurons in the ganglion prior to culture would then be expected to die or remain adrenergic in culture. A third possibility is that one population of sympathetic neurons survives in culture, and all of these SCGN assume cholinergic properties in culture after a period of exposure to some inducing factor *in vitro*. This third possibility raises the additional question of whether the assumption of acetylcholine as a neurotransmitter excludes all adrenergic properties in the same cell or whether components of two neurotransmitter systems can coexist in one cell.

8.2. Evidence That the Sympathetic Neuron "Shifts" Transmitter Production

Several observations indicate that there is initially only one population of adrenergic cells in these cultures and that some factor in the culture environment, which can be transmitted through the bathing medium, induces these cells to undergo a "shift" to synthesize acetylcholine and to make cholinergic synapses.

There is no selective development or proliferation of a new neuron *in vitro* because nearly all of the SCGN in the animal are postmitotic at the time they are taken for culture (Hendry, 1977) and because no mitoses have been seen in these cultures except among the nonneuronal cell population. Furthermore, in many of these preparations, an antimitotic agent is added to the bathing medium or the cultures are irradiated with cobalt to prevent proliferation.

There is probably no selective survival of particular (cholinergic) neurons because the number of neurons per culture dish remains fairly stable after the cells attach to the collagen substrate (Patterson and Chun, 1977a,b; Johnson *et al.*, 1980a). In addition, more than 25% of the total number of neurons present in the ganglia that are taken for culture may survive (Johnson *et al.*, 1980a). This percentage is greater than the expected (5%) population of cholinergic neurons believed to exist in the rat SCG (Yamauchi *et al.*, 1973). Consequently, many adrenergic neurons must be present in the cultures initially. In some older cultures it is possible to demonstrate that nearly every recorded neuron is capable of synaptically activating some other neuron on the culture dish (O'Lague *et al.*, 1978c; Wakshull *et al.*, 1979a). This high percentage of driver neurons is probably a consequence of a substantial number of the neurons shifting to acetylcholine production.

Under optimal conditions the shift that occurs must affect nearly all of the neurons in a culture. There are two series of experiments to support this conclusion. First, in parallel cultures prepared from the same initial tissues, appropriate conditions (see Section 8.3) in one series will result in cholinergic synapses and increased acetylcholine synthesis, but few cholinergic synapses and persistence of catecholamine synthesis in untreated sister cultures (O'Lague *et al.*, 1974; Patterson and Chun, 1974, 1977a,b; Reichardt and Patterson, 1977). Reichardt and Patterson (1977) have directly demonstrated with biochemical assays of single cells that acetylcholine synthesis can be switched on in individual neurons, that the number of neurons shifted per culture is a function of the concentration of a conditioned medium used to feed the cultures, and that most cells assume the synthesis of only one neurotransmitter even when they are exposed to intermediate concentrations of conditioned medium. In addition, systematic analyses of synaptic vesicle cytochemistry in a parallel series of cultures studied at different *in vitro* ages revealed that nearly all synaptic profiles contain large numbers of small agranular vesicles after 8 weeks *in vitro,* whereas fewer than 10% are agranular prior to 3 weeks *in vitro* (Johnson *et al.*, 1976, 1980a). This result is presumably due to a shift in the cytochemistry of all of the neurons once they become cholinergic. Coincident with the time course of altered cytochemistry, Johnson *et al.* (1976) showed a 1000-fold increase in choline acetyltransferase and a marked increase in both the complexity and the frequency of cholinergic interactions.

8.3. The Inducing Factor for the Cholinergic Shift

In the initial biochemical studies of SCGN in culture, Mains and Patterson (1973a–c) showed that catecholamine synthesis persisted *in vitro* when the neurons were grown with or without ganglionic nonneuronal cells. Small amounts of acetylcholine could be found in cultures grown without supporting cells only after the SCGN were cultured for more than 4 weeks. In contrast, ACh synthesis increased from 30- to 1000-fold in the presence of nonneuronal cells (Patterson and Chun, 1974). The increased ACh production could be caused by a heterogenous population of ganglionic supporting cells or C6 glioma cells. Subsequent work by Ross and Bunge (1976) showed that the addition of either Schwann cells or periosteal fibroblasts could also increase choline acetyltransferase (ChAc) activity in a culture of neurons, whereas these nonneuronal cells alone did not demonstrate any ChAc activity. An important corollary to these observations was the demonstration that cholinergic synapses could be found with ease in cultures containing nonneuronal cells (O'Lague *et al.,* 1974, 1976).

Two separate observations indicated that the physical presence of the nonneuronal cells *per se* was not responsible for these effects. First, large numbers of synapses (Wakshull *et al.,* 1978, 1979a) and a 1000-fold increase in ChAc activity (Ross and Bunge, 1976) could be observed in pure neuronal cultures fed with medium containing human placental serum and embryo extract. Second, Patterson and his collegaues (Patterson *et al.,* 1977; Patterson and Chun, 1977a,b; Reichardt and Patterson, 1977) showed that the ratio of ACh to norepinephrine (NE) in pure neuronal cultures could be manipulated by altering the medium to include various proportions of medium taken from cultures of nonneuronal cells. Consequently, high levels of ACh synthesis could be induced by a diffusible factor produced by primary rat cells from blood vessels, heart, skeletal muscle, and embryo fibroblasts and from rat cell lines of fibroblasts, L8 skeletal muscle, and C6 glioma (Patterson and Chun, 1977a) or found in serum or embryo extract (Ross and Bunge, 1976; Wakshull *et al.*, 1978, 1979a). The nature of this diffusible factor is presently under investigation (Patterson, 1978).

Additional data, however, suggest that the amount of ACh produced by a cell can be influenced by other factors. For example, Ross and Bunge (1976) showed that the average ChAc activity per neuron is directly related to neuronal density and that chronically blocking cholinergic receptors with hexamethonium reduces ChAc activity. One possible explanation of these data is that the level of synaptic activity expressed by a cell influences its production of ACh. Consequently, more ACh is synthesized per neuron in high density cultures because a greater frequency of spontaneous activity may occur, whereas less ACh is produced when no synaptic activity is possible in the presence of hexamethonium. These possible effects of synaptic activity on level of ACh synthesis may, however, only be expressed after a neuron has been induced to become cholinergic.

In contrast, Walicke *et al.* (1977) have suggested that chronic depolarization (induced by growing the cells in a high-K^+ medium) of the SCGN decreases ACh synthesis and tends to keep the cells adrenergic even in the presence of conditioned medium. SCGN will still shift toward cholinergicity when grown in high K^+ and conditioned medium if the Mg^{2+} concentration is also increased or if Ca^{2+} entry into the cells is specifically blocked. According to Walicke *et al.* (1977), the induction of a shift toward cholinergicity may involve Ca^{2+} modulation of the metabolism of cyclic nucleotides. These studies by Ross and Bunge (1976) and Walicke *et al.* (1977) indicate that induction and subsequent levels of ACh synthesis may not be related; the former may be related to changes in the metabolism of the cell, and the latter may be controlled by the cell's total level of synaptic activity.

8.4. Time Course of Inducibility

The time course of the induction of acetylcholine synthesis in dissociated SCGN cultures was followed by determining the sensitivity of the cells to conditioned medium at various times after the cells were grown in culture (Patterson and Chun, 1977b). These experiments suggested that exposure to a conditioned medium yielded an increase in ACh synthesis during the initial 10–20 days and less increase when conditioned medium was first added after 30–40 days *in vitro*. In addition, the cells needed only a minimal exposure to the conditioned media to remain permanently committed to a cholinergic phenotype. Similarly, Hill and Hendry (1977) and Ross *et al.* (1977) showed that a decision to remain adrenergic persists when the sympathetic cells were allowed to develop in the animal for more than 3 weeks postnatally prior to explantation to culture.

Whether the commitment to be adrenergic is irrevocable in all cells could not be determined in these experiments with explant cultures because the behavior of individual cells was not assayed biochemically or physiologically. Similarly, whether a cell that has shifted to become cholinergic gives up all adrenergic characteristics is uncertain. All biochemical studies of these neurons show a decline in the level of synthetic enzymes for catecholamine production when ACh synthesis rises (Patterson and Chun, 1977a,b; Patterson *et al.*, 1976; Ross and Bunge, 1976). However, the adrenergic pump for the uptake of catecholamines persists in individual cholinergic neurons (Wakshull *et al.*, 1978). In addition, the uptake and release of NE in a Ca^{2+}-dependent manner has been demonstrated in mass cultures at an age *in vitro* when ACh synthesis predominates (Burton and Bunge, 1975; Patterson *et al.*, 1976). In order to determine whether the specific adrenergic uptake system is totally independent of the shift to ACh synthesis, it will still be necessary to specify the K_m value for uptake as a function of different proportions of conditioned medium and then in neurons in which varying degrees of acetylcholine synthesis have been induced.

8.5. Two Neurotransmitters in One Cell

In a recent study of dissociated SCGN cultures prepared from more mature neurons (up to 12.5-week-old rats), cholinergic synapses were seen between neurons and between neurons and skeletal muscle (Wakshull *et al.*, 1979b). These experiments with young adults again raise the issue of selective survival *in vitro* of cholinergic neurons that are known to be present *in vivo*. The problem is greater for cultures derived from postnatal tissues because more of these neurons die. Another alternative is that upon dissociation even postnatal neurons can regain susceptibility to cholinergic conditioning factors (Wakshull *et al.*, 1979b). A third possibility is that these experiments demonstrate the presence of small but physiologically detectable quantities of acetylcholine in any adrenergic sympathetic neuron.

Furshpan *et al.* (1976) have convincingly demonstrated that embryonic SCGN can release both ACh and NE onto co-cultured heart muscle cells in culture (Fig. 6). This dual transmitter behavior was detected physiologically; however, in other experiments it was found biochemically in only 2 cells (Reichardt and Patterson, 1977). Previous findings in mass cultures of intermediate ratios of ACh to NE synthesis have been interpreted to indicate relative numbers of cells committed to one or another transmitter system as a function of culture conditions and not to refer to differences in the ratio of ACh to NE within individual cells (Patterson and Chun, 1977a,b; Reichardt and Patterson, 1977; Patterson, 1978). However, the studies with adult SCGN (Wakshull *et al.*, 1979a,b) and with embryonic SCGN plus heart (Furshpan *et al.*, 1976) suggest that there may be no final commitment to one neurotransmitter. What is not known is whether the physiologically studied cells from embryonic SCGN–heart cultures are at an intermediate, transitional stage of development and are, therefore, at the point where cytochemical analyses find a substantially mixed population of snyaptic vesicles (cf. Landis, 1976; Landis *et al.*, 1976; Johnson *et al.*, 1980a,b). Alternatively, some sympathetic neurons may be capable of always maintaining a dual or a flexible phenotype for neurotransmitter synthesis, as has been suggested recently by Burnstock (1978). Several experiments are needed to test these ideas. First, longitudinal physiological studies of single embryonic neurons and heart muscle as functions of time and conditioned medium concentration would show whether some cells maintained dual transmitter release over substantial periods of time or only during a transitory stage of induction. Second, physiological and cytochemical studies with adult sympathetic neurons and heart muscle in culture would demonstrate whether some, any, or all sympathetic neurons can release two transmitters. In both of these studies, it is likely that only the more sensitive techniques (i.e., physiological) can be used to detect the minute quantities of transmitter, because this release may represent only a portion of the transmitter from a single cell.

The question of the existence of more than one neurotransmitter in autonomic neurons *in vivo* has been considered at length in several recent reviews (Burn, 1977; Burnstock, 1976, 1978). Of particular interest is the theory developed by Burn and Rand (1965) that acetylcholine release is linked to the subsequent release of norepinephrine from sympathetic terminals. According to this theory, the ACh released by a sympathetic cell acts to enhance the additional release of NE. The evidence supporting this cholinergic link theory of autonomic transmission has been subject to critical evaluations (Ferry, 1963, 1966; Campbell, 1970). In most circumstances, alternative explanations are available (e.g., nonspecific pharmacological effects on nerve terminals or the existence of separate populations of adrenergic and cholinergic sympathetic fibers).

Recently, Burn (1977) has cited the experiments in tissue culture as providing significant support for his dual transmitter theory. However, no studies in culture demonstrate that the release of acetylcholine is a prerequisite for the release of norepinephrine from the same cell (Furshpan *et al.*, 1976). In fact, according to the Burn model, an inhibitory phase should be seen prior to any acceleration in muscle activity caused by norepinephrine release. Many cells shown to release NE onto cardiac muscle did not cause a prior inhibition (Furshpan *et al.*, 1976), and only a small subset of the SCGN demonstrate inhibitory and excitatory effects on cardiac myocytes (Fig. 6).

9. DISCUSSION

The cells of the neural crest that give rise to the various neurons of the autonomic nervous system also give rise to a variety of other cells including such diverse types as cartilage, melanocytes, and Schwann cells (for review see Weston, 1970). The mechanisms by which this wide diversity of cell types is generated are not known. One possibility for the generation of the different kinds of neuron in this system would be specific genetic restrictions imposed on each neuron at (or before) the time of its terminal mitosis, thus removing it from further environmental influence. This model would clearly imply that at least certain autonomic neurons are specified prior to their formation of synapses with target tissue. An example is provided by neurons of the rat SCG that cease mitosis before birth but that complete their innervation of the iris some time after birth (see R. Bunge *et al.*, 1978).

An alternate model would separate the differentiation of autonomic neurons into several stages. During an initial step the cells would acquire properties permitting their migration to specific positions within the embryo. Subsequent differentiation would be delayed so that the cells might interact with the environment at the final point of residence (of either the cell soma or axon terminals) and thus initiate final steps of differentiation appropriate to that environment.

A large number of studies have shown that the environment of neural crest cells does indeed influence their differentiation (LeDouarin and Teillet, 1974; LeDouarin *et al.*, 1975, 1977, 1978). In general, these studies have utilized experimental alteration of the developing embryo and have, of necessity, been conducted on cells that are still capable of dividing. Consequently, these results cannot distinguish between restrictions of a given cell's potentiality by the new environment and selective survival of particular classes of cells that are produced in the new setting. The most prominent result of the studies on adrenergic neurons grown in tissue culture is that plasticity with respect to neurotransmitter synthesis is retained in postmitotic cells. This potentiality may even occur in some adult adrenergic neurons grown under certain culture conditions as has been discussed. Previous evidence for plasticity in neurotransmitter synthesis in postmitotic neurons has been rare (Ceccarelli *et al.*, 1971; Kemplay and Garrett, 1976; Fujiwara and Kurahashi, 1976) and has generally been open to alternative explanations (O'Lague *et al.*, 1978c).

The propensity of adrenergic neurons grown in culture to express cholinergic properties raises the important issue of what keeps these cells adrenergic in the animal. Certainly the environment seen by adrenergic neurons *in vivo* contains many of the cells that so potently induce acetylcholine synthesis in culture. Several considerations (e.g., temporal course of development, concentration gradients of trophic factors, presence of countervailing conditions) may be involved in determining the response of these neurons to their environment.

The concept that cholinergic influences are dominant in development and, therefore, provide primary influence until a subset of developing neurons "capture" these influences and make them inaccessible to later developing neurons has been presented by Bunge *et al.* (1978).

The recent experiments by Walicke *et al.* (1977) raise the additional question of whether alterations in the culture environment that prevent the action of conditioning factors are related to events in the animal that create or maintain adrenergic properties in neurons. Walicke *et al.* (1977) have shown that chronic depolarization in the presence of conditioned medium, which normally induces cholinergic mechanisms, keeps neurons adrenergic and that this effect is mediated by the transport of Ca^{2+} into the cell. In the animal, Black (1977) and colleagues (Cochard *et al.*, 1978) have also shown that the preganglionic innervation of sympathetic ganglia influences the level of tyrosine hydroxylase activity in these cells. However, the relationship between the depolarization caused by neural events *in vivo* and the chronic depolarization studied in culture is not known, although in both instances Ca^{2+} conductance changes occur. In the animal, these conductance changes are of short duration, and prolonged alteration in the cell's metabolism may not be present, whereas in culture the opposite may occur. It may also be unreasonable to expect the preganglionic input to be responsible for instructing all sympathetic cells to be adrenergic because adrenergic and cholinergic neurons are often intimately mixed within the same autonomic ganglion, e.g., the pelvic ganglion of the rat (Dail *et al.*, 1975).

Another issue related to comparisons between induction of SCGN in the animal and in culture is the site where the inducing factor is taken into the cell. In the animal, the peripheral axons within the target tissue probably take up inducing, as well as trophic, agents. In neurons with long axons, the question must then be raised whether the environment at the site of residence of the nerve cell body or the site of axon termination is more critical (Hill and Hendry, 1976). Under culture conditions these two sites are not in separate tissue compartments as they may be in the animal. Thus, in culture, manipulations would tend to influence both parts of the neuron equally, whereas in the animal they may not. It is known that the growing axon is provided with uptake mechanisms for certain macromolecules, with some suggestion that uptake at the growth cone is more active than that at the cell body. It is also known that uptake and retrograde transport occur from synaptic regions and that the extent of this uptake may be related to the amount of activity at the synapse (for discussion, see M. Bunge, 1977).

10. SUMMARY

Neurons from the sympathetic, parasympathetic, and enteric portions of the autonomic nervous system have been successfully cultured. Extensive studies of the adrenergic sympathetic neuron have demonstrated that its basic physiological properties are similar to those found *in vivo.* In tissue culture preparations these neurons exhibit acetylcholine sensitivity, can receive nicotinic cholinergic synapses from spinal cord, and can provide active synapses on heart muscle, smooth muscle, and fat cells. In addition, under certain culture conditions, these neurons form synaptic contacts on one another (and on themselves) that are cholinergic. It has been established that the development of these cholinergic mechanisms is not by the selection in culture of a subpopulation of cholinergic neurons, but by the addition of cholinergic properties in neurons initially expressing adrenergic characteristics. The ability of the sympathetic neuron to "shift" transmitter choices is related to experimental embryological studies providing similar observations. These results suggest the conclusion that certain aspects of the differentiation of autonomic neurons are critically influenced by environmental factors at the site of final residence of the neuron cell body and/or its axonal terminals.

ACKNOWLEDGMENTS. Work from the authors' laboratories has been supported by National Institutes of Health Grants NS09809 to Dr. Harold Burton, NS09923 and NS14416 to Dr. Richard P. Bunge, and NS11888 to both Drs. Harold Burton and Richard P. Bunge.

We are especially grateful to our colleagues (Drs. Ko, Johnson, Ross, and Wakshull) who have at various times provided critical discussions that led to some of the ideas expressed in this review. We also wish to thank Dr. Dennis Bray for critically reading the manuscript.

REFERENCES

Aiken, J.W., and Reit, E.A., 1969, A comparison of the sensitivity to chemical stimuli of adrenergic and cholinergic neurons in the cat stellate ganglion, *J. Pharmacol. Exp. Ther.* **169:**211.

Betz, W., 1976, The formation of synapses between chick embryo skeletal muscle and ciliary ganglia grown *in vitro, J. Physiol. (Lond.)* **254:**63.

Black, I.B., 1977, Regulation of the growth and development of sympathetic neurons *in vivo,* in: *Progress in Clinical and Biological Research, Vol. 15, Cellular Neurobiology* (Z. Hall, R. Kelly, and C.F. Fox, eds.), pp. 61–71, Alan R. Liss, New York.

Blackman, J.G., Ginsborg, B.L., and Ray, C., 1963, Some effects of changes in ionic concentration on the action potential of sympathetic ganglion cells in the frog, *J. Physiol. (Lond.)* **167:**374.

Bray, D., 1970, Surface movements during the growth of single explanted neurons, *Proc. Natl. Acad. Sci. USA* **65:**905.

Bray, P., 1973, Branching patterns of individual sympathetic neurons in culture, *J. Cell Biol.* **56:**702.

Brown, D.A., and Fumagalli, L., 1977, Dissociation of α-bungarotoxin binding and receptor block in the rat superior cervical ganglion, *Brain Res.* **129:**165.

Bunge, M.B., 1973, Fine structure of nerve fibers and growth cones of isolated sympathetic neurons in culture, *J. Cell Biol.* **56:**713.

Bunge, M.B., 1977, Initial endocytosis of peroxidase or ferritin by growth cones of cultured nerve cells, *J. Neurocytol.* **6**:407.

Bunge, R., Johnson, M., and Ross, C.D., 1978, Nature and nurture in development of the autonomic neuron, *Science* **199**:1409.

Bunge, R.P., Rees, R., Wood, P., Burton, H., and Ko, C.-P., 1974, Anatomical and physiological observations on synapses formed on isolated autonomic neurons in tissue culture, *Brain Res.* **66**:401.

Burn, J.H., 1977, The function of acetylcholine released from sympathetic fibres, *Clin. Exp. Pharmacol. Physiol.* **4**:59.

Burn, J.H., and Rand, M.J., 1965, Acetylcholine in adrenergic transmission, in: *Annual Review of Pharmacology* (W.S. Cutting, R.H. Dreisbash, and H.W. Elliott, eds.), pp. 163–182, Annual Reviews, Palo Alto, California.

Burnstock, G., 1976, Do some nerve cells release more than one transmitter? *Neuroscience* **1**:239.

Burnstock, G., 1978, Do some sympathetic neurons synthesize and release both noradrenaline and acetylcholine? *Prog. Neurobiol.* **10**:1.

Burnstock, G., and Bell, C., 1974, Peripheral autonomic transmission, in: *The Peripheral Nervous System* (J.I. Hubbard, ed.), pp. 277–327, Plenum Press, New York.

Burnstock, G., and Costa, M., 1975, *Adrenergic Neurons, Their Organization, Function and Development in the Peripheral Nervous System,* Chapman and Hall, London.

Burton, H., and Bunge, R.P., 1975, A comparison of the uptake and release of [^{3}H] norephinephrine in rat autonomic and sensory ganglia in tissue culture, *Brain Res.* **97**:157.

Campbell, G., 1970, Autonomic nervous supply to effector tissues, in: *Smooth Muscle* (E. Bülbring, A.F. Brading, A.W. Jones, and T. Tomita, eds.), pp. 451–495, The Williams & Wilkins Co., Baltimore.

Carbonetto, S.T., and Fambrough, D.M., 1978, Rates of metabolism and site of insertion of a plasma membrane protein in rapidly growing neurons, *Neurosci. Abstr.* **4**:510.

Carbonetto, S.T., Fambrough, D.M., and Muller, K.J., 1977, α-Bungarotoxin binding to chick sympathetic neurons, *Neurosci. Abstr.* **3**:454.

Carbonetto, S.T., Fambrough, D.M., and Muller, K.J., 1978, Nonequivalence of α-bungarotoxin receptors and acetylcholine receptors in chick sympathetic neurons, *Proc. Natl. Acad. Sci. USA* **75**:1016.

Ceccarelli, B., Clementi, F., and Mantegazza, P., 1971, Synaptic transmission in the superior cervical ganglion of the cat after reinnervation by vagus fibres, *J. Physiol. (Lond.)* **216**:87.

Chamley, J., Mark, G., Campbell, G., and Burnstock, G., 1972, Sympathetic ganglia in culture. I. Neurons, *Z. Zellforsch. Mikrosk. Anat.* **135**:287.

Chamley, J., Campbell, G., and Burnstock, G., 1973, An analysis of the interactions between sympathetic nerve fibers and smooth muscle cells in tissue culture, *Dev. Biol.* **33**:344.

Chamley, J., Campbell, G., and Burnstock, G., 1974, Dedifferentiation, redifferentiation and bundle formation of smooth muscle cells in tissue culture: The influence of cell number and nerve fibres, *J. Embryol. Exp. Morphol.* **32**:297.

Christ, D.D., and Nishi, S., 1971a, Site of adrenaline blockade in the superior cervical ganglion of the rabbit, *J. Physiol. (Lond.)* **213**:107.

Christ, D.D., and Nishi, S., 1971b, Effects of adrenaline on nerve terminals in the superior cervical ganglion of the rabbit, *Br. J. Pharmacol.* **41**:331.

Chun, L.L.Y., and Patterson, P.H., 1977a, Role of nerve growth factor in the development of rat sympathetic neurons *in vitro.* I. Survival, growth, and differentiation of catecholamine production, *J. Cell Biol.* **75**:694.

Chun, L.L.Y., and Patterson, P.H., 1977b, Role of nerve growth factor in the development of rat sympathetic neurons *in vitro.* II. Developmental studies, *J. Cell Biol.* **75**:705.

Chun, L.L.Y., and Patterson, P.H., 1977c, Role of nerve growth factor in the development of rat sympathetic neurons *in vitro.* III. Effect on acetylcholine production, *J. Cell Biol.* **75**:712.

Claude, P., 1973, Electron microscopy of dissociated rat sympathetic neurons in culture, *J. Cell Biol.* **59**:57a.

Cochard, P., Goldstein, M., and Black, I.B., 1978, Ontogenetic appearance and disappearance of tyrosine hydroxylase and catecholamines in the rat embryo, *Proc. Natl. Acad. Sci. USA* **15**:2986.

Coughlin, M.D., 1975, Target organ stimulation of parasympathetic nerve growth in the developing mouse submandibular gland, *Dev. Biol.* **43**:140.

Crain, S.M., 1968, Development of functional neuromuscular connections between separate explants of fetal mammalian tissues after maturation in culture, *Anat. Rec.* **160**:466.

Crain, S.M., 1976, *Neurophysiologic Studies in Tissue Culture,* Raven Press, New York.

Dail, W.G., and Evan, A.P., 1978, Ultrastructure of adrenergic terminals and SIF cells in the superior cervical ganglion of the rabbit, *Brain Res.* **148**:469.

Dail, W.G., Jr., Evan, A.P., Jr., and Eason, H.R., 1975, The major ganglion in the pelvic plexus of the male rat: A histochemical and ultrastructural study, *Cell Tissue Res.* **159**:49.

Dennis, M.J., Harris, A.J., and Kuffler, S.W., 1971, Synaptic transmission and its duplication by focally applied ACh in parasympathetic neurons in the heart of the frog, *Proc. R. Soc. Lond. B* **177**:509.

Dichter, M.A., and Fischbach, G.D., 1977, The action potential of chick dorsal root ganglion neurones maintained in cell culture, *J. Physiol. (Lond.)* **267**:281.

Dreyfus, C.L., Bornstein, M.B., and Gershon, M.D., 1977a, Synthesis of serotonin by neurons of the myenteric plexus *in situ* and in organotypic tissue culture, *Brain Res.* **128**:125.

Dreyfus, C.L., Sherman, D.L., and Gershon, M.D., 1977b, Uptake of serotonin by intrinsic neurons of the myenteric plexus grown in organotypic tissue culture, *Brain Res.* **128**:109.

Dun, N., and Karczmar, A.G., 1977, The presynaptic site of action of norepinephrine in the superior cervical ganglion of guinea pig, *J. Pharmacol. Exp. Ther.* **200**:328.

Dun, N.J., and Karczmar, A.G., 1978, Involvement of an interneuron in the generation of the slow inhibitory postsynaptic potential in mammalian sympathetic ganglia, *Proc. Natl. Acad. Sci. USA* **75**:4029.

Dun, N., and Nishi, S., 1974, Effects of dopamine on the superior cervical ganglion of the rabbit, *J. Physiol. (Lond.)* **239**:155.

Elfvin, L.-G., 1971, Ultrastructural studies on the synaptology of the inferior mesenteric ganglion of the cat, *J. Ultrastruct. Res.* **37**:411.

Eränkö, L., 1972, Ultrastructure of the developing sympathetic nerve cell and the storage of catecholamines, *Brain Res.* **46**:159.

Ferry, C.B., 1963, The sympathomimetic effect of acetylcholine on the spleen of the cat, *J. Physiol. (Lond.)* **167**:487.

Ferry, C.B., 1966, Cholinergic link hypothesis in adrenergic neuroeffector transmission, *Physiol. Rev.* **46**:420.

Fujiwara, M., and Kurahashi, K., 1976, Cholinergic nature of the primary afferent vagus synapsed in cross anastomosed superior cervical ganglia, *Life Sci.* **19**:1175.

Furness, J.B., and Costa, M., 1974, The adrenergic innervation of the gastrointestinal tract, *Ergeb. Physiol.* **69**:1.

Furshpan, E.J., MacLeish, P.R., O'Lague, P.H., and Potter, D.D., 1976, Chemical transmission between rat sympathetic neurons and cardiac myocytes developing in microcultures: Evidence for cholinergic, adrenergic, and dual-function neurons, *Proc. Natl. Acad. Sci. USA* **73**:4225.

Greengard, P., 1978, *Cyclic Nucleotides, Phosphorylated Proteins, and Neuronal Function,* Raven Press, New York.

Grillo, M.A., 1966, Electron microscopy of sympathetic tissues, *Pharmacol. Rev.* **18**:387.

Harrison, R., 1910, The outgrowth of the nerve fiber as a mode of protoplasmic movement, *J. Exp. Zool.* **9:**787.

Hashiguchi, T., Ushiyama, N.S., Kobayashi, H., and Libet, B., 1978, Does cyclic GMP mediate the slow excitatory synaptic potential in sympathetic ganglia? *Nature* **271:**267.

Hendry, I.A., 1977, Cell division in the developing sympathetic nervous system, *J. Neurocytol.* **6:**299.

Hild, W., and Tasaki, I., 1962, Morphological and physiological properties of neurons and glial cells in tissue culture, *J. Neurophysiol.* **25:**277.

Hill, C.E., and Hendry, I.A., 1976, Differences in sensitivity to nerve growth factor of axon formation and tyrosine hydroxylase induction in cultured sympathetic neurons, *Neuroscience* **1:**489.

Hill, C.E., and Hendry, I.A., 1977, Development of neurons synthesizing noradrenaline, and acetylcholine in the superior cervical ganglion of the rat *in vivo* and *in vitro, Neuroscience* **2:**741.

Hill, C.E., Purves, R.D., Watanabe, H., and Burnstock, G., 1976, Specificity of innervation of iris musculature by sympathetic nerve fibres in tissue culture, *Pflügers Arch.* **361:**127.

Hirst, G.D.S., and Spence, I., 1973, Calcium action potentials in mammalian peripheral neurons, *Nature New Biol.* **243:**54.

Hooisma, J., Slaaf, D.N., Meeter, E., and Stevens, W.F., 1975, The innervation of chick striated muscle fibers by the chick ciliary ganglion in tissue culture, *Brain Res.* **85:**79.

Horn, J.P., and McAfee, D.A., 1978, Norepinephrine antagonizes calcium-dependent potentials in rat sympathetic neurons, *Neurosci. Abstr.* **4:**235.

Jacobowitz, D., 1970, Catecholamine fluorescence studies of adrenergic neurons and chromaffin cells in sympathetic ganglia, *Fed. Proc.* **29:**1929.

Jacobowitz, D.M., 1974, The peripheral autonomic system, in: *The Peripheral Nervous System* (J.I. Hubbard, ed.), pp. 87–110, Plenum Press, New York.

Jessen, K., McConnell, J., Purves, R., Burnstock, G., and Chamley-Campbell, J., 1978, Tissue culture of mammalian enteric neurons, *Brain Res.* **152:**573.

Johnson, M., Ross, D., Meyers, M., Rees, R., Bunge, R.P., Wakshull, E., and Burton, H., 1976, Synaptic vesicle cytochemistry changes when cultured sympathetic neurons develop cholinergic interactions, *Nature* **262:**308.

Johnson, J.I., Ross, C.D., Meyers, M., Spitznagel, E.L., and Bunge, R.P., 1980a, Morphological and biochemical studies on the development of cholinergic properties in cultured sympathetic neurons. I. Correlative changes in choline acetyltransferase and synaptic vesicle cytochemistry, *J. Cell Biol.* **84:**680.

Johnson, J.I., Ross, C.D., and Bunge, R.P., 1980b, Morphological and biochemical studies on the development of cholinergic properties in cultured sympathetic neurons. II. Dependence on postnatal age, *J. Cell Biol.* **84:**692.

Joó, F., Lever, J.D., Ivens, C., Mottram, D.R., and Presley, R., 1971, A fine structural and electron histochemical study of axon terminals in the rat superior cervical ganglion after acute and chronic preganglionic denervation, *J. Anat.* **110:**181.

Katz, B., and Thesleff, S., 1957, A study of the "desensitization" produced by acetylcholine at the motor end plate, *J. Physiol. (Lond.)* **138:**63.

Kemplay, S.K., and Garrett, J.R., 1976, Effects of a heterologous cross-culture between the postganglionic sympathetic and the preganglionic parasympathetic nerve trunks of submandibular glands in cats, *Cell Tissue Res.* **167:**197.

Ko, C.-P., 1975, Electrophysiological Studies of Synaptic Transmission in Nerve Tissue Cultures of Rat Spinal Cord and Dissociated Superior Cervical Ganglion Neurons, Ph.D. Thesis, Washington University, St. Louis, Missouri.

Ko, C.-P., Burton, H., and Bunge, R.P., 1976a, Synaptic transmission between rat spinal cord explants and dissociated superior cervical ganglion neurons in tissue culture, *Brain Res.* **117:**437.

Ko, C.-P., Burton, H., Johnson, M.I., and Bunge, R.P., 1976b, Synaptic transmission between rat superior cervical ganglion neurons in dissociated cell cultures, *Brain Res.* **117:**461.

Kobayashi, H., Hashiguchi, T., and Ushiyama, N.S., 1978, Postsynaptic modulation of excitatory process in sympathetic ganglia by cyclic AMP, *Nature* **271:**268.

Koketsu, K., and Nishi, S., 1969, Calcium and action potentials of bullfrog sympathetic ganglion cells, *J. Gen. Physiol.* **53:**608.

Kuffler, S.W., and Yoshikami, D., 1975, The distribution of acetylcholine sensitivity at the postsynaptic membrane of vertebrate skeletal twitch muscles: Ionotophoretic mapping in the micron range, *J. Physiol. (Lond.)* **244:**703.

Landis, S.C., 1976, Rat sympathetic neurons and cardiac myocytes developing in micro-cultures: Correlation of the fine structure of endings with neurotransmitter function in single neurons, *Proc. Natl. Acad. Sci. USA* **73:**4220.

Landis, S.C., 1977, Morphological properties of the dendrites and axons of dissociated rat sympathetic neurons, *Neurosci. Abstr.* **3:**525.

Landis, S.C., MacLeish, P.R., Potter, D.D., Furshpan, E.J., and Patterson, P.H., 1976, Synapses formed between dissociated sympathetic neurons: The influence of conditioned medium, *Neurosci. Abstr.* **2:**197.

LeDouarin, N.M., and Teillet, M.A., 1974, Experimental analysis of the migration and differentiation of neuroblasts of the autonomic nervous system and of neuroectodermal mesenchymal derivatives, using a biological cell marking technique, *Dev. Biol.* **41:**162.

LeDouarin, N.M., Renaud, D., Teillet, M.A., and LeDouarin, G., 1975, Cholinergic differentiation of presumptive adrenergic neuroblasts in interspecific chimeras after heterotropic transplantations, *Proc. Natl. Acad. Sci. USA* **72:**728.

LeDouarin, N.M., Teillet, M.A., and LeLievre, C., 1977, Influence of the tissue environment on the differentiation of neural crest cells, in: *Cell and Tissue Interactions* (J. Lash and M. Burger, eds.), pp. 11–28, Raven Press, New York.

LeDouarin, N.M., Teillet, M.A., Ziller, C., and Smith, J., 1978, Adrenergic differentiation of cells of the cholinergic ciliary and Remak ganglia in avian embryo after *in vivo* transplantation, *Proc. Natl. Acad. Sci. USA* **75:**2030.

Levi, G., and Delorenzi, E., 1934, Transformazione degli elementi dei gangli spinali e simpatici cultivati *in vitro, Arch. Ital. Anat. Embriol.* **33:**443.

Levi-Montalcini, R., and Angeletti, P., 1963, Essential role of the nerve growth factor in the survival and maintenance of dissociated sensory and sympathetic nerve cells *in vitro, Dev. Biol.* **7:**658.

Levi-Montalcini, R., Meyer, H., and Hamburger, V., 1954, *In vitro* experiments on the effects of mouse sarcomas 180 and 37 on the spinal and sympathetic ganglia of the chick embryo, *Cancer Res.* **14:**49.

Lewis, W.H., and Lewis, M.R., 1912, The cultivation of sympathetic nerves from the intestine of chick embryos in saline solutions, *Anat. Rec.* **6:**7.

Libet, B., 1975, The SIF cell as a functional dopamine-releasing interneuron in the rabbit superior cervical ganglion, in: *SIF Cells: Structure and Function of the Small, Intensely Fluorescent Sympathetic Cells, Fogarty International Center Proceedings, No. 30* (O. Eränkö, ed.), pp. 163–177, U.S. Government Printing Office, Washington, D.C.

Lichtman, J.W., 1977, The reorganization of synaptic connexions in the rat submandibular ganglion during postnatal development, *J. Physiol. (Lond.)* **273:**155.

Lichtman, J.W., Yip, J.W., and Purves, D., 1978, Segmentally selective innervation of mammalian sympathetic ganglia: Relative roles of intraganglionic position and postganglionic targets, *Neurosci. Abstr.* **4:**119.

Mains, R.E., and Patterson, P.H., 1973a, Primary cultures of dissociated sympathetic neurons. I. Establishment of long-term growth in culture and studies of differentiated properties, *J. Cell Biol.* **59:**329.

Mains, R.E., and Patterson, P.H., 1973b, Primary cultures of dissociated sympathetic neurons. II. Initial studies on catecholamine metabolism, *J. Cell Biol.* **59:**346.

Mains, R.E., and Patterson, P.H., 1973c, Primary cultures of dissociated sympathetic neurons. III. Changes in metabolism with age in culture, *J. Cell Biol.* **59:**361.

Mark, G.E., Chamley, J.H., and Burnstock, G., 1973, Interactions between autonomic nerves and smooth and cardiac muscle cells in tissue culture, *Dev. Biol.* **32:**194.

Masurovsky, E., and Benitez, H., 1967, Apparent innervation of chick cardiac muscle by sympathetic neurons in corganized culture, *Anat. Rec.* **157:**285.

Matsuda, Y., Yoshida, S., and Yonezawa, T., 1978, Tetrodotoxin sensitivity and Ca component of action potentials of mouse dorsal root ganglion cells cultured *in vitro, Brain Res.* **154:**69.

Matthews, M.B., and Raisman, G., 1969, The ultrastructure and somatic efferent synapses of small granule containing cells in the superior cervical ganglion, *J. Anat.* **105:**255.

Matthews, M.R., 1974, Ultrastructure of ganglionic junctions, in: *The Peripheral Nervous System* (J.I. Hubbard, ed.), pp. 111–150, Plenum Press, New York.

Merrillees, N.C.R., 1968, The nervous environment of individual smooth muscle cells of the guinea pig *vas deferens, J. Cell Biol.* **37:**794.

Murray, M., and Stout, A., 1947, Adult human sympathetic ganglion cells cultivated *in vitro, Am. J. Anat.* **30:**225.

Neale, E.A., MacDonald, R.L., and Nelson, P.G., 1978, Intracellular horseradish peroxidase injection for correlation of light and electron microscopic anatomy with synaptic physiology of cultured mouse spinal cord neurons, *Brain Res.* **152:**265.

Nishi, R., and Berg, D.K., 1977, Dissociated ciliary ganglion neurons *in vitro:* Survival and synapse formation, *Proc. Natl. Acad. Sci. USA* **74:**5171.

Nishi, R., and Berg, D.K., 1979, Survival and development of ciliary ganglion neurones grown alone in cell culture, *Nature* **277:**232.

Nishi, S., 1974, Ganglionic transmission, in: *The Peripheral Nervous System* (J.I. Hubbard, ed.), pp. 225–255, Plenum Press, New York.

Njå, A., and Purves, D., 1977a, Specific innervation of guinea-pig superior cervical ganglion cells by preganglionic fibres arising from different levels of the spinal cord, *J. Physiol. (Lond.)* **264:**565.

Njå, A., and Purves, D., 1977b, Re-innervation of guinea-pig superior cervical ganglion cells by preganglionic fibres arising from different levels of the spinal cord, *J. Physiol. (Lond.)* **272:**633.

Njå, A., and Purves, D., 1978, Specificity of initial synaptic contacts made on guinea-pig superior cervical ganglion cells during regeneration of the cervical sympathetic trunk, *J. Physiol. (Lond.)* **281:**45.

Nurse, C.A., and O'Lague, P.H., 1975, Formation of cholinergic synapses between dissociated sympathetic neurons and skeletal myotubes of the rat in cell culture, *Proc. Natl. Acad. Sci. USA* **75:**1955.

Obata, K., 1974, Transmitter sensitivities of some nerve and muscle cells in culture, *Brain Res.* **73:**71.

Okun, L.M., 1972, Isolated dorsal root ganglion neurons in culture: Cytological maturation and extension of electrically active processes, *J. Neurobiol.* **3:**111.

O'Lague, P.H., Obata, K., Claude, P., Furshpan, E.G., and Potter, D.D., 1974, Evidence for cholinergic synapses between dissociated rat sympathetic neurons in cell culture, *Proc. Natl. Acad. Sci. USA* **71:**3602.

O'Lague, P.H., MacLeish, P.R., Nurse, C.A., Claude, P., Furshpan, E.J., and Potter, D.D.,

1976, Physiological and morphological studies on developing sympathetic neurons in dissociated cell culture, *Cold Spring Harbor Symp. Quant. Biol.* **XL**:399.

O'Lague, P.H., Potter, D.D., and Furshpan, E.J., 1978a, Studies on rat sympathetic neurons developing in cell culture. I. Growth characteristics and electrophysiological properties, *Dev. Biol.* **67**:384.

O'Lague, P.H., Furshpan, E.J., and Potter, D.D., 1978b, Studies on rat sympathetic neurons developing in cell culture. II. Synaptic mechanisms, *Dev. Biol.* **67**:404.

O'Lague, P.H., Potter, D.D., and Furshpan, E.G., 1978c, Studies on rat sympathetic neurons developing in cell culture. III. Cholinergic transmission, *Dev. Biol.* **67**:424.

Olson, M.I., and Bunge, R.P., 1973, Anatomical observations on the specificity of synapse formation in tissue culture, *Brain Res.* **59**:19.

Patterson, P.H., 1978, Environmental determination of autonomic neurotransmitter functions, *Ann. Rev. Neurosci.* **1**:1.

Patterson, P.H., and Chun, L.L.Y., 1974, The influence of non-neuronal cells on catecholamine and acetylcholine synthesis and accumulation in cultures of dissociated sympathetic neurons, *Proc. Natl. Acad. Sci. USA* **71**:3607.

Patterson, P.H., and Chun, L.L.Y., 1977a, The induction of acetylcholine synthesis in primary cultures of dissociated rat sympathetic neurons. I. Effects of conditioned medium, *Dev. Biol.* **56**:263.

Patterson, P.H., and Chun, L.L.Y., 1977b, The induction of acetylcholine synthesis in primary cultures of dissociated rat sympathetic neurons. II. Developmental aspects, *Dev. Biol.* **60**:473.

Patterson, P.H., Reichardt, L.F., and Chun, L.L.Y., 1976, Biochemical studies on the development of primary sympathetic neurons in cell culture, *Cold Spring Harbor Symp. Quant. Biol.* **XL**:389.

Patterson, P.H., Chun, L.L.Y., and Reichardt, L.F., 1977, The role of non-neuronal cells in the development of sympathetically derived neurons, in: *Progress in Clinical and Biological Research, Vol. 15, Cellular Neurobiology* (Z. Hall, R. Kelly, and C.F. Fox, eds.), pp. 95–103, Alan R. Liss, New York.

Perri, V., Sacchi, O., and Casella, C., 1970, Electrical properties and synaptic connections of the sympathetic neurons in the rat and guinea pig superior cervical ganglion, *Pflügers Arch.* **314**:40.

Phillis, J.W., 1970, *The Pharmacology of Synapses,* Wheaton and Co., Exeter.

Purves, D., 1976, Competitive and non-competitive re-innervation of mammalian sympathetic neurones by native and foreign fibres, *J. Physiol. (Lond.)* **261**:453.

Purves, D., and Lichtman, J.S., 1978, Formation and maintenance of synaptic connections in autonomic ganglia, *Physiol. Rev.* **58**:821.

Purves, D., and Njå, A., 1976, Effect of nerve growth factor on synaptic depression after axotomy, *Nature* **260**:535.

Purves, D., Lichtman, J.W., and Yip, J.W., 1978, Segmentally selective innervation of mammalian sympathetic ganglia: Further evidence for the importance of postganglionic target position, *Neurosci. Abstr.* **4**:124.

Purves, R.D., Hill, D.E., Chamley, J.H., Mark, G.E., Fry, D.M., and Burnstock, G., 1974, Functional autonomic neuromuscular junctions in tissue culture, *Pflügers Arch.* **350**:1.

Raisman, G., Field, P.M., Östberg, A.-J.C., Iversen, L.L., and Zigmond, R.E., 1974, A quantitative ultrastructural and biochemical analysis of the process of reinnervation of the superior cervical ganglion in the adult rat, *Brain Res.* **71**:1.

Rees, R., and Bunge, R.P., 1974, Morphological and cytochemical studies of synapses formed in culture between isolated rat superior cervical ganglion neurons, *J. Comp. Neurol.* **157**:1.

Rees, R.P., Bunge, M.B., and Bunge, R.P., 1976, Morphological changes in the neuritic growth cone and target neuron during synaptic junction development in culture, *J. Cell Biol.* **68:**240.

Reichardt, L.F., and Patterson, P.H., 1977, Neurotransmitter synthesis and uptake by isolated sympathetic neurones in microcultures, *Nature* **270:**147.

Roper, S., 1976a, An electrophysiological study of chemical and electrical synapses on neurones in the parasympathetic cardiac ganglion of the mud puppy, *Necturus maculosus:* Evidence for intrinsic ganglionic innervation, *J. Physiol. (Lond.)* **254:**427.

Roper, S., 1976b, The acetylcholine sensitivity of the surface membrane of multiply-innervated para-sympatehtic ganglion cells in the mudpuppy before and after partial denervation, *J. Physiol. (Lond.)* **254:**455.

Ross, D., and Bunge, R.P., 1976, Choline acetyltransferase in cultures of rat superior cervical ganglion, *Neurosci. Abstr.* **2:**769.

Ross, D., Johnson, M., and Bunge, R., 1977, Development of cholinergic characteristics in adrenergic neurones is age dependent, *Nature* **267:**536.

Sargent, P.B., and Dennis, M.J., 1977, Formation of synapses between parasympathetic neurones deprived of preganglionic innervation, *Nature* **268:**456.

Shimada, Y., and Fischman, D., 1973, Morphological and physiological evidence for the development of functional neuromuscular junctions *in vitro, Dev. Biol.* **31:**200.

Silberstein, S., Johnson, D., Jacobowitz, D., and Kopin, I., 1971, Sympathetic reinnervation of the rat iris in organ culture, *Proc. Natl. Acad. Sci. USA* **68:**1121.

Sjöqvist, F., 1963, The correlation between the occurrence and localization of acetylcholinesterase-rich cell bodies in the stellate ganglion and the outflow of cholinergic sweat secretory fibers to the forepaw of the cat, *Acta Physiol. Scand.* **57:**339.

Szantroch, Z., 1933, Beobachtung an den Kulturen des Sympathetikus. Ergebnisse der Zuchtung des Remakschen Darmnerven, *Arch. Exp. Zellforsch.* **14:**442.

Tamarind, D.L., and Quilliam, J.P., 1971, Synaptic organization and other ultrastructural features of the superior cervical ganglion of the rat, kitten, and rabbit, *Micron* **2:**204.

Tashiro, N., and Nishi, S., 1972, Effects of alkali-earth cations on sympathetic ganglion cells of the rabbit, *Life Sci.* **11:**941.

Thompson, W., Njå, A., and Purves, D., 1978, Effects of postganglionic axotomy on the regeneration of selective synaptic connections in the guinea-pig superior cervical ganglion, *Neurosci. Abstr.* **4:**128.

Tuttle, J., Ard, M., and Suszkiw, J., 1978, Neuronal survival and cat activity in dissociated cell cultures of ciliary ganglion, *Neurosci. Abstr.* **4:**596.

Uehara, Y., Campbell, G.R., and Burnstock, G., 1976, *Muscle and its Innervation,* Edward Arnold, London.

Varon, S., and Raiborn, C., 1971, Excitability and conduction in neurons of dissociated ganglionic cell cultures, *Brain Res.* **30:**83.

Volle, R.L., 1969, Ganglionic transmission, *Annu. Rev. Pharmacol.* **9:**135.

Wakshull, E.M., 1978, Studies on the *in Vitro* Development of Rat Sympathetic Neurons, Ph.D. Thesis, Washington University, St. Louis, Missouri.

Wakshull, E.M., Johnson, J.I., and Burton, H., 1978, Persistence of an amine uptake system in cultured rat sympathetic neurons which use acetylcholine as their transmitter, *J. Cell Biol.* **79:**121.

Wakshull, E.M., Johnson, M.I., and Burton, H., 1979a, Postnatal rat sympathetic neurons in culture. I. A comparison with embryonic neurons, *J. Neurophysiol.* **42:**1410.

Wakshull, E.M., Johnson, M.I., and Burton, H., 1979b, Postnatal rat sympathetic neurons in culture. II. Synaptic transmission by postnatal neurons, *J. Neurophysiol.* **42:**1426.

Walicke, P.A., Campenot, R.B., and Patterson, P.H., 1977, Determination of transmitter function by neuronal activity, *Proc. Natl. Acad. Sci. USA* **74:**5767.

Weight, F.F., 1971, Mechanisms of synaptic transmission, in: *Neurosciences Research, Vol. 4* (S. Ehrenpreis and O.C. Solnitzky, eds.), pp. 1–27, Academic Press, New York.

Weston, J., 1970, The migration and differentiation of neural crest cells, *Adv. Morphol.* **8**:41.

Williams, T.H., and Palay, S.L., 1969, Ultrastructure of the small neurons in the superior cervical ganglion, *Brain Res.* **15**:17.

Yamauchi, A., Lever, J.D., and Kemp, K.W., 1973, Catecholamine loading and depletion in the rat superior cervical ganglion: A formol fluorescence and enzyme histochemical study with numerical assessments, *J. Anat.* **114**:271.

Yarowsky, P., and McAfee, D.A., 1977, Calcium dependent potentials in the rat superior cervical ganglion, *Neurosci. Abstr.* **3**:25.

Yip, J.W., Purves, D., and Lichtman, J.W., 1978, Segmentally selective innervation of mammalian sympathetic ganglia: Comparative innervation of cervical and thoracic ganglia, *Neurosci. Abstr.* **4**:130.

Yokota, R., and Yamauchi, A., 1974, Ultrastructure of the mouse superior cervical ganglion with particular reference to the pre- and postganglionic elements covering the soma of its principal neurons, *Am. J. Anat.* **140**:281.

2

Electrophysiological and Structural Studies of Neurons in Dissociated Cell Cultures of the Central Nervous System

PHILLIP G. NELSON, ELAINE A. NEALE, and ROBERT L. MACDONALD

1. INTRODUCTION

The dissociated cell culture approach to the analysis of central nervous system (CNS) function begins with a radical disassembly of the system into its single-cell components. Reconstitution of the system must be done in such a way that the reassembled components convincingly capture those neurobiological properties or processes that are of concern to the investigator. In the most general scheme, this would involve separation of the single cell suspension into pure subpopulations corresponding to each major CNS cell type. The intrinsic properties of each cell type and their development with time could be determined by appropriate electrophysiological, morphological, and biochemical studies. Interactions among the different cell types could then be studied in a variety of cultures formed from combinations of the different cell types. Some of these interactions would presumably be specific to particular cell combinations and would contribute to an understanding of the assembly of specifically organized neuronal systems in the brain.

PHILLIP G. NELSON and ELAINE A. NEALE • Laboratory of Developmental Neurobiology, National Institute of Child Health and Human Development, National Institutes of Health, Bethesda, Maryland 20205. ROBERT L. MACDONALD • Department of Neurology, University of Michigan, Ann Arbor, Michigan 48109.

Although such a comprehensive scheme is far from realization, less elaborate reconstitution has generated a number of experimental neural systems that are attractive for an analysis of important apsects of CNS development and function. Specific, highly differentiated neuronal morphologies develop in cell cultures, and the organization of a variety of neuronal types into complex and active synaptic circuits is a prominent feature of such cultures. Thus, central neuronal morphogenesis and synaptogenesis are captured in preparations which allow a high degree of experimental accessibility and manipulation. Because of this great accessibility, detailed neuropharmacological studies are feasible, and correlative physiological and anatomical analysis of central synaptic mechanisms can be approached more readily then in intact systems. Biochemical analysis of cellular interactions related to neuronal specification and development is facilitated *in vitro,* since diffusible substances mediating such interactions can be identified by conditioned medium experiments. Exploitation of such possibilities is the primary goal in tissue culture studies of the nervous system, although technical efforts to improve the culture systems still occupy a substantial part of the energies of many laboratories engaged in the art of neural tissue culture.

It should be noted that, while the visibility and accessibility of dissociated cell cultures greatly facilitate intracellular recordings, the pioneering intracellular studies on cultured dorsal root ganglion (DRG) cells were done on explanted preparations (Crain, 1956). Intracellular recordings from neurons in hippocampal explants have demonstrated characteristic paroxysmal depolarizing bursts (Zipser *et al.,* 1973), and spike and synaptic potentials have been recorded from hypothalamic explants (Gähwiler *et al.,* 1978). Thus, even if intracellular recordings were required for a study, explanted or reaggregated preparations might be compatible with that requirement. Explant cultures of cerebral cortex and cerebellum exhibit distinctive patterns of electrical activity (Calvet, 1974; Calvet *et al.,* 1974; Schlapfer *et al.,* 1972), and cortical and deep nuclear interactions have been documented in cerebellar explants (Hendelman *et al.,* 1977; Seil and Leiman, 1977). The usefulness of cerebellar explants as neuropharmacological tools has been well shown (Gähwiler, 1975, 1976; Geller and Woodward, 1974), and functional interactions between inferior olive and cerebellar explants demonstrated (Gähwiler, 1978). Distinctive burst patterns in hypothalamic explants suggest that some electrical properties characteristic of the tissue *in vivo* are maintained *in vitro* (Gähwiler *et al.,* 1978; Geller, 1975). Morphological evidence for synapse formation (Masurovsky *et al.,* 1971) and peptide synthesis (Toran-Allerand, 1978) in hypothalamic explants further indicates the high degree of differentiated functions expressed by these preparations. For the many valuable papers outside the scope of the present chapter that illustrate the broad range of neurobiological phenomena captured in explants and cell cultures from the CNS, previous reviews should be consulted (Crain, 1976; Fedoroff and Hertz, 1977; Fischbach and Nelson, 1977; Nelson, 1975).

The present chapter will focus largely on primary dissociated cell cultures. It is encouraging that it has proven possible to prepare dissociated cell cultures from essentially all levels of the neuraxis including cerebral cortex (Dichter, 1978; Godfrey *et al.,* 1975; Varon and Raiborn, 1969; Yavin and Menkes, 1973; Yavin and Yavin, 1977), cerebellum (Lasher and Zagon, 1972; Messer, 1977; Nelson and Peacock, 1973; Stefanelli *et al.,* 1977; Trenkner and Sidman, 1977), hypothalamus (Benda *et al.,* 1975), and hippocampus (Banker and Cowan, 1979; Peacock, 1979; Peacock *et al.,* 1979). A description of three preparations is given in an appendix to the present chapter to illustrate the

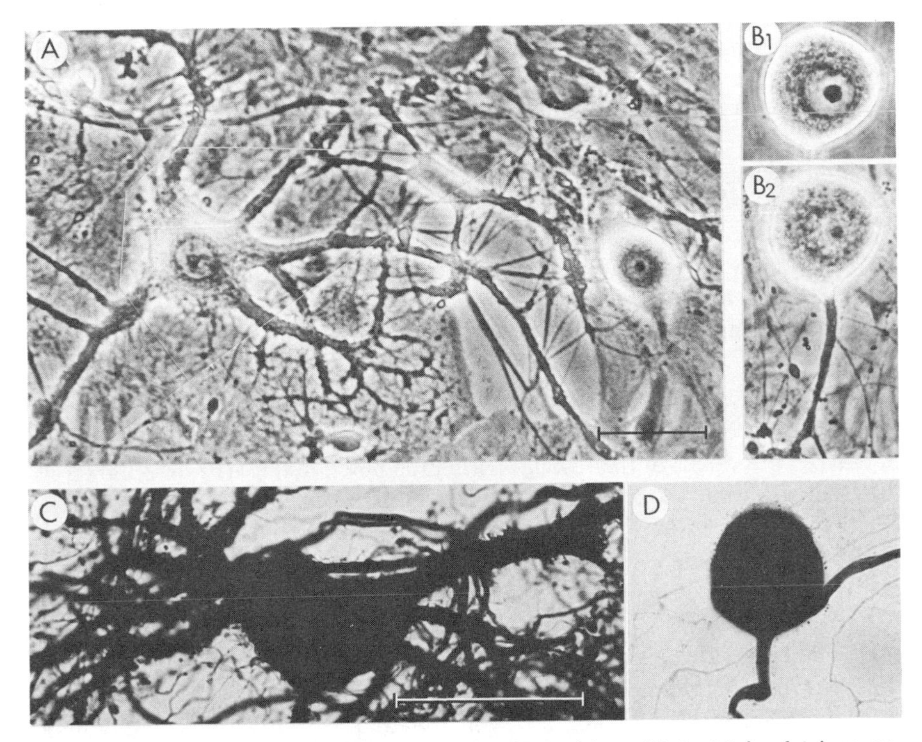

FIGURE 1. Neurons in dissociated cell culture prepared from 13- to 14-day fetal mouse spinal cords and dorsal root ganglia and maintained > 4 weeks. (A) Photomontage showing the soma and processes of a large multipolar spinal cord (SC) neuron (left) and the phase bright soma and process of a dorsal root ganglion (DRG) neuron (right). A small SC neuron is seen between the two larger neurons. The neurons lie upon flattened background cells which cover the surface of the tissue culture substrate. (B) DRG neuron photographed in two focal planes. In B_1, the sharp outline of the soma and distinct nucleus and nucleolus are shown, while in B_2 a single process is seen to emerge from the soma. (C) and (D) Neurons injected with horseradish peroxidase (HRP), fixed, reacted histochemically, and embedded in Epon. (C) Soma and dendrites of an HRP-injected SC neuron. Fine axonal branches and swellings are also visible in the field. (D.) HRP-injected DRG neuron. A and B, phase optics. C and D, bright-field optics. Magnification bars = 50 μm. Bar in C applies to B–D.

variables that have been considered important in optimizing different cell culture systems. For a detailed discussion of methodological issues, a previous review (Fischbach and Nelson, 1977) should be consulted. It seems likely that in order to capture *in vitro* as full a CNS functional repertoire as possible, the accessibility of a variety of brain areas will be essential. Certainly region-specific morphology is expressed in the cell cultures (see Section 3.1), and it is reasonable to suppose that other specific properties will also be preserved.

Chick and rodent spinal cord (SC) and dorsal root ganglion (DRG) cell cultures are discussed here in most detail since intracellular electrophysiological techniques have been used extensively in these preparations (Fischbach, 1970; Fischbach and Dichter, 1974; Peacock *et al.*, 1973; Ransom *et al.*, 1977a–c; Varon and Raiborn, 1971). The two cell types are clearly distinguishable, and it has been possible to study systematically the differentiated properties peculiar to each. Certain classes of synaptic interactions in the DRG–SC cell cultures have been examined with combined electrophysiological and anatomical methods (E. Neale *et al.*, 1978a). Figure 1 demonstrates the cellular resolution afforded by neurons in a two-dimensional system and illustrates morphological features typical of SC and DRG cell types.

2. ELECTRICAL PROPERTIES OF CULTURED NEURONS

2.1. Resting Membrane Potential

Absolute values of the resting membrane potential (RMP) determined in different studies must be compared with caution, because the impalement of cells with intracellular recording microelectrodes produces variable and unknown amounts of damage and depolarization. Criteria for selecting those cells that will be considered "normal" enough to be included in the data base vary from study to study, undoubtedly influencing the average value of RMP that is determined. This appeared to be the case in two studies that measured the RMP of DRG and SC cells. In one case (Peacock *et al.*, 1973), a significant difference was found, i.e., RMP of −40 and −51 mV for SC and DRG cells, respectively, while in the other case (Ransom *et al.*, 1977c), the average RMPs for the two cell types were not significantly different, each being about −50 mV. This discrepancy in result is probably attributable to a less restrictive selection criterion in the earlier study.

There is substantial agreement that the RMP of DRG cells is −40 to −50 mV (Dichter and Fishbach, 1977; Matsuda *et al.*, 1978; Peacock *et al.*, 1973; Scott *et al.*, 1969). The membrane potential of these cells was significantly increased by 4.4 mV when the concentration of potassium ions (K^+) in the bathing medium was decreased from 5.4 mM to 3 mM. Such a decrease in external K^+ did not affect the RMP of SC cells (Ransom *et al.*, 1977c), but substantial increases in external K^+ did produce depolarization of spinal cord cells in explant cultures (Hösli *et al.*, 1972) and in cell cultures (Ransom and Blank, unpublished data). A large decrease in extracellular chloride ions (Cl^-) did not produce a significant change in the RMP of SC cells (Ransom *et al.*, 1977c).

The resting potentials of cerebral cortical cells were substantially more negative (−65 mV) (Dichter, 1978) than those of SC and DRG cells. While the technical uncertainties mentioned above must be borne in mind, a difference this large (15 mV) would seem to reflect a biological difference, particularly since the higher value of RMP was obtained in smaller, more fragile cells that presumably would be more depolarized by the recording microelectrode. Data selection criteria would not explain the difference, since the mean RMP value for cortical cells lay nearly outside the distribution of RMP values obtained from DRG cells (Scott *et al.*, 1969). A useful comparison of electrical properties of brain cells *in vivo* and *in vitro* was compiled by Dichter (1978) and is reproduced in Table I.

If the difference between brain and SC cells were of biological origin, it would suggest that the permeability of the surface membrane to Na^+ (P_{Na}) relative to that of K^+ (P_K) is lower in brain cells than in SC cells; that is, the $P_{Na} : P_K$ ratio is lower in brain cells than in SC cells. In turn, this would suggest that brain cells would be more responsive to changes in extracellular K^+ than the SC cells. An alternative interpretation involves an electrogenic sodium pumping mechanism. SC cells depolarize by about 5 mV when exposed to strophanthidin, an inhibitor of the Na : K ATPase responsible for active sodium transport (Ransom *et al.*, 1975). It is possible that the sodium pump is more active or more electrogenic in brain cells than in SC cells, accounting for the greater negativity of the brain cell RMP. No estimates of intracellular K^+ concentration are available, nor have possible contributions of Cl^- permeability or active transport to the resting membrane potential been extensively studied.

The uncertainties evident in this brief discussion reflect the fact that few studies have been done on the influence of a changed ionic milieu on central neuronal RMP. In view of the significant changes in extracellular K^+ that occur in the CNS in different states of activation (Somjen, 1975), neuronal sensitivity to such changes must be considered an important determinant of CNS function; further study is surely indicated.

2.2. Cell Resistance and Capacitance

The input resistance (R_{in}) of neurons is measured by injecting an appropriate current pulse (ΔI) across the surface membrane through an intracellular electrode and measuring the voltage change (ΔV) produced by the current; the ratio $\Delta V/\Delta I$ is equal to R_{in}. The R_{in} of cultured spinal cord and brain neurons is 2- to 10-fold higher than that measured in mature cat motoneurons *in vivo*.

TABLE I. Electrophysiological Properties of Cortical Neurons[a]

Reference	Preparation	No. of neurons	Resting potential (mV)	Action potentials: Overshoot (mV)	Action potentials: Size (mV)	Duration[b] (msec)	Input resistance (MΩ)	Time constant (msec)
Dichter (1978)	Rat cortex in tissue culture	31[c]	65 ± 1	15 ± 2	78 ± 2 (up to 108)	1.1 ± 0.1	50 ± 4	5 (2–12)
Takahashi (1965)	Cat (pentobarbital)	90	58–62	—	60–80 (up to 100)	0.5–1.5	2–15	3–15
Creutzfeldt *et al.* (1966)	Cat (pentobarbital)	—	55–70	—	60–75 (up to 100)	0.3–1.0	28 ± 12	8
Lux and Pollen (1966)	Cat (pentobarbital)	64	50–70	—	—	0.75	4–15	8–10
Li *et al.* (1971)	Cat (pentobarbital)	ca. 100	60 ± 7	—	—	—	9 ± 3	6
Kandel *et al.* (1961), Spencer and Kandel (1961)	Cat hippocampus (hexobarbital)	40	48	13	63 (up to 110)	1.8	13	12
Schwartzkroin (1975)	Guinea pig hippocampal slice	215	54 ± 12	—	60 ± 10	—	16 ± 5	10 ± 3

[a]From Dichter (1978).
[b]Duration of action potentials measured at one-half height by Dichter and at foot in other reports.
[c]Consecutive impalements with membrane potential larger than or equal to −40 mV in two cultures.

Single electrodes in a bridge circuit configuration have been most widely used in studies of neuronal electrical properties. Nonlinear behavior of the electrodes can pose serious problems in such studies, although useful data can be obtained (see Fig. 2B for comparison of records obtained inside and outside the cell illustrating excellent bridge balance). Two-electrode recordings can be obtained from the cultured neurons (Fig. 2A), and these data, with independent voltage sensing and current passage, are undoubtedly more reliable. The membrane time constant (T_m), is quite comparable *in vivo* and *in vitro*. The membrane time constant is a useful indicator in that it is independent of total cell surface area since it is equal to the specific membrane capacitance (C_m) times the specific membrane resistance (R_m); these latter are the capacitance and resistance of a unit area of membrane. If it were assumed that C_m is relatively invariant from preparation to preparation, the similarity of T_m between *in vivo* and *in vitro* conditions would imply that the specific neuronal membrane resistances are also similar. The higher R_{in} obtained *in vitro* therefore implies a lesser total surface membrane area in the cultured cells, since $R_{in} = R_m$/cell surface area. It is not unreasonable that cultured cells would be smaller than their counterparts in intact mature systems. This is but one example of the discrepancies, demonstrated or potential, that must be kept in mind when interpreting results from the model system.

2.3. Dendritic Characteristics

Despite their smaller overall size, the cultured cells do develop an impressive dendritic apparatus that is reflected in their passive electrical behavior. Compared with immature pyramidal neurons of comparable age *in vivo,* hippocampal neurons in culture show a remarkable similarity in general form, although with somewhat smaller somata and less extensive dendritic trees (Banker and Cowan, 1979). The analysis of voltage transients elicited by transmembrane current steps indicates that the total electrotonic length of the dendrites of large SC cells in culture is about one characteristic length constant (Ransom *et al.,* 1977c) (Fig. 2A,B). This is only about half that of the total electrotonic length of motoneuronal dendrites in the mature cat spinal cord (Nelson and Lux, 1970), again consistent with a lesser degree of development for cultured cells. Distal dendritic synaptic inputs *in vitro* are thus less removed electrically from the cell body than is the case *in vivo*. DRG cells give no electrical evidence of a dendritic tree, consonant with their simple morphology.

2.4. Action Potentials

Essentially all cells identified as neuronal on morphological grounds and from which stable intracellular recordings could be obtained are electrically excitable. Substantial differences among neuronal types with respect to action

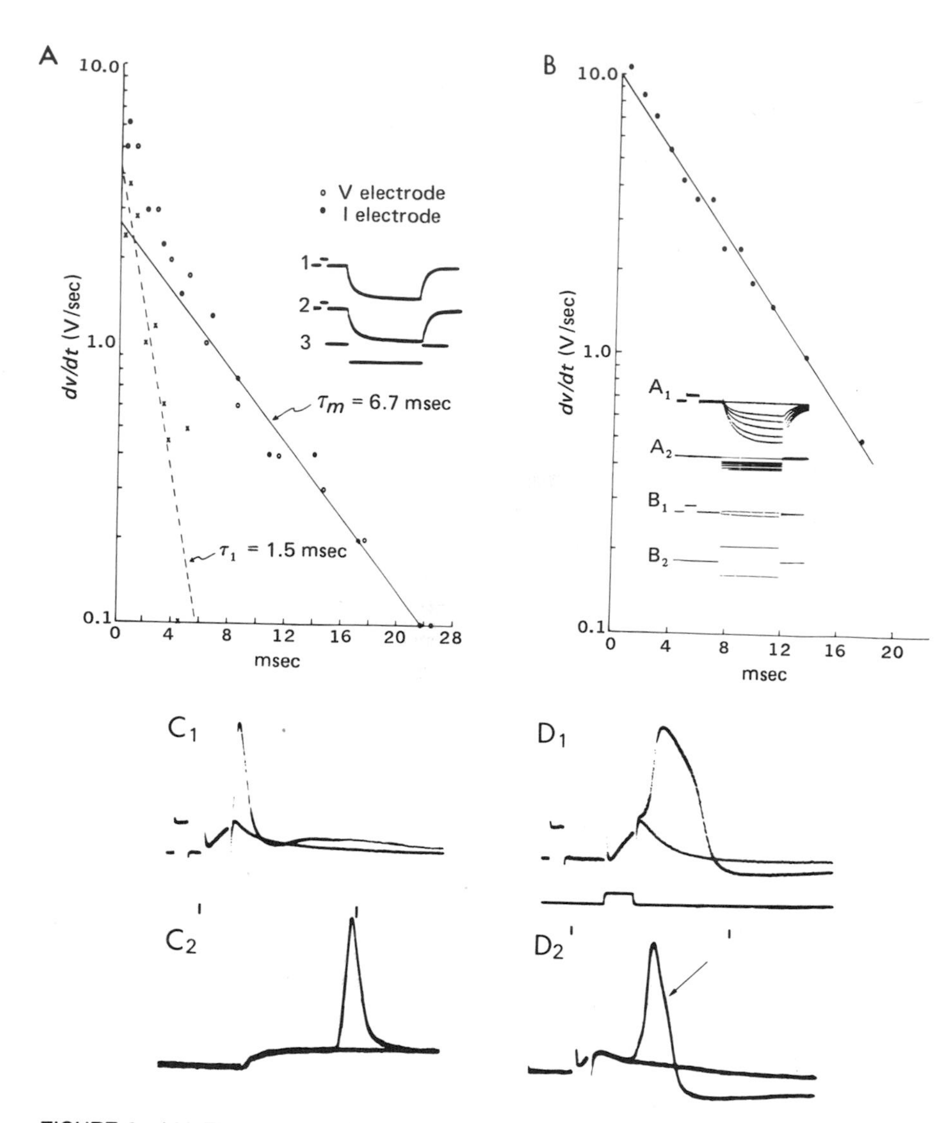

FIGURE 2. (A) Time course of the transmembrane potential change elicited by a current step in an SC neuron. Two electrodes were inserted into the same cell. The current step was passed through one electrode, and the transmembrane voltage response was measured both with a second electrode which carried no current (trace 1 of inset) and with the current-carrying electrode by use of a bridge circuit (trace 2 of inset). The calibration pulses represent 10 mV and 5 msec. The current step of approximately 1.5 nA is shown in trace 3. The bridge circuit recording is indistinguishable from that obtained with the second electrode. The electrical transient response of the membrane (——) is plotted as the rate of change of potential with respect to time as a function of time after the beginning of the current step. The negative slope of the late portion of this graphical representation of the transient is related to the membrane time constant. At early times the data points deviate

potential mechanism have been noted, however, and Fig. 2 illustrates the different action potential configurations of SC and DRG neurons. In both spinal cord (Dichter and Fischbach, 1977; Ransom and Holz, 1977) and cerebral cortical (Dichter, 1978) cultures, all neurons exhibit action potentials that are completely blocked by tetrodotoxin (TTX), an agent that acts specifically on voltage-dependent Na^+ channels. Since SC neurons are inexcitable in low-Na^+ solutions, the depolarizing phase of the action potential would appear to involve a permeability increase largely for Na^+. In the presence of tetraethylammonium ions, however, substantial calcium (Ca^{2+}) currents can be detected in SC cells (Bergey and Nelson, unpublished data). In DRG cells, by contrast, both Na^+ and Ca^{2+} make substantial contribution to the upstroke of the action potential. The sodium component of the DRG action potential may be sensitive to TTX [chick (Dichter and Fischbach, 1977)] or at least partially insensitive to TTX [mouse (Matsuda *et al.*, 1978; Ransom and Holz, 1977)]. A developmental shift from Ca^{2+}-dependent action potentials to a predominantly Na^+-dependent mechanism was suggested for amphibian neurons in culture (Spitzer and Lamborghini, 1976) and *in vivo* (Baccaglini, 1978).

That a similar developmental progression may be characteristic of mammalian DRG cells as well has been proposed by Matsuda *et al.* (1978) who have described in some detail the variation among DRG action potentials. Dichter and Fischbach (1977) discussed the relationship of the DRG action potential as observed *in vitro* to the situation *in vivo* and suggested that some Ca^{2+} involvement in the action currents may persist in adult DRG cells. (See Burton and Bunge, this volume, for further discussion). A particularly interesting aspect of the Ca^{2+} component of the DRG action potential is that it can

from the solid line. The dashed line and × points represent the difference between data points and the extrapolated solid line. (B) Time course of the transmembrane potential change elicited by a current step in a DRG neuron. Inset A shows membrane voltage response (1) to various currents (2). Maximum current pulse is about 1 nA. Inset B shows voltage records (B_1) taken with the electrode outside the cell with depolarizing and hyperpolarizing current pulses being passed through the electrode. Calibration pulses represent 10 mV and 5 msec. The graph is produced in the same way as in A. The negative slope of this plot is a measure of the membrane time constant. No significant deviation from the single semilogarithmic relationship is evident. No detectable alteration in the graph resulted if control extracellular records were subtracted from the intracellular records. (C_1) and (C_2) Action potentials (AP) recorded from mouse SC cells showing relatively simple, uninflected configuration and lesser afterhyperpolarization as compared with DRG APs. Calibration pulse in C_1 represents 20 mV and 0.5 msec in both C_1 and C_2. Resting membrane potential (RMP) was −60 to −65 mV in both C_1 and C_2. (D_1) and (D_2) Action potential recorded from mouse DRG cells. Early slow falling phase of the AP is followed by a more rapid phase with a hyperpolarizing afterpotential. Slow phase is more obvious in D_1; arrow shows inflection in D_2. Calibration pulse in D_1 is 20 mV and 0.5 msec for D_1 and D_2. RMP was −60 to −65 mV in both. Current pulse (lowest trace of D_1) represents about 0.5 nA. (Panels A and B reproduced with permission from Ransom *et al.*, 1977c.)

be modulated by a number of neuroactive compounds (Dunlap and Fischbach, 1978). GABA, serotonin, norepinephrine, and enkephalin (Mudge *et al.*, 1979) produce striking decreases in calcium currents, while glutamate has no effect. These workers raised the possibility that Ca^{2+} modulation may be at least in part responsible for presynaptic inhibitory effects. Conversely, the increase in calcium spike currents occurring as a result of serotonin application in *Aplysia* neurons has been implicated in a reversal of desensitization (Castellucci *et al.*, 1978).

Postactivation potentials occur in both SC and DRG cells but appear to be mediated, at least in part, by different mechanisms (Ransom *et al.*, 1975). In DRG cells, both the afterhyperpolarization (AHP) following single spikes and posttetanic hyperpolarization (PTH) following repetitive action potentials were due to a similar conductance increase mechanism, probably involving K^+. In SC cells, however, postactivation potentials, particularly those hyperpolarizations following large glutamate-induced depolarizations, appeared to involve an electrogenic pump mechanism.

The chemical excitability of central neurons and the topographical distribution of various receptors over the neuronal surface can be studied readily in the cultured preparations (Barker and Ransom 1978a,b; Ransom *et al.*, 1977a). This topic is dealt with in detail elsewhere (Macdonald and Barker, this volume) and will not be developed here.

3. CENTRAL SYNAPTIC TRANSMISSION

The single cell suspension that is inoculated onto collagen-coated culture dishes as the starting material for the cell culture systems undergoes an impressive development into functional synaptic networks. A variety of patterns of spontaneous electrical activity is seen in mature (>1 month *in vitro*) cultures. A typical pattern recorded from spinal cord cell cultures is an irregular sequence of mixed excitatory and inhibitory synaptic potentials leading to frequent action potentials (Fig. 3A). More complex combinations of quiescence and paroxysmal bursts of synaptic and spike activity are also seen, sometimes with sustained depolarization (Fig. 3B). Dorsal root ganglion cells in DRG–SC co-culture, by contrast, are electrically silent, showing no spontaneous activity (Fig. 3C).

The synaptic connection between two specific cells can be analyzed by stimulating and recording from a pair of cells impaled with separate electrodes. In a substantial proportion of cell pairs, action potentials generated in one cell will elicit simple monosynaptic postsynaptic potentials in the other (Fig. 3D). Powerful, self-exciting synaptic circuits frequently give prolonged bursts of synaptic activity as a result of brief stimulation of a single cell (Fig. 3E).

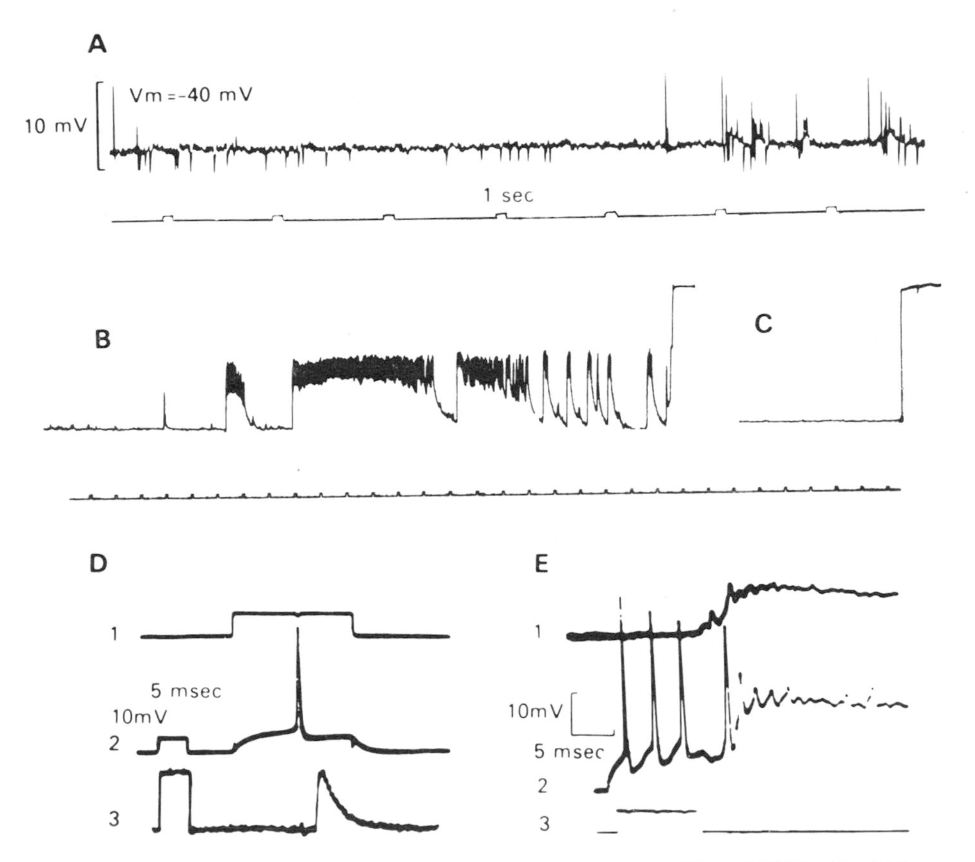

FIGURE 3. Spontaneous and evoked synaptic activity in mouse SC and DRG cells. (A) Chart recording showing downgoing inhibitory postsynaptic potentials (IPSPs) and depolarizing excitatory postsynaptic potentials (EPSPs) and spikes. (B) Quiescence and paroxysmal activity alternate in this SC cell. (C) DRG cell exhibits no spontaneous synaptic or spike activity; microelectrode is withdrawn at end of record showing the resting potential. Calibration bar in A represents 40 mV in B and C. (D) Current pulse of approximately 1 nA in line 1 elicits an action potential in one SC cell (line 2) which produces a monosynaptic EPSP in a second SC cell (line 3). (E) Current pulse of approximately 1 nA (line 3) elicits 3 action potentials in one SC cell (line 2), which are followed by a sustained burst of synaptic and action potentials in that cell and in an adjacent cell (line 1).

Evoked inhibitory activity is also commonly seen (Fig. 4). The diagnostic reversal of the inhibitory postsynaptic potential (IPSP) at approximately −80 mV is readily demonstrated (Fig. 4B,C).

With this sort of accessible and synaptically competent mammalian central neuronal system available, three sorts of questions may be profitably

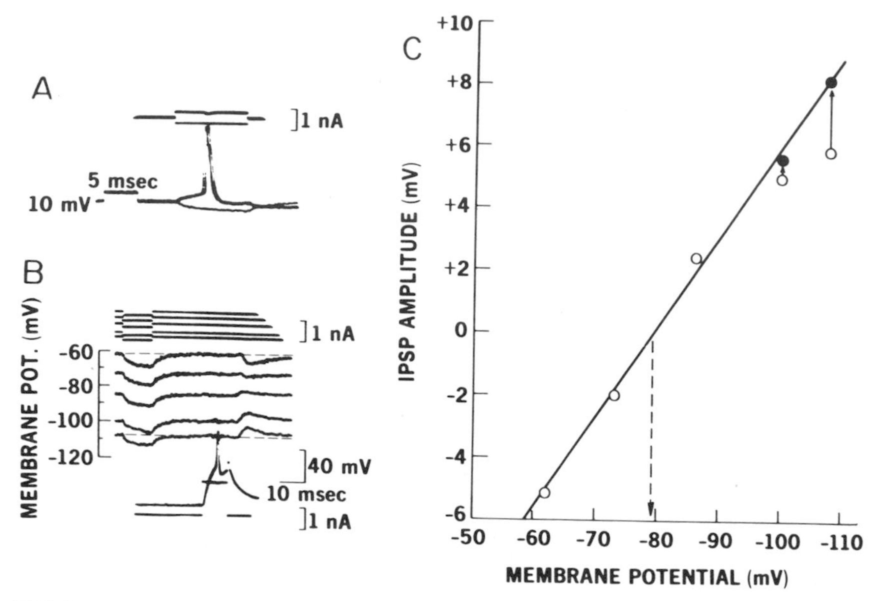

FIGURE 4. Inhibitory synaptic interaction between two cells. (A) Behavior of the postsynaptic cell (lower trace) to current pulses (upper trace) of both polarities injected through the recording pipette. (B) Presynaptic action potentials produced by depolarizing current pulses (bottom two traces) evoked IPSPs in the postsynaptic cell (middle set of traces). The membrane potential of the postsynaptic cell was varied by the passage of steady polarizing current (top set of traces), and the passive properties of the cell were monitored by the injection of constant-current pulses (hyperpolarizing transients seen at the left in the middle set of traces). The hyperpolarizing potential passively elicited by the current pulse becomes smaller at the more negative membrane potentials indicating some degree of anomalous rectification. (C) Amplitudes of the evoked IPSPs in panel B are plotted versus membrane potential, and the broken line with arrow indicates the reversal potential at about −79 mV. The two values in the upper right hand corner of this plot have been corrected for changes in membrane resistance associated with hyperpolarization (i.e., anomalous rectification). The resting potential of the postsynaptic cell was −62 mV. (Reproduced with permission from Ransom *et al.*, 1977b.)

addressed: (1) What is the appearance of specific central synapses? Are there morphological correlates of the physiology and/or biochemistry of a given synaptic interaction? (2) How do central synapses work? The excellent visibility and accessibility of the neurons and the relative ease of obtaining stable recordings from two or more synaptically linked cells make attractive the detailed analysis of basic membrane and cellular mechanisms involved in central synaptic transmission. (3) How do central synapses develop? Since the entire developmental process required for the assemblage of powerfully active synaptic networks occurs while the preparations are under a considerable degree of experimental control, manipulation of the synaptogenic process should be possible and informative.

3.1. Morphophysiological Aspects

Because of the visibility of individual neurons in a monolayer culture, it is feasible to analyze a neuron electrophysiologically and to obtain morphological data on the same neuron by phase microscopy, various light microscopic staining techniques, and electron microscopy. Both by phase microscopy of living cultures and examination of silver-stained preparations, it is apparent that a variety of morphological cell types survives in dissociated SC–DRG cell cultures (Giller *et al.*, 1977; Ransom *et al.*, 1977c). The neurons settle upon flattened "background" cells which may include fibroblasts, glial and ependymal cells. Additional networks of small processed biopolar cells (presumably Schwann cells) frequently lie in association with thick cables of neuronal processes. The neurons are phase bright and are classified (SC versus DRG) by the size and shape of their somata and by the number and branching pattern of emerging processes (see Fig. 1).

By phase microscopy, however, it is often difficult to follow individual neurites for any great distance from the soma, and silver-stained neurons frequently appear in complex networks. In order to better demonstrate the anatomical complexity of individual neurons, various substances (fluorescent or opaque markers, radioactive precursors) have been injected into selected cells. After intracellular injection of the enzyme horseradish peroxidase (HRP) (Cullheim and Kellerth, 1976; Jankowska *et al.*, 1976; Kitai *et al.*, 1976; Muller and McMahan, 1976; Snow *et al.*, 1976) and a time interval to allow movement of the tracer throughout the cells, followed by fixation and histochemical reaction, an injected cell is opaque and in many cases is evidently stained in its entirety. The HRP reaction product further allows one to visualize, with the light microscope, apparent contacts between the injected cell axon and neighboring unlabeled cells (Brown *et al.*, 1977; Brown and Fyffe, 1978; Burke *et al.*, 1979; Cullheim *et al.*, 1977; Cullheim and Kellerth, 1978; Hendelman and Marshall, 1978; Rastad *et al.*, 1977). By injection of the presynaptic cell in a pair of synaptically connected, physiologically studied neurons, it is possible to attempt to correlate gross morphological features with physiological action (E. Neale *et al.*, 1977, 1978a; Nelson *et al.*, 1978) and to obtain information on the number and location of apparent contacts in relation to a particular evoked postsynaptic potential (Christensen and Teubl, 1978). The electron density of the HRP reaction product allows one to identify those synaptic boutons derived from a single neuron and thus to approach a number of questions with electron microscopy. What is the range in the ultrastructural appearance of those boutons formed by a single neuron? Are there one or more fine structural features of presynaptic terminals that appear consistently associated with a given postsynaptic response? What is the quantitative relationship between synaptic potential amplitude and bouton number? Such a physiology–morphology experimental paradigm has been applied to pairs of synaptically linked neurons growing in culture.

3.1.1. DRG Neurons

The DRG neuron, observed with phase optics, is morphologically defined as a cell with a phase bright bulbous soma about 40 μm in diameter with a distinct nucleus and one or more nucleoli and a small number of emerging slender processes (see Figs. 1A,B,D). This cell type is further defined on electrophysiological grounds, as discussed in Section 2.4. DRG neurons generally receive no synaptic input (but see Dichter and Fischbach, 1977), and the soma surface appears free of synaptic investment (Ransom *et al.,* 1977c). Stimulation of a DRG cell and simultaneous recording from a synaptically linked SC cell reveal excitatory postsynaptic potentials (EPSPs) as shown in Fig. 9. Although their size varies, these recorded EPSPs are, on average, smaller than those produced by SC cells (Fig. 3D). An HRP-injected DRG neuron is shown in Fig. 5A. From a collection of such injected neurons, it appeared that emerging processes commonly numbered from one to three, showed diameters that averaged 2 μm at the point of emergence and could be traced for distances up to 3 mm from the soma. Whether there are structural features that might discriminate among the processes of these cultured DRG neurons is not known. Each process appeared to branch and form a limited number of swellings along its length.

The morphological counterpart of electrophysiological analysis has shown that the synaptic investment of large SC cells by DRG neurons was composed of several afferent branches that formed a number of swellings primarily, but not exclusively, in contact with the dendrites of the postsynaptic cell (Fig. 7A). Occasionally, smaller rounded neurons, frequently clustered, were contacted by DRG fibers that enwrapped and formed boutons on their somata. Although variation was seen in the branching complexity and the number of swellings formed along a collateral, a terminal collateral branch usually included about ten swellings. In eight cells studied, there appeared a consistent relationship between the amplitude of the recorded EPSP (2–20 mV) and the number of swellings (20–70) thought to contact the recorded postsynaptic cell.

In general, the synaptic structures formed by DRG neurons in these cultures are not unlike those dorsal root terminations described by Proshansky and Egger (1977) in the frog, by Brown *et al.* (1977) in the cat dorsal horn, and by Iles (1973, 1976), Burke *et al.* (1979), and Brown and Fyffe (1978) in the cat ventral horn. It has been demonstrated *in vivo* with HRP injection techniques that collaterals of cutaneous afferents exhibit more complex terminal arborizations than do Ia muscle spindle afferents and that the morphology of the latter vary in complexity with the lamina in which they terminate (Brown

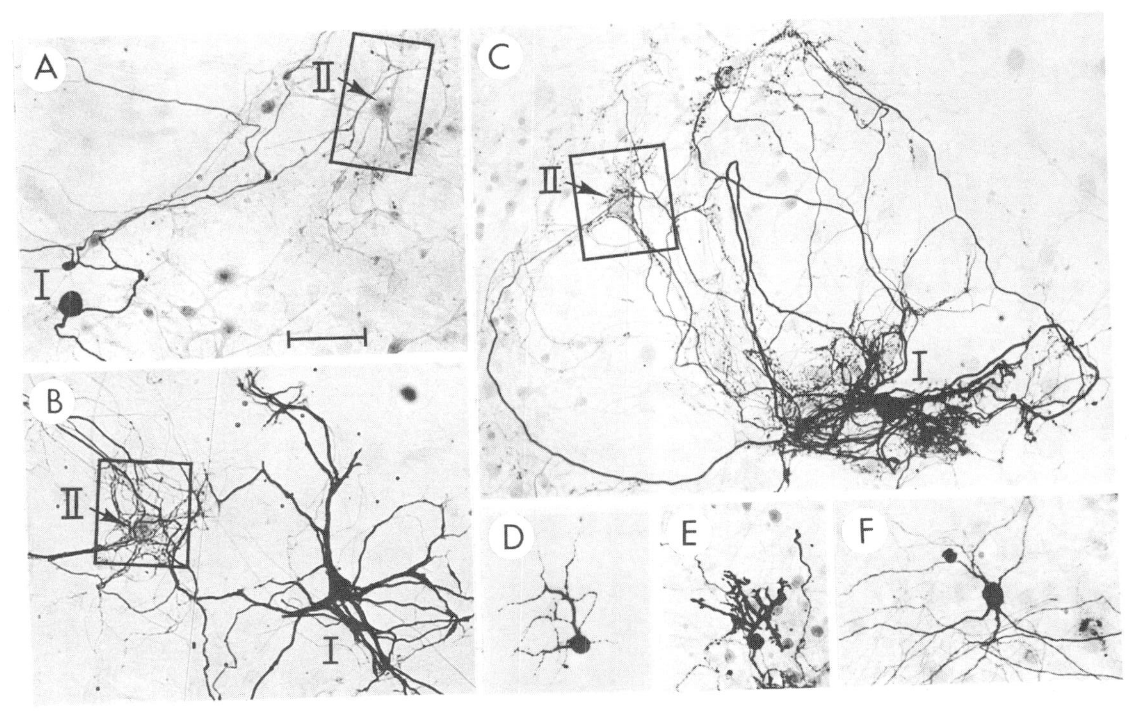

FIGURE 5. Bright-field photomicrographs showing morphologic prototypes among neurons in culture. The stained neurons (except that shown in D) had been injected with horseradish peroxidase (HRP), fixed, reacted for the enzyme, and processed and embedded for electron microscopy. The neuron labeled "I" in A, B, and C was found by intracellular recordings, to be presynaptic to the neuron labeled "II." The postsynaptic cells (boxed areas) are shown at higher magnification in Fig. 7. (A) Pair of neurons in which Cell I is a DRG neuron that when stimulated by intracellular current injection evoked an EPSP in a large SC neuron. (B) Pair of synaptically connected mouse SC neurons in which stimulation of Cell I produced an EPSP in Cell II. (C) Pair of mouse SC neurons in which stimulation of Cell I elicited an IPSP in Cell II. (These cells were reciprocally connected; stimulation of Cell II caused an EPSP to be recorded in Cell I.) (D) Pyramidal neuron in culture prepared from fetal mouse hippocampus. This cell had been injected with Lucifer Yellow CH and is shown in reversed photographic image. (E) HRP-injected neuron in a culture prepared from fetal mouse cerebellum. This cell resembles a Purkinje neuron in the form of its dendritic arbor and the spiny covering of the dendrite surfaces (see Fig. 6D). (F) Neuron in a culture prepared from fetal rat cerebellum. Because of the soma size, the long tapering dendrites and extensive axonal arborization, and the occurrence of similar cells in small groups, this cell is thought to be a neuron derived from one of the deep nuclei. Bar in A = 100 μm; same magnification in all. (The authors are grateful to Drs. J. Peacock, W. Gibbs, and G. Moonen for providing the micrographs D, E, and F, respectively.)

et al., 1977; Brown and Fyffe, 1978). Direct comparisons with these *in vivo* data are difficult since it has not been possible to specify a given DRG neuron in culture as to receptor type nor yet to associate a given SC cell with a particular lamina of the *in vivo* mouse cord. As illustrated in Fig. 7A, it appears, however, that the morphology of the DRG–SC synaptic investment is relatively simple and, in this respect, is consistent with that described *in vivo*.

HRP-labeled DRG swellings that contacted the dendrites of a large SC cell were, in fact, synaptic as defined by ultrastructural criteria (Fig. 8B). Extensive active zones with prominent postsynaptic densities frequently formed around a spinelike protrusion of the postsynaptic dendrite. Synaptic vesicles were relatively large and rounded and showed a marked tendency to aggregate. With the EM examination of boutons from a larger number of identified DRG neurons and the further characterization, perhaps by enzyme histochemistry (Spater *et al.*, 1977) and immunocytochemistry (Hökfelt *et al.*, 1977; Matthew *et al.*, 1979; J. Neale *et al.*, 1978; Nelson *et al.*, 1980; Pickel *et al.*, 1977; Schultzberg *et al.*, 1978) of the DRG neurons in these cultures, it may be possible to make additional structure–function correlations and to compare these data with the ultrastructure of DRG terminations within the intact SC (Beattie *et al.*, 1978).

3.1.2. SC Neurons

Spinal cord (SC) neurons show both excitatory (Fig. 3) and inhibitory (Fig. 4) synaptic responses. We have uniformly observed that stimulation of a given neuron elicits the same monosynaptic response in synaptically linked neurons, and the stimulated presynaptic neuron may thus be considered an excitatory or inhibitory cell.

Stimulation of the largest SC cells (average soma, 40 × 45 μm; see Figs. 5B and 6A) in these cultures has evoked without fail excitatory responses in postsynaptic neurons. As a rule, "excitatory" neurons had at least three main dendrite trunks that tapered in diameter from about 2–10 μm near the soma to delicate tips about 350 μm and at least two branch points distant. Dendritic surfaces were generally smooth (Fig. 6B). The axon originated directly from the soma or, more commonly, from the base of a dendrite, maintained a rather consistent diameter over a long distance (Fig. 6A), and divided by forming collaterals and branches that ultimately formed swellings, usually *en passant*, that contacted the soma and processes of neighboring cells. Somewhat smaller excitatory neurons with rounded somata 35 μm in diameter and delicately radiating dendrites have also been visualized with the HRP technique.

In some cases the EPSPs produced by SC cell activation were quite large and were capable of eliciting postsynaptic action potentials even when the postsynaptic membrane potential had been hyperpolarized with large currents passed through the impaling microelectrode. The range of intensity of synaptic

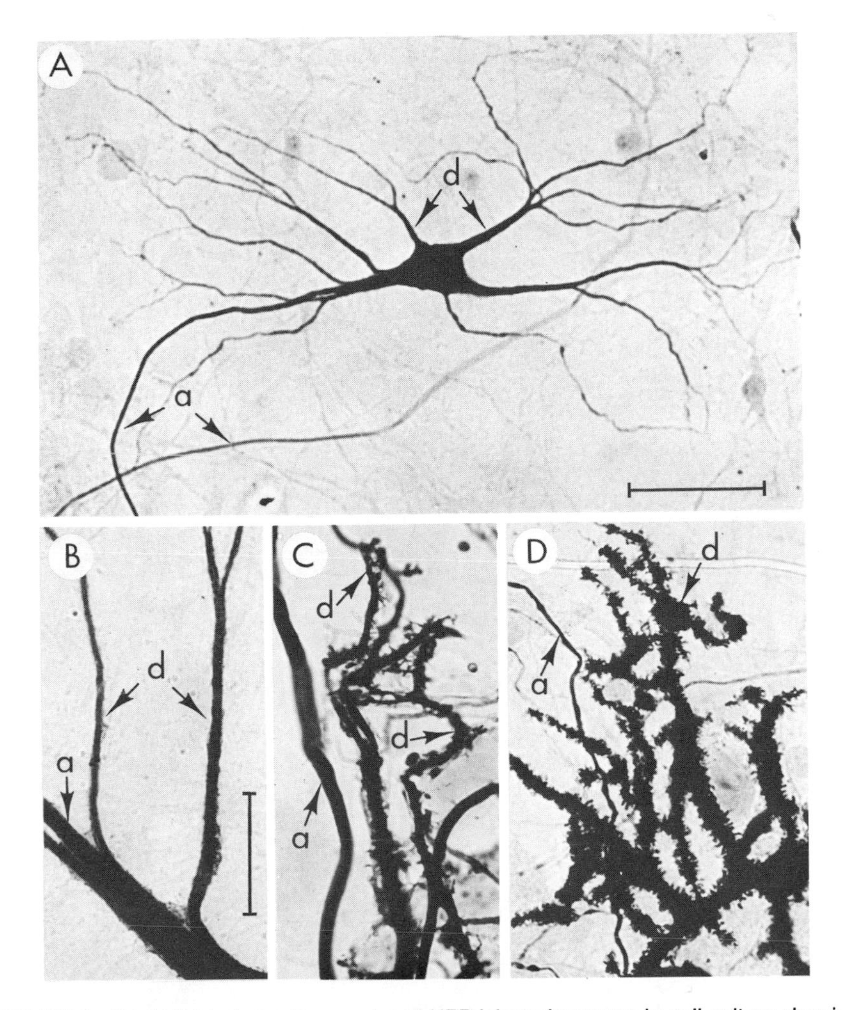

FIGURE 6. Bright-field photomicrographs of HRP-injected neurons in cell culture showing details of axonal (a) and dendritic (d) processes. (A) Several large dendritic trunks radiate from the soma of this neuron, branch several times, and extend for about 300 μm. The axon emerges from the base of a dendrite and courses a greater distance (traceable for 2.5 mm) maintaining a rather constant diameter. Bar = 100 μm. (B) Relatively smooth-surfaced dendrites of an SC "excitatory" neuron. (C) Rough-surfaced dendrites of an SC "inhibitory" neuron. (D) Spine-covered dendrites of a cerebellar Purkinje neuron. B–D, bar = 25 μm.

action from extremely weak connections to those powerful suprathreshold inputs was much wider in SC–SC pairs than in DRG–SC pairs. Excitatory synaptic investments have likewise shown a range of morphological forms from a few boutons derived from one or two delicate collaterals (Fig. 7B), as has been described in the intact SC (Rastad *et al.*, 1977; Cullheim and Kellerth,

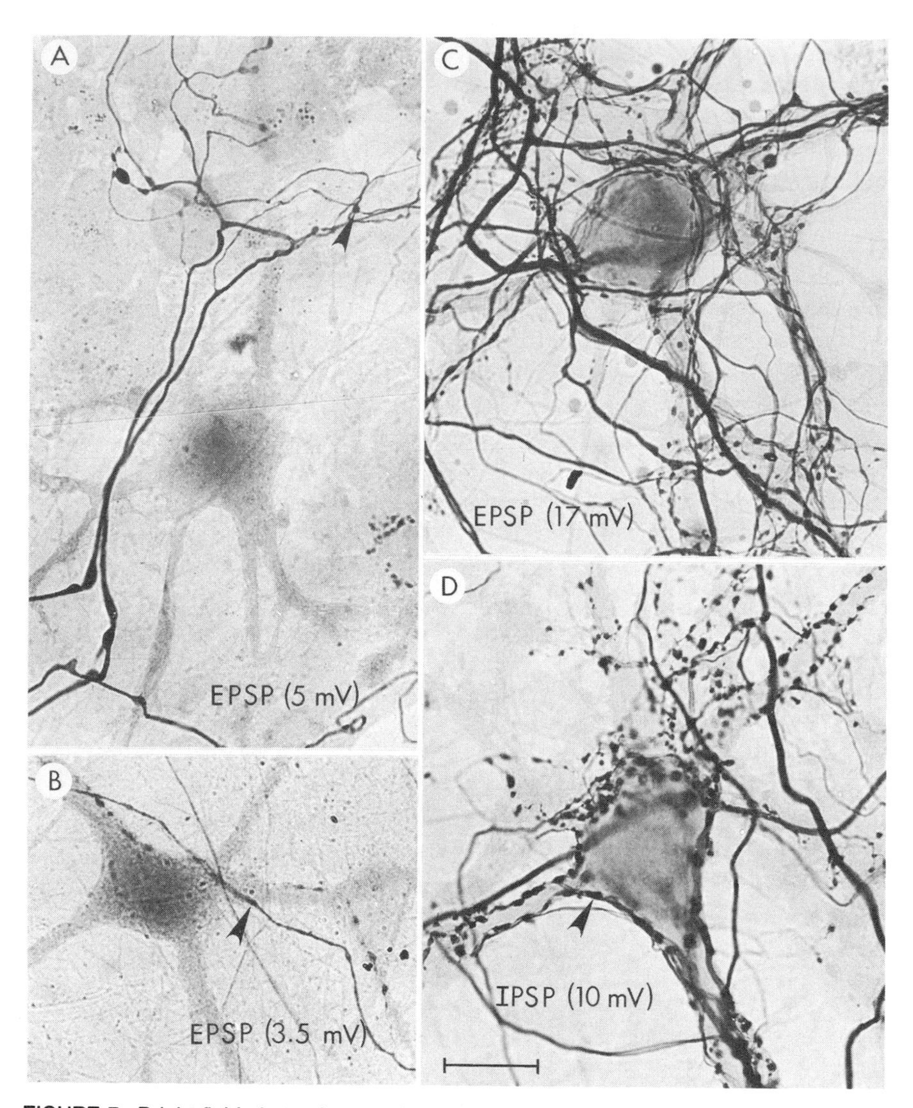

FIGURE 7. Bright-field photomicrographs of mouse spinal cord neurons in dissociated cell culture. Each of the cells shown was the postsynaptic neuron in a synaptically connected, physiologically studied pair. The peak amplitude of the postsynaptic potential recorded after stimulation of the presynaptic cell is noted on each panel. The anatomy of the synaptic connection was demonstrated by labeling the presynaptic neuron with intracellular HRP. The swellings marked with arrowheads are shown in the electron micrographs of Fig. 8. (A) SC neuron which was synaptically linked to a DRG neuron and which responded to stimulation of the DRG cell with a small (5 mV) depolarizing EPSP. The synaptic investment formed by the DRG cell consisted of, at most, 30 *en passant* axonal swellings. In general, DRG cell processes formed relatively few swellings, a large proportion of which appeared to contact

1978), to connections that apparently involve several large diameter presynaptic fiber branches forming numerous swellings on somatic and dendritic surfaces (Fig. 7C). Such swellings have been shown to correspond ultrastructurally to synaptic boutons. Data from two different excitatory neurons indicated that within these swellings vesicles were rounded and tended to cluster at multiple active zones; postsynaptic dense material was not always evident (Fig. 8A).

With the HRP injection technique to delineate the synaptic scale of a presynaptic neuron, it has been possible to relate quantal number (see Section 3.2.1) to number of boutons or, at the EM level, to a number of release sites (E. Neale *et al.,* 1978a). For an SC cell producing an EPSP in another SC cell, the number of boutons contacting the postsynaptic cell was roughly equivalent to the number of quanta released by each action potential. Since there were several active zones per bouton, it appeared that quantal (vesicular) release occurred at one in five release sites with each trial.

Generalizations on structure are more difficult for "inhibitory" SC neurons. The somata were rounded (mean diameter, 30 μm) (Fig. 5C), and dendrites were frequently thick and rough-surfaced (Fig. 6C), extending 150 μm from the soma. The axonal arborization was in most cases quite extensive and usually confined to within less than a millimeter of the soma. The pattern of synaptic investment was generally that of rather massive encrustation of the soma and proximal dendrites of postsynaptic cells (Fig. 7D). Occasionally, a lesser number of swellings was seen to contact postsynaptic neurons, but the site of the interaction was consistently somatic.

The swellings derived from an inhibitory neuron, studied in some detail with the electron microscope (Fig. 8C), have been shown to contain rather diffusely disposed synaptic vesicles. Many of the vesicles were pleomorphic in shape and showed a minimal tendency to cluster. There were usually several small active zones within the boutons, and postsynaptic density at these sites was rarely evident.

---

the dendrites of any given postsynaptic cell. (B) Neuron in an SC–SC pair from which was recorded a small (3.5 mV) EPSP following stimulation of the presynaptic cell. The contact between the two neurons was distinguished as a small number of *en passant* swellings of a branch of a delicate axon collateral. These swellings were seen to lie upon the soma and dendrites of the postsynaptic neuron. (C) Neuron in an SC–SC pair which responded to stimulation of the presynaptic cell with a relatively large (approaching 20 mV) EPSP. The demonstrated anatomical contact was correspondingly large with many labeled swellings impinging on the soma and proximal dendrites of this postsynaptic neuron. (D) Neuron whose surface is contacted by hundreds of HRP-labeled swellings. Synaptic investments formed by numerous boutons lying predominantly along the soma and proximal dendrites were found in five of seven neuronal pairs where the response to stimulation of the presynaptic neuron was a hyperpolarizing IPSP. Bar = 25 μm, same magnification throughout.

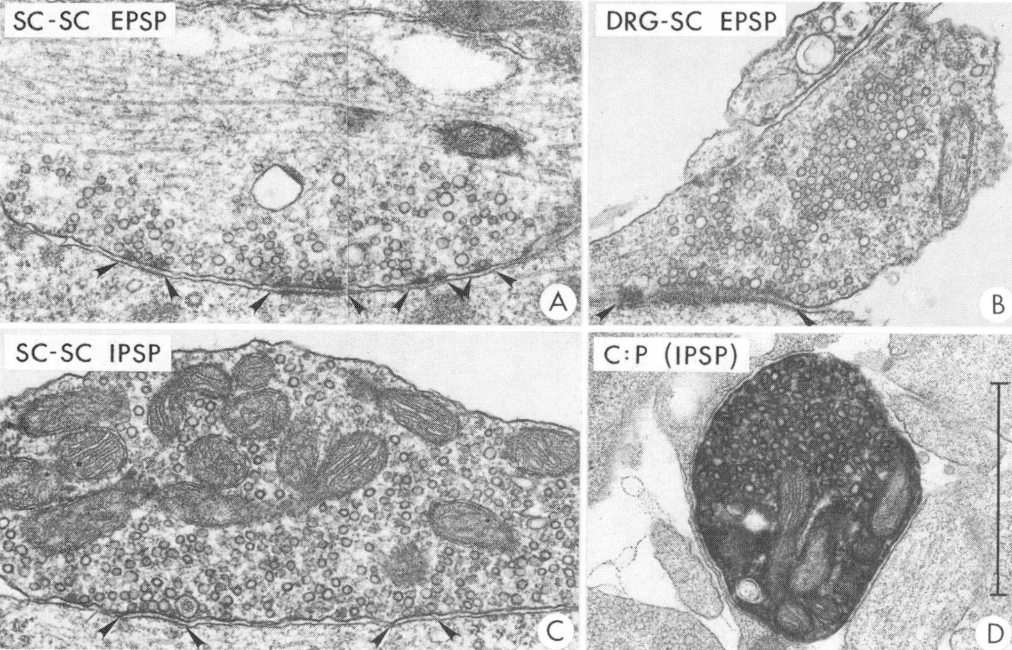

FIGURE 8. Electron micrographs of labeled axonal swellings of HRP-injected cultured neurons. (A) Swelling (arrow in Fig. 7B) of an axonal collateral of an SC ''excitatory'' neuron. Within the swelling, rounded synaptic vesicles tend to cluster around multiple small active zones (marked with arrowheads). The occurrence of postsynaptic dense material is variable. (B) Swelling (arrow in Fig. 7A) of a DRG axonal process. A synaptic contact (arrowheads) is formed by this swelling in which rounded vesicles show a pronounced tendency to cluster, and the active zone is extensive and marked by prominent postsynaptic dense material. (C) Swelling (arrow in Fig. 7D) of an axonal branch of a SC ''inhibitory'' neuron. Mitochondria and somewhat pleomorphic vesicles fill this swelling. Active zones (arrowheads) are apparent at two sites; postsynaptic dense material is minimal. (D) Swelling formed along an axonal branch of a cerebellar Purkinje neuron. Although the HRP reaction product is dense, it is possible to discern mitochondria and rather pleomorphic vesicles within the swelling. No definitive active zone can be discriminated. Bar = 1 μm, same magnification throughout.

3.1.3. Supraspinal Neurons

Dissociated cell cultures prepared from whole brain (Godfrey *et al.*, 1975), cerebral cortex (Dichter, 1978), hippocampus (Banker and Cowan, 1977, 1979; Peacock, 1979; Peacock *et al.*, 1979), and cerebellum (Lasher and Zagon, 1972; Macdonald *et al.*, 1978 and in preparation; Messer, 1977; Moonen *et al.*, in preparation; E. Neale *et al.*, 1978b; Trenkner and Sidman, 1977) have been described. It is apparent that, despite the dissociation procedures, such cultures contain neurons that develop specific electrophysiological and morphological properties consistent with their tissue of origin. Pyramidal and stellate neurons showing spontaneous electrical activity are seen in long-term (>1 month) (see Fig. 6D) cultures of fetal rat cortex (Dichter, 1978). Neurons in 1- to 2-month cultures prepared from fetal mouse hippocampus display complex patterns of electrical activity, a basic pyramidal form, and a marked degree of dendritic differentiation and synapse development (Peacock, 1979; Peacock *et al.*, 1979). Cultures prepared from fetal rat and mouse cerebellar tissue (Moonen, in preparation; Macdonald *et al.*, 1978; E. Neale *et al.*, 1978b) contain a number of cell types tentatively identified after HRP injection and/or EM examination. The most commonly occurring macroneuron is judged to be the culture equivalent of a Purkinje cell (Fig. 5E), while other large neurons are apparently derived from the deep nuclei (Fig. 5F). EM studies provided evidence for cerebellar granule cells and other small neurons, some the size of stellate and basket cells. Purkinje cell dendrites were thick and blunt-ended and marked by large numbers of spiny protruberances (Fig. 6D), features not noted on any other cell type in these culture studies. With electron microscopy, a rich synaptic neuropil and a large variety of bouton morphologies were found. HRP labeling of axonal swellings of presumed Purkinje cells indicated the presence of vesicles that were quite pleomorphic (Fig. 8D). Morphologically distinct active zones were not found in these swellings, possibly because rather subtle features of presynaptic membrane specialization were obscured by the dense HRP reaction.

3.1.4. Additional Comments

Neurons maintained in dissociated cell culture will undoubtedly show deviations in morphology from what would have existed had the neurons developed in intact, undisturbed tissue. HRP injection of neurons in culture has revealed potentially "aberrant" features of morphology, e.g., connections between large spinal cord cells, extensive branching of axonal processes resulting in thick, matlike enwrapments of postsynaptic neurons, and DRG neurons with multiple emerging processes. The determination as to which of these features can be attributed to the culture environment and which actually occur, but have not been reported *in vivo,* awaits the accumulation of additional data from intact systems. Although the technical difficulties involved in such experiments are not trivial, the results have been extremely informative. After intracellular injection of HRP, direct synaptic connections between α-motoneurons via recurrent collaterals have, in fact, been visualized in the intact cat spinal cord (Cullheim *et al.,* 1977).

Additional characterization of the properties of individual neurons in culture might be combined with physiological and morphological data. Radioautographic methods have been used to identify those neurons which take up specific neurotransmitters, e.g., [^{3}H]GABA (Burry and Lasher, 1975, 1978; Farb *et al.,* 1979; Lasher, 1974; Sotelo *et al.,* 1972; White *et al.,* 1980), or presumed transmitter precursors, e.g., [^{3}H]choline (Barald and Berg, 1978, 1979a,b; Woodward and Lindström, 1977). Radioactive or fluorescent ligands and specific antibodies directed against neuronal cell surface markers might differentially label neuronal cell types. Such techniques have been used to visualize toxin binding sites (Dimpfel *et al.,* 1975; Raff *et al.,* 1979), antigenic sites specific for neuronal or nonneuronal cells (Fields *et al.,* 1978; Raff *et al.,* 1979), and receptor sites related to neurotransmitter or neuromodulator chemicals (Chan-Palay *et al.,* 1978; Hazum *et al.,* 1979). Immunocytochemical approaches applied to cultures have demonstrated enkephalin- and substance-P-containing neurons in dissociated SC–DRG cell cultures (J. Neale *et al.,* 1978, Matthew *et al.,* 1979, Nelson *et al.,* 1980). These radioautographic and immunocytochemical techniques are useful for both light and electron microscopic analysis. The further application of statistical methods to such combined findings (E. Neale *et al.,* 1979) in order to discriminate populations of neurons (or boutons) will be most helpful in understanding the synaptic neuropil that is found in mature cultures (Bird, 1978; Nelson, 1976). This information, then, will ultimately serve as the basis for studies of the specificity of synaptic connections that occur in culture and of the factors that might influence the development of these connections.

3.2. Synaptic Mechanisms

The morphological evidence of synaptic development has its counterpart in intracellular microelectrode recordings which show powerful synaptic interactions between the cultured neurons. Some 50% of cell pairs examined in one study showed evidence of excitatory synaptic coupling (Ransom *et al.,* 1977b), and essentially all the neurons within a small (several hundred micrometers in diameter) area may be involved in the generation of rhythmic network activity (Nelson *et al.,* 1977; Peacock, 1979). Both GABA- and glycine-mediated inhibitions probably occur in the SC cultures, and network activity is affected strongly by these inhibitory actions (Nelson *et al.,* 1977). The cultures are technically favorable so that intracellular recordings can be obtained from one or more cells for periods of over an hour. Thus a variety of central synapses are captured in a preparation that lends itself to physiological analysis. Such analysis involves the study of mechanisms of synaptic transmission which may be usefully considered under the two categories of presynaptic transmitter-release processes and postsynaptic responses to transmitter chemicals (see Takeuchi, 1977; Martin, 1977, for discussion). We will address presynaptic issues first.

3.2.1. Presynaptic Mechanisms

It has become increasingly clear that the presynaptic transmitter release apparatus is extremely complex, and the operation of its various elements undoubtedly underlies many important CNS phenomena. The quantal model for transmitter release developed at the neuromuscular junction is the working basis for analysis of central synaptic function (Kuno, 1971), although critical tests of its validity in the CNS are still lacking. The cell culture systems should be useful in this regard in that extensive quantitative data suitable for statistical analysis are obtainable, and necessary morphophysiological correlations are feasible.

If the quantal model were to apply to central synapses, then the quantal size (q) and quantal content (m) for a given connection could be obtained by determining the mean and variance of a series of 100 or so evoked responses (Martin, 1977). This has been done for excitatory connections between DRG and SC cells and between pairs of SC cells. In all cases analyzed, the PSP amplitude distributions have been Gaussian and indicative of relatively high (>5) mean quantal contents (Fig. 9). The quantal size has been in the range of 100–200 μV both for the DRG–SC and the SC–SC connection, about the same as for the DRG–SC synapse measured *in vivo* (Kuno, 1971). The conductance change responsible for this small quantal event in cells of relatively high resistance is about 10^{-10} mho, only about 1/10 that of the quantal conductance increase for the Ia excitatory synapse in the intact spinal cord. Either the amount of transmitter released at synapses in culture is less than *in vivo* or the postsynaptic receptor concentration is lower. Some regulatory process may be operating to match pre- and postsynaptic mechanisms with the result that quantal size is relatively constant (see Ransom *et al.,* 1977b, for discussion).

A quantal event of 200 μV is at or below the noise level of the recording system, and spontaneous miniature quantal synaptic events are generally not recognizable in cultures in which spike activity is blocked by TTX. Similarly, because of this low signal-to-noise ratio, it has been difficult to study evoked PSPs with an *m* equal to 2 or fewer. In this situation, the quantal model predicts a relationship between mean quantal content and the incidence of failures for any postsynaptic response to occur. Such data and Poisson amplitude distributions with connections exhibiting a low *m* have constituted dramatic quantitative confirmation of the quantal model at the neuromuscular junction (Katz, 1966). Powerful methods are now available for dealing with low-level biological signals embedded in relatively large system noise levels (Edwards *et al.*, 1976a–c). Application of these techniques to the cultured material is clearly needed in order to obtain more discriminating tests of the quantal model.

The statistical analysis of synaptic action is useful in distinguishing pre- and postsynaptic actions of pharmacological agents. Very substantial specific opiate-induced reductions in the dorsal horn potentials produced by DRG stimulation were observed in SC explant cultures (Crain *et al.*, 1978). A presynaptic locus of action for an opiate agonist, etorphine, was found in experiments

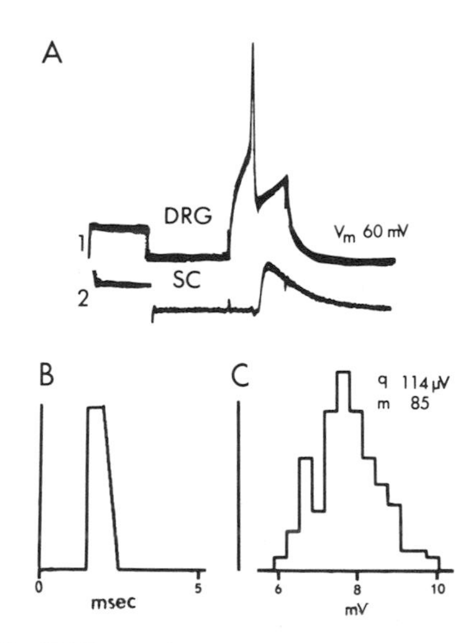

FIGURE 9. Quantitative features of synaptic transmission between a DRG cell and an SC cell in culture. (A) Simultaneous intracellular recordings were obtained from a DRG cell (A1) and an SC cell (A2). Excitation of the DRG cell was followed by a short-latency (2 msec) EPSP in the SC cell. Two superimposed trials are illustrated. Calibration pulses are 5 mV by 10 msec. (B) Latency histogram demonstrates that a series of EPSPs evoked by DRG stimulation had a uniform latency distribution. Maximum value on ordinate represents 120 trials and the histogram bin width is 0.25 msec. (C) Amplitude histogram shows a unimodal distribution of EPSP amplitudes with a mean amplitude of 7.5 mV. The mean quantal content (*m*) and mean quantal amplitude (*q*) were calculated using the coefficient of variation method, and the resulting values were corrected for nonlinear summation of the postsynaptic responses assuming a resting membrane potential of −60 mV and an equilibrium potential for the EPSP of 0 mV. The maximum ordinate value represents 30 trials, and the histogram bin width is 0.25 mV. (Reproduced with permission from Nelson *et al.*, 1978.)

involving the excitatory synaptic linkage between DRG and SC cells. A naloxone-reversible diminution in EPSP amplitude was demonstrated, and this change was quantitatively attributable to a decreased quantum content of the PSP in the absence of any change in quantal size (Macdonald and Nelson, 1978). By contrast, a postsynaptic mechanism of action of the benzodiazepines was indicated, since both IPSPs and iontophoretically elicited GABA responses were augmented by this class of drugs (Choi *et al.*, 1977; Macdonald and Barker, 1978).

3.2.2. Postsynaptic Mechanisms

Postsynaptic responses to transmitter chemicals can involve surface membrane permeability changes or more complex biochemical processes in which cytoplasmic components participate. A large number of candidate neurotransmitters have been proposed (Krnjević, 1974). One of the necessary steps in establishing that a given candidate is in fact a natural transmitter is to demonstrate identical postsynaptic responses to presynaptically released neurotransmitter and to the iontophoretically applied candidate. The many difficulties that beset the establishment of this identity *in vivo* are by no means eliminated by using the cell culture model system. However, the idealized but realistic experiment (using cultured material) as described below would contribute to the resolution of this kind of question. A parallel description of the data available at present will illustrate the large gaps in our knowledge.

At least four physiological tests can be applied to the question of identity of membrane action for two agents:

1. Are the reversal potentials identical for the permeability changes produced by the two agents?
2. Do changes in the ionic environment produce similar changes in reversal potential for both agents? These two criteria reduce to the single criterion that identical specific ionic permeability changes be produced by the two agents.
3. Do antagonists affect the two responses similarly, i.e., are the same receptors involved in both responses?
4. Are the kinetics of ionic channel opening and closing measured for the candidate chemical compatible with the time course of the naturally occurring synaptic event?

To approach these questions, the following experimental sequence is possible.

An HRP-filled electrode, inserted into a presynaptic neuron, is used either to stimulate the neuron or to inject HRP. Two electrodes, one for current passage and one for recording, are inserted into the postsynaptic cell. The two electrodes can be arranged in either a voltage clamp or constant current configuration. An agonist-containing (e.g., glutamate or aspartate) micropipette is positioned immediately outside the postsynaptic cell for iontophoretic application. This arrangement allows the determination of agonist and synaptic reversal potentials. Following HRP injection and appropriate histochemistry, it is possible to obtain information on the number and distribution of synaptic contacts responsible for the recorded PSP. The equivalent total electrotonic length of the dendrites can be measured electrically, and thus the electrotonic distance of the synapses from the neuronal soma can be derived (Ransom *et al.*, 1977c). The effect of a dendritic synaptic location on the reversal potential of the synaptic response can be solved analytically. A more direct approach would be to restrict functioning synaptic contacts to the neuronal soma. This could be achieved by bathing the culture in a low-Ca^{2+}, high-Mg^{2+} medium to block synaptic function generally and by applying locally to the neuronal soma a high-Ca^{2+} solution from a 2- to 5-μm-tip pipette. With HRP labeling of presynaptic terminals, features of synaptic ultrastructure might be analyzed vis-à-vis a given neurotransmitter.

IPSPs, GABA responses, and glycine responses all have closely similar reversal potentials (Ransom *et al.*, 1977a,b). Noise analysis experiments with the appropriate agonist can provide the kinetic constants for channel opening and closing for comparison with the synaptic potential (or current) time course (see Lecar and Sachs, this volume). Measurements of glycine and GABA

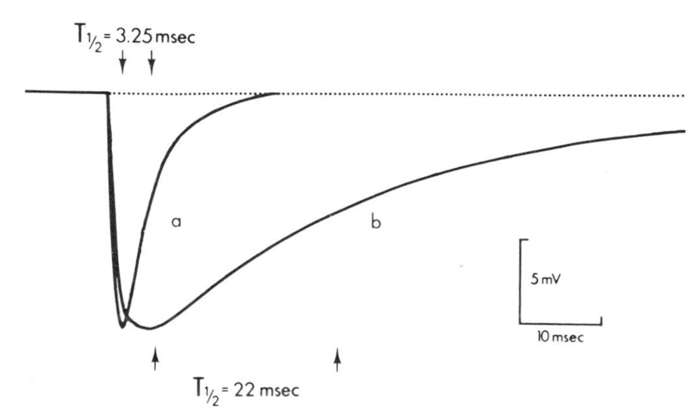

FIGURE 10. Line drawings of IPSPs obtained from different cells showing markedly different time courses. The longer IPSP is a graphically constructed average waveform of two slow IPSPs ($T_{1/2}$ = 17 msec and 35 msec).

average channel open times give values of 5.2 and 20 msec, respectively (Barker and McBurney, 1979; McBurney and Barker, 1978). Inhibitory postsynaptic potentials in spinal cord cultures are of at least two distinct time courses with a five- to tenfold difference in half time (Fig. 10). Most IPSPs are of the more rapid time course, and the channel open time measurements suggest that these would be glycinergic. Even more direct measures of channel properties may be possible using the techniques developed by Neher and Sakmann (1976; Neher *et al.,* 1978) on skeletal muscle (Jackson and Lecar, personal communication).

Perfusion experiments can be used to assess the effects of either ion changes or antagonist actions. Changes in external concentrations of K^+ and Cl^- can be effected over a considerable range without introducing serious problems. Excessive reduction in external Na^+ concentration would block presynaptic action potentials, but focal neurite stimulation could still induce transmitter release. Iontophoretic application of many antagonists is feasible, although the distributed nature of synaptic contacts poses a difficulty for this approach. The effective restriction of active synapses to the neuron soma would be useful in this regard.

Essentially all components of these idealized experiments have been achieved, but results are still fragmentary. The conclusion of one set of experiments was that a substantial discrepancy existed between the behavior of the glutamate response and the EPSPs that linked SC cells (Ransom *et al.,* 1977a). The reversal potential for the EPSP was several millivolts more positive than that for the glutamate response (Fig. 11). Neither the glutamate response nor the EPSP was actually reversed in these experiments, however, and only the extrapolated values for the reversal potentials were obtained. As pointed out in other systems (Dionne and Stevens, 1975; Dudel, 1974), the voltage dependence of chemically activated channels can lead to substantial discrepancies between extrapolated values of reversal potential values and the potential at which drug synaptic responses actually do reverse. Clearly, the difficulties inherent in complex, combined experiments documenting all aspects of postsynaptic responsiveness must be faced in order to obtain unambiguous results.

3.3. Developmental Studies

As stated earlier, it is an attractive feature of the cell cultures that highly complex neuronal morphologies develop over a period of several days or a few weeks, and extensive synaptic networks come to link the neurons during this period. The environment in which neurodifferentiation is occurring is under experimental control, and it would seem natural to ask what environmental manipulations modify neural development. Two types of studies in this area can be described.

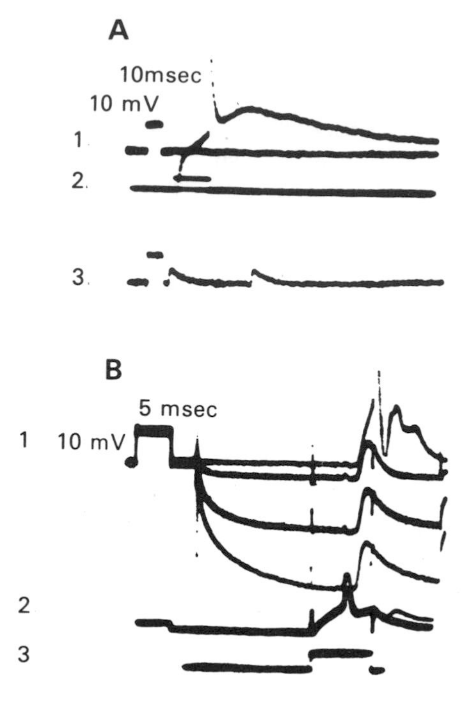

FIGURE 11. Comparison of glutamate-induced responses and EPSPs. (A) Depolarization of SC cell (1), produced by glutamate pulse of approximately 30 nA (2). "Spontaneous" EPSPs in same cell are shown in (3). (B) Stimulus pulse of approximately 0.5 nA (lowest trace) elicits spike in one cell (second lowest trace) and EPSP in second cell (upper multiple traces). Hyperpolarizing current pulses were delivered to this second cell to demonstrate the effects on EPSP amplitude of such hyperpolarizations. In the absence of hyperpolarization, the EPSP elicits an action potential in the second cell (uppermost trace). Note complex PSP following this action potential and the late EPSP on record of first cell. (C) Graph of relationship between steady membrane potential and the amplitude of glutamate potentials (○) and EPSPs (●). Note 20-mV disparity in extrapolated equilibrium potentials. (Reproduced with permission from Nelson, 1976.)

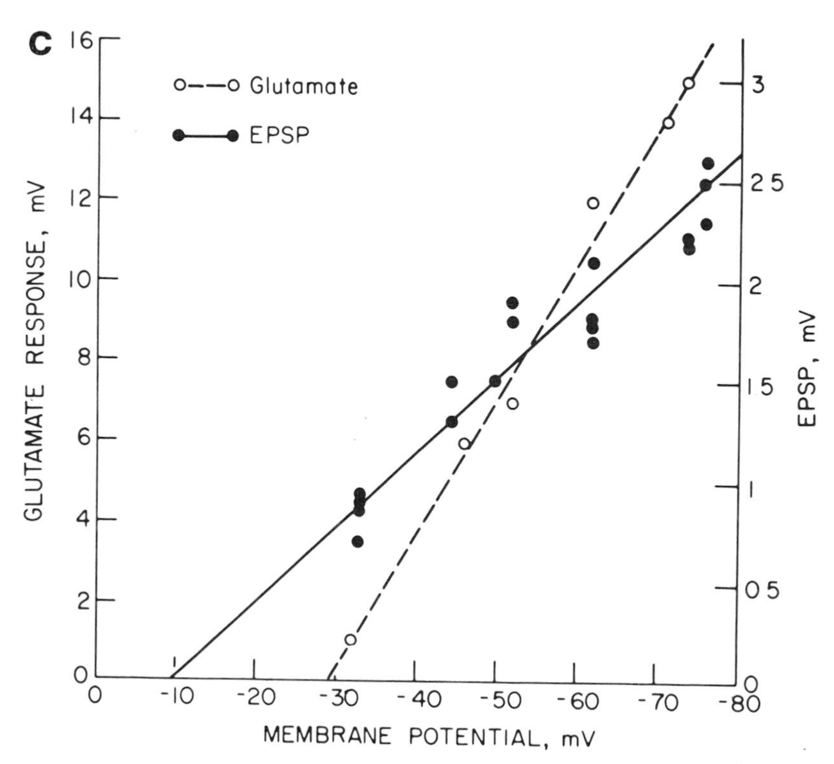

3.3.1. Trophic Materials

Co-culture of neural material with other cell types may influence the development of the neurons. Spinal cord cultures grown with muscle cells develop several-fold higher levels of choline acetyltransferase (CAT), the enzyme responsible for acetylcholine synthesis, than do SC cultures grown in the absence of muscle cells (Fig. 12). Conditioned medium from muscle cultures, when added to SC cultures, also produces an increase in CAT activity (Giller *et al.*, 1973, 1977). Presumably material synthesized and released into the medium by muscle cells has this inductive effect, and the experimental data suggest that this material is of a high molecular weight. Efforts to purify the material have not been successful to date, however. A similar but more dramatic effect has been demonstrated in rat sympathetic ganglion cell preparations and is described elsewhere (Burton and Bunge, this volume).

An induction of catecholamine synthesis in SC cells by co-culture with liver cells has been described (Bird and James, 1975). This is particularly surprising in view of the findings in the intact spinal cord that all catecholamine-containing structures are fibers descending into the spinal cord from cell bodies located above the spinal cord. A number of workers have reported effects on muscle fibers produced by the application of soluble material obtained from nerve (Engelhardt *et al.*, 1976; Hasegawa and Kuromi, 1977; Kuromi and Hasegawa, 1975; Oh and Markelonis, 1978).

These sorts of studies would serve, in principle, to define the molecular basis for some of the cellular interactions that regulate neurodifferentiation. A different orientation underlies experiments seeking to define the role of neuronal activity in modulating maturation of the nervous system.

3.3.2. Effects of Activity

Rather disparate data have been obtained regarding the degree to which nervous system development is affected by the functional state imposed upon it. It is clear that synapse formation can occur in the face of a variety of procedures that block synaptic function. Crain and co-workers (1968; Model *et al.*, 1971) have shown impressive functional and morphological maturation in nervous system explant cultures grown in solutions containing lidocaine which blocked organized functional activity. On the other hand, the importance of functional activation for the normal development of the visual system is well documented (Wiesel and Hubel, 1965).

In dissociated SC cell cultures it has been shown that blockade of functional activity by TTX has a drastic but selective effect on neuronal maturation and survival (Bergey *et al.*, 1978). Only 5–10% of SC neurons survive in TTX-treated cultures, while DRG cells are unaffected by such treatment. The effect on SC cell cultures is stage-dependent; early treatment (beginning on day 1 or

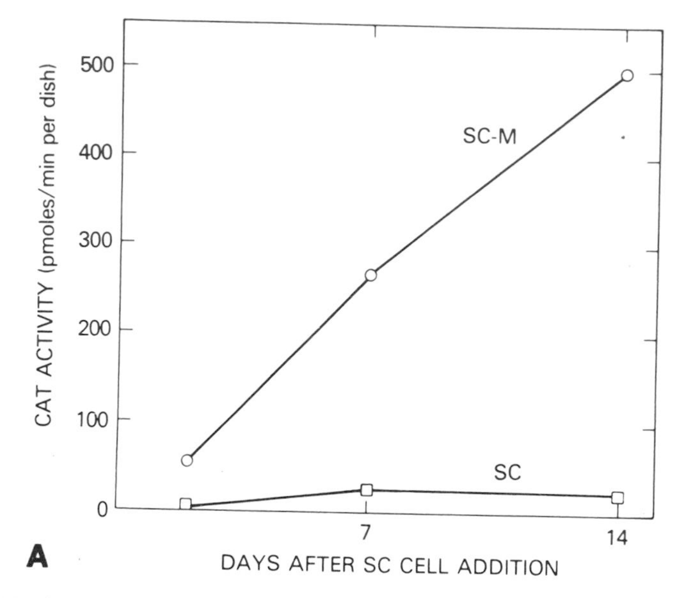

FIGURE 12. (A) Time course of the development of choline acetyltransferase (CAT) in culture. SC cells from 16-day mouse embryos were inoculated onto collagen-coated dishes (SC, □—□) or onto similar dishes containing mouse myotubes that had been cultured in replicate for 2 weeks (SC-M, ○—○). Cultures were harvested at the times shown. Each datum shows CAT activity per dish in a single homogenate made from the contents of two or three dishes. Muscle cultures without SC cells had no detectable CAT activity. (B) Effect of dose of muscle conditioned medium (CM) on CAT activity in SC cell cultures. SC cells from 14-1/2-day mouse fetuses were inoculated into collagen-coated dishes and grown as in A except that on days 2, 5, 7, 9, 11, and 13 the medium changes included the indicated proportions of CM. Cultures were harvested on day 14 and assayed for CAT activity and protein. Each datum represents CAT activity per dish (○—○), and per milligram of protein (□—□). (Reproduced with permission from Giller *et al.*, 1977.)

day 8 of culture) produces the result, while treatment started after 1 month of culture does not. The stage- and cell-type specificity argues against a nonspecific toxic effect of the TTX, but further work with other blockers of activity is clearly needed.

4. GENERAL COMMENTS

A comment is probably necessary as to the validity of the culture systems as models in which membrane and cellular mechanisms important for the intact central nervous system may be analyzed. It is abundantly clear that deviations from the *in vivo* nervous system are to be expected in cell culture. It is equally apparent that many complex features of the intact nervous system are captured by these preparations. There is no clear-cut general way to decide

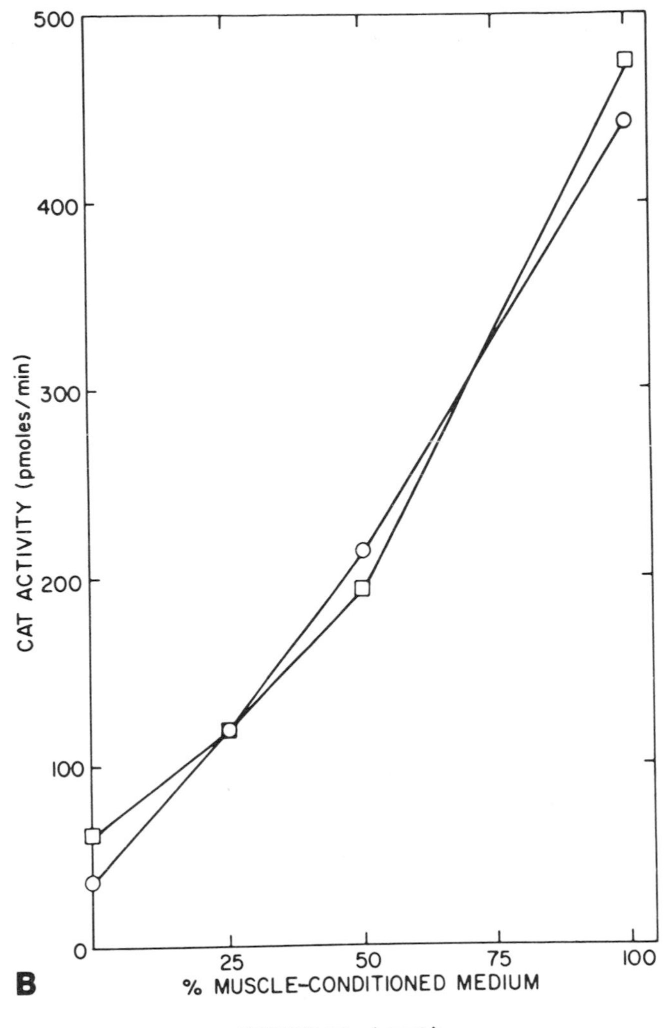

FIGURE 12. (*cont.*)

in a given situation whether one is dealing with deviation or faithful simile. It seems reasonable to argue, however, that basic cellular and membrane mechanisms occurring and demonstrable in cultured cells will have some counterpart *in vivo.*

A critical requirement for coping with these reservations is the identification of cell types in the cultures. Cell purification techniques have shown considerable promise (Berg and Fischbach, 1978; Masuko *et al.,* 1979), and a number of radioautographic and cytochemical methods should also prove useful in this regard. Monolayer culture systems offer the technical advantages of accessibility of individual cells to exogenously applied compounds along with a degree of cellular resolution for visualization of specific uptake or binding of such compounds. In these respects, cell cultures may be the system of choice for certain binding studies.

When quantitative characterization of well-identified counterparts of constituent cell types from the intact nervous system has been accomplished, it will still remain to document relevance to the *in vivo* situation. The necessity for this validation does not minimize the enormous benefit to be derived from having functional, differentiated central nervous system components revealed and accessible to us in all of their intricate beauty.

APPENDIX

Rat Cerebellum (Lasher and Zagon, 1972)

This method involves postnatal (2-day-old) animals, a mechanical dissociation (by trituration), and growth in enriched high-K^+ medium.

Cerebellums were dissected from 2-day postnatal rats, placed in a small petri dish containing saline G (Puck *et al.*, 1958), and cleaned of any adhering tissue. Excised cerebellums were cut into small pieces, transferred to a round-bottom tube containing 2 ml of saline G, and agitated. Following aspiration of the saline, the procedure was repeated with medium containing 10% fetal calf serum (MFCS). The pieces were then resuspended in 4 ml of MFCS and pipetted 30–40 times with a 1 ml plastic pipette (Falcon). This suspension was allowed to settle for several minutes, and then 2 ml of the supernatant were transferred to a conical centrifuge tube. Two ml of MFCS were then added to the pieces, and the procedure was repeated. The supernatants from 4 to 5 pipettings were collected, spun down, and resuspended in several ml of MFCS. This method yielded a cell suspension containing single cells and some small clumps of 5–10 cells. If sufficient time was not allowed for the pieces to settle after pipetting, or if the cell density at plating was too high, large aggregates of up to 100 cells were found. Other dissociation methods were also tried [e.g., incubation in 0.1% trypsin (beef pancreas, Grade V, Miles) for 15–20 min, followed by gentle pipetting] but were not as satisfactory as the method described.

Approximately 5×10^5 cells were plated into 35 mm plastic dishes (Falcon or Linbro) containing a substrate of rat tail collagen (20–30 μg, dried under UV, and rinsed with saline G or MFCS). Collagen-coated coverslips (Corning) were also used when cells were to be fixed for staining at a later time. Two ml of the following medium were placed in each dish: 90% Ham's F-12 (Ham, 1965) modified to contain twice the concentration of amino acids, choline, and pyruvate; 50 μg/ml ascorbate, 600 mg% glucose, 0.5% bovine albumin (Fraction V, Miles), and 10% FCS (Grey Industries, Inc.). This medium contained approximately 3.5 mM K^+ ion and had an osmolarity of 320 mOsm (measured by freezing-point depression). In medium adjusted to 24.5 mM K^+ ion, the osmolarity was kept constant by lowering the NaCl concentration. Cultures were maintained at 35.5°C in a highly humid atmosphere of 5% CO_2 in air.

Single cells and small clumps took 1–3 days to attach and spread out. After 7–10 days, most of the neurons in cultures fed with medium containing 3.5 mM K^+ ion were dead or dying. In comparison, the majority of neurons in cultures fed with medium containing 24.5 mM K^+ ion were still healthy. These latter cultures were maintained for over a month with good neuronal survival and differentiation.

Mouse Spinal Cord (Ransom *et al.*, 1977c)

This method includes mechanical dissociation, plating on collagen-coated substrate, and treatment with fluoro-2′-deoxyuridine to retard background growth. Horse serum is used as the supplement in the growth medium.

Spinal cords were taken from 12- to 14-day-old mouse (C57BL/6J) embryos. The age of the fetuses used was critical in that younger or older ones did not yield good, healthy cultures. The pregnant mice were killed by placing them in a CO_2 atmosphere for 2 or 3 min and then fracturing their necks. The CO_2 narcosis produced unconsciousness without the possible damage to tissue of chloroform or ether anesthesia. The embryos in their sacs were removed and placed in a 60-mm culture dish containing DISGH solution. Each spinal cord took about 1 min to remove.

The DISGH was made up of 50 ml of 20× concentrated Puck's D1 salt solution (Colorado Serum), 6 g glucose, and 15 g sucrose per liter and 10 mM HEPES buffer, pH 7.3. The complete solution was adjusted to 330 mOsm with water, if necessary, in order to bring it to the same osmolarity as that of the nutrient medium to be used.

Following dissection, up to four or five spinal cords were placed in an empty, sterile 35-mm petri dish and minced with iridectomy scissors until the mass of tissue appeared almost gelatinous. This tissue was taken up in 1.5 ml of nutrient medium with a Pasteur pipette and transferred to a sterile 15 ml centrifuge tube. The nutrient medium (MEM 10/10) consisted of 80% Eagle's minimal essential medium (MEM, Gibco) and added glucose (600 mg/100 ml), 10% heat-inactivated (56°C for 30 min) horse serum (HS, Gibco), and 10% fetal calf serum (FCS, Gibco). $NaHCO_3$ (1.5 g/liter) was added to the medium to increase its buffering capacity so that is could be incubated in a 90% air–10% CO_2 atmosphere. No antibiotics were used. The tissue fragments in the centrifuge tube were mechanically dissociated by trituration, using the Pasteur pipette to take up and gently expel the suspension 5–10 times. The resulting suspension was allowed to settle for a few minutes, and the supernatant was removed and saved. One milliliter of MEM 10/10 was added to the remaining pellet, and the trituration procedure was repeated using a Pasteur pipette with a slightly narrowed orifice (achieved by flaming the tip). After again allowing the suspension to settle, the supernatant was again removed and added to that previously collected. This cycle of resuspension and trituration was continued until the supernatant volume reached about 1 ml per spinal cord (i.e., 4–5 ml). This final suspension was plated on collagen-coated (see below) 35-mm-diameter tissue culture plates (Falcon) by adding 0.5 ml of the suspension (about 1.5×10^6) cells to plates containing 1 ml of MEM 10/10 which had been preincubated for about an hour so that the pH and temperature of the medium were equilibrated to the incubation conditions (10% CO_2–90% air and 36°C).

Collagen-coated culture dishes were prepared in the following manner: a collagen solution was made by adding 50 mg of calf skin collagen (Calbiochem; this comes in 50 mg vials, and the collagen may be treated as though it were sterile) to 100 ml of sterile, 1 : 1000 glacial acetic acid solution. This mixture was stirred continuously at room temperature. Dissolving the collagen could take as much as 12 hr, and even after this period the solution sometimes appeared slightly cloudy. One drop of the collagen solution from a Pasteur pipette is sufficient to spread out and coat a 35-mm culture plate. The covered plates can be air dried or, if necessary, dried rapidly in an oven at a temperature no greater than 60°C.

The freshly plated cultures were incubated for 3–5 days before their first medium change with 1.5 ml of MEM 10/10. The second medium change was done after 2 days and if, as was usually the case, the nonneuronal background cells were confluent at this time, the change was made using medium from which FCS was deleted (MEM 10) and to which 5′-fluoro-2′-deoxyuridine (FdU) and uridine (final concentrations of 15 and 35 μg/ml, respectively) were added. Occasionally the FdU treatment was postponed until the third change if the background cells were not confluent at the time of the second change. The FdU served to control the rapid division of the background cells and possibly enhanced neural differentiation (Godfrey *et al.*, 1975). The MEM with FdU and uridine was changed after 2 days, and all subsequent changes were made twice a week using MEM 10 alone. Cultures prepared in the manner described above were allowed to mature for about 3 weeks prior to study and could routinely be maintained for periods up to and sometimes exceeding 2 months. At about 2–3 weeks, some loss of neurons regularly occurred after which the cultures stabilized.

Rat Cerebral Cortex (Dichter, 1978)

Trypsinization and trituration provide for the dissociation in this method. Plating on a layer of nonneuronal cells and treatment with a mitotic inhibitor, cytosine arabinoside (Ara–C), produce well visualized, stable populations of neurons.

Mammalian cortical neurons are grown in dissociated cultures using a method modified from that described by Fischbach (1972) for growing dissociated chick nervous system. Pregnant CD rats, 15 days post conception, are stunned, and embryos are removed with sterile technique. The cortices are removed, gently minced into small pieces, and collected in 0.7 ml of 0.027% solution of trypsin (Armour) in MEM. The tissue is incubated at 37°C for approximately 2 hr, and then 4.5 ml MEM plus 5% fetal calf serum (5F) is added. The tissue is triturated with a Pasteur pipette ten times and then again after the pipette lumen is fire polished until it is almost closed. The cell suspension is diluted to 10 ml with 5F, and clumps of cells are removed by filtering through a double layer of washed sterilized lens paper. The cells are counted (with viability estimated by trypan blue exclusion) and plated at between 1.2–1.6 $\times 10^5$ viable cells per 35-mm plastic Falcon tissue culture dish or occasionally onto sterile glass coverslips. In either case, the growing surface has been previously covered with a layer of nonneuronal cells (see below). The cortical cells are grown in a medium which consists of Eagle's MEM, 5% heat-inactivated rat serum (prepared in our laboratory), 100 mg% glucose, 2 mM glutamine, and 20 U/ml penicillin and streptomycin. This medium is completely changed with fresh medium three times per week, and the cells are kept in a 5% CO_2 incubator at 36–37°C. The cells are allowed to grow and differentiate until the dividing nonneuronal cells from the cortical dissociation begin to become confluent. At that time (usually at 6–8 days), the cultures are treated with cytosine arabinoside (Ara–C, 10^{-5} M for 24 hr) to inhibit dividing cells. Neuronal viability continues for periods up to 8–12 weeks and occasionally longer.

Although the cortical cells can be grown on plastic alone or on collagen-coated plastic, cells plated onto a background monolayer of previously inhibited nonneuronal cells will survive with a much lower inoculum and will show much less tendency to form clumps of reaggregating cells. The background layer is formed by dissociating cells from 15-day rat embryos whose heads, spinal cords, hearts, livers, limbs, and other visceral organs had been removed. These cells are plated onto the plastic plates at between 1–2 $\times 10^5$ cells per plate and after several days, when they have almost reached confluence, are inhibited for 24 hr with Ara–C (10^{-5} M). The cells can then be kept for several weeks before cortical cells are placed on top of them.

The cultures require serum for survival. The neurons survive longer and become larger if grown with 5–10% rat serum as compared with commercially available fetal calf or horse serum. Adult rats are bled via intracardiac puncture under ether anesthesia. The blood is allowed to clot (and must be kept away from the black rubber plunger of plastic syringes), and the serum is carefully drawn off and centrifuged. It is then heat inactivated at 56°C for 30 min and filtered through a 0.4 μm Millipore filter. Neither chick embryo extract nor rat embryo extract appeared to enhance neuronal growth or survival.

ACKNOWLEDGMENTS. It is a pleasure to acknowledge the expert assistance of Ms. Sandra Fitzgerald in the tissue culture work and Ms. Rita Lohse and Ms. Nancy Garvey in typing and editing the manuscript.

REFERENCES

Baccaglini, P.I., 1978, Action potentials of embryonic dorsal root ganglion neurones in *Xenopus* tadpoles, *J. Physiol. (Lond.)* **283:**585–604.

Banker, G.A., and Cowan, W.M., 1977, Rat hippocampal neurons in dispersed cell culture, *Brain Res.* **126:**397–425.

Banker, G.A., and Cowan, W.M., 1979, Further observations on hippocampal neurons in dispersed cell culture, *J. Comp. Neurol.* **187:**469–494.

Barald, K.F., and Berg, D.K., 1978, Labeling cholinergic neurons in cell culture, *Neurosci. Abstr.* **4:**589.

Barald, K.F., and Berg, D.K., 1979a, Autoradioautographic labeling of spinal cord neurons with high affinity choline uptake in cell culture, *Dev. Biol.* **72:**1–14.

Barald, K.F., and Berg, D.K., 1979b, Ciliary ganglion neurons in cell culture: High affinity choline uptake and autoradiographic choline labeling, *Dev. Biol.* **72:**15–23.

Barker, J.L., and McBurney, R.N., 1979, GABA and glycine may share the same conductance channel on cultured mammalian neurones, *Nature* **277:**234–236.

Barker, J.L., and Ransom, B.R., 1978a, Amino acid pharmacology of mammalian central neurones grown in tissue culture, *J. Physiol. (Lond.)* **280:**331–354.

Barker, J.L., and Ransom, B.R., 1978b, Pentobarbitone pharmacology of mammalian central neurones grown in tissue culture, *J. Physiol. (Lond.)* **280:**355–372.

Beattie, M.S., Bresnahan, J.C., and King, J.S., 1978, Ultrastructural identification of dorsal root primary afferent terminals after anterograde filling with horseradish peroxidase, *Brain Res.* **153:**127–134.

Benda, P., De Vitry, F., Picart, R., and Tixier-Vidal, A., 1975, Dissociated cell cultures from fetal mouse hypothalamus: Patterns of organization and ultrastructural features, *Exp. Brain Res.* **23:**29–47.

Berg, D.K., and Fischbach, G.D., 1978, Enrichment of spinal cord cell cultures with motoneurons, *J. Cell Biol.* **77:**83–98.

Bergey, G.K., Macdonald, R.L., and Nelson, P.G., 1978, Adverse effects of tetrodotoxin on early development and survival of postsynaptic cells in spinal cord cultures, *Neurosci. Abstr.* **4:**601.

Bird, M.M., 1978, Presynaptic and postsynaptic organelles of synapses formed in cultures of previously dissociated mouse spinal cord, *Cell Tissue Res.* **94:**503–511.

Bird, M.M., and James, D.W., 1975, The culture of previously dissociated embryonic chick spinal cord cells on feeder layers of liver and kidney, and the development of paraformaldehye induced fluorescence upon the former, *J. Neurocytol.* **4:**633–646.

Brown, A.G., and Fyffe, R.E.W., 1978, The morphology of group Ia afferent fibre collaterals in the spinal cord of the cat, *J. Physiol. (Lond.)* **274:**111–127.

Brown, A.G., Rose, P.K., and Snow, P.J., 1977, The morphology of hair follicle afferent fibre collaterals in the spinal cord of the cat, *J. Physiol. (Lond.)* **272:**779–797.

Burke, R.E., Walmsley, B., and Hodgson, J.A., 1979, HRP anatomy of group Ia afferent contacts on alpha motoneurones, *Brain Res.* **160:**347–352.

Burry, R.W., and Lasher, R.S., 1975, Uptake of GABA in dispersed cell cultures of postnatal rat cerebellum: An electron microscope autoradiographic study, *Brain Res.* **88**:502–507.

Burry, R.W., and Lasher, R.S., 1978, Electron microscopic autoradiography of the uptake of [^{3}H]GABA in dispersed cell cultures of rat cerebellums. I. The morphology of the GABAergic synapse, *Brain Res.* **151**:1–17.

Calvet, M.C., 1974, Patterns of spontaneous electrical activity in tissue cultures of mammalian cerebral cortex vs. cerebellum, *Brain Res.* **69**:281–295.

Calvet, M.C., Drian, M.J., and Privat, A., 1974, Spontaneous electrical patterns in cultured Purkinje cells grown with an antimitotic agent, *Brain Res.* **79**:285–290.

Castellucci, V.F., Carew, T.J., and Kandel, E.R., 1978, Cellular analysis of long-term habituation of the gill-withdrawal reflex of *Aplysia californica, Science* **202**:1306–1308.

Chan–Palay, V., Yonezawa, T., Yoshida, S., and Palay, S., 1978, γ-Aminobutyric acid receptors visualized in spinal cord cultures by [^{3}H]muscimol autoradiography, *Proc. Natl. Acad. Sci. USA* **75**:6281–6284.

Choi, D.W., Farb, D.H., and Fischbach, G.D., 1977, Chlordiazepoxide selectively augments GABA action in spinal cord cultures, *Nature* **269**:342–344.

Christensen, B.N., and Teubl, W.P., 1978, A comparison of synaptic junction localization on lamprey spinal cord neurons by physiological and morphological methods, *Neurosci. Abstr.* **4**:387.

Crain, S.M., 1956, Resting and action potentials of cultured chick embryo spinal ganglion cells, *J. Comp. Neurol.* **104**:285–329.

Crain, S.M., 1976, *Neurophysiologic Studies in Tissue Culture,* Raven Press, New York.

Crain, S.M., Bornstein, M.B., and Peterson, E. R., 1968, Maturation of cultured embryonic CNS tissues during chronic exposure to agents which prevent bioelectric activity, *Brain Res.* **8**:363–372.

Crain, S.M., Crain, B., Peterson, E.R., and Simon, E.J., 1978, Selective depression by opioid peptides of sensory-evoked dorsal-horn network responses in organized spinal cord cultures, *Brain Res.* **157**:196–201.

Creutzfeldt, O., Lux, H., and Watanabe, S., 1966, Electrophysiology of cortical nerve cells, in: *The Thalamus* (D. Purpura and M. Yahr, eds.), Columbia University Press, New York

Cullheim, S., and Kellerth, J.–O., 1976, Combined light and electron microscopic tracing of neurons, including axons and synaptic terminals, after intracellular injection of horseradish peroxidase, *Neurosci. Lett.* **2**:307–313.

Cullheim, S., and Kellerth, J.–O., 1978, A morphological study of the axons and recurrent axon collaterals of cat sciatic α-motoneurons after intracellular staining with horseradish peroxidase, *J. Comp. Neurol.* **178**:537–558.

Cullheim, S., Kellerth, J.–O., and Conradi, S., 1977, Evidence for direct synaptic interconnections between cat spinal α-motoneurons via the recurrent axon collaterals: A morphological study using intracellular injection of horseradish peroxidase, *Brain Res.* **132**:1–10.

Dichter, M.A., 1978, Rat cortical neurons in cell culture: Culture methods, cell morphology, electrophysiology, and synapse formation, *Brain Res.* **149**:279–293.

Dichter, M.A., and Fischbach, G.D., 1977, The action potential of chick dorsal root ganglion neurones maintained in cell culture, *J. Physiol. (Lond.)* **267**:281–298.

Dimpfel, W., Neale, J.H., and Habermann, E., 1975, ^{125}I-labelled tetanus toxin as a neuronal marker in tissue cultures derived from embryonic CNS, *Naunyn-Schmiedebergs Arch. Pharmacol.* **290**:329–333.

Dionne, V.E., and Stevens, C.F., 1975, Voltage dependence of agonist effectiveness at the frog neuromuscular junction: Resolution of a paradox, *J. Physiol. (Lond.)* **251**:245–270.

Dudel, J., 1974, Nonlinear voltage dependence of excitatory synaptic current in crayfish muscle, *Pflügers Arch.* **352**:227–241.

Dunlap, K., and Fischbach, G.D., 1978, Neurotransmitters decrease the calcium component of sensory neurone action potentials, *Nature* **276**:837–839.

Edwards, F.R., Redman, S.J., and Walmsley, B., 1976a, Statistical fluctuations in charge transfer at Ia synapses on spinal motoneurons, *J. Physiol. (Lond.)* **259:**665–688.

Edwards, F.R., Redman, S.J., and Walmsley, B., 1976b, Non-quantal fluctuations and transmission failures in charge transfer at Ia synapses on spinal motoneurones, *J. Physiol. (Lond.)* **259:**689–704.

Edwards, F.R., Redman, S.J., and Walmsley, B., 1976c, The effect of polarizing currents on unitary Ia excitatory post-synaptic potentials evoked in spinal motoneurones, *J. Physiol. (Lond.)* **259:**705–723.

Engelhardt, J.K., Ishikawa, K., Lisbin, S.J., and Mori, J., 1976, Neurotrophic effects on passive electrical properties of cultured chick skeletal muscle, *Brain Res.* **110:**170–174.

Farb, D.H., Berg, D.K., and Fischbach, G.D., 1979, Uptake and release of [^{3}H]γ-aminobutyric acid by embryonic spinal cord neurons in dissociated cell culture, *J. Cell Biol.* **80:**651–661.

Federoff, S., and Hertz, L. (eds.), 1977, *Cell, Tissue, and Organ Cultures in Neurobiology,* Academic Press, New York.

Fields, K.L., Brockes, J.P., Mirsky, R., and Wendon, L.M.B., 1978, Cell surface markers for distinguishing different types of rat dorsal root ganglion cells in culture, *Cell* **14:**43–51.

Fischbach, G.D., 1970, Synaptic potentials recorded in cell cultures of nerve and muscle, *Science* **169:**1331–1333.

Fischbach, G.D., 1972, Synapse formation between dissociated nerve and muscle cells in low density cultures, *Dev. Biol.* **28:**407–429.

Fischbach, G.D., and Dichter, M.A., 1974, Electrophysiologic and morphologic properties of neurons in dissociated chick spinal cord cell cultures, *Dev. Biol.* **37:**100–116.

Fischbach, G. D., and Nelson, P.G., 1977, Cell culture in neurobiology, in: *Handbook of Physiology, Sect. 1, The Nervous System, Vol. I* (E.R. Kandel, vol. ed.), pp. 719–774, American Physiological Society, Bethesda.

Gähwiler, B.H., 1975, The effects of GABA, picrotoxin and bicuculline on the spontaneous bioelectric activity of cultured cerebellar Purkinje cells, *Brain Res.* **99:**85–95.

Gähwiler, B.H., 1976, Inhibitory action of noradrenaline and cyclic AMP in explants of rat cerebellum, *Nature* **259:**483–484.

Gähwiler, B.H., 1978, Mixed cultures of cerebellum and inferior olive: Generation of complex spikes in Purkinje cells, *Brain Res.* **145:**168–172.

Gähwiler, B.H., Sandoz, P., and Dreifuss, J.J., 1978, Neurones with synchronous bursting discharges in organ cultures of the hypothalamic supraoptic nucleus area, *Brain Res.* **151:**245–253.

Geller, H. M., 1975, Phasic discharge of neurons in long-term cultures of tuberal hypothalamus, *Brain Res.* **93:**511–515.

Geller, H.M., and Woodward, D.J., 1974, Responses of cultured cerebellar neurons to iontophoretically applied amino acids, *Brain Res.* **74:**67–80.

Giller, E. L., Jr., Schrier, B. K., Shainberg, A., Fisk, H.R., and Nelson, P.G., 1973, Choline acetyltransferase activity is increased in combined cultures of spinal cord and muscle cells from mice, *Science* **182:**588–589.

Giller, E.L., Jr., Neale, J.H., Bullock, P.N., Schrier, B.K., and Nelson, P.G., 1977, Choline acetyltransferase activity of spinal cord cell cultures increased by co-culture with muscle and by muscle-conditioned medium, *J. Cell Biol.* **74:**16–29.

Godfrey, E.W., Nelson, P.G., Schrier, B.K., Breuer, A.C., and Ransom, B.R., 1975, Neurons from fetal rat brain in a new cell culture system: A multidisciplinary analysis, *Brain Res.* **90:**1–21.

Ham, R.G., 1965, Clonal growth of mammalian cells in a chemically defined, synthetic medium, *Proc. Natl. Acad. Sci. USA* **53:**288–293.

Hasegawa, S., and Kuromi, H., 1977, Effects of spinal cord and other tissue extracts on resting and action potentials of organ-cultured mouse skeletal muscle, *Brain Res.* **119:**133–140.

Hazum, I., Chang, K.-J., and Cuatrecasas, P., 1979, Opiate (enkephalin) receptors of neuroblastoma cells: Occurrence in clusters on the cell surface, *Science* **206:**1077–1079.

Hendelman, W.J., and Marshall, K.C., 1978, HRP characterization of neurons in organized cultures of cerebellum, *Neurosci. Abstr.* **4:**591.

Hendelman, W.J., Marshall, K.C., Aggerwal, A.S., and Wojtowicz, J.M., 1977, Organization of pathways in cultures of mouse cerebellum, in: *Cell, Tissue, and Organ Cultures in Neurobiology* (S. Fedoroff and L. Hertz, eds.), pp. 539–554, Academic Press, New York.

Hökfelt, T., Johansson, O., Kellerth, J.-O., Ljungdahl, Å., Nilsson, G., Nygårds, A., and Pernow, B., 1977, Immunohistochemical distribution of substance P, in: *Substance P* (U.S. von Euler and B. Pernow, eds.), pp. 117–145, Raven Press, New York.

Hösli, L., Andres, P.F., and Hösli, E., 1972, Effects of potassium on the membrane potential of spinal neurones in tissue culture, *Pflügers Arch.* **333:**362–365.

Iles, J.F., 1973, Demonstration of afferent terminations in the cat spinal cord, *J. Physiol. (Lond.)* **234:**22–24P.

Iles, J.F., 1976, Central terminations of muscle afferents on motoneurones in the cat spinal cord, *J. Physiol. (Lond.)* **262:**91–117.

Jankowska, E., Rastad, J., and Westman, J., 1976, Intracellular application of horseradish peroxidase and its light and electron microscopical appearance in spinocervical tract cells, *Brain Res.* **105:**557–562.

Kandel, E., Spencer, W., and Brinley, F., 1961, Electrophysiology of hippocampal neurons. I. Sequential invasion and synaptic organization, *J. Neurophysiol.* **24:**225–242.

Katz, B., 1966, *Nerve, Muscle and Synapse,* McGraw–Hill, New York.

Katai, S.T., Kocsis, J.D., Preston, R.J., and Sugimori, M., 1976, Monosynaptic inputs to caudate neurons identified by intracellular injection of horseradish peroxidase, *Brain Res.* **109:**601–606.

Krnjević, K., 1974, Chemical nature of synaptic transmission in vertebrates, *Physiol. Rev.* **54:**418–540.

Kuno, M., 1971, Quantum aspects of central and ganglionic synaptic transmission in vertebrates, *Physiol. Rev.* **51:**647–678.

Kuromi, H., and Hasegawa, S., 1975, Neurotrophic effect of spinal cord extract on membrane potentials of organ-cultured mouse skeletal muscle, *Brain Res.* **100:**178–181.

Lasher, R.S., 1974, The uptake of [^{3}H]GABA and differentiation of stellate neurons in cultures of dissociated postnatal rat cerebellum, *Brain Res.* **69:**235–254.

Lasher, R.S., and Zagon, I.S., 1972, The effect of potassium on neuronal differentiation in cultures of dissociated newborn rat cerebellum, *Brain Res.* **41:**482–488.

Li, C., Okujava, V., and Bak, A., 1971, Some electrical measurements of cortical elements and neuronal responses to direct stimulation with particular reference to input resistance, *Exp. Neurol.* **31:**263–276.

Lux, H., and Pollen, D., 1966, Electrical constants on the neurons in the motor cortex of the cat, *J. Neurophysiol.* **29:**207–220.

Macdonald, R.L., and Barker, J.L., 1978, Benzodiazepines specifically modulate GABA-mediated postsynaptic inhibition in cultured mammalian neurones, *Nature* **271:**563–564.

Macdonald, R.L., and Nelson, P.G., 1978, Specific-opiate-induced depression of transmitter release from dorsal root ganglion cells in culture, *Science* **199:**1449–1451.

Macdonald, R.L., Moonen, G., and Nelson, P.G., 1978, Postsynaptic pharmacology of cerebellar neurons in cell culture, *Neurosci. Abstr.* **4:**447.

Martin, A.R., 1977, Junctional transmission. II. Presynaptic mechanisms, in: *Handbook of Physiology, Sect. 1, The Nervous System, Vol. I* (E.R. Kandel, vol. ed.), pp. 329–355, American Physiological Society, Bethesda.

Masuko, S., Kuromi, H., and Shimada, Y., 1979, Isolation and culture of motoneurons from embryonic chicken spinal cords, *Proc. Natl. Acad. Sci. USA* **76:**3537–3541.

Masurovsky, E.B., Benitez, H.H., and Murray, M.R., 1971, Synaptic development in long-term organized cultures of murine hypothalamus, *J. Comp. Neurol.* **143:**263–278.

Matsuda, Y., Yoshida, S., and Yonezawa, T., 1978, Tetrodotoxin sensitivity and Ca component of action potentials of mouse dorsal root ganglion cells cultured in vitro, *Brain Res.* **154:**69–82.

Matthew, E., Neale, E.A., Nelson, P.G., and Zimmerman, E.A., 1979, Immunocytochemical studies on neuronal peptides in dissociated spinal cord–dorsal root ganglion cell cultures, *Neurosci. Abstr.* **5:**533.

McBurney, R.N., and Barker, J.L., 1978, GABA-induced conductance fluctuations in cultured spinal neurones, *Nature* **274:**596–597.

Messer, A., 1977, The maintenance and identification of mouse cerebellar granule cells in monolayer culture, *Brain Res.* **130:**1–12.

Model, P.G., Bornstein, M.B., Crain, S.M., and Pappas, G.D., 1971, An electron microscopic study of the development of synapses in cultured fetal mouse cerebrum continuously exposed to xylocaine, *J. Cell Biol.* **49:**362–371.

Mudge, A.W., Leeman, S.E., and Fischbach, G.D., 1979, Enkephalin inhibits release of substance P from sensory neurons in culture and decreases action potential duration, *Proc. Natl. Acad. Sci. USA* **76:**526–530.

Muller, K.J., and McMahan, U.J., 1976, The shapes of sensory and motor neurones and the distribution of their synapses in ganglia of the leech: A study using intracellular injection of horseradish peroxidase, *Proc. R. Soc. Lond. B* **194:**481–499.

Neale, E.A., Macdonald, R.L., and Nelson, P.G., 1977, Morphology of peroxidase injected and electrophysiologically studied spinal cord neurons in culture, *J. Cell Biol.* **75:**114a.

Neale, E.A., Macdonald, R.L., and Nelson, P.G., 1978a, Intracellular horseradish peroxidase injection for correlation of light and electron microscopic anatomy with synaptic physiology of cultured mouse spinal cord neurons, *Brain Res.* **152:**265–282.

Neale, E.A., Moonen, G., Macdonald, R.L., Gibbs, W., and Nelson, P.G., 1978b, Morphology and electrophysiology of cerebellar neurons in cell culture, *Neurosci. Abstr.* **4:**592.

Neale, E.A., Bowers, L.M., and Malley, J.D., 1979, Synaptic ultrastructure of identified neurons: A quantitative analysis, *J. Cell Biol.* **83:**142a.

Neale, J.H., Barker, J.L., Uhl, G.R., and Snyder, S.H., 1978, Enkephalin-containing neurons visualized in spinal cord cell cultures, *Science* **201:**467–469.

Neher, B., and Sakmann, B., 1976, Single-channel currents recorded from membrane of denervated frog muscle fibers, *Nature* **260:**799–802.

Neher, E., Sakmann, B., and Steinbach, J.H., 1978, The extracellular patch clamp: A method for resolving currents through individual open channels in biological membranes, *Pflügers Arch.* **375:**219–228.

Nelson, P.G., 1975, Nerve and muscle cells in culture, *Physiol. Rev.* **55:**1–61.

Nelson, P.G., 1976, Central nervous system synapses in cell culture, *Cold Spring Harbor Symp. Quant. Biol.* **40:**359–371.

Nelson, P.G., and Lux, H.D., 1970, Some electrical measurements of motoneuron parameters, *Biophys. J.* **10:**55–73.

Nelson, P.G., and Peacock, J.H., 1973, Electrical activity in dissociated cell cultures from fetal mouse cerebellum, *Brain Res.* **61:**163–174.

Nelson, P.G., Ransom, B.R., Henkart, M., and Bullock, P.N., 1977, Mouse spinal cord in cell culture. IV. Modulation of inhibitory synaptic function, *J. Neurophysiol.* **40:**1178–1187.

Nelson, P.G., Neale, E.A., and Macdonald, R.L., 1978, Formation and modification of synapses in central nervous system cell cultures, *Fed. Proc.* **37:**2010–2015.

Nelson, P.G., Neale, E.A., Matthew, E., and Zimmerman, E.A., 1980, A presynaptic locus of action for the opiates, in: *The Role of Peptides in Neuronal Function* (J.L. Barker and T.S. Smith, eds.), pp. 727–739, Marcel Dekker, New York.

Oh, T.H., and Markelonis, G.J., 1978, Neurotrophic protein regulates muscle acetylcholinesterase in culture, *Science* **200:**337–339.

Peacock, J.H., 1979, Electrophysiology of dissociated hippocampal cultures from fetal mice, *Brain Res.* **169:**247–260.

Peacock, J.H., Nelson, P.G., and Goldstone, M.W., 1973, Electrophysiologic study of cultured neurons dissociated from spinal cords and dorsal root ganglia of fetal mice, *Dev. Biol.* **30:**137–152.

Peacock, J.H., Rush, D.F., and Mathers, L.H., 1979, Morphology of dissociated hippocampal cultures from fetal mice, *Brain Res.* **169:**231–246.

Pickel, V.M., Reis, D.J., and Leeman, S.E., 1977, Ultrastructural localization of substance P in neurons of rat spinal cord, *Brain Res.* **122:**534–540.

Proshansky, E., and Egger, M.D., 1977, Staining of the dorsal root projection to the cat's dorsal horn by anterograde movement of horseradish peroxidase, *Neurosci. Lett.* **5:**103–110.

Puck, T.T., Cieciura, S.J., and Robinson, A., 1958, Genetics of somatic mammalian cells. III. Long-term cultivation of euploid cells from human and animal subjects, *J. Exp. Med.* **108:**945–956.

Raff, M.C., Fields, K.F., Hakomori, S., Mirsky, R., Pruss, R.M., and Winter, J., 1979, Cell-type-specific markers for distinguishing and studying neurons and the major classes of glial cells in culture, *Brain Res.* **174:**283–308.

Ransom, B.R., and Holz, R.W., 1977, Ionic determinants of excitability in cultured mouse dorsal root ganglion and spinal cord cells, *Brain Res.* **136:**445–453.

Ransom, B.R., Barker, J.L., and Nelson, P.G., 1975, Two mechanisms for poststimulus hyperpolarisations in cultured mammalian neurones, *Nature* **265:**424–425.

Ransom, B.R., Bullock, P.N., and Nelson, P.G., 1977a, Mouse spinal cord in cell culture. III. Neuronal chemosensitivity and its relationship to synaptic activity, *J. Neurophysiol.* **40:**1163–1177.

Ransom, B.R., Christian, C.N., Bullock, P.N., and Nelson, P.G., 1977b, Mouse spinal cord in cell culture. II. Synaptic activity and circuit behavior, *J. Neurophysiol.* **40:**1151–1162.

Ransom, B.R., Neale, E., Henkart, M., Bullock, P.N., and Nelson, P.G., 1977c, Mouse spinal cord in cell culture. I. Morphology and intrinsic neuronal electrophysiologic properties, *J. Neurophysiol.* **40:**1132–1150.

Rastad, J., Jankowska, E., and Westman, J., 1977, Arborization of initial axon collaterals of spinocervical tract cells stained intracellularly with horseradish peroxidase, *Brain Res.* **135:**1–10.

Schlapfer, W.T., Mamoon, A.–M., and Tobias, C.A., 1972, Spontaneous bioelectric activity of neurons in cerebellar cultures: Evidence for synaptic interactions, *Brain Res.* **45:**345–363.

Schultzberg, M., Ebendal, T., Hökfelt, T., Nilsson, G., and Pfenniger, K., 1978, Substance P-like immunoreactivity in cultured spinal ganglia from chick embryos, *J. Neurocytol.* **7:**107–117.

Schwartzkroin, P., 1975, Characteristics of CA1 neurons recorded intracellularly in the hippocampal *in vitro* slice preparation, *Brain Res.* **85:**423–436.

Scott, B.S., Engelbert, V.E., and Fisher, K.C., 1969, Morphological and electrophysiological characteristics of dissociated chick embryonic spinal ganglion cells in culture, *Exp. Neurol.* **23:**230–248.

Seil, F.J., and Leiman, A.L., 1977, Spontaneous versus driven activity in intracerebellar nuclei: A tissue culture study, *Exp. Neurol.* **54:**110–127.

Snow, P.J., Rose, P.K., and Brown, A.G., 1976, Tracing axons and axon collaterals of spinal neurons using intracellular injection of horseradish peroxidase, *Science* **191:**312–313.

Somjen, G., 1975, Electrophysiology of neuroglia, *Annu. Rev. Physiol.* **37:**163–190.

Sotelo, C., Privat, A., and Drian, M.–J., 1972, Localization of [^{3}H]GABA in tissue culture of rat cerebellum using electron microscopy radioautography, *Brain Res.* **45:**302–308.

Spater, H.W., Novikoff, A.B., Spater, S., and Quintana, N., 1977, Pyridoxal-5-phosphatase: A new enzyme activity of GERL in neurons, *J. Cell Biol.* **75:**114a.

Spencer, W., and Kandel, E., 1961, Electrophysiology of hippocampal neurons. III. Firing level and time constant, *J. Neurophysiol.* **24:**260–271.

Spitzer, N.C., and Lamborghini, J.E., 1976, The development of the action potential mechanism of amphibian neurons isolated in culture, *Proc. Natl. Acad. Sci. USA* **73:**1641–1645.

Stefanelli, A., Cataldi, E., and Ieradi, L.A., 1977, Specific synaptic systems in reaggregated spherules from dissociated chick cerebellum cultivated in vitro, *Cell Tissue Res.* **182:**311–325.

Takahashi, K., 1965, Slow and fast groups of pyramidal tract cells and their respective membrane properties, *J. Neurophysiol.* **28:**908–924.

Takeuchi, A., 1977, Junctional transmission. I. Postsynaptic mechanisms, in: *Handbook of Physiology, Sect. 1, The Nervous System, Vol. I* (E.R. Kandel, vol. ed.), pp. 295–327, American Physiological Society, Bethesda.

Toran–Allerand, C.D., 1978, The luteinizing hormone-releasing hormone (LH–RH) neuron in cultures of the newborn mouse hypothalamus/preoptic area: Ontogenetic aspects and responses to steroid, *Brain Res.* **149:**257–265.

Trenkner, E., and Sidman, R.L., 1977, Histogenesis of mouse cerebellum in microwell cultures, *J. Cell Biol.* **75:**915–940.

Varon, S., and Raiborn, C.W., Jr., 1969, Dissociation, fractionation, and culture of embryonic brain cells, *Brain Res.* **12:**180–199.

Varon, S., and Raiborn, C., 1971, Excitability and conduction in neurons of dissociated ganglionic cell cultures, *Brain Res.* **30:**83–98.

White, W.F., Snodgrass, S.R., and Dichter, M., 1980, Identification of GABA neurons in rat cortical cultures by GABA uptake and autoradiography, *Brain Res.* **190:**139–152.

Wiesel, T.N., and Hubel, D.H., 1965, Comparison of the effects of unilateral and bilateral eye closure on cortical unit responses in kittens, *J. Neurophysiol.* **28:**1029–1040.

Woodward, W.R., and Lindstrom, S.H., 1977, A potential screening technique for neurotransmitters in the CNS: Model studies in the cat spinal cord, *Brain Res.* **137:**37–52.

Yavin, E., and Menkes, J.H., 1973, The culture of dissociated cells from rat cerebral cortex, *J. Cell Biol.* **57:**232–237.

Yavin, Z., and Yavin, E., 1977, Synaptogenesis and myelinogenesis in dissociated cerebral cells from rat embryo on polylysine coated surfaces, *Exp. Brain Res.* **29:**137–147.

Zipser, B., Crain, S.M., and Bornstein, M.B., 1973, Directly evoked "paroxysmal" depolarizations of mouse hippocampal neurons in synaptically organized explants in long-term culture, *Brain Res.* **60:**489–495.

3

Neuropharmacology of Spinal Cord Neurons in Primary Dissociated Cell Culture

ROBERT L. MACDONALD and JEFFERY L. BARKER

1. INTRODUCTION

Investigation into the physiological properties of central nervous system (CNS) neurons and the pharmacological actions of clinically important drugs on the CNS have progressed in parallel over the past two decades. The suggestion that most interneuronal communication is mediated by release of chemical neurotransmitters (Krnjevic, 1974) has stimulated research into all aspects of synaptic transmission including regulation of synthesis and release of neurotransmitter from presynaptic terminals, properties of pre- and postsynaptic neurotransmitter receptors, properties of neurotransmitter-coupled ionic channels, and neuroanatomical distribution of specific neurotransmitters.

The importance of comprehensive understanding of neurotransmitter pharmacology is equally apparent in the study and management of an increasing number of neurological diseases, the most well-established being Parkinson's disease and Huntington's disease, where alteration of dopaminergic, GABAergic, and cholinergic synaptic transmission are associated with marked motor abnormalities (Hornykiewicz, 1972; Hornykiewicz *et al.,* 1976; Chase, 1976). While clinically relevant centrally acting drugs such as anticonvulsants, anesthetics, or antidepressants may have direct actions on neuronal or axonal

ROBERT L. MACDONALD • Department of Neurology, University of Michigan, Ann Arbor, Michigan 48109. JEFFERY L. BARKER • Laboratory of Neurophysiology, National Institute of Neurological and Communicative Disorders and Stroke, National Institutes of Health, Bethesda, Maryland 20205.

membrane properties, much recent research has involved investigation of possible effects of these drugs on synaptic transmission.

While numerous immunohistochemical and neurochemical techniques can be applied to the CNS to determine the distribution of putative neurotransmitter pathways, neurophysiological techniques for investigation of the cellular bases of neurotransmitter and drug action have more restricted application. Two approaches have been commonly applied to the intact mammalian nervous system. One involves extracellular recording from neurons which are either unidentified or are "identified" by a characteristic firing pattern and/or via antidromic or orthodromic activation from known neuroanatomical pathways. Once a neuron is selected, a drug or putative neurotransmitter is applied either by intravenous administration or by iontophoresis from a micropipette coupled to the recording microelectrode (Curtis and Eccles, 1958a,b; Curtis *et al.,* 1959). These techniques have also been extended to include intracellular recording coupled with either intravenous or iontophoretic application of drugs (Curtis *et al.,* 1968; Krnjevic *et al.,* 1977). The second approach involves sucrose gap recording from the dorsal and/or ventral roots of the isolated spinal cord (Curtis *et al.,* 1961; Koketsu *et al.,* 1969). This allows characterization of the action of drugs and neurotransmitters on various "root potentials" which themselves reflect the algebraic sum of excitability changes in many neurons.

Both of these techniques have clear limitations. Neither extracellular recording nor sucrose gap recording permits direct observation of the membrane events that underlie the neurotransmitter or drug action. It is often difficult if not impossible to determine whether the effects result from actions at pre- or postsynaptic sites or whether there are direct actions on neuronal membrane properties. Because of these limitations and the complexity of the mammalian CNS, many investigators have chosen to use "simple" *in vitro* model systems to study synaptic events in the CNS. Such preparations include synapses peripheral to the CNS such as those found at neuromuscular junctions, in isolated invertebrate ganglia, or in peripheral vertebrate autonomic ganglia. Intact, but isolated and perfused, amphibian spinal cord has also been employed as an *in vitro* preparation. The peripheral synaptic systems are less complex than intact CNS and permit routine application of intracellular recording for study of synaptic physiology and pharmacology. A significant drawback to these preparations as models for synaptic transmission in the CNS is that the postsynaptic pharmacology is either limited in chemosensitivity to a few neurotransmitters or else is overly complex. For example, at the neuromuscular junction, the postjunctional membrane most likely contains only receptors for acetylcholine (Hubbard, 1975), while neurons from *Aplysia* ganglia often respond to individual neurotransmitters with both excitatory and inhibitory responses of varying durations. Thus despite having technical advantages, such systems often appear to manifest pharmacology that is quite dissimilar to that in the intact mammalian CNS, thus limiting their use as model systems for mammalian CNS pharmacology.

To achieve the technical advantages of *in vitro* preparations while retaining important features of mammalian CNS physiology, two mammalian *in vitro* preparations have evolved. The "slice technique" involves the acute preparation of thin tissue slices from specific nervous system regions such as hippocampus (Yamamoto and Kawai, 1967, 1968; Skrede and Westgaard, 1971), cortex (Courtney and Prince, 1977), or cerebellum (Yamamoto, 1974; Okamoto *et al.,* 1976). Such preparations remain viable for many hours and permit rigorous control over the bathing or extracellular medium as well as perfusion with drug-containing solutions. While primarily extracellular recording has been used with this technique, intracellular recordings have been obtained from specific cell types (Schwartzkroin, 1975; Yamamoto, 1972). A major advantage of the slice preparation is that normal organization is preserved, while a disadvantage is that the preparation is viable only for a limited period of time.

Another approach involves growth and maintenance of CNS neurons in organotypic or explant culture (Crain, 1975) or in primary dissociated cell (PDC) culture (Nelson, 1976; Ransom *et al.,* 1977a). Explants of fetal CNS maintained in tissue culture provide chronic or long-term preparations and develop or retain some normal organizational features. Mammalian neurons in PDC culture offer significant advantages and overcome many limitations of other *in vivo* and *in vitro* preparations since they permit direct visualization and accessibility of neurons while achieving differentiation of neuronal morphology and electrophysiological properties (Nelson, 1976; Ransom *et al.,* 1977a; Neale *et al.,* 1978). Thus it is routinely possible to impale cultured neurons with one or two microelectrodes and apply conventional current or voltage clamp techniques for the study of membrane events. Neurotransmitters and/or drugs can be applied to discrete somatic or dendritic locations or using either iontophoresis or "miniperfusion." Thus this culture system offers a favorable preparation for study of the cellular aspects of CNS pharmacology. In the following sections, we will discuss the culture system in more detail, methodology for electrophysiological and pharmacological studies, results of studies on postsynaptic pharmacology of putative neurotransmitter amino acids, and results of investigations on the mechanisms of action of convulsant and anticonvulsant drugs.

2. MAMMALIAN SPINAL CORD NEURONS IN PRIMARY DISSOCIATED CELL CULTURE

Fetal murine spinal cord co-cultured with dorsal root ganglia in PDC cultures contain spinal cord (SC) neurons which make excitatory or inhibitory synapses with other SC neurons, and dorsal root ganglion (DRG) cells which make exclusively excitatory contacts with SC cells (Ransom *et al.,* 1977b; Neale *et al.,* 1978). SC and DRG neurons can be unambiguously identified using morphological as well as electrophysiological criteria. Under phase contrast microscopy, DRG cells have smooth, round, phase-bright cell bodies, sharply defined nuclei and nucleoli, and usually fewer than two thin processes. Large SC neurons (>20-μm-diameter soma) have a rough surface with several large tapering and branching processes (Fig. 1). At the electron microscopic level, DRG cell bodies appear devoid of synaptic contacts, while SC cell surfaces (including somatic and dendritic regions) contain numerous synaptic terminals (Ransom *et al.,* 1977a). The fine structure of these synapses is varied, with dense-core vesicles and variable types of membrane thickenings characterizing the different classes of synapses.

Electrophysiological properties of these neuronal cell types are consistent with those of mature differentiated vertebrate neurons (Ransom *et al.,* 1977a). DRG and SC cells have average resting membrane potentials of about −50

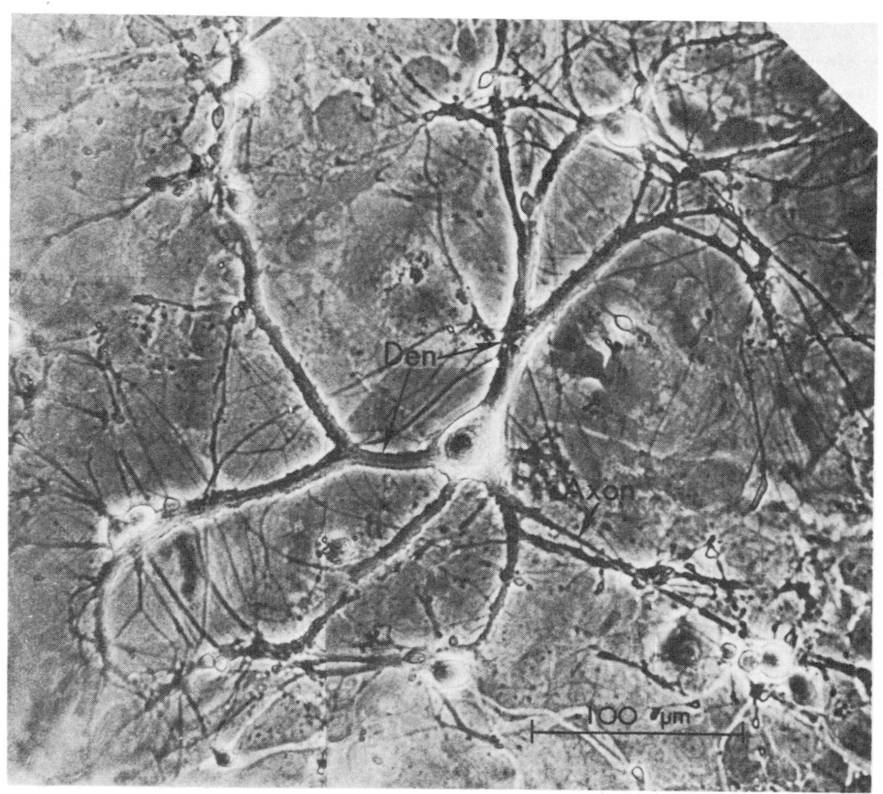

FIGURE 1. A spinal cord neuron that was maintained in dissociated cell culture for 8 weeks and was representative of those employed for electrophysiological study; × 450, reproduced at 53%. (From Macdonald and Barker, 1979b.)

mV (5.4 mM K^+) and produce active responses when depolarized. SC, but not DRG, neurons characteristically undergo repetitive firing, and both cell types exhibit afterhyperpolarization (Ransom *et al.*, 1975, 1977a). SC action potentials are blocked by tetrodotoxin and thus are secondary to regenerative increase in sodium conductance, whereas DRG action potentials have both sodium and calcium components (Matsuda *et al.*, 1976; Dichter and Fischbach, 1977; Ransom and Holtz, 1977). SC cells are spontaneously active with random firing of action potentials superimposed upon mixtures of excitatory and inhibitory synaptic potentials, while DRG cells are electrically silent and receive no synaptic input (Ransom *et al.*, 1977a).

Monosynaptic connections between DRG and SC neurons or SC and SC neurons have received considerable study (Ransom *et al.*, 1977a,b; Nelson *et al.*, 1978; Neale *et al.*, 1978). Excitatory postsynaptic potentials (EPSPs) evoked by stimulation of DRG or SC neurons are frequently large (5–10 mV) with only a small variation in amplitude and no preferred amplitudes (as would be expected with low mean quantum contents). Thus, most of these synaptic connections are well-developed and relatively secure with high quantum contents (Ransom *et al.*, 1977b). Inhibitory postsynaptic potentials are readily reversible with hyperpolarization of membrane potential and appear to be produced by extensive ramification of inhibitory terminal branches over the postsynaptic cell (Neale *et al.*, 1978). Additional discussion of the electrophysiology of neurons in PDC culture can be found in another chapter in this volume (Nelson *et al.*, this volume).

The postsynaptic pharmacology of SC neurons will be discussed in more detail in Section 4, but, briefly, the majority of SC neurons respond to iontophoretically applied inhibitory neutral amino acids such as γ-aminobutyric acid (GABA) and glycine (GLY) and to excitatory acidic amino acids such as glutamic acid (GLU) (Ransom *et al.*, 1977c). Neutral amino acids responses are mediated primarily by increases in chloride conductance, while GLU responses are likely to result from increases in sodium and potassium conductances (Barker and Ransom, 1978a).

Thus spinal cord neurons in PDC culture have been demonstrated to have differentiated neuronal morphology, inherent excitability, rich synaptic connectivity, and extensive amino acid sensitivity. They would thus appear to be appropriate for study of the physiology of central synaptic transmission and the amino acid pharmacology of central mammalian neurons.

3. METHODOLOGY

3.1. Primary Dissociated Cell Cultures

Cultures are prepared from spinal cords obtained from 12.5- to 14-day-old fetal mice according to the method developed in the laboratory of Phillip

Nelson (Ransom *et al.*, 1977a). Briefly, the spinal cords with attached dorsal root ganglia are minced with iridectomy scissors, suspended in a 0.25% trypsin solution, and incubated at 37°C for 30 min, following which the trypsin is inactivated by the addition of a solution containing 10% heat-inactivated horse serum (10 HS) and 10% fetal calf serum (10 FCS). The resultant single cell suspension is then plated in collagen-coated 35-mm culture dishes at a density of 106 cells/ml in 1.5 ml of a medium consisting of 80% Eagle's minimal essential medium (MEM), 10 FCS, and 10 HS. Following 6–8 days of incubation at 36°C in 90% air/10% CO_2, uridine and 5′-fluoro-2′-deoxyuridine (FUdr) (35 μg/ml and 15 μg/ml) are added to the culture medium to suppress growth of rapidly dividing nonneuronal background cells, and the medium is changed to 90% MEM/10 HS. After 2 days, the medium is totally changed to 90% MEM/10% HS, and the cultures are incubated for 5–12 weeks prior to electrophysiological study.

3.2. Intracellular Recording

Electrophysiological experiments are performed on the stage of an inverted phase microscope modified to allow heating of the culture medium to 35–37°C. Cultures are bathed either in normal growth medium (90% MEM/10% HS) superfused by a CO_2/air mixture (bicarbonate/CO_2 buffer system) or in a phosphate-buffered saline with pH being maintained in physiological range (7.25–7.4). Intracellular recordings are made under visual control using glass micropipettes (25–50 MΩ) filled with either 3 M KCl or 4 M K-acetate positioned using Leitz micromanipulators. A conventional bridge system with a high input impedance allows simultaneous recording and current injection when a single recording microelectrode is used. Data are displayed on an oscilloscope and recorded on a 6-channel polygraph. In addition, data are digitized on-line by a small laboratory computer for on- and off-line analysis.

The voltage clamp technique is reliably applied to SC neurons by using two independently positioned intracellular microelectrodes. The voltage clamp system which has been successfully applied to cultured neurons is considered in detail elsewhere in this volume (Smith *et al.*, this volume). The membrane potential is sensed by the "voltage" electrode using a unity gain head stage amplifier. The potential is then amplified by an inverting amplifier with variable gain and frequency response. The signal from this second amplification stage is then fed to a differential amplifier of unity gain and finally to the "current" electrode. Current flows through the neuronal membrane, through the bathing medium, and to the summing point of the "current monitor" which holds the bath at virtual ground. The instrumentation time constant need not be fast to record pharmacologically induced membrane currents.

Application of the voltage clamp technique allows one to study in a more direct manner membrane conductance events underlying the generation of membrane potential changes. With the membrane potential held constant, the ionic driving force acting on the conductance to produce the membrane current is kept constant, and contributions of membrane current derived from capacitative and voltage change sources are nullified. The technique permits considerable control of membrane potential in regions of potential where this is impossible using a single micropipette or two micropipettes without a feedback (or clamp) circuit.

3.3. Iontophoresis and Miniperfusion

Putative neurotransmitters and drugs are applied to either somatic or dendritic locations using either iontophoresis or miniperfusion. In most experiments, the magnesium concentration is increased to 12 mM in the bathing medium to suppress spontaneous synaptic activity and to allow a clearer recording of responses elicited by iontophoresis or miniperfusion of drugs or neurotransmitters. Iontophoretic currents are applied using a custom built constant current iontophoretic unit (Smith and Hoffer, 1978), and with neurotransmitters, short current pulses (10 to 100 msec) are used to avoid desensitization of the responses and to permit analysis of the time course of the responses. Under visual control, iontophoretic pipette tips can be positioned within 2 μm of a selected portion of the soma or dendritic tree. When the interaction between two agents is being investigated, the tips of both electrodes can be placed within 2 μm of each other on the neuronal surface. With four micromanipulators placed about the microscope stage, four micropipettes can be positioned independently providing one recording pipette and three iontophoretic pipettes. In addition, use of multibarrel pipettes increases the number of agents available for iontophoresis.

Neurotransmitters, anticonvulsants, and convulsants are also applied to individual neurons using a miniperfusion technique. Glass micropipettes are prepared using a conventional microelectrode puller, and large tips (5–15 μm) are produced by adjusting heat and tension appropriately. Alternatively high-resistance, small-tip (<1 μm) microelectrodes can be converted to miniperfusion pipettes by mechanical fracture under microscopic visual control. The electrodes are then filled with solutions of the desired composition which have been adjusted to the pH of the bathing medium. The miniperfusion pipettes are attached to micromanipulators and positioned under visual control with the pipette tip within 20 μm of the neuronal membrane. Neurotransmitters and drugs are ejected from the miniperfusion pipette using regulated pressure applied to the open end of the pipette via a tight-fitting polyethylene tube. The advantage of miniperfusion is that known concentrations of neurotransmitters and drugs can be applied to neurons, thus allowing more quantitative analysis of neurotransmitter and drug action. It also permits determination of whether the pharmacological actions being studied are occurring at physiological or therapeutic concentrations. The relatively rapid time course of application also allows study of neurotransmitters that are rapidly desensitizing. Since drugs are applied at physiological or therapeutic concentrations, it is possible to study the actions of agents that act at very low concentrations as well as those that are relatively insoluble. Iontophoresis of such drugs is always difficult and unreliable because of their low and variable current-carrying ability. The drugs can also be applied at physiological pH, whereas with iontophoresis, the pipettes contain neurotransmitters and drugs that have had acid or base added to provide the agent with a net positive or negative charge, and this often results in pHs as low as 1.5 or as high as 11.0. While pH effects on neuronal membrane properties are not present when there is another ionized species at high concentration in the pipette (such as GABA), possible pH artifacts might occur when low concentrations of ionized drug are iontophoresed from a high- or low-pH solution. In this case, a higher percentage of current might be carried by hydrogen or hydroxyl ions, and local alterations in pH might be possible. With miniperfusion, no pH alterations occur.

Miniperfusion is not, however, without disadvantages. Since large portions of the cell are perfused, the responses obtained are an aggregate of responses obtained from regions of high as well as low sensitivity. Mechanical artifacts also can occur when miniperfusion is used, although they can be minimized by careful control of perfusion pressure and pipette location. Fewer concentrations of neurotransmitter can be applied to each given cell because of space limitations for pipettes. Finally, miniperfusion often requires long-duration application of drugs (seconds) that precludes study of rapid-time-course kinetics.

In summary, miniperfusion increases the quantitative capability of the present neuronal system by permitting application of known quantities of drugs. Iontophoresis, however, remains an important technique. Short pulses of neurotransmitter can be applied iontophoretically (<10 msec), allowing time-course kinetics to be investigated. Iontophoretic studies are less demanding technically, and thus, more neurons can be studied in each experiment. This is especially important when examining descriptive questions (such as specificity of action of a drug) where the action of a single drug on responses to many different neurotransmitters is being recorded. Iontophoresis is also useful when reversal potential is determined, since a fixed neurotransmitter response is recorded at a number of membrane potentials. Dose–response data obtained using iontophoresis, while limited in value, may have use if results can be validated with miniperfusion.

3.4. Perfusion

Medium containing the desired composition is warmed to 35–37°C, and pH is adjusted to that of the bathing medium prior to perfusion into the culture dish. Perfusion can be achieved during intracellular recordings which can be maintained for several hours with several changes in bathing medium.

4. AMINO ACID PHARMACOLOGY OF SPINAL NEURONS IN PDC CULTURE

Iontophoresis of either GABA, GLY, β-alanine (BALA), or GLU onto the surface of cultured spinal neurons evoked membrane potential and conductance changes in virtually every cell tested. Typical examples of membrane responses to these four amino acids are shown in Fig. 2. Iontophoresis of all three neutral amino acids typically led to hyperpolarizing inhibitory responses, while iontophoretically applied glutamate uniformly evoked depolarizing excitatory response. Focal iontophoresis of the amino acids onto the surface of spinal cord cells revealed nonuniform distributions of the sensitivities over the cell surface (Fig. 3). The nonuniform distribution in sensitivity for acidic and neutral amino acids did not coincide, suggesting that lack of responsivity at certain sites was not due to the presence of overlying tissue preventing agonist molecules from diffusing to their receptors. Rather, the results indicated a topography of amino acid receptors distributed in clusters, or "hot spots," over the cell surface. The relationship between these "hot spots" and sites of synaptic innervation is a line of investigation for future consideration.

Closer inspection of Fig. 3 reveals the presence of complex responses to iontophoresis of GABA on some but not all processes. The hyperpolarizing phase of these responses inverted at about the same membrane potential as monophasic hyperpolarizing events, suggesting that they were due to similar mechanisms (Barker and Ransom, 1978a). The depolarizing phases of the complex responses extrapolated to a slightly positive inversion (0 to +20 mV) and indicated that GABA can apparently activate more than one conductance mechanism. The depolarizing response desensitized far more rapidly than did the hyperpolarizing response (Barker and Ransom, 1978a). Similar complex responses with nonuniform distribution were observed with the other neutral

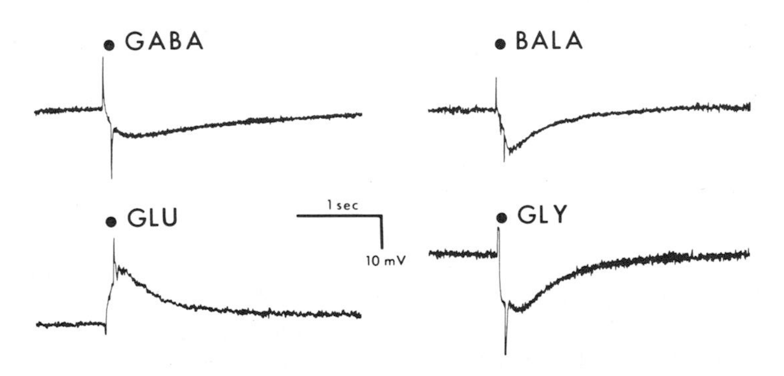

FIGURE 2. The response of a single spinal cord neuron to iontophoretic application of GABA (30 nA), β-alanine (60 nA), glutamate (40 nA), and glycine (30 nA) for 100 msec. Membrane potential ranged from −40 to −44 mV.

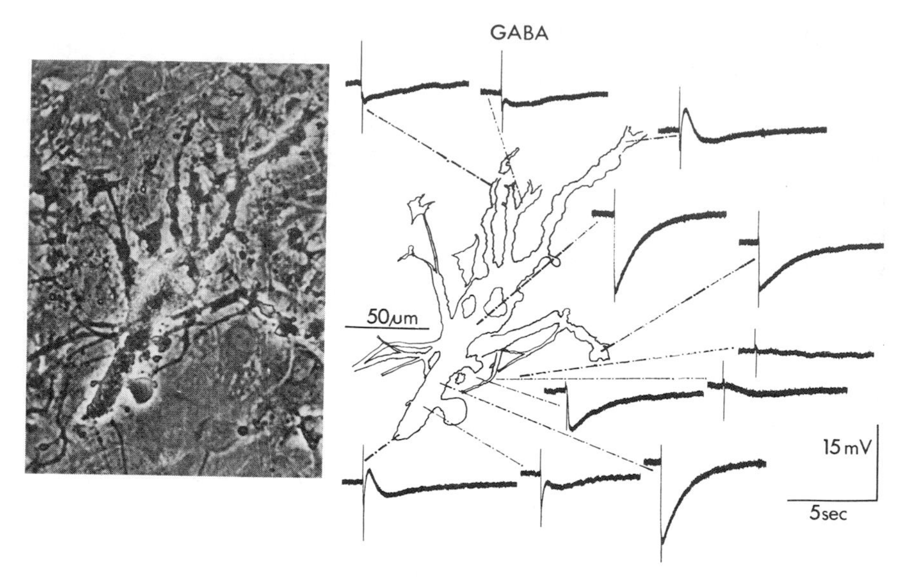

FIGURE 3. Iontophoresis of (22 nA, 100 msec) GABA onto the surface of a cultured spinal cord neuron showed both nonuniform topography of sensitivity and response complexity. Phase contrast photomicrograph of cell on left with intracellular recording of GABA response on right. Monophasic, hyperpolarizing responses were present at the cell body, while multiphasic responses containing a depolarizing component were present on several processes. Withdrawing the iontophoretic pipette from the somal surface in 10-μm steps (lower right) showed a progressive depression in response amplitude and slowing of response time course. The responses illustrated were representative of three evoked at each site. Resting membrane potential: −53 mV. (Modified from Barker and Ransom, 1978a.)

amino acids studied (GLY and BALA). GLY and BALA hyperpolarizing responses inverted at potentials similar to, or identical with, inversion potentials of GABA responses (Fig. 4), suggesting that all three neutral amino acids utilized similar conductance mechanisms (Barker and Ransom, 1978a). The shift in inversion potential for the neutral amino acids from about −60 mV to about −20 mV when KCl recording pipettes (rather than KAc) were used (Fig. 4) indicated a role for Cl^- ions in the membrane responses.

That the hyperpolarizing responses were due to activation of Cl^- conductance was indicated from experiments in which the extracellular Cl^- ion concentration was changed. For these experiments, spinal cord cells were bathed in a medium containing a low concentration of Cl^- ions, and a blunt micopipette filled with a high concentration of Cl^- ions was placed near the iontophoretic pipette close to the cell surface to observe what effects Cl^- ions had on resting membrane properties and amino acid-evoked responses (Barker and Ransom, 1978a). Placement of the Cl^--ion-filled pipette had small and variable effects on resting membrane properties (Fig. 5). GABA responses were, however, inverted from depolarizing to hyperpolarizing, and conductance was increased upon placement of the Cl^- ion diffusion pipette (Fig. 5). GLU responses remained unaltered. Under these conditions GABA responses sometimes reverted to approximate the voltage change seen initially in the continued presence of high extracellular Cl^- ion concentration; however, the conductance increase remained much higher than that obtained in low-Cl^--ion medium (Fig. 5A). The variable reversion of voltage responses probably reflects the labile distribution of Cl^- ions across the membrane. The results have been

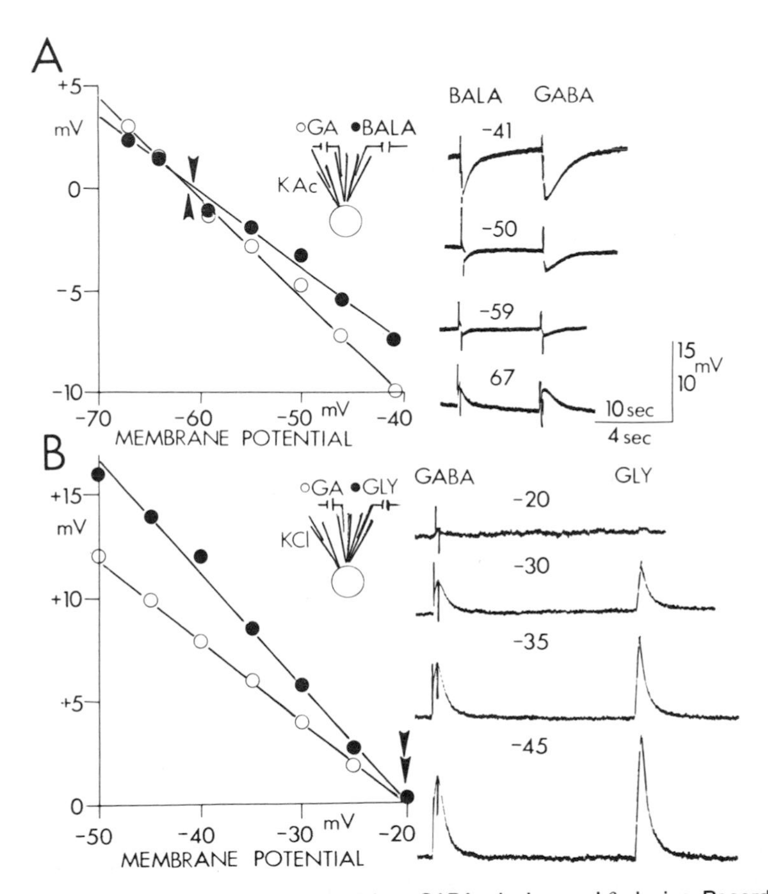

FIGURE 4. Comparison of inversion potentials to GABA, glycine, and β-alanine. Recordings from two different spinal cord neurons. Specimen records on right. (A) K-acetate recording pipette. Inversion potential of β-alanine response was similar to that of GABA (−60 mV). β-Alanine pulse: 30 nA, 100 msec; GABA pulse: 20 nA, 100 msec. (B) KCl recording pipette. Inversion potential of glycine responses was similar to that of GABA (−20 mV). Glycine pulse: 16 nA, 100 msec; GABA pulse: 24 nA, 100 msec. Lower calibrations apply to B.

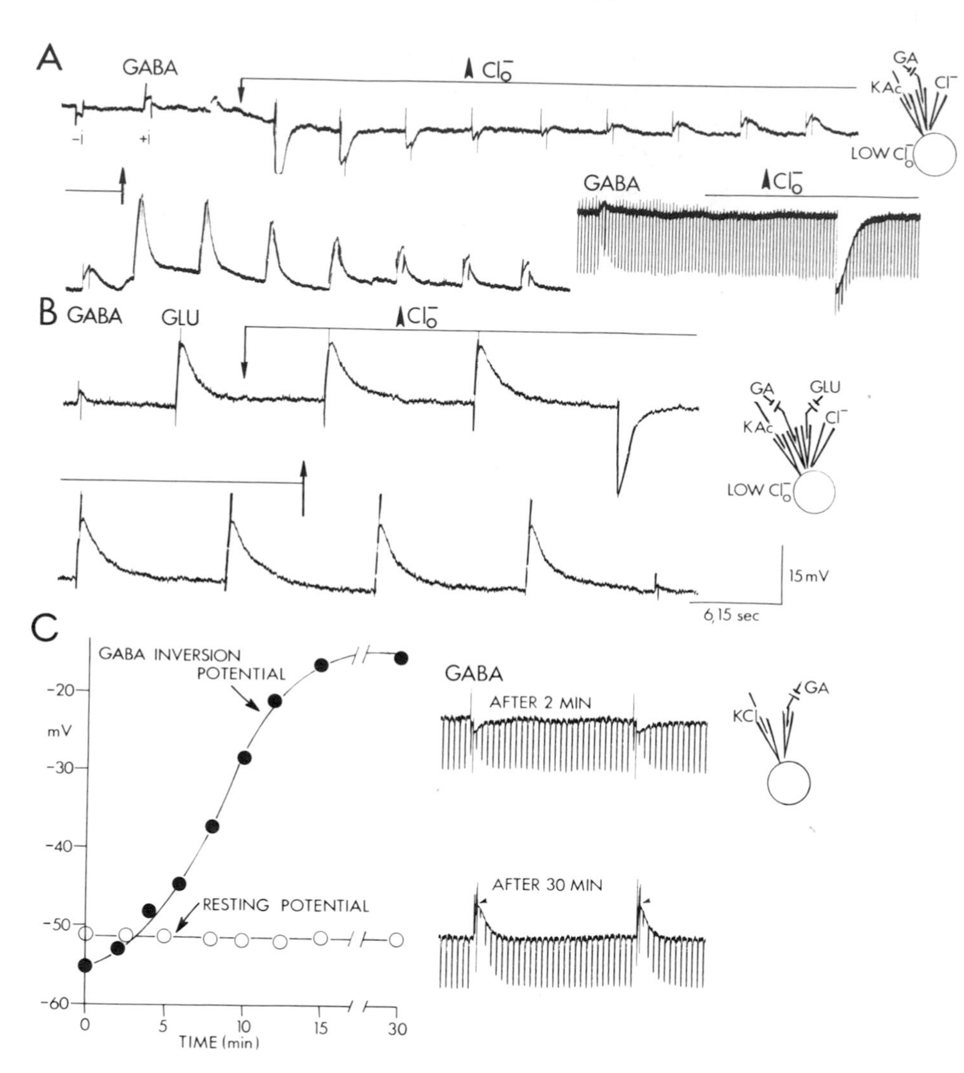

FIGURE 5. Cl^- ion dependency of GABA responses. Recordings from three different spinal cord cells. Cells in A and B were recorded from in a low-Cl^--ion solution containing 9 mM $[Cl^-]_o$. (A) GABA was applied iontophoretically with 30-nA, 100-msec cationic pulses ("+i" below trace). Current pulses of the opposite polarity ("−i" below trace) produced a coupling artifact but no detectable response. Placement of a blunt micropipette containing 159 mM Cl^- close to the surface of the cell ("Cl^-_o") caused a transient increase in membrane potential and transient inversion of the GABA response from depolarizing to hyperpolarizing. Continued application led to a gradual reversal in the polarity of the response. Withdrawal of the diffusion pipette was associated with a transient depolarization of the membrane potential and concomitant, transient enhancement of the depolarizing response to GABA. Trace on right shows GABA-induced increase in membrane conductance before and after placement of Cl^- diffusion pipette. A clear enhancement of the GABA-induced conductance increase as well as a change in response polarity without change in resting

interpreted to indicate a primary role for Cl^- ions in hyperpolarizing responses to GABA (and, by inference, to GLY and BALA).

We have recently begun to apply voltage clamp techniques to the study of pharmacological responses on cultured spinal neurons and have been able to record membrane current responses to iontophoretic application of neutral and acidic amino acids (Fig. 6; Barker *et al.*, 1977). The voltage clamp technique

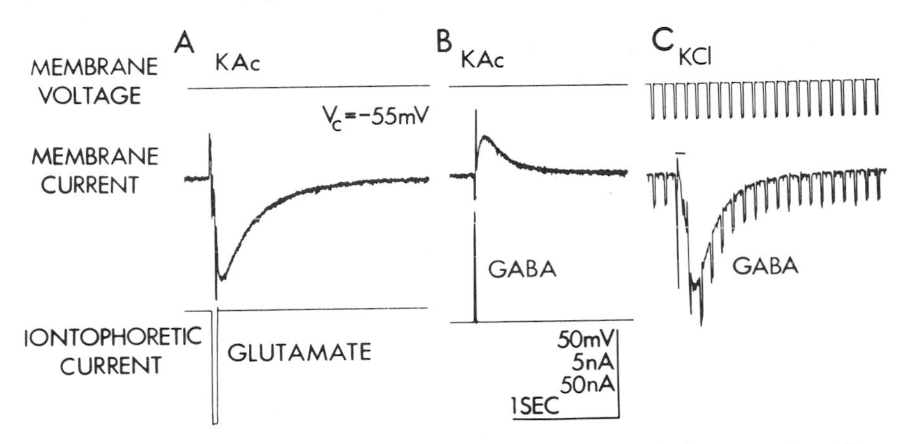

FIGURE 6. Application of voltage clamp techniques to cultured spinal neurons. A and B are recordings from the same spinal cord cell voltage-clamped to −55 mV using two KAc microelectrodes. In the unclamped state at −55 mV, the membrane response to a 55-nA, 50-msec pulse of glutamate was a depolarization, while that to a 60-nA, 10-msec pulse of GABA was hyperpolarization. Under voltage clamp, glutamate evoked an inward current response, and GABA an outward current response. C was a recording from another cell voltage-clamped to −55 mV using two KCl microelectrodes. Under these conditions in the unclamped state a 40-nA, 100-msec pulse of GABA elicited a depolarizing response. With the potential clamped, GABA evoked an inward current response. Superimposed in the holding potential were 20-mV, 50-msec hyperpolarizing commands which produced current responses, the size of which were directly related to membrane conductance. At the peak of the GABA response, the current response to the constant voltage commands was increased relative to control, indicating an increase in membrane conductance during the response.

---

membrane potential or conductance was evident. Conductance was assessed using 0.5-nA, 50-msec constant-current hyperpolarizing pulses. Resting potential: −56 mV. (B) Placement of the Cl^- diffusion pipette inverted the response to a 35-nA, 100-msec iontophoretic pulse of GABA but did not affect depolarizing responses to 18-nA, 100-msec pulses of glutamate. Resting potential: −58 mV. (C) KCl recording from a spinal cord cell in normal Cl^- (125 mM). Inversion potential of response to an iontophoretic pulse (26 nA, 100 msec) of GABA recorded as a function of time shifted from −55 mV immediately after penetration to −17 mV after 30 min, at which time the depolarizing GABA response produced excitation (spikes marked by arrowheads in lower righthand specimen record). Conductance assessed with 0.5-nA, 50-msec constant-current hyperpolarizing pulses. 15-sec time calibration applies to traces with conductance pulses. (J. L. Barker and B. R. Ransom, unpublished observations.)

has allowed us to reach membrane potential regions not easily attained with single microelectrodes because of potential-dependent, high-membrane-conductance states. Thus, we have reexamined the inversion potentials of the neutral and excitatory amino acids while recording with two KCl microelectrodes. The neutral amino acid responses inverted at about −15 mV, and GLU inverted at about 0 mV (Barker *et al.*, 1977). The shift in the inversion potential of GLU responses is similar to the extrapolated inversion potential of certain excitatory postsynaptic potentials recorded in cultured SC cells (Macdonald and Nelson, unpublished observations).

The voltage clamp technique has also allowed us to confirm our earlier observations on estimates of the number of agonist molecules participating in particular receptor-coupled conductance events. By using the limiting slopes of agonist-induced dose–response curves as estimates of the number of agonist

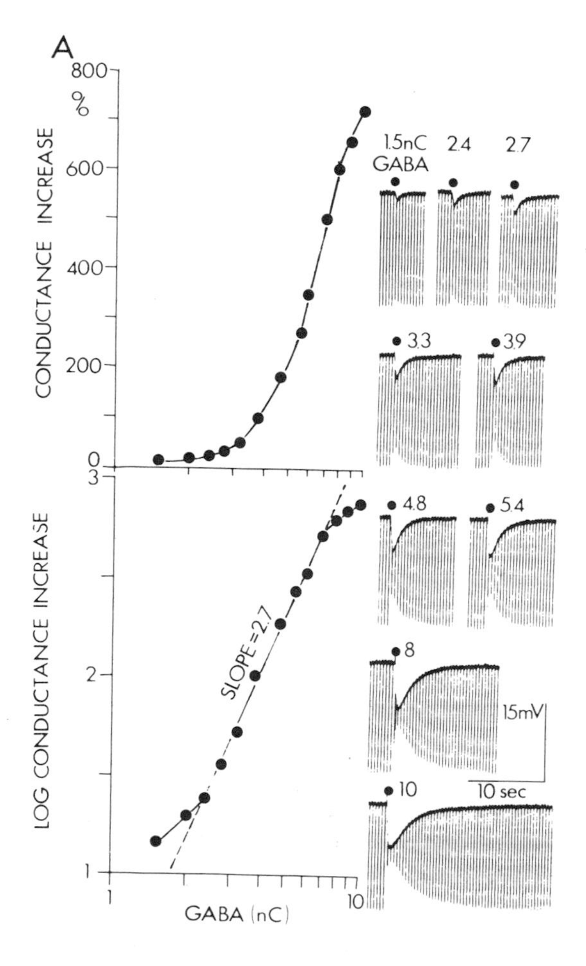

FIGURE 7. (A) Dose-response curve of membrane conductance increase as a function of charge applied to the GABA iontophoretic pipette. Semilogarithmic plot of extrapolated percentage conductance increase as a function of charge (upper graph) and log-log plot of data (lower graph) from a spinal cord cell. Specimen records illustrate slope of log–log curve (drawn by eye) is 2.7. GABA iontophoretic pulse: 29 nA, 100 msec. Resting potential: −54 mV. Constant current pulses: −0.6 nA. (B) Dose-response curve of depolarizing response to glutamate. Semilogarithmic plot (upper) and log–log plot of data (lower) from another spinal cord cell. Specimen records illustrate some of the data points. Limiting slope, extrapolated from linear part of log–log curve (drawn by eye), is 0.95. Glutamate iontophoretic pulse: 20 nA, 50 msec. Resting potential: −70 mV. (From Barker and Ransom, 1978a.)

molecules involved in pharmacological responses (Werman, 1969), we have found that one GLU and probably two GABA molecules were involved in generating their respective responses (Fig. 7; Barker and Ransom, 1978a). Similar results for GLU and GABA current responses have been observed under voltage clamp (Barker *et al.*, 1977). In addition, consideration of glycine dose–response curves suggested that at least three GLY molecules cooperate to produce the pharmacologic response. If such cooperativity contributes to physiologically elaborated postsynaptic events, then excitation events mediated by glutamate would be linearly related to concentration, while inhibitory events mediated by neutral amino acids would be nonlinearly related to concentration. Superficially, the results suggested that the functional differences between these synaptic physiologies go beyond the simple contrast between excitation and inhibition.

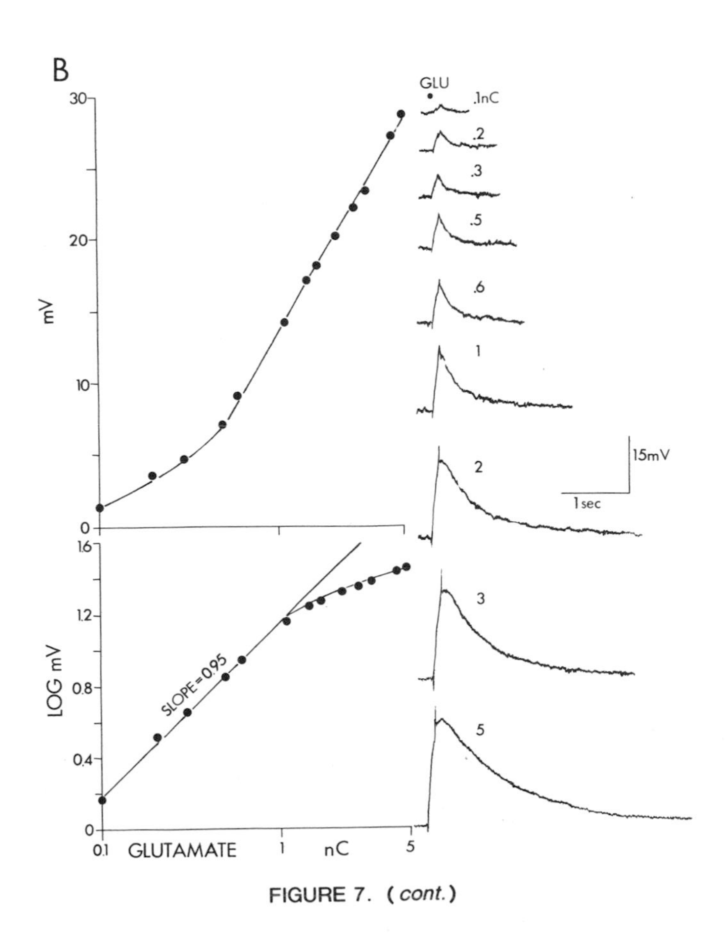

FIGURE 7. (*cont.*)

The voltage clamp technique, in conjunction with computer analysis of current fluctuations set by application of amino acid agonists, has permitted estimation of the elementary events associated with amino acid receptor activation. The results showed that GABA and GLY activate single channels with distinctly different properties, but that each amino acid could, under certain conditions, interfere with the membrane response to the other agonist (McBurney and Barker, 1978; Barker and McBurney, 1978, 1979). This latter observation was interpreted as evidence for a common channel with two receptors activatable by either amino acid rather than one receptor and one channel as appears to occur at cholinergic neuromuscular junctions. Estimates of single channel kinetic properties pharmacologically induced by transmitter agonists showed a close correspondence to quantal synaptic events at both invertebrate and vertebrate neuromuscular junctions (Katz and Miledi, 1972; Anderson and Stevens, 1973; Crawford and McBurney, 1976). We observed several populations of quantal synaptic events which were clearly distinguishable on the basis of their time constant of decay (J.L. Barker and R.N. McBurney, unpublished observations). One population had a time constant of decay which approximated that observed during pharmacological activation of GABA receptors. Furthermore, these physiological events were sensitive to drugs which specifically affect GABA pharmacological responses, suggesting that the former were responses to naturally occurring quantal release of GABA.

Thus, pharmacological approaches to amino acid receptors in cultured central neurons have provided both a degree of resolution not presently attainable *in vivo* and a basis for understanding and studying physiological events mediated by amino acids.

5. CONVULSANT AND ANTICONVULSANT ACTIONS ON SPINAL CORD NEURONS IN PDC CULTURE

Another major potential use of mammalian neurons in PDC is in the study of mechanisms of drug action. In this section we will review our studies on the actions of convulsants and anticonvulsants on spontaneous neuronal activity, on neuronal membrane properties, and on postsynaptic response to iontophoretically applied amino acids.

SC neurons in normal growth medium had a characteristic spontaneous activity which consisted of a random mixture of excitatory and inhibitory postsynaptic potentials and action potentials (Fig. 8). Short bursts of action potentials lasting less than one second occasionally occurred but were somewhat uncommon. Thus the normal spontaneous activity of SC neurons in culture represented the randomly summated synaptic input from multiple neurons (Ransom *et al.*, 1977b). This pattern of activity was in marked contrast to that recorded from neurons following addition of convulsant agents to the bathing medium. Strychnine, penicillin (PCN), pentylenetetrazol (PTZ), picrotoxin

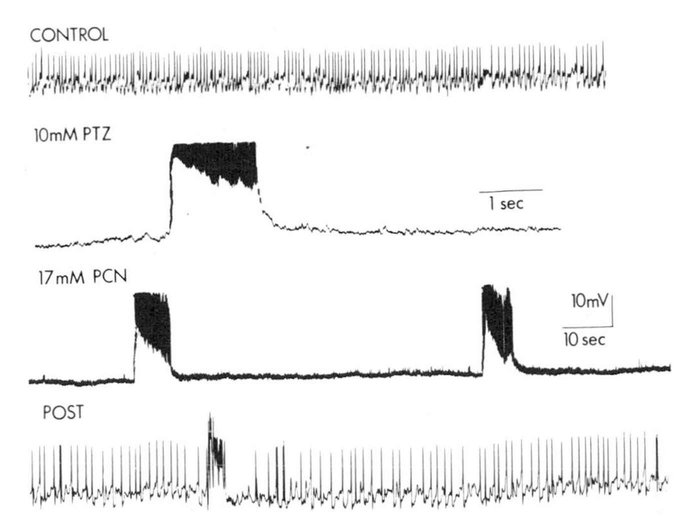

FIGURE 8. Intracellular recordings from spinal cord neurons taken from a pen recorder. Addition of 10 mM pentylenetrazol or 17 mM penicillin altered baseline spontaneous activity (control) and produced paroxysmal depolarizing events, an action which was reversible (post). The 10-sec time bar applies to all traces except that for 10 mM pentylenetetrazol (see 1-sec time bar). (From Macdonald and Barker, 1978a.)

(PICRO), or bicuculline (BICUC) added to the bathing medium produced large depolarizing potentials associated with bursts of action potentials (Fig. 8) (Macdonald and Barker, 1978a). The depolarizing potentials had durations of 1 to 8 sec and occurred randomly. Removal of the convulsant abolished these paroxysmal depolarizing events (PDE) and restored the normal baseline activity. Topical application of these convulsants to the mammalian cortex also produced abnormal electrical activity (Prince, 1968; Matsumoto *et al.*, 1969) in surface electroencephalographic (EEG) recordings and during intracellular recordings from neurons near the point of convulsant application. Surface EEG recordings revealed the development of focal interictal spikes prior to the development of typical electroencephalographic seizure activity. Simultaneous intracellular recordings from cortical neurons in the region of convulsant application revealed large, randomly occurring depolarizations of membrane potential associated with volleys of action potentials. These paroxysmal depolarizing shifts (PDS) thus represented the cellular neurophysiological correlate of interictal spikes recorded in the surface EEG. PDS appeared to be sterotypical whether chronic or acute techniques were employed to produce the seizure activity, and thus such paroxysmal bursts have been accepted as the hallmark of epileptiform activity as recorded with intracellular techniques. The similarity of PDE to PDS suggests that PDE represents an *in vitro* equivalent of PDS, and thus investigation of PDE may provide insight into the basic mechanisms underlying PDS.

The basis for PDS produced by acute application of convulsants remains controversial, but two basic hypotheses have been dominant: (1) that PDS is the result of an increase of or summation of excitatory synaptic potentials (Futamachi and Prince, 1975) and/or a decrease in inhibitory synaptic potentials (Davidoff, 1972a,b; Curtis *et al.*, 1972; Takeuchi and Takeuchi, 1969; Curtis and Felix, 1971; Nicoll and Padjen, 1976; Macdonald and Barker, 1978a; Davidoff and Hackman, 1978) resulting in a "giant" EPSP (Prince, 1968; Matsumoto *et al.*, 1969; Dichter and Spencer, 1969) or (2) that nonsynaptic factors, such as intrinsic or convulsant-induced voltage-dependent regenerative processes or alterations in the neuronal ionic microenvironment, form the basis for PDS (David *et al.*, 1974; Williamson and Crill, 1976a,b; Prince, 1978). These alternatives are not mutually exclusive, and alterations in synaptic input could lead to populations of neurons which then reach threshold for endogenous or intrinsic PDS.

Such a mechanism has been proposed for burst generation in CA3 neurons in hippocampal slices when exposed to penicillin (Wong and Prince, 1978). In these neurons, orthodromic stimulation in stratum radiatum produced EPSP–IPSP sequences in dendrites but did not produce burst responses while intracellularly injected depolarizing current pulses triggered intrinsic burst discharges in dendrites. Application of penicillin led to generation of spike bursts by interfering with the dendritic inhibition that normally served to suppress dendritic burst generation during orthodromic activation. Thus in this model system, both intrinsic and synaptic mechanisms appear to be involved in convulsant-induced PDS.

To eludicate the basic mechanisms underlying convulsant-induced PDE, we investigated the action of the convulsants PCN, PTZ, PICRO, and BICUC on postsynaptic responses to iontophoretically applied amino acid putative neurotransmitters. All four convulsants rapidly and reversibly antagonized the voltage response to GABA (Fig. 9) in a dose-dependent manner (Macdonald and Barker, 1977, 1978a). However, these convulsants did not alter responses elicited by other neutral inhibitory amino acids, GLY and BALA, or by the acidic excitatory amino acid, GLU (Fig. 10). Furthermore, when applied at currents effective in antagonizing GABA responses, the convulsants did not alter resting membrane potential, input resistance, GABA equilibrium potential, or properties of the action potential (except for PTZ which occasionally slightly increased neuronal input resistance). Thus, the convulsants PCN, PTZ, PICRO, and BICUC all produced PDE when applied to cultured mammalian spinal cord neurons and specifically antagonized GABA-mediated postsynaptic inhibition, suggesting that the convulsant activity of these agents derives in part from selective antagonism of GABA-mediated postsynaptic inhibition.

Since convulsant application to the medium bathing mammalian spinal cord neurons in cell culture readily produced paroxysmal alterations in membrane potential, it seemed reasonable to examine the action of anticonvulsant

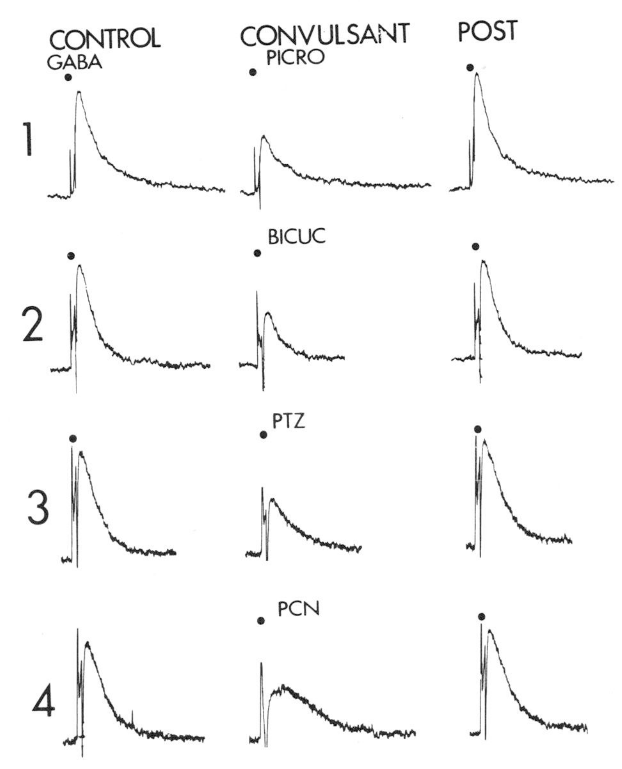

FIGURE 9. GABA responses were antagonized by iontophoresis of picrotoxin, bicuculline, pentylenetetrazol, and penicillin. All records were obtained using 3 M KCl micropipettes with membrane potential hyperpolarized to about −90 mV. Calibration: 10 mV × 1 sec. (From Macdonald and Barker, 1978a.)

agents on convulsant-induced PDE and on postsynaptic responses to putative neurotransmitter amino acids. Addition of 0.2 mM phenobarbital (PhB) to the cell culture during an intracellular recording produced minimal alteration of synaptic activity or of spontaneous action potentials (Fig. 11B) (Macdonald and Barker, 1978b, 1979a,b). Following addition of 0.08 mM PICRO to the bathing medium, however, paroxysmal activity was demonstrable (Fig. 11E). Addition of 0.2 mM PhB abolished the paroxysmal activity and restored the spontaneous activity to the random pattern characteristic of the control recordings (Fig. 11F). Thus, it was clearly demonstrated that the anticonvulsant PhB had pharmacological anticonvulsant activity against PICRO-induced PDE *in vitro*.

The mechanisms underlying this action were investigated by determining the effect of the anticonvulsants PhB and mephobarbital (MB) as well as the benzodiazepines [diazepam (DZ) and chlordiazepoxide (CDZ)] and valproic acid (VPA) on responses elicited by putative neurotransmitter amino acids.

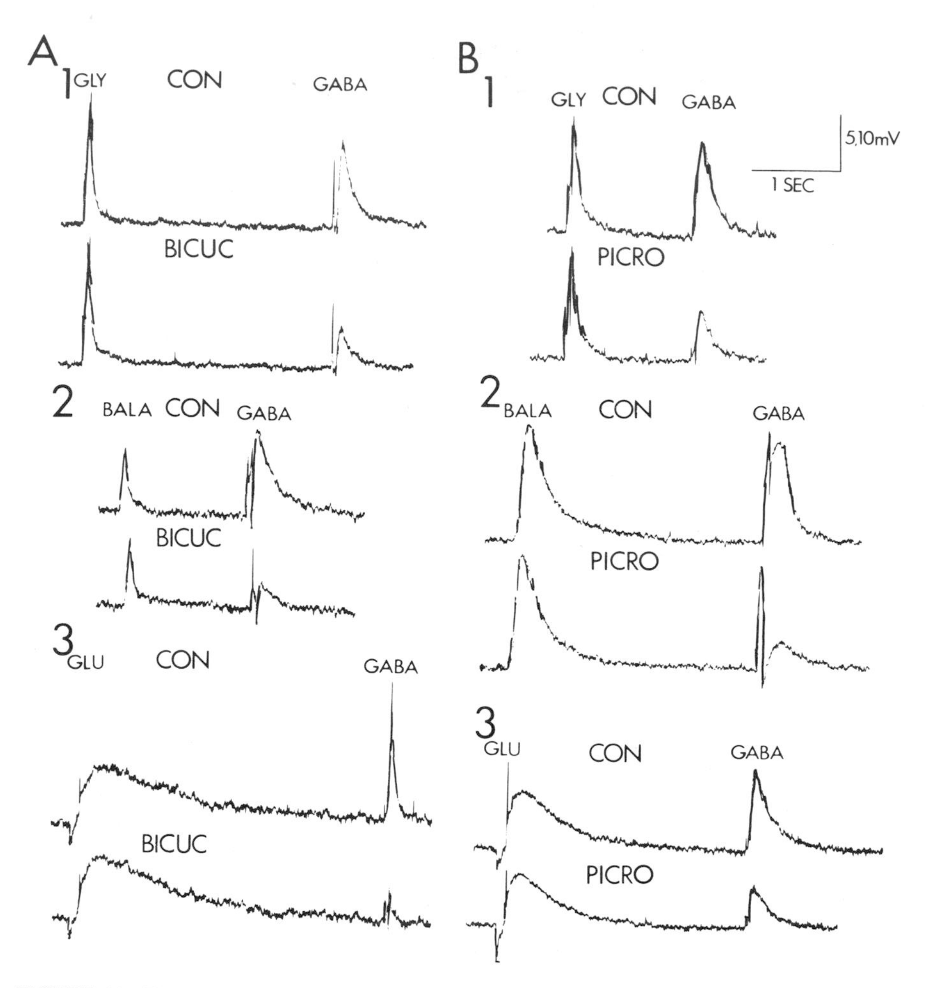

FIGURE 10. Bicuculline (A) and picrotoxin (B) selectively antagonized GABA responses without altering GLY, BALA, or GLU responses. See caption to Fig. 9 for recording details.

PhB and MB rapidly and reversibly augmented the voltage response (Fig. 12) and conductance change elicited by GABA without altering responses elicited by GABA or GLY (Fig. 12) (Macdonald and Barker, 1979a,b). Both barbiturates also antagonized responses elicited by GLU in a dose-dependent and reversible manner. Similar modulation of GABA and GLU responses by the anesthetic barbiturate pentobarbital have been observed (Barker and Ransom, 1978b; Macdonald and Barker, 1978b, 1979b) but at lower iontophoretic current applications of the anesthetic compared to the anticonvulsant barbiturate (Macdonald and Barker, 1978b, 1979a,b). Further, anesthetic, but not anticonvulsant, barbiturates directly increased neuronal membrane conductance at low iontophoretic currents, an action which was antagonized by the GABA

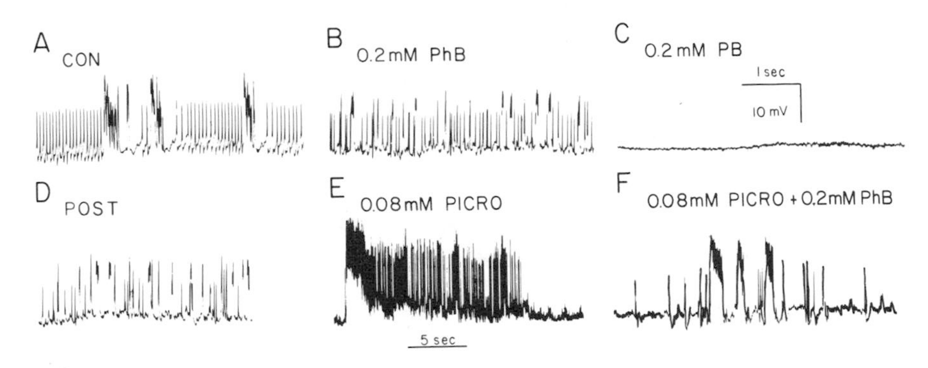

FIGURE 11. Intracellular recordings from a spinal cord neuron during barbiturate and picrotoxin perfusion. Drugs in perfusate are indicated over each specimen record. All records were obtained from the same neuron. The 1-sec time bar applies to all traces except E (see 5-sec time bar). (From Macdonald and Barker, 1979b.)

antagonists PICRO and PCN. The benzodiazepines (Fig. 13) (Choi *et al.*, 1977; Macdonald and Barker, 1978c, 1979a) and valproic acid (Fig. 14) (Macdonald and Bergey, 1979) also rapidly and reversibly augmented GABA-mediated postsynaptic inhibition without altering responses elicited by GLY or GLU.

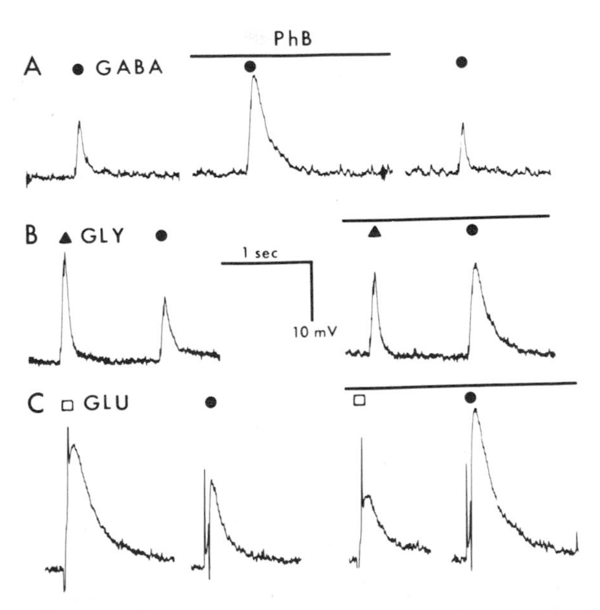

FIGURE 12. Phenobarbital augmented GABA responses (A) and reduced GLU responses (C) without altering GLY responses (B). See caption to Fig. 9 for recording details.

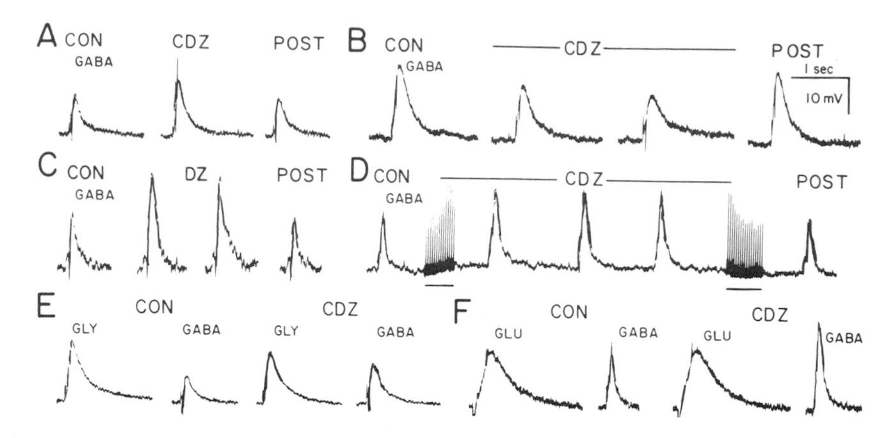

FIGURE 13. Chlordiazepoxide (A,D) and diazepam (C) augmented GABA responses without altering GLY (E) or GLU responses (F). At high iontophoretic currents (B), the GABA response was occasionally reduced. See caption to Fig. 9 for recording details. (From Macdonald and Barker, 1978c.)

Thus, these findings suggested that the level of GABA-mediated inhibition present in the neuronal cell culture would be dependent on the relative levels of convulsant and anticonvulsant drugs in the bathing medium. If the convulsant concentration was relatively higher, antagonism of GABA-mediated inhibition occurred along with the development of PDE. Alterna-

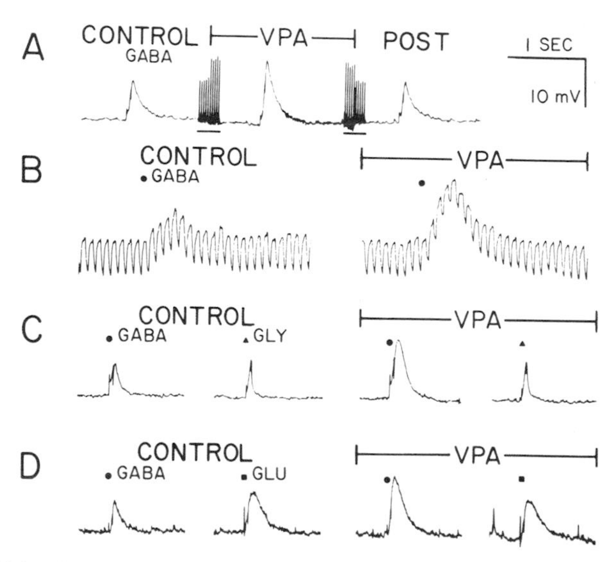

FIGURE 14. Valproic acid (VPA) augmented GABA responses (A,B) without alteration of GLY (C) or GLU (D) responses. (From Macdonald and Bergey, 1979.)

tively, if anticonvulsant concentration was relatively higher and within therapeutic range, GABA-mediated inhibition would be restored to control levels or augmented, and PDE would therefore be disrupted.

Extrapolation of these findings to the pathogenesis and therapy of epilepsy in man implies that the level of GABA-mediated inhibition present in nervous system regions with low threshold for paroxysmal activity may determine the presence or absence of interictal and ictal activity. Diminished GABA-mediated inhibition could be induced by specific drugs, metabolites, or toxins via several alternative mechanisms including (1) selective loss of GABAergic neurons, (2) diminished release of GABA because of reduced glutamic acid decarboxylase activity and, thus, reduced synthesis of GABA, (3) antagonism of GABA release, (4) postsynaptic antagonism of GABA binding, or (5) reduction of GABA-coupled chloride channel function. GABA-mediated inhibition could then be restored by administration of an anticonvulsant that augmented the postsynaptic action of GABA, as demonstrated in these experiments, or alternatively one that increased presynaptic release of GABA. Clearly these mechanisms require intact GABAergic neurons for their action, and if all such neurons were destroyed, such drugs would not be expected to be active. These findings also document a difference in action between anticonvulsant and anesthetic barbiturates which may help to explain their different clinical actions. The distinction between an anesthetic and anticonvulsant barbiturate may depend on the relative concentrations producing GABA augmentation and GABAmimetic action. If both effects occur at similar concentrations, the barbiturate would have anesthetic properties, since postsynaptic GABA receptors would be activated throughout the CNS with a resultant decrease in neuronal excitability. On the other hand, if the threshold for GABA augmentation is substantially lower than that for GABAmimetic action, the agent would selectively augment GABAergic synaptic transmission and thus produce anticonvulsant action without anesthesia.

The mechanism whereby these convulsants and anticonvulsants modulate postsynaptic amino acid responses remains uncertain, but several alternative actions have been proposed. For the convulsants, competitive (Hori *et al.*, 1978; Shank *et al.*, 1974), noncompetitive (Takeuchi and Ondera, 1972; Takeuchi and Takeuchi, 1969; Shank *et al.*, 1974; Constanti, 1978; Hochner *et al.*, 1976), or mixed (Constanti, 1978) dose–response curves in crustacean muscle or stretch receptor neurons have been reported. In addition, PCN and PTZ antagonize choride-mediated responses in *Aplysia* neurons (Pellmar and Wilson, 1977a,b), suggesting that the convulsants combine with choride channels. Binding studies (Olsen *et al.*, 1978a) have shown that BICUC, PCN, and PTZ, but not PICRO, displace GABA binding from mouse brain synaptosomes. Based on these studies, Olsen and Ticku (Olsen *et al.*, 1978a,b; Ticku *et al.*, 1978) have suggested that PICRO binds to and modulates the GABA-coupled chloride channel.

Similarly, Young and Snyder (1974a,b) investigated the displacement of [^{3}H]strychnine by GLY and strychnine from synaptic membrane fractions of rat spinal cord. They concluded that strychnine binds to a site different from the GLY binding site and proposed that the strychnine binding site was associated with a chloride channel. Binding of a convulsant to the chloride channel would decrease postsynaptic GABA- or GLY-elicited conductance changes without altering the binding of GABA or GLY to their receptor. In the experiments reported above using PDC neurons in culture, it was clear that a "nonspecific" combination of convulsants with chloride channels could not explain the clear selectivity of these agents. Since all active neutral amino acids studied produced increases in chloride conductance, a simple combination of the convulsants with chloride channels should result in antagonism of responses to GLY, BALA, and taurine in addition to antagonism of GABA-mediated inhibition. Our results, then, suggested that the convulsants must bind to a sterospecific site on the GABA receptor–chloride channel complex producing (1) decreased affinity of GABA for its receptor, (2) antagonism of GABA binding by competing for GABA for its receptor, or (3) decreased GABA-coupled chloride channel function by decreasing either single channel open time or unitary conductance. To determine which of these alternatives is correct will require additional experiments involving analysis of GABA-induced noise in the presence and absence of convulsant, dose–response analysis using GABA and convulsant perfusion, and direct measurement of GABA binding with and without convulsants.

Possible sites for anticonvulsant action are clearly different. Since the barbiturates and benzodiazepines augmented GABA responses, it was clear that they could not compete with GABA for receptor binding. However, they could act at a modulator site separate from the active site of GABA binding to increase the affinity of the receptor for GABA binding, enhance the coupling between GABA binding and production of a chloride conductance increase, or increase the function of the chloride channel. Studies of GABA binding to rat brain synaptic membrane fragments have demonstrated no alteration in GABA binding by barbiturates (Olsen *et al.,* 1978a; Peck *et al.,* 1976) or benzodiazepines (Olsen *et al.,* 1978a), suggesting that the former alternative is unlikely. Benzodiazepine binding has been actively investigated (Braestrup *et al.,* 1977; Mohler and Okada, 1977; Squires and Braestrup, 1977), and benzodiazepines have been found to bind to sites which cannot be displaced by GABA, GABA antagonists, or barbiturates (Braestrup and Squires, 1978). Thus from these studies and our own in culture, it would appear that the anticonvulsant barbiturates and benzodiazepines bind to sterospecific sites related to the GABA receptor. Binding to these sites most likely either increases coupling between GABA binding and the chloride conductance mechanism or increases GABA-coupled chloride channel function. More detailed information should be obtained through study of the action of barbiturates and benzodiazepines on GABA-induced noise to determine whether ionophore function is altered by these drugs.

6. SUMMARY

The preceding studies have demonstrated that mammalian neurons in primary dissociated cell culture can be used to investigate neurotransmitter and drug actions. In the studies discussed above, spinal cord neurons were shown to have two inhibitory neutral amino acid receptors, one GABAlike and the other GLYlike, and an excitatory acid amino acid receptor for GLU. The inhibitory amino acid responses were chloride-mediated, while the excitatory amino acid responses were sodium- (and probably potassium-) mediated. Dose–response data suggested that inhibitory amino acids had cooperative interactions with their receptors, whereas GLU did not. Finally, analysis of current fluctuations evoked by amino acids also demonstrated that GABA and GLY activated single channels with distinctly different properties.

Several conclusions were also drawn from the studies of convulsant and anticonvulsant actions. All of the convulsants studied antagonized inhibitory amino acid responses, an action which may underlie, at least in part, their convulsant mechanism. The anticonvulsants, in contrast, augmented GABA-mediated inhibition, and thus they could reverse convulsant-induced reduction in GABA-mediated inhibition, an action which may be in part responsible for their anticonvulsant action. Barbiturate anesthetics, however, had another action at anesthetic concentrations. The anesthetics had a direct chloride-mediated inhibitory action that appeared to involve GABA receptors. Thus, anesthetic barbiturates may produce anesthesia by the summed action of augmented GABAergic synaptic transmission and by direct GABAmimetic inhibition.

7. FUTURE DIRECTIONS

Investigation of mammalian neurons in primary dissociated cell culture has numerous advantages over the study of the nervous system *in vivo,* and these include the ability to (1) control the extracellular medium in acute experiments, (2) reliably obtain intracellular recordings, and (3) control the extracellular medium during neuronal development. There are, however, also significant disadvantages to this culture system. Since the nervous system is disrupted prior to plating, the normal relationships among neurons and between neurons and glia are lost. Thus while genetically determined neuronal maturation continues, the highly ordered structural relationships are lost, and the synaptic physiology is undoubtedly considerably altered. Therefore, one must devise research strategies that exploit the advantages of this preparation and minimize the disadvantages. Future investigations using this technique will employ an increasing number of central nervous system regions including forebrain, hippocampus, cerebellum, and brainstem in culture. The approach to pharmacological study of these neurons will involve more quantitative approaches, including miniperfusion of known concentrations of drugs and neurotransmitters. Use of other culture systems will expand the study of neurotransmitter postsynaptic pharmacology to nonamino acid neurotransmitters such as acetylcholine, norepinephrine, serotonin, and dopamine and to the study of peptides suspected of being neurotransmitters or neuromodulators. Immunocytochemical studies of peptidergic neurons are currently in progress in several laboratories. The actions of traditional neurotransmitters as well as peptides will be studied using voltage clamp and noise analysis techniques as well as by traditional intracellular recording. Important advances will occur with investigation of the neurochemistry of neurotransmitter synthesis, uptake, and release and comparison with the physiology of synaptic transmission. Finally, the developmental aspects of neurons in culture will be investigated so that the factors involved in synaptogenesis can be described.

The ability of investigators to approach specific questions of neuronal function using physiological, pharmacological, biochemical, and anatomical perspectives suggests that study of mammalian neurons in primary dissociated cell culture will continue to provide exciting insights into the properties of mammalian neurons that cannot be developed using other preparations.

REFERENCES

Anderson, C.R., and Stevens, C.F., 1973, Voltage clamp analysis of acetylcholine produced end-plate current fluctuations at frog neuromuscular junction, *J. Physiol. (Lond.)* **235:**655–691.

Barker, J.L., and McBurney, R.N., 1978, Different properties of single channels activated by GABA and glycine on cultured mouse neurones, *J. Physiol. (Lond.)* **284:**127p.

Barker, J.L., and McBurney, R.N., 1979, GABA and glycine may share the same conductance channel on cultured mammalian neurones, *Nature* **277:**234–236.

Barker, J.L., and Ransom, B.R., 1978a, Amino acid pharmacology of mammalian central neurones grown in tissue culture, *J. Physiol. (Lond.)* **280:**331–354.

Barker, J.L., and Ransom, B.R., 1978b, Pentobarbitone pharmacology of mammalian central neurones grown in tissue culture, *J. Physiol. (Lond.)* **280:**355–372.

Barker, J.L., Macdonald, R.L., and Smith, T.G., 1977, Voltage clamp analysis of amino acid currents in cultured mammalian neurons, *J. Gen. Physiol.* **70:**1a.

Braestrup, C., and Squires, R.F., 1978, Pharmacological characterization of benzodiazepine receptors in the brain, *Eur. J. Pharmacol.* **48:**263–270.

Braestrup, C., Albrechtsen, R., and Squires, R.F., 1977, High densities of benzodiazepine receptors in human cortical areas, *Nature* **269:**702–704.

Chase, T.N., 1976, Rational approaches to the pharmacotherapy of chorea, in: *The Basal Ganglia,* Association for Research in Nervous and Mental Disease, pp. 337–349, Raven Press, New York.

Choi, D.W., Farb, D.H., and Fischbach, G.D., 1977, Chlordiazepoxide selectivity augments GABA action in spinal cord cell cultures, *Nature* **269:**342–344.

Constanti, A., 1978, The "mixed" effect of picrotoxin on the GABA dose/conductance relation recorded from lobster muscle, *Neuropharmacology* **17:**159–167.

Courtney, K.R., and Prince, D.A., 1977, Epileptogenesis in neocortical slices, *Brain Res.* **127:**191–196.

Crain, S.M., 1975, Physiology of CNS tissues in culture, in: *Metabolic Compartmentation and Neurotransmission* (S. Berl, D.D. Clarke, and D. Schneider, eds.), pp. 273–303, Plenum Press, New York.

Crawford, A.C., and McBurney, R.N., 1976, On the elementary conductance event produced by L-glutamate and quanta of the natural transmitter at the neuromuscular junction of *Maia squinado, J. Physiol. (Lond.)* **258:**205–255.

Curtis, D.R., and Eccles, R.M., 1958a, The excitation of Renshaw cells by pharmacological agents applied electrophoretically, *J. Physiol. (Lond.)* **141:**435–445.

Curtis, D.R., and Eccles, R.M., 1958b, The effect of diffusional barriers upon the pharmacology of cells within the central nervous system, *J. Physiol. (Lond.)* **141:**446–463.

Curtis, D.R., and Felix, D., 1971, The effect of bicuculline upon synaptic inhibition in the cerebral and cerebellar cortices of the cat, *Brain Res.* **34:**301–321.

Curtis, D.R., Phillis, J.W., and Watkins, J.C., 1961, Actions of amino acids on the isolated hemisected spinal cord of the toad, *Br. J. Pharmacol.* **16:**262–281.

Curtis, D.R., Phillis, J.W., and Watkins, J.C., 1959, The depression of spinal neurones by γ-amino-*n*-butyric acid and β-alanine, *J. Physiol. (Lond.)* **146:**185–203.

Curtis, D.R., Hosli, L., Johnston, G.A.R., and Johnston, I.H., 1968, The hyperpolarization of spinal motoneurones by glycine and related amino acids, *Exp. Brain Res.* **5:**235–258.

Curtis, D.R., Game, C.J.A., Johnston, G.A.R., McCulloch, R.M., and MacLachlan, R.M., 1972, Convulsive action of penicillin, *Brain Res.* **43:**242–245.

David, R.J., Wilson, W.A., and Escueta, A.V., 1974, Voltage clamp analysis of pentylenetetrazol effects on *Aplysia* neurons, *Brain Res.* **67:**549–544.

Davidoff, R.A., 1972a, Penicillin and inhibition in the cat spinal cord, *Brain Res.* **45:**638–642.

Davidoff, R.A., 1972b, Penicillin and presynaptic inhibition in the amphibian spinal cord, *Brain Res.* **36:**218–222.

Davidoff, R.A., and Hackman, J.C., 1978, Pentylenetetrazol and reflex activity of isolated frog spinal cord, *Neurology (Minneap.)* **28:**488–494.

Dichter, M.A., and Fischbach, G.D., 1977, The action potential of chick dorsal root ganglion neurones maintained in cell culture, *J. Physiol. (Lond.)* **267:**281–298.

Dichter, M.A., and Spencer, W.A., 1969a, Penicillin-induced interictal discharges from the cat hippocampus. I. Characteristics and topographical features, *J. Neurophysiol.* **32:**649–662.

Dichter, M.A., and Spencer, W.A., Penicillin-induced intericatal discharges from the cat hippocampus. II. Mechanisms underlying origin and restriction, *J. Neurophysiol.* **32:**663–687.

Futamachi, K., and Prince D., 1975, Effect of penicillin on an excitatory synapse, *Brain Res.* **100:**589–597.

Hochner, B., Spira, M.E., and Werman, R., 1976, Penicillin decreases chloride conductance in crustacean muscle: A model for the epileptic neuron, *Brain Res.* **107:**85–103.

Hori, N., Ikeda, K., and Roberts, E., 1978, Muscimol, GABA and picrotoxin: Effects on membrane conductance of a crustacean neuron, *Brain Res.* **141:**364–370.

Hornykiewicz, O., 1972, Neurochemistry of parkinsonism, in: *Handbook of Neurochemistry, Vol. 7* (A. Lajtha, ed.), pp. 465–501, Plenum Press, New York.

Hornykiewicz, O., Lloyd, K.G., and Davidson, L., 1976, The GABA system, function of the basal ganglia, and Parkinson's disease (E. Roberts, T.N. Chase, and D.B. Tower, eds.), pp. 479–485, Raven Press, New York.

Hubbard, J.I., 1975, Microphysiology of vertebrate neuromuscular transmission, *Physiol. Rev.* **53:**674–723.

Katz, B., and Miledi, R., 1972, The statistical nature of the acetylcholine potential and its molecular components, *J. Physiol. (Lond.)* **224:**665–699.

Koketsu, K., Karczmar, A.G., and Kitamura, R., 1969, Acetylcholine depolarization of the dorsal root nerve terminals in the amphibian spinal cord, *Int. J. Neuropharmacol.* **8:**329–336.

Krnjevic, K., 1974, Chemical nature of synaptic transmission in vertebrates, *Physiol. Rev.* **54:**418–450.

Krnjevic, K., Puil, E., and Werman, R., 1977, Bicuculline, benzylpenicillin and inhibitory amino acids in the spinal cord of the cat, *Can. J. Physiol. Pharmacol.* **33:**670–680.

Macdonald, R.L., and Barker, J.L., 1977, Pentylenetetrazol and penicillin are selective antagonists of GABA-mediated postsynaptic inhibition in cultured mammalian neurons, *Nature* **267:**720–721.

Macdonald, R.L., and Barker, J.L., 1978a, Specific antagonism of GABA-mediated postsynaptic inhibition in cultured spinal cord neurons: A common mode of convulsant action, *Neurology (Minneap.)* **28:**325–330.

Macdonald, R.L., and Barker, J.L., 1978b, Different actions of anticonvulsant and anesthetic barbiturates revealed by use of cultured mammalian neurons, *Science* **200:**775–777.

Macdonald, R.L., and Barker, J.L., 1978c, Benzodiazepines specifically modulate GABA-mediated postsynaptic inhibition in cultured mammalian neurones, *Nature* **271:**563–564.

Macdonald, R.L., and Barker, J.L., 1979a, Enhancement of GABA-mediated postsynaptic inhibition in cultured mammalian spinal cord neurons: A common mode of anticonvulsant action, *Brain Res.* **167:**323–336.

Macdonald, R.L., and Barker, J.L., 1979b, Anticonvulsant and anesthetic barbiturates have different postsynaptic actions in cultured mammalian neurons, *Neurology (Minneap.)* **29:**432–447.

Macdonald, R.L., and Bergey, G.K., 1979, Valproic acid augments GABA-mediated postsynaptic inhibition in cultured mammalian neurons, *Brain Res.* **170:**558–562.

Matsuda, Y., Yoshida, S., and Yonezawa, T., 1976, A Ca-dependent regenerative response in rodent dorsal root ganglion cells cultured *in vitro, Brain Res.* **115:**334–338.

Matsumoto, H., Ayala, G.F., and Gumnit, R.J., 1969, Neuronal behavior and triggering mechanism in cortical epileptic focus, *J. Neurophysiol.* **32:**688–703.

McBurney, R.N., and Barker, J.L., 1978, GABA-induced conductance fluctuations in spinal neurons, *Nature* **274:**596–597.

Mohler, H., and Okada, T., 1977, Benzodiazepine receptor: Demonstration in the central nervous system, *Science* **198:**849–851.

Neale, E.A., Macdonald, R.L., and Nelson, P.G., 1978, Intracellular horseradish peroxidase injection for correlation of light and electron microscopic anatomy with physiology of cultured mouse spinal cord neurons, *Brain Res.* **152:**265–282.

Nelson, P.G., 1976, Central nervous system synapses in cell culture, *Cold Spring Harbor Symp. Quant. Biol.* **XL:**359–371.

Nelson, P.G., Neale, E.A., and Macdonald, R.L., 1978, Formation and modification of synapses in central nervous system cell cultures, *Fed. Proc.* **37:**12–17.

Nicoll, R.A., and Padjen, A., 1976, Pentylenetetrazol: An antagonist of GABA at primary afferents of the isolated frog spinal cord, *Neuropharmacology* **15:**69–71.

Okamoto, K., Quastel, D.M.J., and Quastel, J.H., 1976, Action of amino acids and convulsants on cerebellar spontaneous action potentials in vitro: Effects of deprivation of Cl^-, K^+ or Na^+, *Brain Res.* **113:**147–158.

Olsen, R.W., Ticku, M.K., VanNess, P.C., and Greenlee, D., 1978a, Effects of drugs on γ-aminobutyric acid receptors, uptake, release and synthesis *in vitro, Brain Res.* **139:**277–294.

Olsen, R.W., Ticku, M.K., and Miller, T., 1978b, Dihydropicrotoxinin binding to crayfish muscle sites possible related to γ-aminobutryic acid receptor–ionophores, *Mol. Pharmacol.* **14:**381–390.

Peck, E.J., Miller, A.L., and Lester, B.R., 1976, Pentobarbital and synaptic high-affinity receptive sites for gamma-aminobutyric acid, *Brain Res. Bull.* **1**:595–597.
Pellmar, T.C., and Wilson, W.A., 1977a, Penicillin effects on ionophoretic responses in *Aplysia californica, Brain Res.* **136**:89–101.
Pellmar, T.C., and Wilson, W.A., 1977b, Synaptic mechanism of pentylenetetrazole: Selectivity for chloride conductance, *Science* **197**:912–914.
Prince, D.A., 1968, The depolarization shift in "epileptic" neurons, *Exp. Neurol.* **21**:467–485.
Prince, D.A., 1978, Neurophysiology of epilepsy, *Annu. Rev. Neurosci.* **1**:395–415.
Ransom, B.R., and Holz, R.W., 1977, Ionic determinants of excitability in cultured dorsal root ganglion and spinal cord cells, *Brain Res.* **136**:444–453.
Ransom, B.R., Barker, J.L., and Nelson, P.G., 1975, Two mechanisms for poststimulus hyperpolarizations in cultured mammalian neurones, *Nature* **256**:424–425.
Ransom, B.R., Neale, E., Henkart, M., Bullock, P.N., and Nelson, P.G., 1977a, Mouse spinal cord in cell culture. I. Morphology and intrinsic neuronal electrophysiologic properties, *J. Neurophysiol.* **40**:1132–1150.
Ransom, B.R., Christian, C.N., Bullock, P.N., and Nelson, P.G., 1977b, Mouse spinal cord in cell culture. II. Synaptic activity and circuit behavior, *J. Neurophysiol.* **40**:1151–1162.
Ransom, B.R., Bullock, P.N., and Nelson, P.G., 1977c, Mouse spinal cord in cell culture. III. Neuronal chemosensitivity and its relationship to synaptic activity, *J. Neurophysiol.* **40**:1163–1177.
Schwartzkroin, P.A., 1975, Characteristics of CA neurons recorded intracellularly in the hippocampal *in vitro* slice preparation, *Brain Res.* **85**:423–436.
Shank, R.P., Pong, S.F., Freeman, A.R., and Graham, L.T., 1974, Bicuculline and picrotoxin as antagonists of γ-aminobutyric and neuromuscular inhibition in the lobster, *Brain Res.* **72**:71–78.
Skrede, K.K., and Westgaard, R.H., 1971, The transverse hippocampal slice: A well-defined cortical structure maintained *in vitro, Brain Res.* **35**:589–593.
Smith, B.M., and Hoffer, B.J., 1978, A gated, high voltage iontophoresis system with accurate current monitoring, *Electroencephalogr. Clin. Neurophysiol.* **44**:398–402.
Squires, R.F., and Braestrup, C., 1977, Benzodiazepine receptors in rat brain, *Nature* **266**:732–734.
Takeuchi, A., and Onodera, K., 1972, Effect of bicuculline on the GABA receptor of the crayfish neuromuscular junction, *Nature New Biol.* **236**:55–56.
Takeuchi, A., and Takeuchi, N., 1969, A study of the action of picrotoxin on the inhibitory neuromuscular junction of the crayfish, *J. Physiol. (Lond.)* **205**:377–391.
Ticku, M.K., Ban, M., and Olsen, R.W., 1978, Binding of [^{3}H]-dihydropicrotoxinin, a γ-aminobutyric acid synaptic antagonist, to rat brain membranes, *Mol. Pharmacol.* **14**:391–402.
Werman, R., 1969, An electrophysiological approach to drug-receptor mechanisms, *Comp. Biochem. Physiol.* **30**:997–1017.
Williamson, T.L., and Crill, W.E., 1976a, The effects of pentylenetetrazol on molluscan neurons. I. Intracellular recording and stimulation, *Brain Res.* **116**:217–229.
Williamson, T.L., and Crill, W.E., 1976b, The effects of pentylenetetrazol on molluscan neurons. II. Voltage clamp studies, *Brain Res.* **116**:231–249.
Wong, R.K.S., and Prince, D.A., 1978, Burst generation and calcium mediated spikes in CA3 neurons, *Neurosci. Abstr.* **3**:148.
Yamamoto, C., 1972, Intracellular study of seizure-like after discharges elicited in thin hippocampal section *in vitro, Exp. Neurol.* **35**:154–164.
Yamamoto, C., 1974, Electrical activity observed *in vitro* in thin sections from guinea pig cerebellum, *Jpn. J. Physiol.* **24**:177–188.
Yamamoto, C., and Kawai, N., 1967, Seizure discharges evoked *in vitro* in thin section from guinea pig hippocampus, *Science* **155**:341–342.

Yamamoto, C., and Kawai, N., 1968, Generation of the seizure discharge in thin sections from the guinea pig brain in chloride-free medium *in vitro, Jpn. J. Physiol.* **18:**620–631.

Young, A.B., and Snyder, S.H., 1974a, The glycine synaptic receptor: Evidence that strychnine binding is associated with ionic conductance mechanism, *Proc. Natl. Acad. Sci. USA* **71:**4002–4005.

Young, A.B., and Snyder, S.H., 1974b, Strychinine binding in rat spinal cord membranes associated with the synaptic glycine receptor: Cooperativity of glycine interactions, *Mol. Pharmacol.* **10:**790–809

4

Voltage Clamp Techniques Applied to Cultured Skeletal Muscle and Spinal Neurons

THOMAS G. SMITH, Jr., JEFFERY L. BARKER, BRUCE M. SMITH, and THEODORE R. COLBURN

1. INTRODUCTION

For many years there has been an apocryphal story circulating around the National Institutes of Health about Dr. K. S. Cole, the developer of the voltage clamp technique (Cole, 1949, 1972). It seems that Cole was being introduced to a distinguished group of scientists in some landlocked European country. The introducer remarked that he had always felt that the voltage clamp was the "devil's own invention" and it was his "pleasure to introduce the devil." Well, if the voltage clamp is the devil's own invention, then microelectrodes are the devil's own tools when it comes to using them in voltage clamping, since they are the main source of the problems and limitations in applying the technique successfully.

Voltage clamping has long been recognized as one of the most powerful techniques for studying the membrane properties of electrically and chemically excitable cells. The power of the technique, when it is working properly, lies in its ability to measure membrane current directly and thereby enable the calculation of ionic membrane conductance which is related to membrane perme-

---

THOMAS G. SMITH, Jr., and JEFFERY L. BARKER • Laboratory of Neurophysiology, National Institute of Neurological and Communicative Disorders and Stroke, National Institutes of Health, Bethesda, Maryland 20205. BRUCE M. SMITH and THEODORE R. COLBURN • Research Services Branch, National Institute of Mental Health, National Institutes of Health, Bethesda, Maryland 20205.

ability. For voltage-independent and time-invariant conductances, constant current and voltage-recording techniques are equally useful. Many membranes of interest, however, often possess voltage- and time-dependent conductances, and constant current techniques are inadequate and potentially lead to erroneous conclusions.

While a number of good papers have been devoted to the theory and practical aspects of voltage clamping axons with axial wires (e.g., Moore and Cole, 1963; Moore, 1971), few have dealt exclusively with the problems that arise when clamping with micropipettes. Most of what has appeared in this area has been in the "Methods" section of papers devoted to physiological experiments. One exception is an excellent paper by Katz and Schwartz (1974) who derived the equations applicable to voltage clamping and provided limiting case solutions that give useful insights into the design and use of voltage clamp systems with microelectrodes. They also introduced a new current feedback technique which improves both the speed and fidelity of voltage clamping. In this chapter on voltage clamp of tissue-cultured muscle and nerve cells, we shall rely heavily on Katz and Schwartz's work, including the use of similar terms and nomenclature. We shall not rederive the mathematics but shall draw on their useful equations. Thus, readers who want to follow the details of this chapter should become familiar with the paper of Katz and Schwartz (1974).

In addition to the theoretical problems analyzed by Katz and Schwartz and in this chapter, there are a number of practical problems introduced by use of the high resistance pipettes required to voltage clamp certain tissue-cultured cells. This chapter deals with these problems in an attempt to assist in the application of rigorous biophysical techniques to the broad range of biological material available in tissue culture.

2. WHY VOLTAGE CLAMP?

Before we turn to the details of voltage clamping, perhaps some justification and motivation should be given for undertaking this always difficult and often frustrating technique. The reader may reasonably ask: why bother, if conventional voltage recording and constant current techniques have been so productive and relatively easy? The main reasons are that one can directly measure conductance currents that are related to membrane ionic flux and calculate membrane conductance, which is related to membrane permeability, even when the conductance varies with voltage and time.

As mentioned above, constant current techniques can be inadequate and potentially lead to erroneous conclusions. The inadequacy is apparent from a consideration of the equation relating membrane current to a particular (the ith) ionic species, I_i, to other membrane properties: membrane voltage, V_m;

membrane capacitance, C_m; membrane conductance of the ith ion, G_i; and its equilibrium potential, E_i:

$$I_i = G_i(V_m - E_i) + C_m(dV_m/dt) \qquad (1)$$

As long as membrane potential is varying, $dV_m/dt \neq 0$, an unknown fraction of the current flows through C_m so that the G_i and C_m portions of the voltage responses to I_i are inseparable. In addition, if a conductance change leads to a change in V_m, as it usually does, the magnitude of the current will depend not only on a changing G_i but also a changing $V_m - E_i$ term. Such factors can lead to quantitative errors in response amplitude and kinetic studies.

A more serious problem arises, however, when one attempts to calculate conductances from current–voltage relations using constant-current pulses. Ignoring C_m and differentiating the ohmic part of equation 1 with respect to V_m gives

$$dI_i/dV_m = G_i + (V_m - E_i)(dG_i/dV_m) - G_i(dE_i/dV_m) \qquad (2)$$

Here $dI_i/dV_m = G_i$ exactly only if G_i and E_i do not vary with membrane potential. The possible hazard here of using constant-current techniques is that if G_i is positive and a strong function of V_m, and V_m and E_i have the right values, the second term in equation 2 can be negative. Thus, should the absolute value of the second term be larger than the first, the result is an incorrect *sign* for the calculated conductance. To make the point, consider the action potential. In the absence of any other information, the observation that a small change in depolarizing current at spike threshold leads to a large increase in V_m using equation 2 alone, would lead to the calculation of a large decrease in G_i. This is clearly incorrect, but the observers of new responses in tissue culture are not blessed with the prior knowledge of those who first recorded the intracellular spike.

A similar problem can arise from the third term of equation 2. For example, if a response is the result of a significant G_i change to two ions with widely separated ionic equilibrium potentials (e.g., Na^+ and K^+), and if the response is large, the driving force, $V_m - E_i$, will vary significantly during the time course of the response. Again, dI_i/dV_m will not only not measure G_i but may also be incorrect in sign. Similar problems arise if large ion fluxes during a response lead to changes in concentration gradients and thus to changes in E_i.

Both the second and third terms will lead to inaccurate calculations unless the response conductance completely dominates total membrane conductance. Thus, if the other, nonresponse membrane conductances are nonlinear, voltages evoked by constant-current pulses will reflect these changes in conductance as well as the changes in response conductance. Likewise, as the response voltage changes from resting potential, total membrane dE_i/dV_m will not be zero.

These problems are largely eliminated with a properly functioning voltage clamp. Those discussed with respect to equation 1 are completely eliminated since, by definition, dV_m/dt and thus I_i through C_m are zero. Moreover, the driving force, $V_m - E_i$, is constant unless ion concentration gradients change (see below). Thus I_i, either steady or time-varying, reflects G_i whenever E_i is constant.

The difficulties discussed with respect to equation 2 can also be eliminated by voltage clamping. First, the current will always move in the direction determined by the conductance change and the driving force—the sign of the change will therefore always be correct. Second, any dG_i/dV_m in a response can be detected, and any underlying potential- and time-dependent changes in nonresponse conductances can be eliminated by stepping to a potential and waiting for the nonresponse currents to come to a steady value before evoking the response. Finally, any changes in E_i can be measured by the use of the two-step command paradigm. This latter technique also allows the calculation of chord conductances, as will now be discussed.

The two-step technique is of sufficient importance to warrant its discussion. The basic assumption in this technique is that V_m can be changed and membrane current (I_m) measured accurately before membrane conductance (G_m) changes. Consider first its use with steady-state current–voltage (I–V) curves (Fig. 1A). The paradigm is to change V_m from a holding value (V_h) to another value (V_s) and wait until I_m reaches a steady state (I_s). Then, on successive sweeps, V_m is stepped to different levels (1–4), and the tail currents (I_t, arrow) determined. The tail current is the one immediately following the capacitative current after the second step. Each second-step potential–tail current pair is plotted, and the interconnecting line is the chord conductance, G_c. A straight line should result. If not, then G_m has changed before I_m could accurately be read. The potential at which the curve crosses the I_m axis ($I_m = 0$) is the driving force, E. These values represent G_m and E for V_s at time t_1. G_c is different from the tangent to the steady-state I–V curve at V_s; the latter is the slope conductance (G_s) and may have no readily interpretable meaning in terms of ionic conductances. Constant current methods reflect the slope conductance. For example, in N-shaped I–V curves, G_s is negative in the negative slope region, but G_c is always positive.

The technique used to measure G_c and E for evoked responses is illustrated in Fig. 1B. Here, V_m is changed from V_h, and I_m is allowed to become constant. At time t_r, the response is evoked, and when response current reaches its peak (I_p), the membrane is stepped to a new voltage (1–4) on successive sweeps. Again the tail currents are measured. For each second-step potential, the currents measured in the absence of a response (Fig. 1A2) are subtracted from those obtained during a response (Fig. 1B2), and the differences are plot-

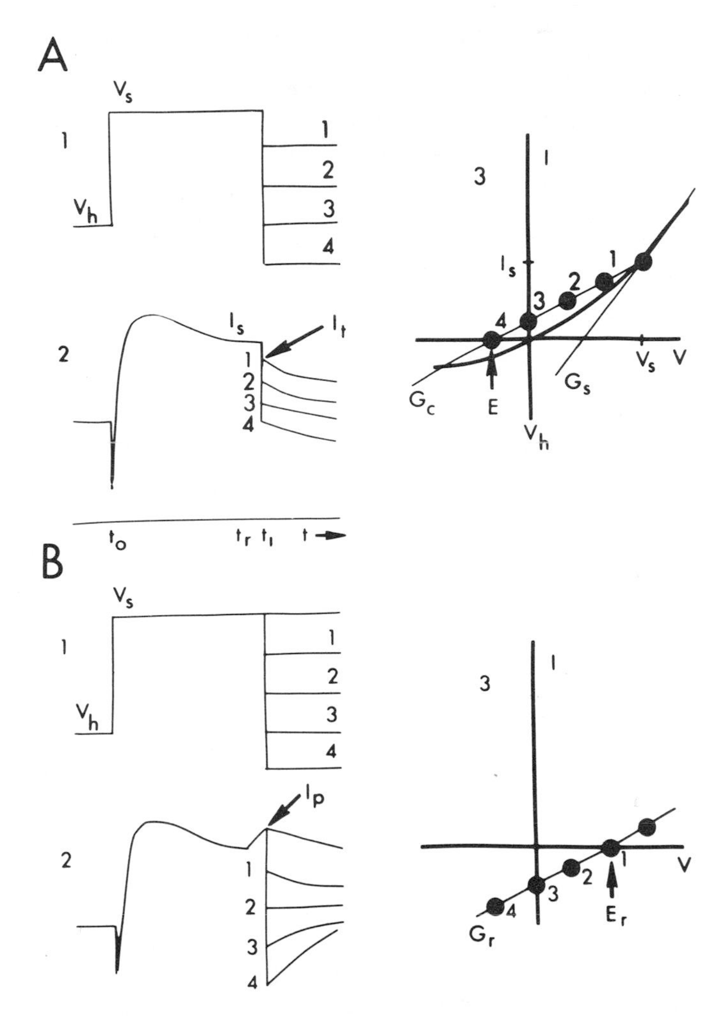

FIGURE 1. Simulated paradigm for determining chord conductance (G_c) and driving force (E) of membrane (A) and of evoked response (B) in voltage clamp experiments. See text.

ted as a function of second-step potential (Fig. 1B3). The chord conductance, G_r, and the E_r that are obtained are the peak conductance and the driving force, respectively, of the response at V_s and t_1.

3. VOLTAGE CLAMP SYSTEM

The essential components of a voltage clamp system are illustrated in Fig. 2, where the cell's passive membrane components are represented by a parallel

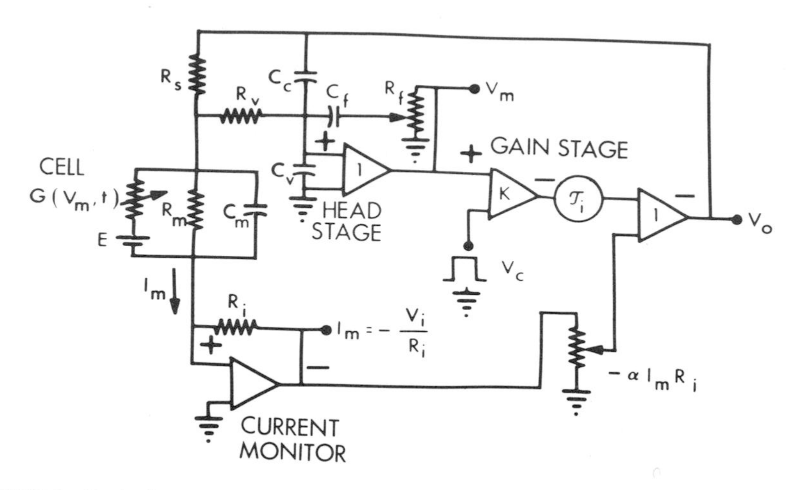

FIGURE 2. Equivalent circuit for voltage clamp system. (Modified from Katz and Schwartz, 1974.) See text.

membrane resistance, R_m, and capacitance, C_m. The battery associated with the resting membrane potential can be omitted with no loss of generality. The excitable conductance branch is assumed to be nonconducting at rest, a function of membrane potential and time $G(V_m,t)$, and driven by a battery, E. The intracellular membrane potential, V_m, is sensed with the microelectrode, R_v, and a unity gain head stage amplifier, the frequency response of which is attenuated by the unwanted stray capacitance to ground, C_v. The effect of C_v can be partially compensated for by positive feedback of part of V_m back to the amplifier's input through the R_f and C_f network. V_m is amplified by an inverting amplifier with variable gain, K, and frequency response with instrumentation time constant, τ_i. The output of the gain–frequency control amplifier is fed to a differential amplifier of unity gain and thence through the current-carrying electrode, R_s. The current then flows through the membrane to the extracellular space of the bathing solution and to the summing point of the current monitor that holds the bath at virtual ground potential. Some of the output of the differential amplifier flows through the unwanted coupling capacitance, C_c, between the two microelectrodes and through R_v and C_v, thereby giving a serious error signal (see Section 5).

Membrane current, I_m, is monitored as an output voltage, V_i, generated by I_m flowing through R_i. As pointed out by Katz and Schwartz (1974), if a fraction of I_m, namely $-\alpha R_i I_m$, is directed into the differential amplifier, the fidelity and speed of clamping can be improved. The result of this maneuver is effectively to reduce the impedance of R_s. This current feedback compensation is different from that employed in clamping axons with axial wires. There, the compensation is an attempt to reduce the effect of a resistance between the voltage monitoring electrode and in series with R_m and C_m (Hodgkin *et al.*, 1952; but cf. Cole, 1972, p. 350).

For purposes of discussion, it is convenient to define

$$\tau_m = R_m C_m$$

$$\tau_s = R_s C_m$$

$$\tau_l = \tau_m \tau_s / (\tau_m + \tau_s)$$

$$\beta = \tau_m / (\tau_m + \tau_s) = R_m / (R_m + R_s) \tag{3}$$

Here, τ_l, the "load" time constant, is what the voltage clamp must attempt to handle and is a composite of the membrane time constant, τ_m, and the "source" time constant, τ_s. The parameter β has the significance that βK is the closed-loop gain of the clamp–load system at zero frequency.

The importance of Katz and Schwartz's contribution to our understanding of voltage clamp systems was their recognition that, in order to achieve critical damping of the response of V_m to a step change in V_c, the important factor is the difference between the load time constant, τ_l, and the instrumentation time constant, τ_i. Critical damping is the fastest response without oscillations and occurs when

$$4K/\tau_i \tau_s = (\tau_l^{-1} - \tau_i^{-1})^2 \tag{4}$$

If this difference between τ_l^{-1} and τ_i^{-1} is too small, the voltage and current responses would display underdamped oscillations or "ringing." If this difference is too large, the responses may be slow or overdamped.

Katz and Schwartz pointed out two limiting case solutions to equation 4, when

$$\tau_l >> \tau_i; \quad \tau_l/\tau_i = 4\beta K \tag{5a}$$

$$\tau_i >> \tau_l; \quad \tau_i/\tau_l = 4\beta K \tag{5b}$$

They also showed that for these two limiting cases, the time (t) for a membrane potential step change to reach 90% of its final value is given by

$$\tau_l >> \tau_i; \quad t = 7.78\tau_i \tag{6a}$$

$$\tau_i >> \tau_l; \quad t = 7.78\,\tau_l \tag{6b}$$

Equations 3–6 were taken from Katz and Schwartz (1974); however, some of the subscripts have been changed. Here the membrane and instrumentation time constants are τ_m and τ_i, respectively. The corresponding time constants in Katz and Schwartz (1974) are τ_i and τ.

These relations form a rational basis for a start in the design and construction of a voltage clamp system for any particular biological preparation. For example, R_m, C_m, and τ_m can be determined from constant current experiments. τ_s is specified by determining the microelectrode resistances (r_s) required for stable recording with two intracellular electrodes and from knowing C_m. τ_l and β can then be calculated. Then, by measuring the speed of recording achievable with the voltage recording microelectrode (R_v) and input amplifier, it can be determined from equation 4 whether the "fast" (equations 5a and 6a) or the "slow" (equations 5b and 6b) case applies. Unless intentionally made otherwise, the input stage of a voltage clamp system employing microelectrodes will be the high-frequency-limiting stage. Ideally the fast system is preferable, since the frequency response is not limited by τ_m (equation 6b) as occurs in the slow system.

The preceding has dealt with the critically damped system. In practice, the system may be made to respond faster by deliberately producing a slightly underdamped oscillation by adjusting the capacitance neutralization (R_f) at the head stage. Dr. Harold Lecar (Lecar and Sachs, this volume; personal communication) has pointed out several useful relationships in assessing this situation. For example, with K large, the natural frequency of the oscillation is given by

$$f_o \simeq (2\pi)^{-1} (K/\tau_i\tau_s)^{1/2} \tag{7}$$

The damping ratio, η, is

$$\eta = [\tau_i\tau_s(\tau_s^{-1} + \tau_m^{-1} + \tau_i^{-1})^2/4K]^{1/2} \tag{8}$$

and is a measure of the settling of the oscillation. For critical damping, $\eta = 1$; for undamped, pure oscillation $\eta = 0$. The best damping ratio is $\eta = 0.5$–0.7, and the time constant of decay of the oscillation, τ_d, is

$$\tau_d = (\tau_s^{-1} + \tau_m^{-1} + \tau_i^{-1})^{-1} \tag{9}$$

With these equations, one can begin to assess the nature and tolerability of the oscillations present in a given system. For clarity we should note that the above equations (3–9) apply to a system simpler than that illustrated in Fig. 2, one that does not include C_v, C_c, C_f, R_f, and current feedback.

Approximate values for these various time constants and parameters are presented in Table I for three different preparations: large molluscan neurons, tissue-cultured chick muscle cells, and tissue-cultured mouse spinal cord cells. As can be seen for the molluscan neurons and for tissue-cultured muscle cells, reasonable values of system gain and response can be obtained with a fast voltage clamp. With gains of (1.5–2.5) × 10^3 in such a system, one would expect, ideally, to have the V_m settle to a step change in V_c within some tens of microseconds (t_{90}), following some high frequency ringing (f_o) that decayed rapidly (η and τ_d). Then one should be able to clamp a spike. In practice, the system does not do as well as suggested by the data in Table I because there are more than two time constants in any voltage clamp system employing two microelectrodes and a preparation in a bath. First, if the frequency response control is not placed at the head stage, there is the time constant of the recording microelectrode, R_v, and its stray capacitance to ground, C_v. Then there is the time constant of the capacity neutralization network which does not match identically the frequency characteristics of R_v and C_v, since C_v is a capacitance that is distributed along the electrode. In addition, there is another complex time constant caused by the coupling capacitance, C_c, between the two microelectrodes across their glass walls and through the bath solution. And, finally, there are apparently one or more time constants whose source may not be known but whose presence is indicated by ringing with voltage steps at a fre-

TABLE I. Approximate Preparation and Instrumentation Variables and Parameters for Critical Damping and for Slightly Underdamped Voltage Clamp for Molluscan Neuron, Tissue-Cultured Muscle Cell, and Spinal Cord Neuron[a]

Parameter	Units	Neuron	Tissue-cultured muscle cell	Tissue-cultured spinal neuron
$\tau_m = R_m C_m$	sec	10^{-1}	10^{-2}	10^{-3}
		$(10^7 \times 10^{-8})$	$(10^7 \times 10^{-9})$	$(10^7 \times 10^{-10})$
$\tau_s = R_s C_m$	sec	10^{-1}	2×10^{-2}	4×10^{-3}
		$(10^7 \times 10^{-8})$	$(2 \times 10^7 \times 10^{-9})$	$(4 \times 10^7 \times 10^{-10})$
$\tau_l = \tau_m \tau_s/(\tau_m + \tau_s)$	sec	5×10^{-2}	7×10^{-3}	8×10^{-4}
$\beta = R_m/(R_m + R_s)$		0.5	0.3	0.2
$K\ (\tau_l >> \tau_i)$		2.5×10^3	1.6×10^3	10^2
$K\ (\tau_i >> \tau_l)$		10^4	10^4	10^4
$\tau_{crit}\ (\tau_l >> \tau_i)$	sec	10^{-5}	10^{-5}	2×10^{-5}
$\tau_{crit}(\tau_i >> \tau_l)$	sec	10^3	84	1.0
f_o	Hz	10^4	1.6×10^4	2×10^4
η		0.7	0.5	0.2
τ_d	sec	10^{-5}	10^{-5}	8×10^{-4}
τ_{90}	sec	8×10^{-5}	8×10^{-5}	6.4×10^{-3}

[a]The values given were derived from unpublished observations by the two senior authors of this paper. They agree, where available, with those obtained in various other laboratories. No claim of priority is intended in this chapter.

quency too high to be accounted for by the known time constants. Nonetheless, adequate clamping can be achieved in molluscan neurons ($K \simeq 3 \times 10^3$, $t_{90} \simeq 10^{-4}$ sec) and in tissue-cultured muscle ($K \simeq 2 \times 10^3$, $t_{90} \simeq 10^{-4}$ sec). An example of spike currents recorded in tissue cultured muscle is shown in Fig. 3 (Fukuda *et al.*, 1976).

When one inspects the data in Table I for tissue-cultured spinal cord neurons, it becomes apparent that the situation is much more difficult indeed. Primarily because of the low C_m (10^{-10} F) and high R_s (4×10^7 Ω), critical damping for a fast clamp allows only a gain of 200, often not enough to clamp synaptic potentials and totally inadequate for spikes. Attempting to increase the gain further leads first to damped oscillations that take considerable time to decay and eventually to undamped oscillations. These values were obtained from equation 5a, where τ_i was 20 μsec, the fastest response obtained with a 40 MΩ electrode with our head stage amplifier. If a gain of 3×10^3 were required, τ_i would have to be 0.3 μsec, an unrealizable value. With a slow clamp, the gain can be increased (10^4), but the frequency response ($t_{90} = 6.4$ msec, $\eta = 0.2$) is poor. Such a system can clamp slow responses (Fig. 4-I, Barker *et al.*, 1978) and allows analysis of current fluctuations (Fig. 4-II, McBurney and Barker, 1978) but not a spike, because the spike conductance

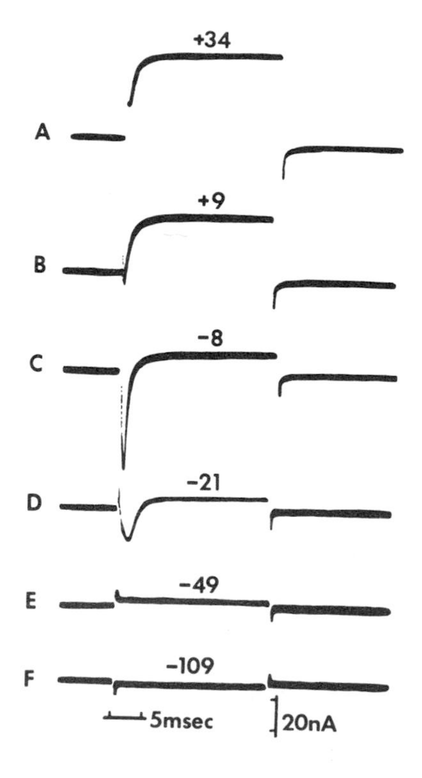

FIGURE 3. Voltage clamp currents from chick striated muscle myosac. Records of current required to clamp membrane potential by a 22-msec pulse to levels indicated by numbers above current traces from a holding potential of −90 mV. Calibrations 20 nA and 5 msec. (Reproduced with permission of *Developmental Biology*.)

occurs before V_m has settled (Fig. 5A). The lack of settling and fidelity of clamping results from the slow frequency response of the system. In these experiments the use of current feedback alone makes an insufficient increase in frequency response to clamp the spike.

In the course of studying our voltage clamp, we found quite by accident that adding another time constant to the system improved the response time markedly and allowed the spike to be clamped moderately well (Fig. 5B). Then, the addition of current feedback further improved clamping fidelity (Fig. 5C). This new addition has a mixture of both fast and slow clamp characteristics. For the fast aspect of the system, gain is controlled by a potentiometer connected between the negative input and the output of an operational amplifier (Fig. 6, inset, R_g). The system is then slowed down by placing a capacitor (C_f) in parallel with R_g. Such a system would be the slow clamp of Katz and Schwartz (1974). This arrangement has the advantage of automatically maintaining the conditions for critical damping (equation 5b) with changes in gain. As gain is increased, τ_i is automatically increased because of the constant gain–bandwidth product of the amplifier circuit. This means that clamp closure can be at low gain with C_f adjusted for critical damping. Then, R_g can be increased to the value needed with the assurance that critical damping will be maintained. In using the slow system, we have found it best to make C_f variable, since there is an optimum C_f for each gain–τ_l condition. In the mixed case, however, C_f only need be large, say 1–10 μF, and fixed. The mixed system results from placing a potentiometer, R_f, in series with C_f. What this lead network does is shown by the Bode plots in Fig. 6A.*

A Bode plot is a graph of log gain versus log frequency and phase shift (in degrees) versus log frequency. More complete discussions of Bode analysis can be found in standard electrical engineering texts (e.g., Graeme *et al.*, 1971). The utility of a Bode plot is that it allows one to see where the important changes in frequency response and phase of a single operational amplifier or a complete system occur. For example, an open loop plot permits one to determine how much gain can be used when the loop is closed while still maintaining stability. The essential condition for stability is that the closed loop gain where the phase shift is $\geqslant 180°$ must be less than unity. Otherwise, in a negative feedback system such as a voltage clamp, those signals become positively fed back to the input, and the system oscillates. For stability all other conditions

*The above discussion and the values given in Table I apply to the resting membrane. When the membrane becomes active through an increase in G_m, τ_m and τ_l decrease. Adequate damping will usually be maintained in a fast system, since the difference between τ_l and τ_i will be increased (equation 4). The same adequacy of damping may not result in a slow or mixed system, since a decreased τ_l may become too near in value to the dominant time constant (τ_1 of Fig. 6A) of such systems. Thus τ_1 should be made sufficiently large to insure adequate damping. This is most conveniently achieved by using much more gain than required for clamping the resting membrane by increasing R_g (Fig. 6A).

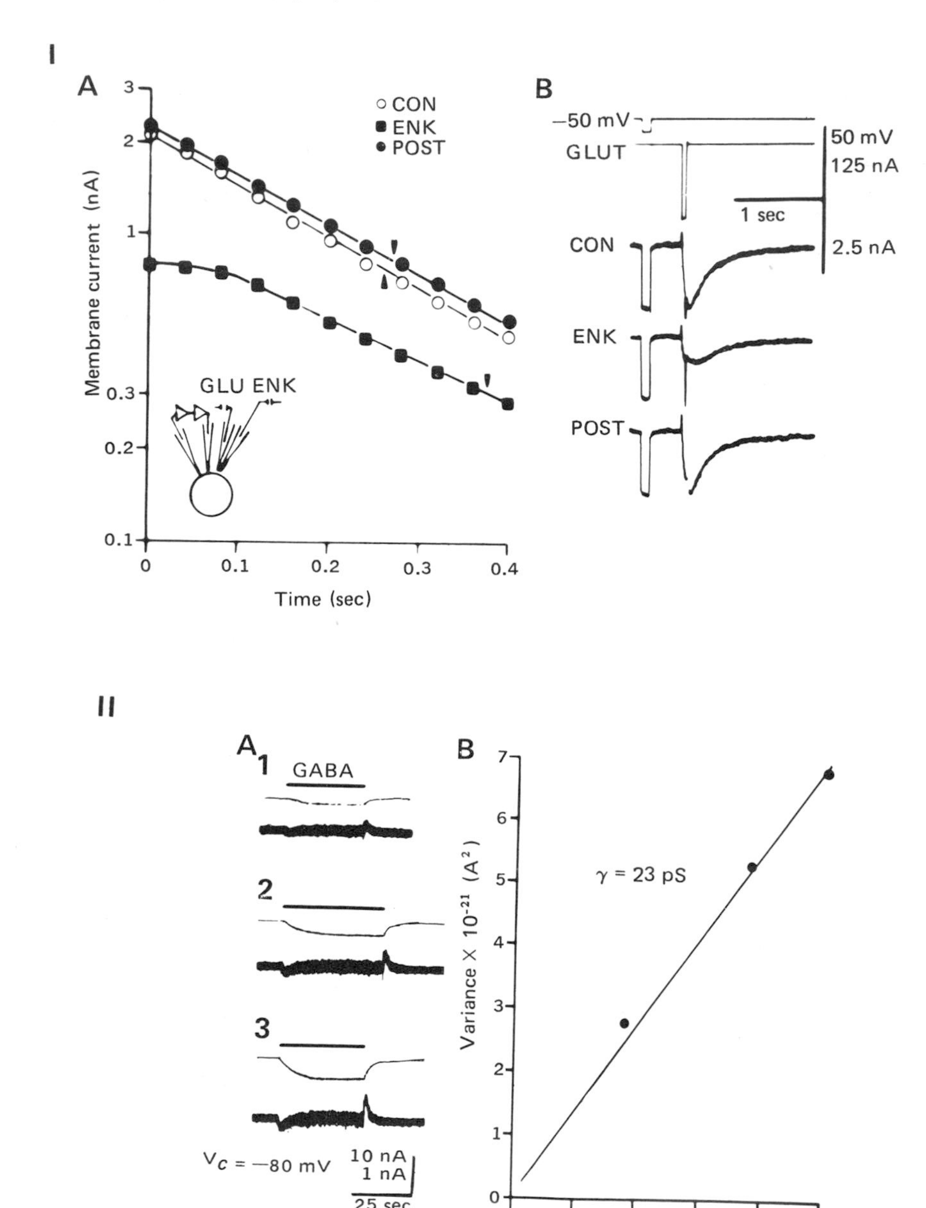

FIGURE 4. (I) Enkephalin depresses glutamate-evoked membrane current response and slows its time course without changing resting membrane conductance. Cell impaled with two KCl microelectrodes, and membrane potential voltage-clamped; glutamate and leu-enkephalin (ENK) iontophoresed from extracellular microelectrodes (schematic insert in A). B shows experimental paradigm with potential clamped to −50 mV and a 10-mV, 100-

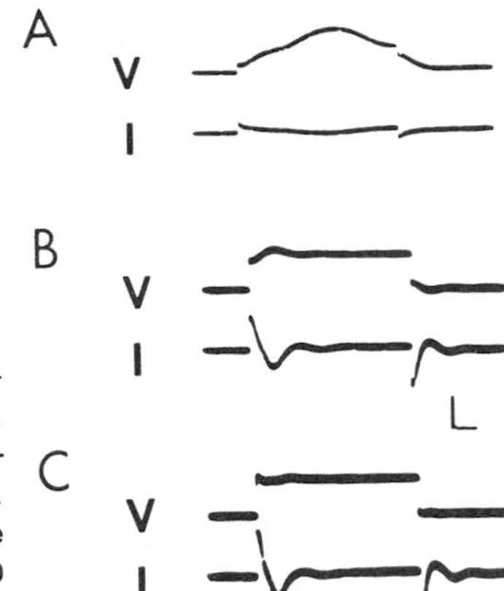

FIGURE 5. Voltage clamp records from tissue-cultured mouse spinal cord cell. V = voltage, I = current. Holding potential −60 mV. Five-msec pulse commands to −40 mV. (A) Slow voltage clamp system. (B) Mixed voltage clamp system. (C) Mixed voltage clamp system plus current feedback. Calibrations: 20 mV and 0.1 μA, 1 msec. See text.

are irrelevant, and any "tricks" one may employ to increase frequency response and fidelity of clamping are legitimate. Having said that, we would add the caveat that one should be sure to have done what was intended, for example, by checking out the system with known electronic "dummies" for the microelectrode and cell membrane.

Curve F_g in Fig. 6A is the Bode plot of our gain control amplifier in the fast operating mode. Its half-power (−3 db) frequency is about 250 kHz, with a τ of just less than a μsec. As shown in curve F_p, the 180° phase point is

---

msec hyperpolarizing command superimposed, followed by a 100-nA, 100-msec glutamate pulse. Membrane current responses to voltage command and glutamate pulse are illustrated before (CON), during (ENK), and after (POST) iontophoresis of 20 nA ENK. ENK depresses the glutamate-evoked current response without hanging the amplitude of current response to the voltage command. Current from peak of response is plotted semilogarithmically in A. Time constant of glutamate current response decay (marked by arrowhead) increases during ENK iontophoresis. (Reproduced with permission of the American Association for the Advancement of Science and of the Elsevier Publishing Company.) (II) (A) Membrane current responses of voltage-clamped spinal cord neurons in cell culture to iontophoretic application of GABA. Steady iontophoretic currents were applied during the times indicated by the black bars above each of the three pairs of records. Each pair contains a DC-coupled trace (above) and a high-gain AC-coupled trace (below). An increase in the thickness or "noise" of the AC-coupled trace accompanied the membrane current response to GABA. The "noise" increased as the intensity of the iontophoretic current and subsequent membrane current response increased (1–3). Clamp potential (V_c): −80 mV. Temperature: 26°C. (B) The variance of the noise induced by GABA was linearly related to the change in membrane current. The increase in variance of the noise for each of the three records shown in (A) was calculated from digitized records of the noise during the plateau phase of each response and plotted against the corresponding change in membrane current. The points were approximated by a straight line whose slope yields a single channel conductance of 23 pS. (Reproduced with permission of *Nature.*)

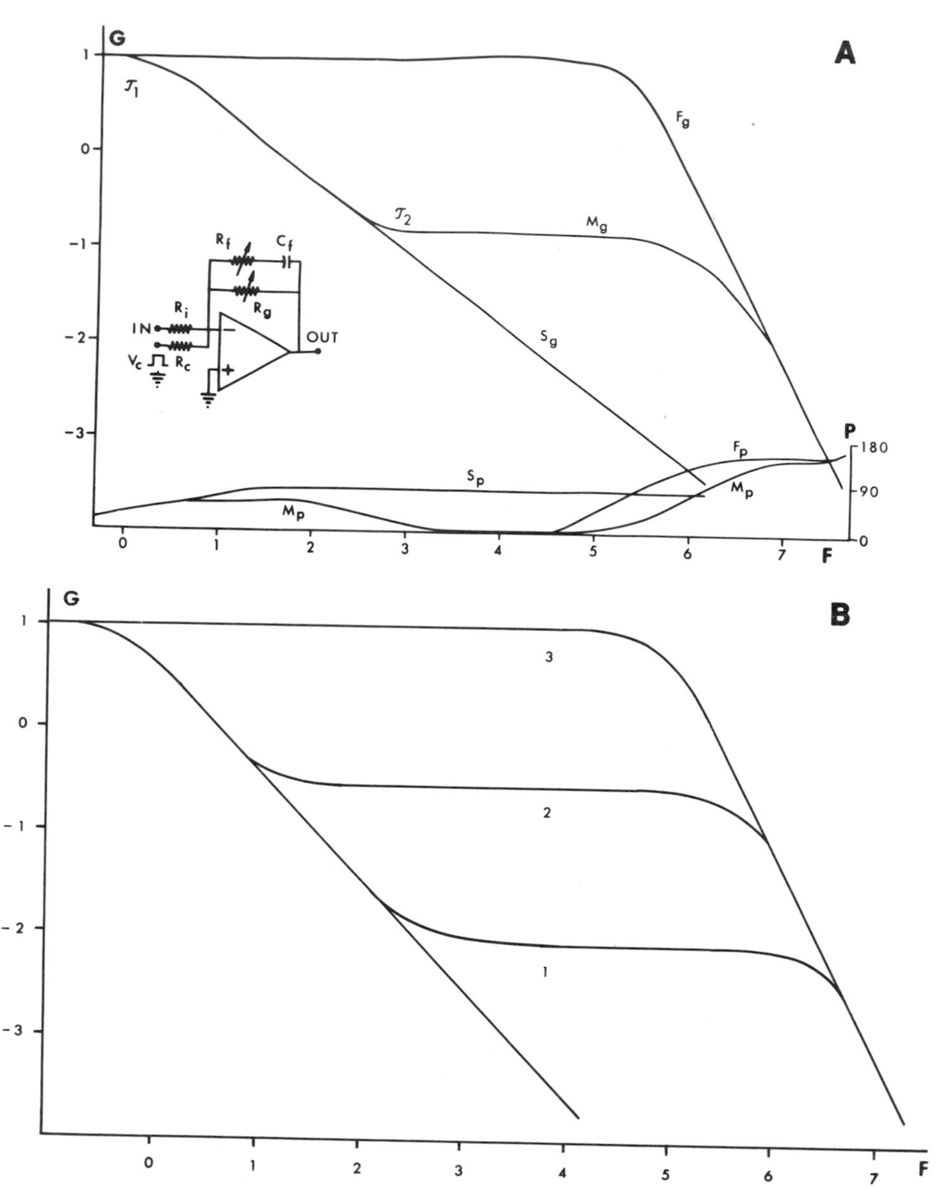

FIGURE 6. (A) Closed-loop Bode plots of gain–frequency control amplifier. Inset: circuit diagram of gain–frequency control amplifier. Pulse into negative input is command pulse. R_i is input resistor which receives the V_m from the head stage; R_c, the input to which the command signal, V_c, is applied; R_g, gain control potentiometer; R_f, frequency response potentiometer; C_f, frequency response capacitor. F_g and F_p are gain and phase curves for fast clamp system, respectively; S_g and S_p for slow system; M_g and M_p for mixed system. Graphs are log gain (G) vs. log frequency (F), with numbers indicating exponents of ten, and phase shift (P) in degrees vs. log frequency (F). (B) Effects of changing R_f. See text.

approximately 2 MHz. Gain control of this stage alone is from 0.1 to 10, which allows an overall system control of from 10^2 to 10^4. Curves S_g and S_p illustrate the gain and phase of the slow mode of operation. Here the low-frequency cutoff, τ_1, is 1 sec, and the phase shift is limited to 90°. Curves M_g and M_p are the gain–frequency plot of the mixed system. The lag, τ_1, of 1 sec remains, but a lead, τ_2, is added, and the gain and phase shift are constant from 10^3 to 2×10^5 Hz. Thereafter, both curves approach the fast system and coincide with it above 5 MHz. The effect of increasing R_f is to move the τ_2 and flat region of the curve from a low gain point on the slow mode curve, Curve 1 of Fig. 6B, to a higher gain point, Curve 2, and eventually to the fast mode curve, Curve 3, when R_f becomes infinite.

In this system,

$$\tau_1 = (R_g + R_f)C_f \tag{10}$$

$$\tau_2 = R_f C_f$$

For the simple slow or fast clamp system, the overall open loop system response is

$$G(s) = K/(\tau_i s + 1)(\tau_l s + 1) \tag{11a}$$

and the relationship between V_m and the command voltage, V_c, with the loop closed, is

$$V_m/V_c = K/[\tau_i \tau_l s^2 + (\tau_i + \tau_l)s + (K + 1)] \tag{11b}$$

Here, s has its usual meaning, namely, the complex frequency, $j\omega$. For the mixed systems,

$$G(s) = K(\tau_2 s + 1)/(\tau_1 s + 1)(\tau_l s + 1) \tag{12}$$

but if $\tau_2 = \tau_l$,

$$G(s) = K/(\tau_1 s + 1) \tag{13a}$$

$$V_m/V_c = K\Big/ (K + 1)\left(\frac{\tau_1 s}{K + 1} + 1\right) \tag{13b}$$

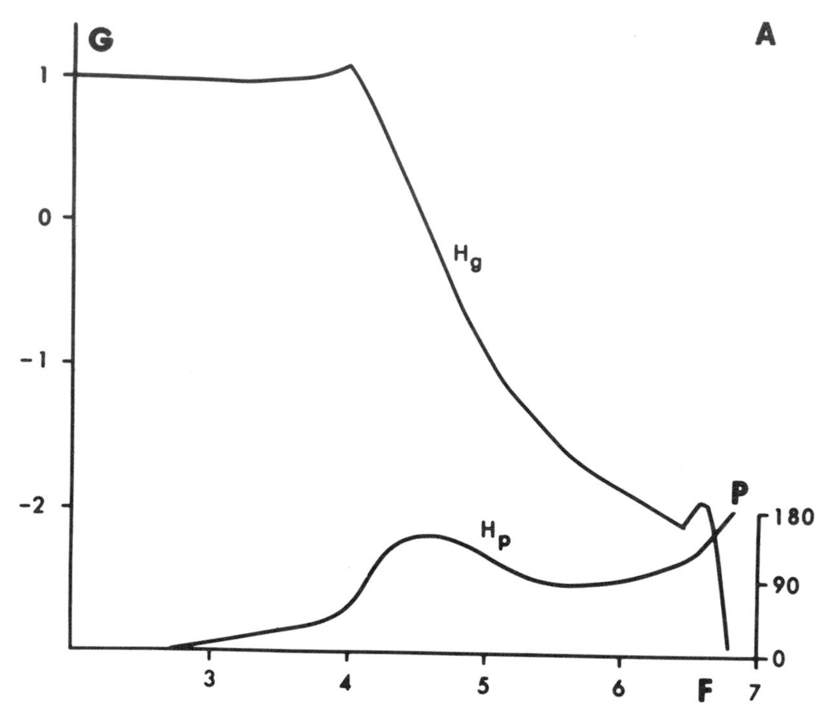

FIGURE 7. Bode plots of log gain (*G*) vs. log frequency (*F*) and phase shift (*P*) vs. log frequency (*F*). Numbers as in Fig. 6. (A) Closed loop head stage curves. (B) Open loop, mixed voltage clamp system curves. See text.

Thus, by making the lead $\tau_2 = \tau_l$, the system is potentially reducible to one with a single time constant (equation 13a), where the effect of τ_1 can be made small with K large (equation 13b). More importantly, the system becomes potentially independent of τ_l.

Unfortunately, in addition to the gain control amplifier, the head stage also plays a role in determining frequency response. Thus, the system still has more than one time constant. The characteristics of the head stage we use, curves H_g and H_p of Fig. 7A, plus the gain control amplifier, curves M_g and M_p of Fig. 6A, and the other very fast amplifiers give the system response, S_g and S_p in Fig. 7B. The practical consequences are to maintain adequate separation between τ_l and τ_i (i.e., τ_1) at high gain but to give a low gain boost at frequencies above τ_2 and retard the increase in phase shift (M_p). The degree to which the high frequency boost in gain can be increased is determined by the phase shift, i.e., the point where the 180° shift exceeds unity gain, and the system becomes unstable and oscillates. In practice, we were able to realize a high frequency gain of about 100 and shift τ_2 to about 10^2 (Fig. 7B), but at the cost of some ringing in the response since $\tau_2 \neq \tau_l$ (Figs. 8B and C).

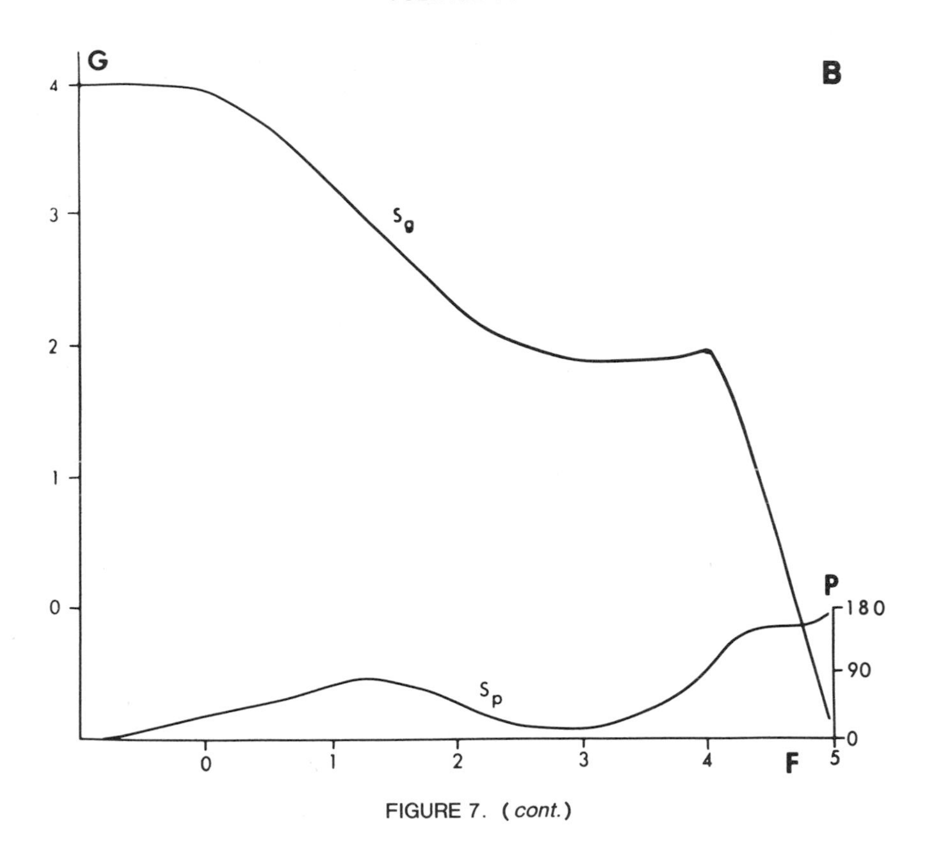

FIGURE 7. (*cont.*)

We should emphasize that a system that has a limiting slope of 3 decades in gain per decade in frequency is potentially very unstable when utilized in closed-loop voltage clamp. Small changes in any component of the total system, including the microelectrodes or cell membrane, can lead to oscillations and to catastrophic destruction of the cell.

4. TAMING THE SPINAL CORD NEURON SPIKE

Using the system just described, we will proceed to illustrate what can be achieved with tissue-cultured spinal cord neurons. After having successfully penetrated a neuron with two microelectrodes and demonstrated the presence of action potentials (Fig. 8A), we remove all holding current and connect the output of the head stage to the input of the voltage clamp system. With the current feedback ($-\alpha R_i I_m$ of Fig. 2) and the frequency response controls (R_f in Figs. 2 and 6) set at minimum values and the gain low, the clamp output (V_o of Fig. 2) is set to zero potential by a DC control potentiometer in the

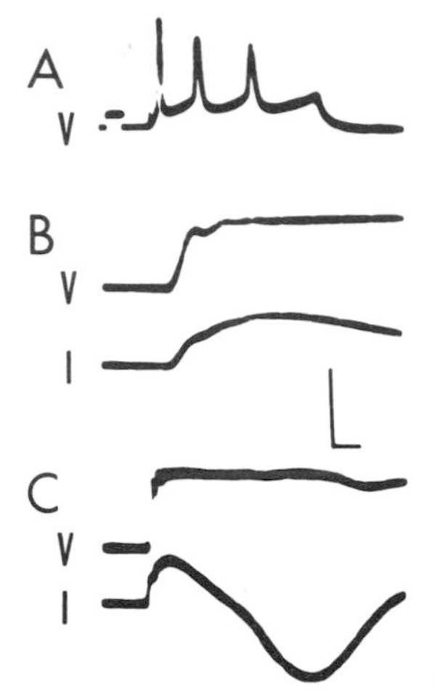

FIGURE 8. Recordings from tissue-cultured mouse spinal cord cell. V = voltage, I = current. (A) Multiple action potentials evoked by 100-msec depolarizing constant-current pulse. (B) and (C) Voltage clamp recordings from cell illustrated in A. Membrane potential stepped from holding potential of −60 mV to −37 mV. Calibrations: 50 mV in A, 25 mV in B and C, 0.125 μA in B and C; 10 msec in A, 20 μsec in B and 100 μsec in C. See text.

clamp (not shown). The loop is then closed by connecting V_o to R_s (Fig. 2). If necessary, the membrane is hyperpolarized to a V_m more negative than the spike threshold with the clamp's DC control. Then a depolarizing command pulse sufficient to evoke a spike is applied, and gain and frequency response are increased (R_g in Fig. 6 and R_f in Figs. 2 and 6) until the best clamp is achieved. Finally, $-\alpha I_m R_i$ is added to increase fidelity and frequency response. With our system, gains of at least 3×10^3 at low frequencies and 50–100 at high frequencies are required to clamp spikes that generated up to 0.1 μA of inward current (Fig. 5C). We have been able to clamp such neurons with a voltage time constant of about 20 μsec (Fig. 8B) using 30–50 MΩ microelectrodes on cells with 1 msec time constants. It appears to take 200–500 μsec for the capacitative current to settle and permit reliable conductance current measurement (Fig. 8C). Part of this initial outward current, which declines before the onset of the inward spike current (Fig. 9A), may represent an active mem-

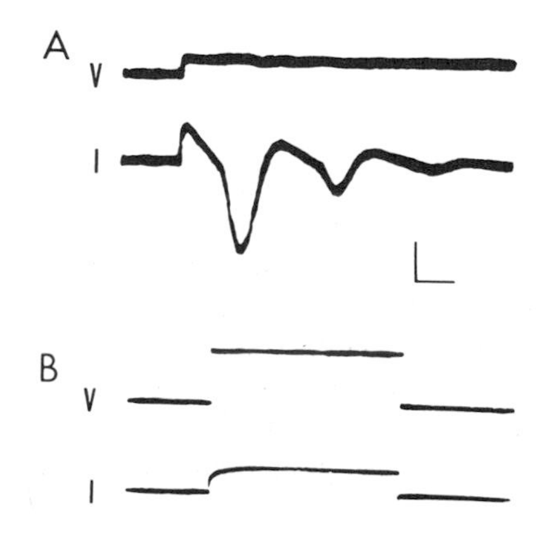

FIGURE 9. Voltage clamp records of spike currents from cell illustrated in Fig. 8. V = voltage, I = current. (A) Membrane potential stepped from −60 mV to −42 mV evoked multiple spike currents. Onset of first spike occurs about 0.5 msec after step command. (B) Membrane potential stepped from −60 mV to 10 mV evoked large outward current. Calibrations: 50 mV and 0.05 μA; 0.5 msec in A, 10 msec in B.

brane conductance, since the current evoked by comparable hyperpolarizing commands settles much faster (not shown).

We have found that most of the spinal cord and dorsal root ganglion cells have significant inward spike currents. Outward currents, presumably potassium, are small and brief at command potentials below zero millivolts in V_m (Figs. 5B and C; Fig. 9A). Only near the reversal potential for the inward spike current (+20 to 40 mV) do the outward currents become appreciable (Fig. 9B).

We have successfully studied spike currents in spinal neurons for periods up to an hour. Stable experiments on dorsal root ganglion (DRG) cells, however, have been infrequent with spike studies, although common with other, less demanding voltage clamp studies. The reasons for this failure are unclear, since DRG cells are as large as, if not larger than, spinal neuronal somata. However, the fact that some DRG cells are swollen following repeated large clamp currents suggests that the considerable charge injection into the cells with large step commands cannot be ejected fast enough. This presumably leads initially to osmotic imbalance and then to water uptake. Thus far we have found no significant differences in spinal and DRG spike currents.

5. MICROELECTRODE SHIELDING AND GUARDING

While all of the aforementioned procedures were necessary to achieve the results illustrated, an additional maneuver was required to deal with the interelectrode coupling capacitance, C_c in Fig. 2. This capacitance results in a positive feedback of the high frequencies of V_o and leads to ringing in the voltage and current traces following step commands (Fig. 10A). This ringing cannot be eliminated without sufficiently lowering the gain and frequency response below that required for adequate clamping of the spike. Applying current feedback only makes matters worse (Fig. 10B).

We have attempted, without success, to circumvent this problem by two

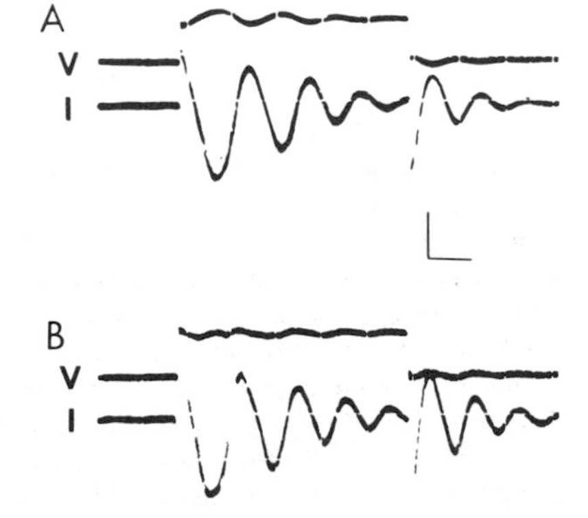

FIGURE 10. Voltage clamp records from spinal cord neuron with microelectrodes inadequately shielded and guarded. Five-msec step from −70 mV to −50 mV. (A) Membrane potential and current trace produce damped oscillations. (B) Effect of adding current feedback. See text. Calibrations: 20 mV and 0.05 μA; 1 msec.

techniques. The first was to place a third microelectrode just outside the cell, to position it so that it would receive the same high frequency V_o signals as R_v, and to record its signals differentially against the intracellular R_v before going to the clamping amplifier. This common mode rejection was unsuccessful because the two signals could not be exactly matched. The second technique was what has been called cross-neutralization (Frank *et al.*, 1959). With this method, one feeds V_o to an inverting amplifier with variable gain and then to the input of the head stage amplifier through a capacitor. It is also a form of common mode rejection and does not work because the distributed capacitance, C_v, cannot be matched with a single capacitor.

The only way we have been able to overcome the C_c problem is to paint the outside of one of the microelectrodes (R_v or R_s) with silver paint (Silver Print, GC Electronics) to within 15–25 μm of the tip of the electrode and then cover the silver with a material (Super Bonder® 420 Adhesive, Loctite Corp.) to insulate it from the extracellular solution. For a coated R_s, the silver is connected to ground. For R_v, the silver is actively driven by the "guard" of the headstage (see Section 7.2). Usually such treatment of one of the two electrodes is sufficient, but more reliable and better results are obtained by treating both.

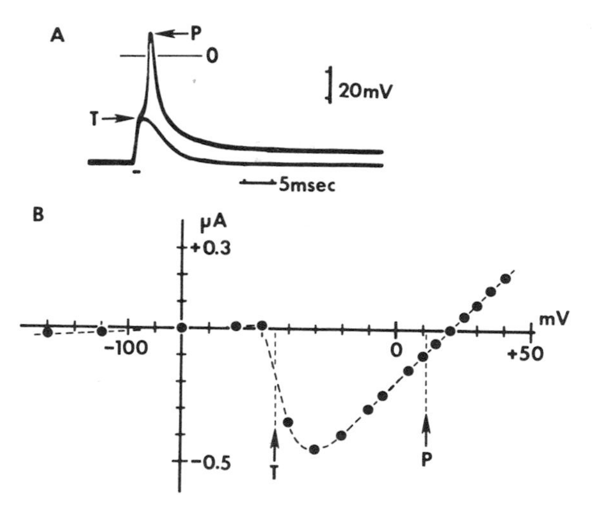

FIGURE 11. Myotube. (A) Determination of threshold and peak potential of Na^+ spike. Two superimposed voltage traces which begin at holding potential of −80 mV. Horizontal line with 0 is zero membrane potential. Constant current pulses applied during time (ca. 1 msec) indicated by bar under voltage traces. One current stimulus was subthreshold, the other evoked an action potential with overshoot. Threshold of −45 mV marked by T, and peak potential of +12 mV marked by P. Calibrations: 20 mV and 5 msec. (B) Voltage clamp current–voltage curve of Na^+ current. Plot of peak inward or minimum currents. Threshold T and peak potential P measured in Fig. 4A indicated on voltage axis. Holding potential of −80 mV. Note lack of correspondence between peak of spike (P) and reversal potential on *I–V* curve because of lack of adequate space clamp. See text. (Reproduced with permission of *Developmental Biology*.)

6. SPACE CLAMPING

In the study of nonspherical cells an important question is whether the cell is isopotential over its total membrane surface, i.e., is total intracellular space isopotential or "space" clamped. This problem has been adequately covered elsewhere (Cole, 1972). The magnitude of the problem depends upon the situation.

With propagated responses such as spikes in nonisopotential cells, adequate space clamping is virtually impossible and can lead to quantitatively erroneous results (Fig. 11). With some preparations, one may be able to

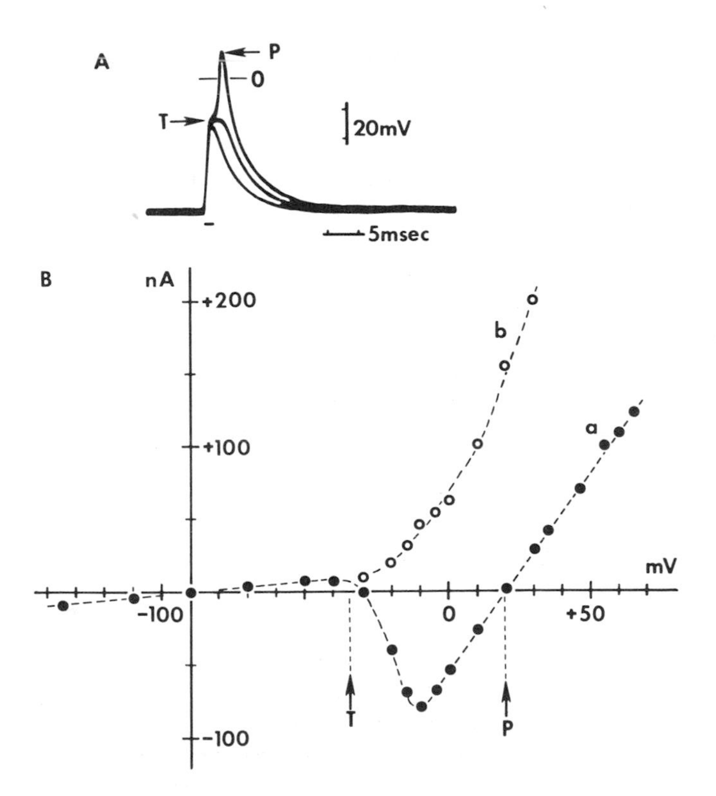

FIGURE 12. Myosac. (A) Determination of threshold and peak potential of Na^+ spike. Three superimposed voltage traces which begin at holding potential of −90 mV. Horizontal line with 0 is zero membrane potential. Constant current pulses applied during time (ca. 1 msec) indicated by bar under voltage trace. Two current stimuli were subthreshold for spike, and one evoked an action potential with overshoot. Threshold of −35 mV marked by T, and peak potential of +20 mV for Na^+ spike marked by P. Calibrations: 20 mV and 5 msec. (B) Voltage clamp current–voltage curves. Data taken from records similar to those shown in Fig. 3. Peak inward-going or minimum currents plotted as filled circles (a) and late currents as open circles (b). Threshold T and peak potential P measured in A are indicated on voltage axis. Holding potential of −80 mV. See text. (Reproduced with permission of *Developmental Biology*.)

achieve isopotentiality by altering the morphology of the cell to be studied as, for example, was done with tissue-cultured muscle by treatment with colchicine (Fukuda *et al.,* 1976). Then, quantitatively meaningful results can be obtained (Fig. 12).

Since dorsal root cells are spherical and have very small axons, an adequate space clamp is probably obtained (Ransom *et al.,* 1977). With spinal cord cells that have extensive dendritic trees, it is unlikely that the entire surface membrane is space clamped. Surprisingly, however, ringing in the current trace, which is the hallmark of an inadequate space clamp, is minimal following a voltage step command. This may be because the total extent of cultured spinal cord neuron's dendritic tree is only one characteristic length constant (Ransom *et al.,* 1977). Still, we doubt that the spike data illustrated in this paper were derived from adequately space-clamped neurons.

When clamping a response that is evoked near the location of the microelectrode tips, e.g., when evoking responses by the brief application of neuroactive substances by iontophoresis at the cell soma, space clamping is apparently not a serious problem. If a drug response or a postsynaptic potential is evoked on a remote dendrite, less significance can be attached to any quantitative analysis of such data.

Focally evoked responses are most accurately measured if the membrane potential is held at resting potential. Changing the membrane potential and then evoking a focal response yields meaningful results only if the "load" of the nonresponse membrane does not interact with the response membrane in a nonisopotential cell. Establishing this as a fact is difficult, if not impossible, and interpretations of data from such experiments should be made with caution.

7. PRACTICAL POINTS

7.1. General

There are a number of good electronic practices that are well applied to all electrophysiology as well as to voltage clamping, such as keeping wires as short as possible, shielding all signal-carrying wires over 2 in. long, and grounding or guarding the shield, as appropriate. Two other good practices are illustrated in Fig. 13, the separation of signal and shield grounds and the "star" method of grounding the signal-measuring system.

In this arrangement, the common or ground connections of the signal system are brought separately to one common point (C) and then connected to the ground of the signal measuring system, usually an oscilloscope. This practice prevents the formation of group loops, which act as antennae for unwanted signals. Similar ground loops in the shield system are unimportant and virtually impossible to avoid. By keeping the signal and shielding grounds separate as

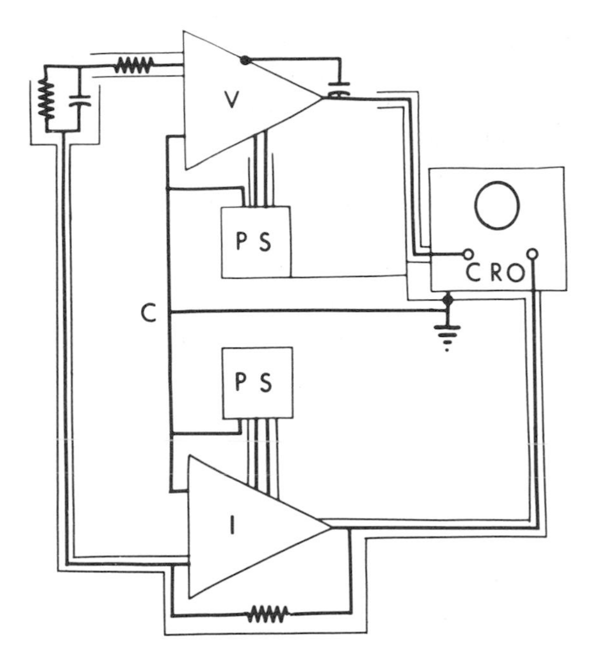

FIGURE 13. Schematic diagram illustrating "star" method of grounding and separation of signal and shielding grounds. Thin lines indicate shielding system; thick lines indicate signal-carrying system. Note common of all signal system components brought to a central connection (C) before being grounded. Note shield of amplifier *V* and guard to microelectrode not grounded but driven by *V* output through a capacitor. See text.

long as possible, the stray signals (e.g., power line frequency and switching transients) picked up by the shields are not injected into the signal common line to generate noise. This is particularly important for low level signals before they have been adequately amplified.

7.2. Microelectrodes

Microelectrodes with the lowest resistances consistent with stable recordings from viable cells should be employed for voltage clamping. For tissue-cultured muscle and spinal neurons, electrodes with values of 20–40 MΩ and 40–60 MΩ, respectively, are required. Since clamping speed and fidelity are limited by the ability of the current-carrying electrode, R_s, it should have the smaller resistance. Also, since a lower-valued R_s requires a smaller V_o to pass a given I_m, the effect of C_c is less. In addition, a smaller R_s means less current noise. The fluid level in the bath should be kept as low as possible to minimize C_c.

Both R_s and R_v and the wires connecting them to the amplifier should be shielded to the extent required to minimize C_c. The R_s shield should be grounded, and the R_v shield guarded. Guarding is achieved by applying the same signal from the feedback control (R_f in Fig. 2) through a separate, large (0.1 μF) capacitor to the R_v shield and headstage chassis (Fig. 13). Up to frequency values limited by amplifier delay, the input signal in the electrode and on the shield will be the same, hence no current flows through the capacitance connecting the electrode and the shield. Since R_f is low compared to R_v, unwanted signals are shunted to ground, and the effect of C_v is minimized.

7.3. Head Stage Amplifier

Since microelectrodes required for tissue culture have such high resistances, the amplifier should have an input impedance of $>10^{10}\ \Omega$ and <4 pF, and high gain–bandwidth product (>10 MHz) and slew rate (>100 V/μsec). Because the microelectrode limits the frequency response of the head stage, one can achieve a gain of ten or better at no cost and thereby reduce the gain requirements of subsequent stages. In all modes of clamping, the microelectrode–head stage combination should be made as fast as possible.

For simplicity of presentation, the head stage amplifier in Fig. 2 is shown with unity gain, but the signals fed back to the guard and to the input through C_f must come from a source with a gain greater than unity if compensation is to be achieved.

7.4. Gain Stage

For all modes of clamping, this stage should have large gain–bandwidth product ($\geqslant 10$ MHz), open loop gain ($\geqslant 100$ db), and slew rate ($\geqslant 100$ V/μsec). Other characteristics are of less importance because the preceding gain ($\geqslant 10$) has boosted the signals to reasonable levels, and one can work with low impedances. The operation of this stage has been discussed previously with Figs. 5 to 7.

7.5. Final Stage

For clamping slow responses only, any fast ($\geqslant 10$ MHz and 100 V/μsec) operational amplifier (op amp) with good (1 decade/decade or 6 decibels/octave) frequency response roll-off will do. For clamping step commands and spikes, however, a high-voltage ($\geqslant \pm 100$ V) amplifier may be required to charge the C_m rapidly and contain the spike. Among commercially available op amps, the choices for this stage are not good. The best we know (Datel Model AM 303) has a gain–bandwidth product of 10 MHz, slew rate of 100 V/μsec, and ± 140 V output. The problem is to make the transition from the ± 10 V output of the gain stage to the high-voltage stage. If only the two amplifiers were used, then the high-voltage stage would require a gain of 14 to match the full dynamic ranges, i.e., 10 V to 140 V. This would give the high-voltage stage a frequency response of only 750 kHz, which would interact with and thus worsen the phase shifts of the headstage in this range (Fig. 7A). We have solved this problem by inserting a discrete component transistor amplifier with a gain of 10, ±75-V output, and a 7-MHz frequency response between the gain stage and the high-voltage amplifier. This allows the latter to be operated at a gain of 2.

7.6. Current Monitor

The type of operational amplifier required for the current monitor depends upon the kind of experiments performed. It has to be fast or quiet—it cannot be both.

7.6.1. Fast Monitor

For clamping spikes and for determining chord conductances the amplifier has to hold the bath at ground potential during rapid and large changes in V_m and G_m that generate large I_ms. The op amp should have large gain–bandwidth product (≥10 MHz) and slew rate (≥100 V/μsec). A relatively small feedback precision (1%) resistor (10–50 kΩ) should be employed to insure speed and accuracy. It is worth noting that the current monitor is operated at a voltage gain much less than unity, as the gain is given by $R_i/(R_m + R_s)$. Thus, the potential frequency response is considerable; however, it is unlikely to be achieved because of microelectrode limitations. This also means that the voltage gain will vary from experiment to experiment. This is of no consequence, however, because I_m, regardless of the source impedance to the current monitor, will be the current fed back through R_i (Fig. 2).

7.6.2. Quiet Monitor

For clamping low-level (pico- to nanoamperes) currents evoked synaptically or iontophoretically and in fluctuation or noise analysis, an electrometer-type op amp is often required. This necessity arises not from the need of high input resistance but for the noise and bias current characteristics available in such amplifiers. The unit should have low input current noise (≤0.1 pA, peak-to-peak, 10 Hz to 10 kHz). A large (10^7 Ω) precision (1%) R_i should be employed to amplify the signal adequately and accurately. Such a unit will have a slow frequency response (≤3 kHz), which often is made slower by further filtering at a later stage of signal amplification.

With both the fast and quiet systems, further amplification of the current signal subsequent to the current monitor is usually needed in order to match the dynamic range of the current signals and recording devices and thus maximize the signal-to-noise ratio. Conventional op amps whose frequency response is much faster and which are quieter than the I_m s recorded are readily available for this task. Precision (1%) input and feedback resistors are used to insure accuracy. If further filtering is required, it is readily achieved by placing a variable capacitor in parallel with the feedback resistor at the final stage of I_m amplification in the way done in the gain stage of the voltage clamp system (C_f in Fig. 6A). R_f is, of course, omitted.

REFERENCES

Barker, J. L., Neale, J. H., Smith, T. G., and MacDonald, R. L., 1978, Opiate peptide modulation of amino acid responses suggests novel form of neuronal communication, *Science* **199:**1451–1453.

Cole, K. S., 1949, Dynamic electrical characteristics of the squid giant axon, *Arch. Sci. Physiol.* **3:**253–258.

Cole, K. S., 1972, *Membranes, Ions and Impulses,* University of California Press, Berkeley.

Frank, K., Fuortes, M. G. F., and Nelson, P. G., 1959, Voltage clamp of motoneuron soma, *Science* **130:**38–39.

Fukuda, J., Fischbach, G. D., and Smith, T. G., Jr., 1976, A voltage clamp study of the sodium, calcium and chloride spikes of chick skeletal muscle cells grown in tissue culture, *Dev. Biol.* **49:**412–424.

Graeme, J. D., Tobey, G. E., and Huelsman, L. P., 1971, *Operational Amplifiers, Design and Application,* McGraw-Hill, New York.

Hodgkin, A. L., Huxley, A. F., and Katz, B., 1952, Measurement of current–voltage relation in the membrane of the giant axon of *Loligo, J. Physiol. (Lond.)* **116:**424–448.

Katz, G. M., and Schwartz, T. L., 1974, Temporal control of voltage-clamped membranes: An examination of principles, *J. Membr. Biol.* **17:**275–291.

McBurney, R. N., and Barker, J. L., 1978, GABA-induced conductance fluctuations in cultured spinal neurones, *Nature* **274:**596–597.

Moore, J. W., 1971, Voltage clamp methods, in: *Biophysics and Physiology of Excitable Membranes* (W. J. Adelman, Jr., ed.), pp. 143–167, Van Nostrand Rheinhold, New York.

Moore, J. W., and Cole, K. S., 1963, Voltage clamp techniques, in: *Physical Techniques in Biological Research, Volume VI, Electrophysiological Methods, Part B* (W. L. Nastuk, ed.), pp. 263–321, Academic Press, New York.

Ransom, B. R., Neale, E., Henkart, M., Bullock, P. N., and Nelson, P. G., 1977, Mouse spinal cord in cell culture. I. Morphology and intrinsic neuronal electrophysiological properties, *J. Neurophysiol.* **40:**1132–1150.

5

Membrane Noise Analysis

HAROLD LECAR and FREDERICK SACHS

1. INTRODUCTION

The fundamental units of excitability are membrane ionic channels. Excitation requires fast switches in membrane permeability, and these are effected by channels that are ion-selective and can be "gated" open and closed. A variety of experimental methods have contributed to establishing the present picture of discrete channels of molecular dimensions, each capable of carrying a rather high ion flux. In this chapter, we discuss an experimental method, membrane noise analysis, which allows one to estimate the ionic currents carried by individual channels. The reason gated ionic channels give rise to a special type of electrical noise is that channels can open and close at random under steady-state conditions. Because the ion flux through a single channel is relatively large, random channel transitions generate observable electrical current fluctuations. These characteristic current fluctuations, often called channel noise, can be distinguished from other types of electrical noise which may be generated in an excitable-cell preparation.

Channel noise can now be recorded in a variety of preparations, and noise analysis has become a useful tool for determining channel conductances and the kinetics of the gating process. This chapter reviews the methods for measurement of membrane noise and discusses some of the results which have been obtained by noise measurement on excitable cells grown in tissue culture. A recent review (DeFelice, 1977) surveys the applications of noise analysis in neurobiology.

HAROLD LECAR • Laboratory of Biophysics, Intramural Research Program, National Institute of Neurological and Communicative Disorders and Stroke, National Institutes of Health, Bethesda, Maryland 20205. FREDERICK SACHS • Department of Pharmacology and Therapeutics, State University of New York, Buffalo, New York 14214.

1.1. Conductance Fluctuations: A Short Historical Overview

The earliest estimate of a unitary conductance from noise analysis was the shot-noise estimate of the response of a photoreceptor to a single photon reported by Hagins and Srebro (Hagins, 1965). Working on slices of excised squid retina, these investigators measured the extracellular current fluctuations induced by steady illumination. From an analysis of the power spectrum, they concluded that the photon-induced conductance event persists for about 100 msec and that the elementary conductance change produced by a single photon is 20 pS. The assumptions underlying this analysis were that all of the single-photon events were identical in shape and that the observed noise was a consequence of a random time series of single-photon "shots." In more recent studies of light-induced noise in photoreceptors, spectra and autocorrelation functions obtained from the noise have been compared to observed and theoretical single-photon responses and have been found to be in general agreement with the notion that the noise originates in conductance events whose shape mirrors the time course of the many-photon flash responses (Dodge *et al.,* 1968; Lamb and Simon, 1977; Schwartz, 1977; Wong, 1978). The single-photon events appear to be the result of the opening or closing of many membrane channels as an indirect consequence of the absorption of a photon in much the same way as a miniature endplate potential results from the opening of many channels in response to the release of a quantal packet of transmitter molecules (Baylor *et al.*, 1974).

Proceeding along a different track, Verveen and Derksen (Verveen and Derksen, 1968, 1969; Derksen and Verveen, 1966; Derksen, 1965; Verveen *et al.,* 1967) examined the spontaneous voltage noise in frog nodes of Ranvier. These investigators were also interested in the sources of noise underlying fluctuations in excitability (Ten Hoopen and Verveen, 1962; Lecar and Nossal, 1971). The early investigations turned up noise having primarily a $1/f$ spectrum, with total power proportional to the square of the departure from the potassium equilibrium potential. Noise spectra with $1/f$ frequency dependence are ubiquitous in electrical systems and have now been observed in various biological and synthetic membranes. As will be shown in the next section, the noise spectrum expected for fluctuations in the gating of ionic channels has a different frequency dependence, falling off as $1/f^2$ at high frequencies. The $1/f^2$ component was apparently masked by the $1/f$ noise in the early studies. However, the channel noise for electrically excitable cells has been studied in recent years by a number of experimenters (Fishman, 1973; Siebenga *et al.,* 1973; Conti *et al.,* 1975, 1976a,b), yielding estimates of the conductances of Na and K channels in the range of 2–10 pS.

Postsynaptic membrane noise was first observed by Katz and Miledi (1972, 1973). They applied steady doses of acetylcholine iontophoretically to the frog muscle endplate and observed increases in voltage noise. As we shall discuss later, voltage fluctuations usually yield only the charge transferred per conductance event rather than the conductance itself. This is because the voltage fluctuations are filtered by the cell membrane capacitance. In order to estimate the dynamic behavior of the channels, Katz and Miledi recorded the extracellular current noise which is not limited by the membrane time constant. From the spectrum of the extracellular fluctuations, they estimated the lifetime of an open channel to be 1 msec. Knowing the charge transferred per event and the average event duration from these noise measurements, they estimated the unit conductance to be approximately 100 pS. Anderson and Stevens (1973) used voltage clamp to study the current fluctuations produced by steady doses of iontophoretically administered acetylcholine (ACh). They estimated the errors carefully and determined the channel conductance to be 30 pS and the lifetime of the open state as 10 msec at a temperature of 10°C. Comparable values have been found for ACh channels in tissue-cultured avian muscle cells (Sachs and Lecar, 1973, 1977) and more recently in excised (Cull-Candy *et al.*, 1978) and cultured human muscle (Bevan *et al.*, 1978).

One development which gave further impetus to the channel noise studies was the direct observation of single-channel fluctuations in synthetic lipid bilayer membranes doped with channel-forming materials. For example, in a lipid bilayer membrane doped with the channel-former EIM, one observes long trains of rectangular current jumps such as those indicated schematically in Fig. 1. These rectangular jumps in current are caused by the opening and closing of individual ionic channels. Since EIM-doped membranes were also known to possess the voltage-dependent conductances needed for electrical excitability (Mueller *et al.*, 1962), the single-channel observations provide an opportunity for discerning the origin of this voltage-dependent conductance. Studies of pulse trains like those of Fig. 1 show that individual open channels are linear conductors but that the rates of channel opening and closing are controlled by the transmembrane electric field. The opening and closing rates vary with applied potential in just the right way to explain the macroscopic voltage-dependent conductance (Ehrenstein *et al.*, 1970; Lecar *et al.*, 1975). Thus, the gated ionic channels produced in bilayers by EIM and other pore-formers now provide model systems for studying membrane excitation caused by channels whose molecular structure can be determined. A recent review (Ehrenstein and Lecar, 1977) summarizes the measurements of single-channel fluctuations in synthetic membranes.

Single-channel fluctuations of the postsynaptic membrane can now also be measured directly (Neher and Sakmann, 1976b, 1978). Neher and Sakmann treated denervated frog muscle fibers with collagenase and protease to remove the basement membrane and then pressed a 2–3 μm diameter pipette containing dilute suberyldicholine in saline against the cell. Currents flowing into the pipette were recorded by a low noise current-to-voltage converter. The single-channel fluctuations caused by agonist interaction with channels under the tip of the pipette were two-state and had the conductance and open time expected from noise measurements on denervated muscle (Neher and Sak-

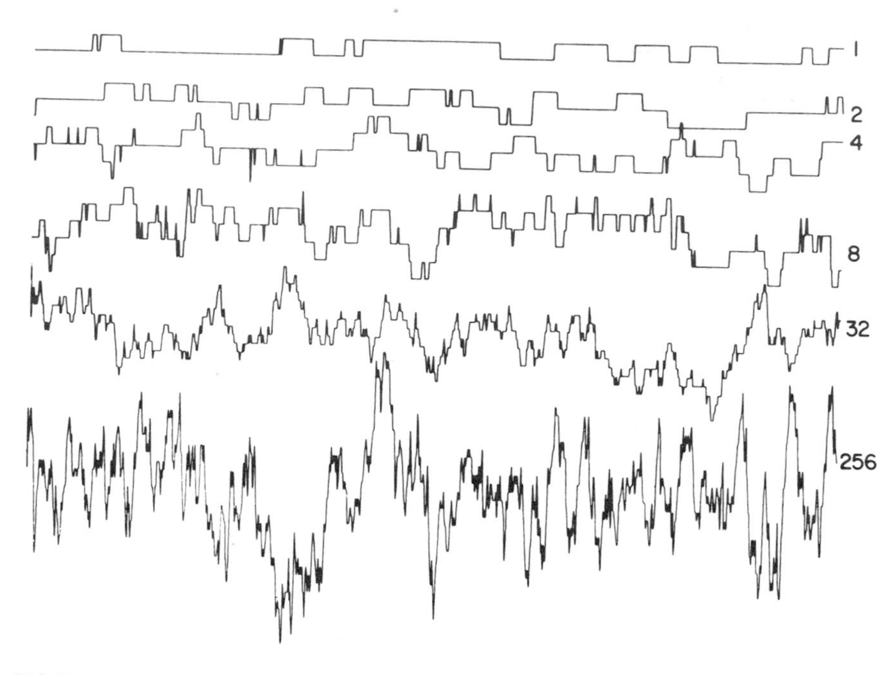

FIGURE 1. A computer simulation of two-state channels showing the transition between single channels and noise. The simulation was made with equal Poisson probabilities for the *on* and *off* transitions, giving an average 50% duty cycle. The simulation was made for different numbers of independent channels as indicated at the right of each trace. Note that the gain has been reduced by 2 on the two lowest traces.

mann, 1976a). Single-channel fluctuations have recently been reported for tissue-cultured muscle, both avian (Nelson and Sachs, 1979) and mammalian (Jackson and Lecar, 1979).

Although this chapter emphasizes noise analysis of gated channels, noise analysis can be applied to nongated channels as well, provided a means can be found for opening and closing the channels. For example, Lindemann and Van Driessche (1977) measured fluctuations caused by the nongated sodium channel in frog skin by using the sodium transport blocker, ameloride, which reversibly blocks the sodium-selective channel. The frequency dependence of this induced channel noise depends on the rates at which the blocking molecules associate and dissociate with the channel. The noise induced by reversible blockage has also been profitably employed to study the kinetics of anesthetic interaction with postsynaptic channels (Katz and Miledi, 1975; Colquhoun and Hawkes, 1977; Ruff, 1976; Colquhoun *et al.,* 1975; Neher and Steinbach, 1978).

Several recent reports show records of quantal jumps occurring as channels form within membranes. Loewenstein *et al.* (1978) observed the opening of cell-to-cell channels in nascent cell junctions, and Krawczyk (1978) observed the formation of single channels on surface layers of spheres of algal protoplasm. Also, channel formation in tissue-cultured muscle was obtained by the addition of the polypeptide antibiotic alamethicin (Sakmann and Boheim, 1979) and by the addition of complement to hapten-treated cells (Stephens *et al.,* 1979).

2. THEORY OF CHANNEL NOISE

The stochastic methods for treating channel noise of various types have been reviewed by several authors (Verveen and DeFelice, 1974; Conti and Wanke, 1975; Neher and Stevens, 1977; Colquhoun and Hawkes, 1977; Hill and Chen, 1972). The stochastic models describe the conductance fluctuations for channels with multiple conductance states as well as channels having a single state of open conductance. Channels with a single conducting state can also have complicated kinetics because they must go through a number of intermediate closed conformations before they open. The differences in all these properties are reflected in different noise spectra.

In order to introduce the reader to noise theory methods, we will consider only the simplest channel, one which merely flips between two states, open and closed. We imagine the channel in a steady state undergoing random transitions between the two conductance states. The observed series of jumps in ionic current at some steady potential constitute a stationary random process, that is, a random process in which statistical parameters such as the average fraction of time spent open or the average rate of transition do not drift in time. The top frame of Fig. 1 shows the time course of the fluctuating current through a single channel. Succeeding frames show the fluctuating current as increasing numbers of channels open and close at random. As can be seen in the figure, when the number of channels becomes sufficiently large, the individual events can no longer be resolved. The many-channel records can be analyzed as a form of electrical noise having statistical properties related to the underlying unitary events.

The simplest way to extract the unit current amplitudes from a noise record is by analysis of the variance of the fluctuating currents. The quantity of interest is the ratio of variance to mean. This ratio can be used to infer the amplitude of the individual rectangular pulses, even though they are not resolvable in the record.

Let the unit amplitude be a, and the channel be open for an average fraction of time, $\Theta = \Sigma t_i / T$ where t_i is the duration of the ith jump and T is the duration of current record. The mean current is then

$$\langle I \rangle = a\Theta \tag{1}$$

and the mean-squared current is

$$\langle I^2 \rangle = a^2\Theta \tag{2}$$

The variance is then given by

$$\mathrm{var}\,(I) = \langle I^2 \rangle - \langle I \rangle^2 = a^2\Theta(1 - \Theta) \tag{3}$$

The ratio of variance to mean is, then,

$$\mathrm{var}\,(I)/\langle I \rangle = a(1 - \Theta) \tag{3a}$$

For small Θ, the right-hand side of equation 3a is approximately equal to the unit amplitude, a.

This argument holds for any number of independent channels, so that the unit current can be measured from the variance even when the unitary fluctuations cannot be resolved. In fact, the validity of equation 3a for a two-state channel is not restricted to stationary processes. Variance measurements can be used to infer the unit amplitude even when the mean conductance is changing in time. A nice experimental example of the measurement of variances for a nonstationary process is a recent determination of the conductance of the sodium channel in frog node of Ranvier (Sigworth, 1977).

So far, the unit events were supposed to be rectangular current jumps, as might be expected for the opening of a simple molecular channel. However, fluctuation analysis is still useful for analyzing the noise emanating from trains of pulses of more complex shape. For example, in the vertebrate photoreceptor, the currents produced by a single photon are thought to be composite membrane events in which a large number of channels are blocked by a substance released from the disks. Thus the noise should be analyzed as a train of identical events having a pulse shape described by some functional form $F(t)$. Another example of noise generated by multichannel events is the neural noise generated by a series of miniature endplate potentials. Each miniature endplate potential is generated by a burst of channel openings initiated by a quantal packet of transmitter molecules.

If the appearance of the events obeys Poisson statistics, the variance is given by Campbell's theorem (Rice, 1954),

$$\mathrm{var}\,(I) = \nu \int_{-\infty}^{\infty} F^2(t)\,dt \tag{3b}$$

where ν is the rate of appearances. Thus for more general pulse shapes, the variance analysis still gives information about the unit event, provided the unit pulse shape is known. Although it is possible to measure variances directly from a time-average of the current noise, it is more common to measure the frequency spectrum of electrical noise. As we shall see in the next section, the variance can be obtained from the integral of the spectrum over all frequencies.

3. SPECTRA AND AUTOCORRELATION

The basic relation underlying the spectral analysis of stationary random processes is the Wiener–Khintchine theorem (Kittel, 1961; Lee, 1960) that relates the power spectrum to the autocorrelation function. The autocorrelation function, $C(T)$, is defined as the average of the product of a stationary random function multiplied by a delayed version of itself. If $X(t)$ is a stationary random process (one whose mean value is not drifting with time), then

$$C(T) = \lim_{A \to \infty} (2A)^{-1} \int_{-A}^{A} X(t)X(t+T)dt \tag{4}$$

$C(T)$ is the average over a long time interval, $2A$, of the product of the function X with a delayed version of itself. The autocorrelation function defined by equation 4 is thus a measure of the persistence of the random events.

The Wiener–Khintchine theorem states that the power spectrum of a stationary random process and the autocorrelation function are Fourier transforms of each other.

$$W(f) = 4 \int_{0}^{\infty} C(T) \cos (2\pi fT) dT \tag{5a}$$

$$C(T) = \int_{0}^{\infty} W(f) \cos (2\pi fT) df \tag{5b}$$

This theorem is significant because it is often convenient to measure the spectrum of electrical noise, but often easier to relate the autocorrelation to theoretical models of the underlying stochastic process. Figure 2 shows several examples of random processes with their autocorrelation functions and spectra.

4. CHANNEL NOISE

As an example of how a power spectrum is derived from a random process, we will consider the fluctuating current $X(t)$ through a single channel which opens and closes at random as shown in Fig. 3A. The derivation follows that of Machlup (1954). We note that this particular random process gives zero for

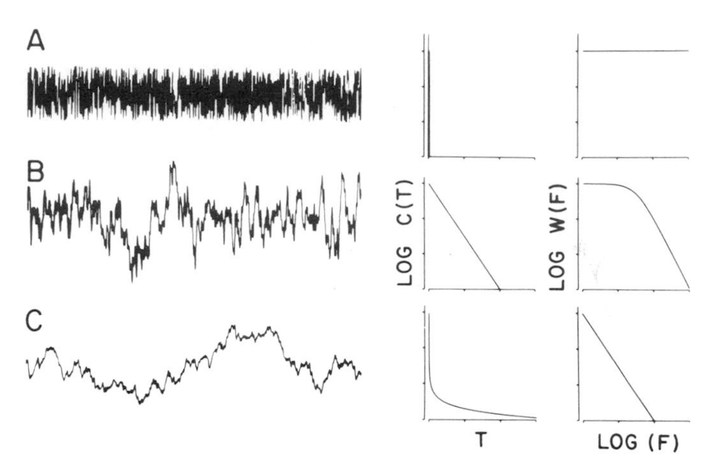

FIGURE 2. Some commonly encountered random time series and their corresponding autocorrelations, $C(t)$, and power spectra, $W(f)$. Part A represents "white" or delta-function-correlated noise. Part B represents a low-pass-filtered white noise or Lorentzian noise, and part C represents $1/f$ or flicker noise.

the product $X(t) \cdot X(t + T)$ when there are an odd number of transitions during the interval between t and $t + T$ and a^2 for the product when there are an even number of transitions. To get the autocorrelation function we must determine the probability of observing an even number of transitions in a given time.

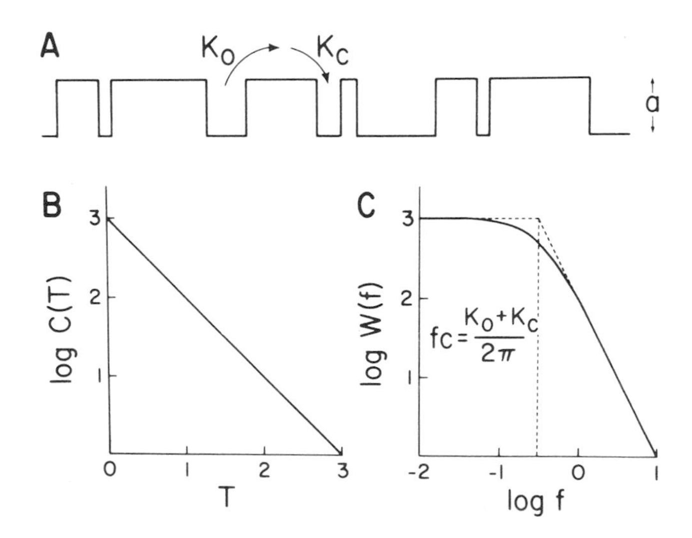

FIGURE 3. The two-state channel as a random process. (A) Sketch of a typical time series. k_o and k_c are the transition rates discussed in the text. (B) Autocorrelation function for process in A. (C) Power spectrum for process in A. The "cutoff" frequency, f_c, is indicated in the figure at the intersection of the high- and low-frequency asymptotes.

The random variable X is the current which flows through the channel; X takes on the values 0 and a. There are transition probabilities per unit time, k_o (closed–open) and k_c (open–closed). We want to find the autocorrelation function of $X(t)$, taken with respect to its mean value. The mean value of X is

$$\langle X \rangle = \lim_{A \to \infty} (2A)^{-1} \int_{-\infty}^{\infty} X(t)dt = \frac{k_o a}{k_o + k_c} \tag{6}$$

which can be seen by merely equating the number of openings over a long time to the number of closings. That is, the rate of opening is equal to k_o times the probability of being closed, and the rate of closing is equal to k_c times the probability of being open. Since

$$P\,(\text{open}) + P\,(\text{closed}) = 1 \tag{6a}$$

and

$$k_o\, P\,(\text{closed}) = k_c P\,(\text{open}) \tag{6b}$$

then

$$P\,(\text{open}) = k_o\,/\,(k_o + k_c) \tag{6c}$$

The autocorrelation function is given by

$$C(T) = \lim_{A \to \infty} (2A)^{-1} \int_{-A}^{A} [X(t) - \langle X \rangle][X(t+T) - \langle X \rangle]\,dt$$

$$= \lim_{A \to \infty} (2A)^{-1} \int_{-A}^{A} X(t)X(t+T)dt - \langle X \rangle^2 \tag{7}$$

Because $X(t)$ is a stationary process, the integrals of $X(t)$ and $X(t + T)$ can both be replaced by $\langle X \rangle$, as in equation 6.

To evaluate the integral in the expression for $C(T)$, we recall that $X(t)$ can have only the values 0 and a, so that the integrand has only the values 0 and a^2. The only situation in which the integrand is equal to a^2 is when the channel is open at t and at $t + T$. Thus, the integral is given by the product of $a^2 \times$ (the probability that the channel is open at any time) $\times$ (the conditional probability that an initially open channel undergoes an even number of jumps in an interval of duration T). That is,

$$\lim_{A \to \infty} \int_{-A}^{A} X(t)X(t+T)dt = \left(\frac{k_o}{k_o + k_c}\right) a^2\, P_e\,(T) \tag{8}$$

Here $P_e(T)$ is the conditional probability that an even number of transitions took place in time T, given the channel was open at time zero.

To obtain $P_e(T)$, consider $P_e(T + dT)$ where dT is such a short interval that the probability of more than one transition in dT is negligible. Letting $P_o(T)$ = the probability of an odd number of jumps in T and given that the channel is open at time zero,

$$P_e\,(T + dT) = P_e(T) \times P\,(0 \text{ jumps in } dT) + P_o(T) \times P\,(1 \text{ jump in } dT)$$

But we know $P_e(T) + P_o(T) = 1$, the probability of an additional closed–open jump in dT is $k_o dT$, and the probability of an additional open–closed jump in dT is $k_c dT$. Hence we obtain

$$P_e(T + dT) = P_e(T)\,(1 - k_c dT) + [1 - P_e(T)]\;k_o dT$$

Upon rearrangement, this becomes

$$dP_e/dT + (k_o + k_c)P_e = k_o \tag{9}$$

The solution of this differential euqation is

$$P_e(T) = k_o/(k_o + k_c) + [P_e\,(0) - k_o/(k_o + k_c)]\,\exp\,[-(k_o + k_c)\,T] \tag{10}$$

Here $P_e(0)$ is tautologically equal to 1, since it is the probability of being open at $t = 0$, given open at $t = 0$. We now have all of the terms in the expression for $C(T)$,

$$\begin{aligned} C(T) &= [k_o a^2/(k_o + k_c)]\,P_e(T) - [k_o a/(k_o + k_c)]^2 \\ &= [k_o k_c a^2/(k_o + k_c)]^2 \exp\,[-(k_o + k_c)\,T] \end{aligned} \tag{11}$$

Recalling that $k_o/(k_o + k_c) = \Theta$ and $k_c/(k_o + k_c) = 1 - \Theta$, and letting $\tau = (k_o + k_c)^{-1}$, we see that

$$C(T) = a^2\Theta(1 - \Theta)\,\exp\,(-T/t)$$

To obtain the power spectrum, we substitute into equation 4, to get

$$\begin{aligned} W(f) &= 4a^2\Theta(1 - \Theta) \int_0^\infty \exp\,(-T/\tau)\,\cos\,(2\pi f T)\;dT \\ &= 4a^2\,\Theta(1 - \Theta)\tau/(1 + 4\pi^2 f^2 \tau^2) \end{aligned} \tag{12}$$

The spectrum of equation 12 is plotted in Fig. 3C. This characteristic spectrum is often called a Lorentzian and, as the spectrum of the simplest kinetic process, plays a central role in the analysis of channel noise. We have dwelt on the details of this derivation because it contains the main ideas of the more powerful and more general stochastic treatments of channel noise (Conti and Wanke, 1975; Neher and Stevens, 1977; Colquhoun and Hawkes, 1977).

In general, one may envisage more complex kinetic schemes, as in electrically excitable channels, in which the channels undergo transitions among various states. In these schemes, the autocorrelation function will be a weighted sum of exponentials and the spectrum a weighted sum of Lorentzians. Although there is now a considerable literature (see reviews cited above) giving predictions for various kinetic models, in practice the multiple-Lorentzian curves are difficult to resolve uniquely.

Not all conductance–fluctuation spectra need be Lorentzian. Consider the example of a Poisson sequence of pulses with shape, $F(t)$, as was discussed for photon noise or noise emanating from secreted packets of transmitter. It can be shown (Rice, 1954) that the spectrum of a Poisson sequence of pulses of identical shape is given by

$$W(f) = 2\nu \mid S(f) \mid^2 \tag{13a}$$

where $S(f)$ is the Fourier transform of the unit pulse shape:

$$S(f) = \int_{-\infty}^{\infty} F(t) \exp(2\pi jft)\, dt \tag{13b}$$

Equation 13 states that the power spectrum depends on the shape of the underlying unit pulse. For the particular case of a unit pulse with exponential shape, the spectrum is in fact Lorentzian, but other pulse shapes will give other spectra. The spectrum for the noise emanating from single-photon responses is an interesting example in which the spectrum may fall off much faster than the Lorentzian at high frequencies (Lamb and Simon, 1977).

As an application of equation 13, we can calculate the channel noise of equation 12 in a different way. Consider a very long train of rectangular pulses. The distribution of pulse durations will be exponential. This can be appreciated by analogy to a radioactive decay process. Each pulse duration is representative of a decay of the open state with no memory of the time of opening. To utilize equation 13, we go through the record and form a subrecord containing only events of duration between T and $T + dT$. If the collection of such subrecords for all values of T were a collection process obeying Poisson statistics, equation 13 would give the spectrum for each subrecord. From equation 13b, the spectrum of a Poisson sequence of rectangular pulses of duration T can be shown to be

$$S_T(f) = 2\nu(a/\pi)^2(\sin \pi fT/f)^2 \tag{14}$$

To get the spectrum of the original process, we sum all the component spectra, making use of the fact that the various pulse durations are exponentially distributed. The spectrum of the overall process is thus the result of integrating equation 14 over the distribution of pulse widths,

$$W(f) = \int_0^\infty S_T(f)[\tau^{-1} \exp(-T/\tau)]\, dT$$
$$= (2\nu/\tau)\,(a/\pi f)^2 \int_0^\infty \sin(\pi f T) \exp(-T/\tau)\, dT \tag{14a}$$

This integral can be evaluated as

$$W(f) = 4\tau \operatorname{var}(I)/[1 + (2\pi\tau f)^2]$$

which is just the Lorentzian spectrum of equation 12.

5. EXPERIMENTAL METHODS FOR MEASURING AND ANALYZING CONDUCTANCE FLUCTUATIONS

5.1. Voltage Fluctuation Measurements

Voltage noise is the simplest type of measurement to make but may be difficult to interpret, since changes in membrane potential can cause changes in more than one ion-conducting system (Poussart, 1969). Some of the problems of simultaneously exciting several types of channels can be eliminated by appropriate pharmacology and ion substitution.

Consider the case of recording voltage fluctuations in response to a chemical agonist (Sachs and Lecar, 1973). A cell is penetrated with a recording microelectrode, and the signal is recorded at low gain to get the mean potential. The AC components (i.e., the fluctuations) are recorded at high gain through appropriate filters which cut off both the DC components and the high frequency instrument noise.

Agonists are introduced either by bath application or, more conveniently, by using iontophoretic pipettes. The voltage variance is measured and compared to the mean voltage displacement to get an estimate of the voltage change per event. Often, the voltage noise is filtered by the membrane capacitance, leading to an underestimate of the variance.

The limitations of voltage noise measurement can be seen by considering the equivalent circuit of Fig. 4. In the figure, $\tilde{g}(t)$ represents the fluctuating channel conductance as a function of time, and E is the reversal potential of the channels (relative to rest). The current noise correlation function is

$$\int_{-\infty}^{\infty} I(t)I(t + T)dt = E^2 \int_{-\infty}^{\infty} \tilde{g}(t)\tilde{g}(t + T)dt$$

so that according to the Wiener–Khintchine theorem, equation 5, the current noise spectrum is proportional to the spectrum of conductance fluctuations,

$$W_I(f) = E^2 W_g(f)$$

Voltage noise, however, can be obtained by solving the circuit equation

$$C_m(dV/dt) + [R_m^{-1} + \tilde{g}(t)]V = \tilde{g}(t)E \qquad (15)$$

Here $\tilde{g}$ is the time-varying conductance induced by the agonist. We can write $\tilde{g} = \bar{g} + \delta g(t)$, where $\bar{g}$ is the mean, and δg represents fluctuations about the mean. In general, $\delta g << \bar{g}$, and $\bar{g}$ will be large or small compared to R_m^{-1}depending on agonist dose.

If for simplicity we assume $\bar{g}(t) << R_m^{-1}$, and transform equation 15 to the frequency domain, we will be able to get an expression for the voltage spectrum as

$$W_v(f) = W_I(f)R_m^2/[1 + (2\pi R_m C_m f)]^2$$
$$= E^2 R_m^2 W_g(f)/[1 + (f/f_m)]^2 \qquad (16)$$

We see that determination of the frequency dependence of the physical quantity of interest, $W_g(f)$, from $W_v(f)$ depends on whether or not the channel fluctuations fall off at frequencies lower than $f_m = 1/2\pi R_m C_m$. If the conductance

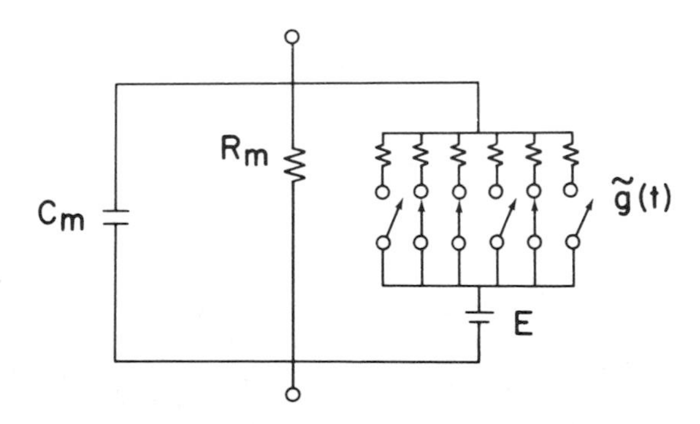

FIGURE 4. Equivalent circuit for membrane containing channels which randomly open and close. The element $\tilde{g}(t)$ is the equivalent randomly varying conductance. (After Lecar and Nossal, 1971.)

fluctuations persist to higher frequencies than f_m, the observed voltage spectrum will not show evidence of the channel time constant.

5.2. Sources of Background Noise

Even in the absence of applied agonist, there can be considerable noise on the amplified voltage record. There are three major sources of background noise: electrode noise, extraneous membrane noise, and amplifier noise. In addition, agonist iontophoresis may introduce extraneous noise, and there is noise in the current recording system. We shall now discuss each of these five sources of extraneous noise in turn.

5.2.1. Electrode Noise

The noise of the voltage recording electrode has two components, thermal (or Johnson) noise and $1/f$ noise. Thermal noise power is proportional to resistance. This noise is found in all conducting materials and is caused by Brownian motion of the charge carriers. The magnitude of thermal noise is given by $V_{\mathrm{rms}} = (4kTRB)^{1/2}$, where V_{rms} is the root mean square of the voltage fluctuations, k is Boltzmann's constant (1.38×10^{-16} erg deg^{-1}), T is absolute temperature, R is resistance, and B is the bandwidth of the recording system. Thermal noise is a general property of any conductor in equilibrium and hence is the irreducible minimum electrical noise. Figure 5 shows the Johnson noise expected for different resistances and bandwidths. A convenient number to remember is that a 10-MΩ resistor produces 5 μV rms of noise in a 100 Hz bandwidth. Reduction of the Johnson noise can be accomplished by reducing either the bandwidth or the electrode resistance. This noise places a theoretical limit on the accuracy of any electrical measurement and serves as a standard for amplifier performance. As long as the later stages of amplification contrib-

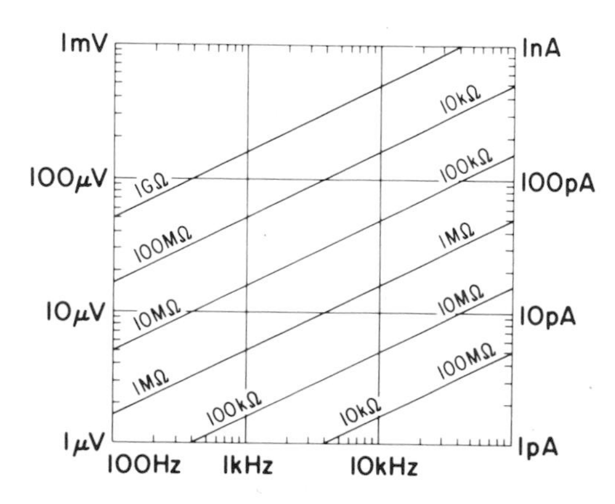

FIGURE 5. Nomogram for calculating rms Johnson noise for resistors at room temperature. Voltage and resistances along the left-hand ordinate refer to voltage noise (for the appropriate resistances). The right-hand ordinate similarly refers to current noise. The bandwidth of the noise is indicated on the abscissa as the cutoff frequency of a single-pole low-pass filter. (After Neher and Stevens, 1977.)

ute significantly less noise than the electrode Johnson noise, no improvement in the amplifier performance can improve the system signal-to-noise ratio.

The $1/f$ component of electrode noise is characteristic of many nonequilibrium systems. An interesting discussion of the properties of $1/f$ noise can be found in a *Scientific American* article by Gardner (1978). DeFelice and Firth (1971) investigated the $1/f$ noise associated with microelectrodes and found that when the electrode-filling solution is of a different concentration than the external solution, $1/f$ noise becomes significant at low frequencies. The noise in excess of the Johnson noise is of the form

$$W_e(f) = b/f^{\alpha}$$

where $W_e(f)$ is the square of the rms noise voltage per cycle across the electrode, f is frequency, and α is constant between 0.7 and 1. The constant b varies as the square of the voltage across the electrode and varies with the ratio of the ionic concentrations inside and outside of the electrode, increasing with an increase in ionic concentration gradient. For example, a 3 M KCl-filled electrode is noisier in 3mM KCl than in 30 mM KCl. The excess noise is not strongly dependent upon the particular ionic species involved.

Interestingly, the electrode noise was not found to be minimum under zero-current conditions, but at a potential between the zero current potential and the potassium Nernst potential. The noise decreased by a factor of 10 for a 3 M/100 mM KCl gradient (see DeFelice and Firth, 1971, Fig. 4) with the electrolyte filling solution 60 mV negative to the zero-current value. The decreases in noise were even larger for more dilute external solutions. DeFelice and Firth (1971) suggest that the observed electrode noise may be reduced by one of two methods: (1) biasing the electrode toward the Nernst potential for the more mobile ion, or (2) decreasing the concentration of the filling solution to match the ionic strength of the external medium.

When the internal and external solutions are identical, the system is in equilibrium and exhibits only Johnson noise. Thus there is a crossover point in the electrode noise versus filling solution ionic strength. As the ionic strength of the electrode solution is reduced toward the external ionic strength, the $1/f$ noise decreases, but the Johnson noise increases because of the increase in resistance. The $1/f$ component, which is most significant at low frequencies, tends to fall beneath the Johnson noise at high frequencies. "High" and "low" are a function of the experimental parameters. For example, using DeFelice and Firth's (1971) Fig. 4 for a 26-MΩ electrode at 10 Hz, the Johnson noise would be about 6×10^{-14} V^2/Hz. For a zero current measure, the $1/f$ component of noise would be about 10 times higher. At 100 Hz, the total electrode noise is only about twice the Johnson noise.

Considerations of electrode noise are important particularly in two experimental conditions: extracellular recording and one-electrode voltage clamp.

Extracellular pipettes are often filled with high salt solutions (2 M NaCl) to lower the resistance and the Johnson noise. The increased $1/f$ noise, however, can defeat the aim of low Johnson noise, and it is often better to use lower concentrations of salt in the electrode.

5.2.2. Extraneous Membrane Noise

The second component of background noise is produced by the cell itself and is predominantly of $1/f$ character. Generally. the cell $1/f$ noise is larger than the electrode noise (Sachs and Lecar, unpublished data). This component can be reduced, along with the electrode and amplifier noise, by taking the difference between the response in the presence and the absence of stimulus. The background noises are uncorrelated with the response, so that variances and spectra are additive. For voltage-activated channels, this procedure is not adequate, since the background noise varies with applied voltage (Fishman and Moore, 1977). Agonist-induced fluctuations (Anderson and Stevens, 1973) do not appear to introduce significant $1/f$ noise. Perhaps this is because the conductance fluctuations are so large that they cover up the $1/f$ noise for frequencies of interest (usually greater than 1 Hz).

5.2.3. Amplifier Noise

A third component of background noise is produced by the input stage of the voltage follower. With the ultra-low-input-current JFET devices currently available, such as the Analog Devices AD515, the input noise is below the thermal noise of even low-resistance microelectrodes at frequencies above 10 Hz and is generally negligible in comparison to other sources. If negative capacity neutralization is applied, however, the electrode amplifier noise may become dominant at several hundred hertz.

5.2.4. Noise Associated with Iontophoresis

One important source of extraneous noise that must be checked in any experiment is the noise produced by iontophoresis itself. The current flow through the drug-filled electrodes will produce large amounts of noise which may be coupled to the voltage electrode, either directly or through inter-electrode capacitance. One check for the influence of iontophoretic noise is to turn off the current, simply letting the drug diffuse out of the tip, to see if the difference spectrum is changed. Another check is to pass the iontophoretic current with the voltage electrode outside the cell and look for an increase in the noise recorded in the voltage-recording amplifier.

5.2.5. Current Noise

In a virtual-ground system of current recording using low-current-noise amplifiers, such as the AD515, the dominant noise will be the thermal noise of the feedback resistor in the current-to-voltage converter. Faster amplifiers may have a significant noise contribution of their own. In extreme circumstances, where noise of the feedback resistor is the limiting factor, the feedback resistor Johnson noise may be reduced (Sachs and Lecar, unpublished data) by cooling the resistor with liquid nitrogen to take advantage of the $T^{1/2}$-dependence of the Johnson noise, noting however that resistance itself may be temperature-dependent.

6. CURRENT NOISE UNDER VOLTAGE CLAMP IN A MICROELECTRODE SYSTEM

In this section we analyze an experiment in which current fluctuations are measured under voltage clamp control. In voltage clamp, a feedback circuit is employed to keep the membrane potential constant as the conductance fluctuates. On small cells, voltage clamp can be accomplished with two microelectrodes, one to monitor membrane potential and the other to deliver current. To simplify the discussion, we assume an idealized experimental arrangement in which the membrane potential recording follows the true membrane potential faithfully. In reality, the membrane potential is filtered by the resistance and capacitance of the voltage electrode, and a capacitative feedback compensation must be employed to regain a rapid response to changes in membrane potential. Figure 6 shows a schematic of a two-microelectrode voltage clamp for measuring the current noise spectrum of agonist-induced conductance fluctuations. The equivalent circuit of Fig. 6 shows capacitative coupling between the micro-

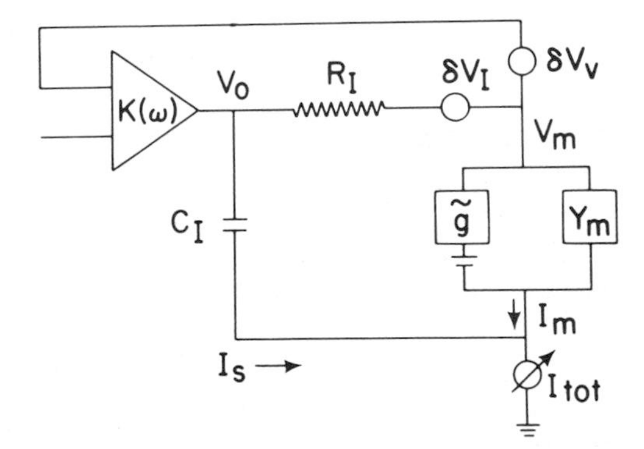

FIGURE 6. Equivalent circuit of a two-microelectrode voltage clamp used for measuring noise in a small cell, showing equivalent noise generators. (After Sachs and Lecar, 1977.) See text for details.

electrodes and ground but omits the stray capacitance between the current and voltage electrodes.

Extraneous noise sources are represented by equivalent noise generators labeled δV_c, δV_v, δV_I. These sources stand for amplifier (or command signal) noise, noise from the voltage microelectrode, and noise from the current electrode, respectively. The physical origin of the extraneous noise sources was discussed in the previous section. The noise source represented by δg is the fluctuating part of the membrane conductance that we regard as the signal of interest, all other noise sources being interferences. The equivalent noise source δV_I, which is meant to model the noise output of the current electrode, is a source of both Johnson noise and $1/f$ noise. When the clamp must deliver significant amounts of current, the $1/f$ noise dominates. In this case, the actual noise voltage of the electrode depends upon the current, and the noise generated will be more properly treated as a resistance fluctuation.

In order to simplify the analysis, we will consider all the signals in the frequency domain. Hence a signal written as δV is taken to mean

$$\delta V(\omega) = \int_{-\infty}^{\infty} \delta V(t) e^{j\omega t} dt$$

Here j is $\sqrt{-1}$, and $\omega = 2\pi f$ is frequency expressed in radians $\sec^{-1}$. The spectrum of the random variable $\delta V(t)$ is given by equations 4 and 5. Spectra, in the following discussion, will be denoted by subscripted variables such as $W_V(\omega)$ which have the dimensions of the subscripted variable squared per second. The spectra can be thought of as being obtained by some long-time averaging process as in equation 4. Thus,

$$W_V(\omega) = \lim_{A \to \infty} A^{-1} \mid \delta V(\omega) \mid^2$$

The purpose of the somewhat lengthy analysis which follows is to show how various noise components with known spectra contribute to the current-noise spectrum observed at the output of the voltage clamp. Readers not interested in the details of the analysis can skip to equation 31 for the major conclusions.

Referring to the schematic of Fig. 6, we see that the output voltage of the feedback system is (again remembering that all voltages are Fourier amplitudes)

$$V_o(\omega) = K(\omega)[V_c + \delta V_c(\omega) - V_m(\omega) - \delta V_v(\omega)] \tag{18}$$

We can equate this to the potential drop across the membrane and its access resistance through the current microelectrodes,

$$V_o(\omega) = V_m(\omega) + R_I I_m(\omega) + \delta V_I(\omega) \tag{19}$$

By equating the two expressions for $V_o(\omega)$, we can write the equation for $V_m(\omega)$,

$$V_m(\omega) = K(\omega)[K(\omega) + 1]^{-1}[V_c + \delta V_c(\omega) - \omega V_v(\omega)] - [R_I I_m(\omega) + \delta V_I(\omega)][K(\omega) + 1]^{-1} \tag{20}$$

A perfect voltage clamp would have very high gain $[K(\omega) \rightarrow \infty]$ for all frequencies of interest. In that case,

$$V_m(\omega) \cong V_c + \delta V_c(\omega) - \delta V_v(\omega) \tag{20a}$$

However, this ideal is not always a good approximation. For example, let us say $K \simeq 10^3$, and that R_I for a small cell can be 10 to 100 MΩ. Then for I_m of the order of 100 nA, the quantity $(R_I I_m)/(K + 1)$ can be of the order of 10 mV, which is not a negligible error. For simplicity, we confine ourselves to a case in which the amplifier gain, $K(\omega)$, can be made very high, so that the membrane potential is approximated by equation 20a. With this approximation, we see that the clamped membrane potential contains amplifier noise, input noise, and noise from the voltage microelectrode. Membrane potential fluctuations caused by noise in the current microelectrode, however, have been suppressed. The total current in the current detector is the sum of membrane current and the shunt current running to ground through the current-electrode capacitance. Thus

$$I = I_m + I_s \tag{21}$$

where

$$I_m = (V_c - E)(\bar{g} + \delta g) + Y_m(\omega)(V_c + \delta V_o) + \bar{g}\delta V_o \tag{22}$$

Here, δV_o is the total output noise produced by the clamp,

$$\delta V_o = \delta V_c - \delta V_v \tag{22a}$$

The agonist-induced conductance is given as $\bar{g} = \bar{g} + \delta g$, where $\bar{g}$ is the mean value and δg is the fluctuating part. The membrane admittance in the absence of agonist is given by $Y_m = G_m + j\omega C_m$.

The current flowing through the capacitative shunt is

$$I_s = j\omega C_I V_o(\omega) \tag{23}$$

where V_o, the output voltage of the clamp, contains the noise term δV_I, as can be seen in equation 19.

The equivalent noise generator of the current microelectrode produces both thermal noise and $1/f$ noise. The $1/f$ noise generally dominates when substantial amounts of current are flowing. The $1/f$ noise voltage generated is proportional to the current flowing through the electrode, so that δV_I can be written as

$$\delta V_I \equiv \bar{I}_m \delta R_I$$

where $\bar{I}_m$ is the average current and δR_I is an equivalent fluctuating resistance. Thus equation 23 can be rewritten as

$$I_s = j\omega C_I \left[V_c + I_m(\omega) R_I + \bar{I}_m \, \delta R_I(\omega) \right] \tag{24}$$

We can write both I_m and I_s as the sum of an average part and a fluctuating part, so that

$$I_m = \bar{I}_m + \delta I_m \tag{25}$$

where

$$\bar{I}_m = (V_c - E)\bar{g} + V_c Y_m(\omega) \tag{25a}$$

$$\delta I_m = (\bar{g} + Y_m)\delta V_o + (V_c - E)\delta g \tag{26}$$

Similarly,

$$I_s = \bar{I}_s + \delta I_s \tag{27}$$

where

$$\bar{I}_s = j\omega C_I (V_c + \bar{I}_m R_I) \tag{27a}$$

$$\delta I_s = j\omega C_I (\bar{I}_m \delta R_I + R_I \, \delta I_m) \tag{28}$$

In the foregoing we have dropped all terms containing the product of two noise terms; since the noise sources are assumed to be uncorrelated, their cross products will not contribute to the noise spectrum.

The total noise current recorded at the detector is given by

$$\begin{aligned} \delta I(\omega) &= \delta I_m + \delta I_s \\ &= (1 + j\omega \tau_I)[(V_c - E)\delta g + j\omega\tau_I \, \delta I_I + (\bar{g} + Y_m) \, \delta V_o] \end{aligned} \tag{29}$$

where $\tau_I = R_I C_I$ is the time constant of the current microelectrode, and δI_I is the current noise generated by the microelectrode, now written as

$$\delta I_I = \bar{I}_m \, \delta R_I / R_I \tag{30}$$

The current noise is seen to be composed of three terms multiplied by a distortion factor $(1 + j\omega\tau_I)$: the membrane conductance fluctuation noise, the shunt noise, and the combined electrode and amplifier noise. To obtain the power spectrum, we take the absolute value squared of $\delta I(\omega)$, recalling, again, that cross terms do not contribute. Thus, the power spectrum of the total current is

$$W_T(\omega) = (1 + \omega^2\tau_I^2)[(V_c - E)^2 W_g(\omega) + \omega^2\tau_I^2 \bar{I}_m^2 W_R(\omega) + |Y_m + \bar{g}|^2 W_0(\omega)] \tag{31}$$

where $W_R(\omega)$ denotes the spectrum appropriate to the resistance fluctuations in the current electrode.

Equation 31 is similar to an equation given in an earlier publication (Sachs and Lecar, 1977). In that treatment, however, the experimental situation envisioned was one in which the agonist-induced conductance was small compared to the resting membrane conductance. In the present discussion, we do not make this restriction. The difference between the two situations is that some of the background noise which would have been canceled by the measurement of a difference spectrum (spectrum in the presence of agonist minus spectrum without agonist) for the small-dose case will not be entirely canceled in the more general case.

In the absence of agonist, we do not expect $W_R(\omega)$ and $W_0(\omega)$ to change, but we do expect the $\bar{I}_m$ and g to be different. With no agonist, the power spectrum of the background is given by

$$W_B(\omega) = (1 + \omega^2\tau_I^2)\,[\omega^2\tau_I^2 \bar{I}_m'^2 \, W_R(\omega) + |Y_m|^2 \, W_0(\omega)]. \tag{31a}$$

Here $\bar{I}_m = Y_m V_c + \bar{g}(V_c - E)$, and $\bar{I}'_m$ is the same expression in the absence of agonist: $\bar{I}'_m = Y_m V_c$. The difference spectrum is obtained by subtracting equation 31a from equation 31:

$$\begin{aligned} W_D(\omega) &= W_I(\omega) - W_B(\omega) \\ &= (1 + \omega^2\tau_I^2)[(V_c - E)^2 W_g(\omega) + (V_c - E)^2\omega^2\tau_I^2 \\ &\quad \cdot g(g + 2G_m)\, W_R(\omega) + \bar{g}\,(\bar{g} + 2G_m)\, W_0(\omega)] \end{aligned} \tag{32}$$

We note that when g is small, the last two terms of equation 32 are small, and $W_D(\omega)$ is just the desired conductance spectrum. Figure 7 shows an exper-

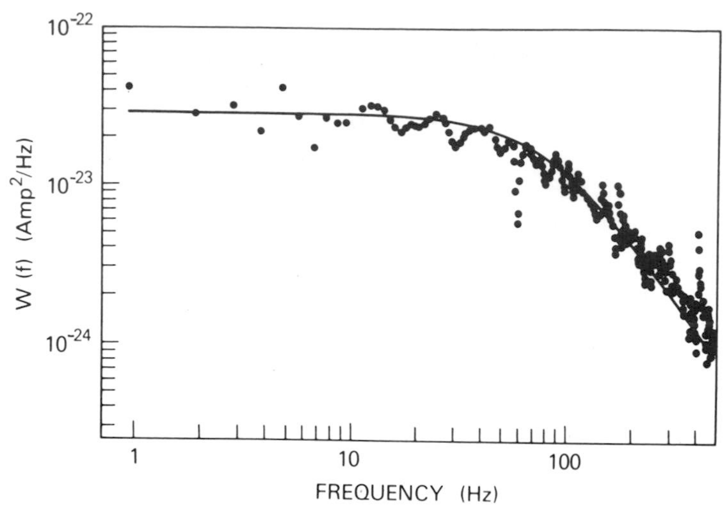

FIGURE 7. Noise spectrum for vinblastine-treated chick pectoral muscle grown in tissue culture. (After Sachs and Lecar, 1977.)

imental difference spectrum obtained when tissue-cultured muscle is exposed to a small steady dose of an agonist. Thus for the small-dose case one can obtain the full benefit of the difference spectrum. In order to evaluate optimum conditions for accurate difference spectra, one must also include the dependence of $W_g(\omega)$ on agonist dosage. From equation 12, we can determine that $W_g(\omega)$ is proportional to $\bar{g}(1 - \bar{g}/\bar{g}_{\max})$. Thus, for nonsaturating agonist dosages, all of the terms in equation 32 have a linear increase. Even when the extraneous noise terms are small, the true spectrum is distorted by the term $(1 + \omega^2\tau_I^2)$ caused by the shunt capacitance. This leads to large noise at high frequencies and limits the bandwidth over which the noise can be accurately measured.

Equation 32 may explain why the $1/f$ noise reported for different preparations is so variable. Whether or not it will cancel depends on how much the average membrane current varies under different conditions. If the analogous calculation were done for an electrically excitable cell instead of for a chemically excitable cell, the complications would have been a bit more severe. The impedances which appear as coefficients in equation 32 would be voltage-dependent and would certainly vary as clamp potential is varied.

In summary, in current-noise measurement under microelectrode voltage control, one must be wary of two sources of difficulty. First, the current noise spectrum will be multiplied by the distortion factor $(1 + \omega^2\tau_I^2)$ because the current microelectrode acts as a high pass filter. This limits stable operation to frequencies below $f = 1/2\pi\tau_I$, which in practice might be 1–10 kHz.

The second difficulty is that the background noise will only be canceled by a difference spectrum when the agonist-induced conductance change is small.

For large conductance changes, one can expect an increase in the residual background noise. The signal-to-noise ratio will tend to improve, however, because the channel switching will be much larger than the background noise.

The foregoing considerations describe a situation in which the experimenter is trying to minimize extraneous sources of noise. However, there is another source of uncertainty in the measurement of a noise spectrum having to do with statistical error in the determination of the power spectrum itself. The statistical error is discussed by Bendat and Piersol (1971). Generally the larger the amount of channel noise power collected, the more precise will be the determination of the spectrum. Difference spectra, when the background noise is large, may not be of much help when there is large statistical error, because the difference spectrum will contain the statistical deviations of the larger powers from which it was obtained.

7. MEASUREMENT OF SINGLE-CHANNEL CURRENTS

Many of the problems and ambiguities arising in noise analysis can be resolved by direct observation of single-channel responses. Single channels have been observed for a number of substances in lipid bilayers (Ehrenstein and Lecar, 1977). These results provided two essential incentives for developing techniques for recording single channels in biological systems: an expectation that single-channel fluctuations might be observable under the right circumstances, and an anticipation that some channels might be more complex than mere two-state conductances. For example, Alvarez *et al.* (1975) found that for EIM in artificial bilayers, the open state increased its conductance with temperature, whereas the low conductance state (which has a measurable conductance and is not really "off") decreases its conductance with increasing temperature. Also, multiple-state channels are common, such as those produced by alamethicin (Eisenberg *et al.*, 1973) and hemocyanin (Latorre *et al.*, 1975).

7.1. Methods of Recording Single Channels from Biological Systems

The essential elements necessary to record single responses are that the background rms noise level be at least three times lower than the channel amplitude and that there be so few channels that individual responses are distinguishable. The first requirment is more difficult to achieve. The diagram in Fig. 8 represents the equivalent circuit of an extracellular pipette pressed against the surface of a cell. Such an arrangement limits the cell area under observation and can provide a low-noise recording system. An excellent discussion of the patch recording technique is given in a review by Neher *et al.* (1978).

The signal, indicated by the current source i in Fig. 8 represents the current flowing through a single channel in the cell membrane. The channel current is divided into two branches, one going to the current-to-voltage converter with resistance R_p, and the other into the shunt with resistance R_s. The current captured by the pipette and thus measured by the amplifier is

$$I_{obs} = iR_s/(R_s + R_p) \tag{33}$$

Equation 33 shows that the observed signal increases with shunt resistance. In practice, single acetylcholine channels have been recorded with a ratio $R_p : R_s$ between 3 and 4, at which ratio 75 to 80% of the channel current is captured. Further increases in R_s would serve to increase the signal-to-noise ratio as long as the thermal noise generated by R_s is the dominant source of background noise.

Assuming the amplifier noise to be insignificant, and assuming that the feedback resistance of the current amplifier is much greater than the sum of patch electrode and shunt resistance, the interfering noise current referred to the amplifier input is approximately

$$I_N = [4kTB/(R_p + R_s)]^{1/2} \tag{34}$$

Dividing equation 33 by equation 34, we obtain an expression for the signal-to-noise ratio of the patch recording,

$$S/N = iR_s/[4kTB(R_p + R_s)]^{1/2} \tag{35}$$

Equation 35 shows that a high seal resistance increases the signal and simultaneously decreases the noise. In practice (Neher *et al.*, 1978), R_p and R_s are

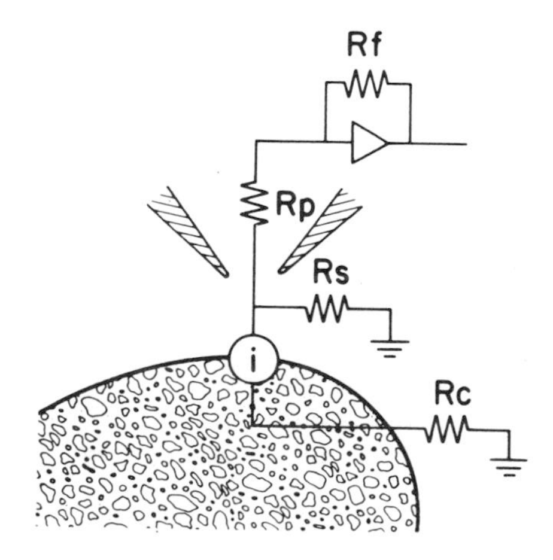

FIGURE 8. Schematic diagram of the electrical equivalent circuit of the patch recording system. See text for details.

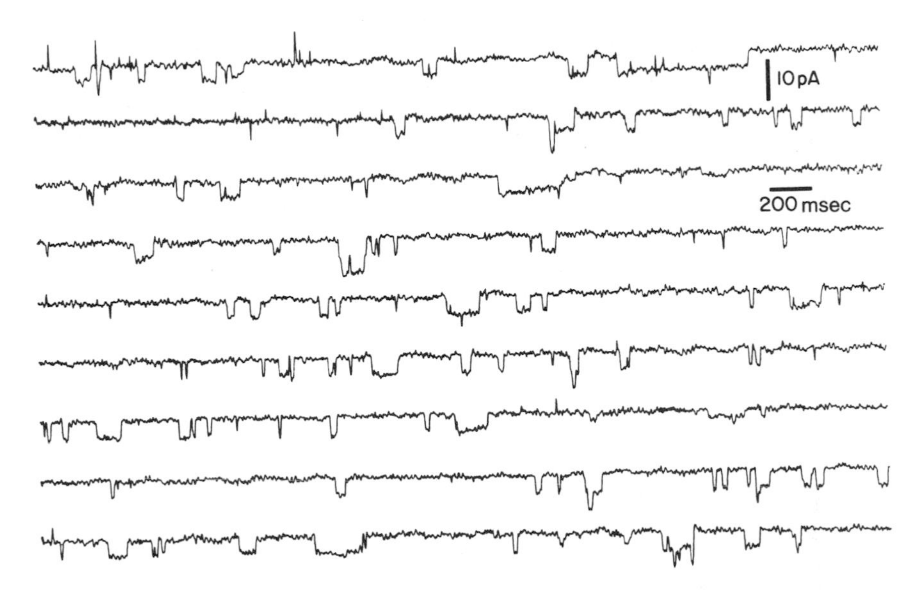

FIGURE 9. Single channel currents recorded from tissue-cultured muscle cells using the technique of Neher and Sakmann (1976a): 0.5 μM suberyldicholine, 20°C, cell unclamped with resting potential approximately −70 mV, sealing ratio approximately 75%, 100 Hz bandwidth established with eight-pole Butterworth filter. Recording made in Dulbecco's phosphate-buffered saline.

of the same order of magnitude, and small changes in either one can spell the difference between success and failure.

Tissue culture cells make good candidates for single-channel experiments. Because these cells can be made essentially free of basement membrane and overlying supporting cells, the pipette can be pressed close to the cell surface, maximizing R_s. Single-channel responses recorded by Sachs and Neher from nicotinic channels in tissue-cultured chick muscle using suberyldicholine as an agonist are shown in Fig 9. Recently, single-channel responses have been studied both in chick pectoral muscle, using suberyldicholine (Nelson and Sachs, 1979) and in rat skeletal muscle, using carbamylcholine (Jackson and Lecar, 1979).

Extracellular pipette construction is discussed fully in the paper by Neher *et al.* (1978). The pipette must be as steeply tapering as possible, so that the pipette resistance is minimal. The tip should be fire-polished to assist in sealing against the cell surface. The orifice should be as small as possible (generally in the range of 1 to 2 μm). To date, surface coating of the pipettes has not helped to increase the quality of the seal. Some of the components tested include Sylgard (catalyzed and uncatalyzed), mineral oils, silicone oils, silanes, petroleum jelly, fluorocarbon waxes and oils, polyethyleneimine, and covalently linked concanavalin A. These discouraging results should not deter enterprising investigators, but are merely meant as a guide. The general experience is that sticky substances pick up debris from solution and from the cell surface, rendering the tip no longer sticky. Hydrophobic coats are unable to displace saline from the cell surface, since the cell surface is hydrophilic. Suction is the one method found to increase the seal resistance. However, for drug-activated channels, the pipette is filled with the activating agonist, and suction causes solution flow into the pipette, removing drug molecules from the region under study.

A few technical hints may be helpful. The figures given for minimum noise assume that: (1) The pipette interior is filled with the same saline as is being used outside, and that any agonist is present in trace amounts. If the interior solution differs from the exterior, the nonequilbrium noise of mixing will dominate the background noise. (2) There is no current flow (above that of the channels) flowing through the pipette. This is extremely important in practical terms, for if a current flows through the pipette, vibration will modulate R_s and cause a great excess of background noise. Thus the current-to-voltage converter should be referenced in the bath and not to system ground. The current-to-voltage converter should incorporate a means for offsetting the input potential. (3) If the patch pipette is applied to a voltage-clamped cell, the extracellular potentials must be extremely stable, since changes of only a few microvolts in the bath potential can drive picoamperes through a patch pipette of a few megohm resistance. A separate voltage clamp of the bath potential may help to reduce artifacts of clamp current (Sachs, unpublished data).

7.2. Analysis of Noise Data

Once one is able to record the fluctuations satisfactorily, the problem then arises of how to analyze them. The simplest procedure is to measure the rms value of the signal with an appropriate meter. Provided that the low frequency cutoff of the meter does not filter too much of the noise, the reading will represent the square root of the total signal variance. If the variance measurement in the presence of the stimulus is subtracted from the measurement in the absence of stimulation, the result will be the net stimulus-induced variance. The variance divided by the mean voltage change gives the unitary voltage change. Voltage noise measurements in which the applied agonist causes a significant change in the membrane potential must be corrected for the change in driving force through the channels as the membrane depolarizes (Sachs and Lecar, 1973). A paper by Stevens (1976) contains insights into the problems of correcting for nonlinear summation resulting from depolarization of the membrane.

Dionne and Ruff (1977) used straightforward variance measurements to determine endplate fluctuations. They showed that the fluctuations go to zero at the reversal potential of the mean current, proving that only one set of channels with one reversal potential is responsible for the endplate current. Crawford and McBurney (1976) used a home-built "variance" meter to record glutamate and aspartate fluctuations at the crab neuromuscular junction. They used a high-gain AC amplifier to isolate the fluctuations and fed the result into an analog squarer followed by an analog integrator. The output, recorded on a pen recorder, gave a running measurement of the variance. This very inexpensive approach can easily be used in laboratories without access to sophisticated signal-analysis equipment.

A serious problem with the total variance measurement is the lack of time or frequency information to guide the data interpretation. For example, power line noise contributes to the total variance, but it cannot be distinguished without spectral or time information. In a total variance measurement, $1/f$ noise is indistinguishable from channel noise whose spectrum falls off as $1/f^2$.

There are two approaches to analyzing the temporal behavior of the noise: autocorrelation and spectral analysis. The two methods are equal in information value since they are mathematically related by the Wiener–Kintchine theorem. The autocorrelation function is easier to visualize for time series, is easier and faster to compute (See Kam *et al.*, 1975, for inexpensive methods), and may be curve-fit somewhat more easily. For first-order relaxation processes, for example, the autocorrelation function is linear on a semilog plot.

Spectral analysis has two main advantages. Because the noise is described in frequency terms, equivalent noise sources may easily be introduced into circuit analysis (Sachs and Lecar, 1977). Since current and voltage noise are related by the transfer function of the system, measurement of one kind of noise can be combined with knowledge of the impedance to predict the other. In addition, the error distribution of the power spectral density (the error in estimating the mean value of the spectral density) is simply *chi*-squared distributed. The errors in the autocorrelation are not so simple to predict. Power spectra are usually displayed on a log–log plot of the variable squared (voltage, current, or conductance versus frequency). This gives, for simple first order relaxation processes, a plot of the form shown in Fig. 3.

There are several ways to perform the experimental operation of measuring the autocorrelation function. Probably the most flexible alternative is to use one of the readily available micro- or minicomputers, preferably with some storage device attached, most preferably, disk storage. Such systems are presently available for well under $10,000. The computer approach will easily permit data acquisition in a frequency range of up to 2 kHz. Small computers allow data manipulation and storage, but the actual calculation time is longer than the processing time of hard-wired signal analyzers. For most applications, however, the execution time for analysis should not be prohibitive, being only a few seconds for a power spectrum calculation of 1024 points. Computer programs for obtaining power spectra invariably make use of the fast Fourier transform algorithm (FFT), which greatly reduces the computation time over older methods (Bendat and Piersol, 1971; Cochran *et al.*, 1967).

A second approach to data gathering uses dedicated devices to measure the power spectra or autocorrelation functions. These devices cost from about \$8000 to \$50,000, depending on options. The dedicated devices are fast and simple to operate, but they often suffer from inconveniences in such data manipulations as taking difference spectra, evaluating correlations, providing storage capability, or fitting curves. An analog tape recorder is a valuable adjunct to these machines, since it provides storage of AC and DC data for off-line analysis.

The most inexpensive method for performing power spectral analysis is to use a narrow-band filter in front of an rms voltmeter. The filter is set at the frequency to be measured, and the rms value at that frequency recorded. Another frequency is then selected, and a new record taken. Alternatively, a tape-recorded version of the signal can be replayed through the system. This procedure suffers only from the slowness of the data processing. However, for relatively featureless data, such as the spectra associated with conductance fluctuations, the filter bandwidth can be opened to accept more power per measurement, and relatively few data points are needed (Sachs and Lecar, 1973). Excellent discussions of the modes and methods of random data acquisition are presented in the book by Bendat and Piersall (1971).

7.3. Applicability of Fluctuation Measurement

Let us summarize the major uses of noise measurement on excitable cells.

7.3.1. Unit Conductance

Foremost, the noise measurement gives an estimate of the conductance of a single channel in a manner that is relatively independent of the kinetic model used. Most models will predict a variance-to-mean ratio close to the average value of the single channel conductance. Reliable values for channel conductance are difficult to obtain by other means, such as titration with channel-blocking toxins (Keynes, 1975).

7.3.2. Channel Kinetics

In the case of channels having multiple conductances, fluctuation analysis is required to sort out the transitions among the various conductance states (Neher and Stevens, 1977). Noise measurements, because they are small signal measurements with a small ratio of peak-to-average power, permit channel kinetics to be measured when there is insufficient voltage clamp control for large signal measurements. This feature may be of considerable use to tissue culture systems where the small cell size and consequent high electrode resistance prevent the use of the large-open-loop-gain amplifiers needed to displace the membrane potential far from rest.

Noise is also of use in measuring kinetics when other macroscopic relaxation techniques are not applicable. For example, one might want to measure the channel open time or relaxation time for a nonphysiological agonist. This could be done by a delivering a brief iontopheretic pulse of agonist and analyzing the subsequent response, but tests would be needed to establish that the observed kinetics do not reflect diffusion to and from the receptor (Dionne and Stevens, 1976). Alternatively, one could use a voltage jump, if the channel kinetics were voltage sensitive (Adams, 1977). However, if the channel were only weakly coupled to the electric field across the membrane, voltage jump experiments would require excessively fast voltage rise times and amplitudes. In the noise measurement, thermal fluctuations represent small, local departures from equilibrium which relax back with the same rate constant as would be obtained from a macroscopic relaxation measurement.

7.3.3. Limitations

The main limitation of this technique providing low-noise instrumentation so that the signal can be seen above background. In most cases, this limits the measurements to a bandwidth of a few kilohertz.

A second limitation is that channel noise must be identified among all the extraneous noise sources. Typically this involves separating ion flows through different types of channels, so that a particular conducting system is studied in isolation. Often uncorrelated noise sources must be removed on the basis of frequency dependence. This comes up in the trivial case of power line interference which shows up as a set of sharp peaks superimposed on the broadband spectrum. Commonly, $1/f$ components must be separated from $1/f^2$-type components in order that a quantative analysis can be done.

8. SUMMARY OF MAJOR RESULTS OBTAINED BY FLUCTUATION ANALYSIS ON CHEMICALLY EXCITED SYSTEMS

1. In vertebrate muscle, the acetylcholine channel conductance is of the order of 25 S. It increases with temperature and is essentially independent of voltage. Glutamate-activated channels have been measured for invertebrate muscle and vertebrate nerve synapse giving a large spread of values. Some representative values of channel conductances determined by noise analysis are given in Table I.
2. The decay of endplate currents is approximately the same as the decay of thermally induced fluctuations implying that the decay rate of the endplate currents represents the closing rate of the channels and not the decay of transmitter in the cleft.
3. With the time resolution presently available, channel gating appears to be an all-or-none phenomenon.
4. Different agonists cause the channels to be open for different periods of time.

TABLE I. Some Recent Channel Conductance Measurements by Fluctuation Analysis

Channel	Conductance (pS)	Temp (°C)	τ (msec)	Potential for τ det. (mV)	Reference
Excised muscle					
Ach, frog muscle endplate	25	12	3.2	−70	Colquhoun *et al.* (1975)
Ach, denervated muscle, extrajunctional	15	8	11	−80	Neher and Sakmann (1976a)[a]
Ach, mouse muscle endplate	22	10	3.8		Dreyer *et al.* (1976a)
Ach, mouse muscle, extrajunctional	9.1	11	4		Dreyer *et al.* (1976a)
GABA, crayfish muscle	14	23	1	−60	Dudel *et al.* (1977)
Glutamate, locust muscle endplate	100	20	2		Anderson *et al.* (1976)
	133	21	2.8	−45	Patlak *et al.* (1979)[a]
Excised nerve					
Na, squid axon	4	6			Conti *et al.* (1975)
Na, frog node	7.9	13			Conti *et al.* (1976a,b)
	7.7	2			Sigworth (1977)
K, squid axon	12	6			Conti *et al.* (1975)
K, frog node	4	15			Beginisich and Stevens (1975)
Ca, *Aplysia* neuron	0.1				Akaike *et al.* (1978)
Tissue culture					
Ach, chick muscle, tissue culture	31	23	7.2	−80	Lass and Fischbach (1976)
	39	25	4	−80	Sachs and Lecar (1977)
	60	22	13 (sub)		Nelson and Sachs (1979)[a]
Ach, rat muscle, tissue culture	48	22	3.2 (carb)	−90	Jackson and Lecar (1979)[a]
GABA, mouse CNS neuron, tissue culture	18	26	20	−50	McBurney and Barker (1978)
Glycine, mouse CNS neuron, tissue culture	34		5		Barker and McBurney (1979)

[a]Measurements by single-channel fluctuations.

8.1. Some Surprises from Noise Analysis

With the initial results on the muscle endplate, it appeared that all postsynaptic channels might be expected to have conductances on the order of 25 pS. Working with glutamate-activated channels in locust muscle, Anderson *et al.* (1976) found channel conductances of 100 to 200 pS, a result that has recently been confirmed by single-channel measurements (Patlak *et al.*, 1979).

One of the most surprising and unexpected findings to emerge from the experiments on noise is the apparent sharp transition in the temperature dependence of channel conductance. Anderson and Stevens (1973) found a rather low temperature dependence in frog muscle (Q_{10} of about 2) over a relatively narrow temperature range. Sachs and Lecar (1977) found that tissue-cultured chick muscle has a higher temperature dependence (9 kcal/mole). In an almost identical tissue-cultured chick pectoral muscle preparation, Lass and Fischbach (1976; Fischbach and Lass, 1978a,b) found abrupt breaks in the Arrhenius plot of conductance. This was interpreted as a phase transition of the conducting channels. The temperature dependences of channel conductance and open time were recently measured for single-channel fluctuations (Nelson and Sachs, 1979), but there was no evidence of a phase transition.

Another surprising finding from fluctuation analysis at the endplate is that the channel open time appears to depend upon the permeating ion. Van Helden *et al.* (1977) found that among the alkali ions, Cs, Rb, and Li, applied as replacements for sodium in the extracellular medium, the effects of the different ions on channel open lifetime were in the opposite order to their permeability through the channel. For example, Li, the least permeable, kept the channel open longest.

Another interesting result derived from fluctuation measurements is that nonpolar aliphatic molecules, which can act as general anesthetics, alter the open time of the channels without significantly changing the conductance. Gage *et al.* (1975) proposed that the anesthetics change the membrane dielectric constant, thus altering the electric field across some part of the channel structure. The altered transmembrane electric field would cause the channels to change their open time. Since vertebrate muscle endplate channels increase their open time with hyperpolarization, and channels in crustacean muscle endplates decrease their duration with hyperpolarization, the effect of general anesthetics should depend upon the type of voltage dependence involved. In this regard, the work of Dionne and Stevens (1975) shows that even if channels are open longer under the influence of hyperpolarization, the average induced endplate currents may decrease. This is explained by them according to the model proposed by Magleby and Stevens (1972) in which the probability of opening and the probability of closing are voltage dependent. If the opening rate is slowed by hyperpolarization, then a short duration dose of agonist, such as produced by nerve stimulation, will open relatively fewer channels than would be opened in a depolarized preparation. The same phenomenon may apply to other synapses, so that anesthetics may be associated with a decreased opening rate rather than decreased channel open time.

The noise results discussed here all were based on the assumption that postsynaptic channels operate independently. In denervated muscle, this seems to be borne out (Neher and Sakmann, 1976b). However, denervated muscles have a relatively low density of channels ($<10^3/\mu m^2$). At the endplate, the packing is much higher, perhaps $10^4/\mu m^2$, so that interaction might be possible (Kuffler and Nicholls, 1976). Scheutze *et al.* (1978) measured the open time of channels in regions of different density in tissue-cultured chick muscle and found no significant changes over three orders of magnitude in channel density. In contrast to these results, the junctional and extrajunctional channels of intact muscle appear to form separate populations with different conductances and open-state lifetimes (Sakmann, 1975; Dreyer *et al.*, 1976b).

8.2. Conductance Fluctuation Studies in Tissue Culture

At this writing, noise analysis has been applied to several tissue-cultured preparations: chick pectoral muscle (Sachs and Lecar, 1973, 1977; Lass and Fischbach, 1976; Fischbach and Lass, 1978a,b; Scheutze *et al.,* 1978), mouse CNS neurons (McBurney and Barker, 1978), and human skeletal muscle (Cull-Candy *et al.,* 1978; Bevan *et al.,* 1978). Similar experiments are probably proceeding in a number of other laboratories, and we will not attempt to review the specific results obtained with tissue-cultured cells. Instead we shall confine ourselves to a discussion of some of the advantages and limitations of membrane noise measurement in tissue culture.

We list six advantages of tissue-cultured cells:

1. In sparse cultures, the precise geometry of cells may be observed, and agonists applied to particular regions of interest, as Scheutze *et al.* (1978) have done.
2. The lack of connective tissue and supportive cells is a distinct advantage for recording the response of single channels.
3. Tissue culture offers the advantage of being able to monitor trophic effects on channels by manipulating growth conditions (temperature, nutrients, etc.).
4. Cell geometry may be manipulated. Sachs and Lecar (1977) used vinblastine to cause muscle cells to become spherical, thus making them more closely isopotential than the normal cablelike myotubes. Fischbach and Lass (1978a) produced round cells by growing chick skeletal muscle culture on nonadhesive substrates.
5. Cultures growing as monolayers have a negligible diffusion barrier for bath-applied drugs, and in properly designed chambers, drugs of known concentration can be introduced in well under 1 sec. This provides a means of examining dose–response relationships.
6. The ability to perform somatic fusion experiments and the develop-

ment of clonal mutants promises to provide some new information on the biosynthesis and expression of ionic channels in cells.

The main limitations of tissue culture techniques in studying electrophysiological mechanisms arise from the small size of the usual cells and the consequent high resistance of the electrodes needed to study them. It is not possible to make an accurate high-speed, low-noise, voltage clamp using high-resistance electrodes. We expect that the ability to grow excitable cells free of connective tissue and supporting cells will lead to voltage clamp experiments using low-impedance extracellular patch voltage clamps such as have been done with Aplysia cells (Akaike *et al.*, 1978). In the meantime, it should be possible with existing techniques to examine fluctuations under voltage clamp conditions when the average currents are in the nanoampere range. In addition, it should be possible to gain information about a variety of ionic channels by using the combination of intracellular voltage recording to give estimates of the charge transfer and extracellular recording to give relaxation time information.

ACKNOWLEDGMENTS. We thank Ms. M. Schaefer for her patience and care in preparing the manuscript. We also wish to thank Drs. D. Nelson, C. Morris, M. Jackson, B. Wong, and T. Smith for their helpful comments.

REFERENCES

Adams, P. R., 1977, Relaxation experiments using bath-applied suberyldicholine, *J. Physiol. (Lond.)* **268**:271–289.

Akaike, N., Fishman, H. M., Lee, K. S., Moore, L. E., and Brown, A. M., 1978, The units of calcium conduction in helix neurones, *Nature* **274**:379–382.

Alvarez, D., Latorre, R., and Verdugo, R., 1975, Kinetic characteristics of the excitability-inducing material channel in oxidized cholesterol and brain lipid membranes, *J. Gen. Physiol.* **65**:421–439.

Anderson, C. R., and Stevens, C. F., 1973, Voltage clamp analysis of acetylcholine produced current fluctuations at frog neuromuscular junction, *J. Physiol. (Lond.)* **235**:655–691.

Anderson, C. R., Cull-Candy, S. G., and Miledi, R., 1976, Glutamate and quisqualate noise in voltage-clamped locust muscle fibres, *Nature* **261**:151–152.

Barker, J. L., and McBurney, R. N., 1979, GABA and glycine may share the same conductance channel on cultured mammalian neurones, *Nature* **277**:234–236.

Baylor, D. A., Hodgkin, A. L., and Lamb, T. D., 1974, The electrical response of turtle cones to flashes and steps of light, *J. Physiol. (Lond.)* **242**:685–727.

Beginisich, T., and Stevens, C. F., 1975, How many conductance states do potassium channels have? *Biophys. J.* **15**: 843–846.

Bendat, J. S., and Piersol, A. G., 1971, *Random Data: Analysis and Measurement Procedures*, Wiley–Interscience, New York.

Bevan, S., Kullberg, R. W., and Rice, J., 1978, Acetylcholine-induced conductance fluctuations in cultured human myotubes, *Nature* **273**:469–470.

Cochran, W. T., Cooley, J. W., Favin, D. L., Helms, H. D., Kaenel, R. A., Lang, W. W., Maling, G. C., Jr., Nelson, D. E., Rader, C. M., and Welch, P. D., 1967, What is the fast Fourier transform?, *Proc. IEEE* **55**:1664–1674.

Colquhoun, D., and Hawkes, A. G., 1977, Relaxation and fluctuation of membrane currents that flow through drug-operated channels, *Proc. R. Soc. London* **B199:**231–262.

Colquhoun, D., Dionne, V. E., Steinbach, J. H., and Stevens, C. F., 1975, Conductance of channels opened by acetylcholine-like drugs in muscle end-plate, *Nature* **253:**204–206.

Conti, F., and Wanke, E., 1975, Channel noise in nerve membrane and lipid bilayers, *Q. Rev. Biophys.* **8:**451–506.

Conti, F., DeFelice, L. J., and Wanke, E., 1975, Potassium and sodium ion current noise in the membrane of the squid giant axon, *J. Physiol. (Lond.)* **248:**45–82.

Conti, F., Hille, B., Neumcke, B., Nonner, W., and Stämpfli, R., 1976a, Measurement of the conductance of the sodium channel from current fluctuations at the node of Ranvier, *J. Physiol. (Lond.)* **262:**699–727.

Conti, F., Hille, B., Neumcke, B., Nonner, W., and Stämpfli, R., 1976b, Conductance of the sodium channel in myelinated nerve fibres with modified sodium inactivation, *J. Physiol. (Lond.)* **262:**729–742.

Crawford, A. C., and McBurney, R. N., 1976, On the elementary conductance event produced by L-glutamate and quanta of the natural transmitter at the neuromuscular junctions of *Maia squindo, J. Physiol. (Lond.)* **259:**205–225.

Cull-Candy, S. G., Miledi, R., and Trautmann, A., 1978, Acetylcholine-induced channels and transmitter release at human endplates, *Nature* **271:**74–75.

DeFelice, L. J., 1977, Fluctuation analysis in neurobiology, *Int. Rev. Neurobiol.* **20:**169–208.

DeFelice, L. J., and Firth, D. P., 1971, Spontaneous voltage fluctuations in glass microelectrodes, *IEEE Trans. Biomed. Eng.* **18:**339–351.

Derksen, H. E., 1965, Axon membrane voltage fluctuations, *Acta Physiol. Pharmacol. Neerl.* **13:**373–466.

Derksen, H. E., and Verveen, A. A., 1966, Fluctuations of resting neural membrane potential, *Science* **151:**1388–1389.

Dionne, V. E., and Ruff, R. L., 1977, Endplate current fluctuations reveal only one channel type at frog neuromuscular junction, *Nature* **266:**263–265.

Dionne, V. E., and Stevens, C. F., 1975, Voltage dependence of agonist effectiveness of the frog neuromuscular junction: Resolution of a paradox, *J. Physiol. (Lond.)* **251:**245–270.

Dodge, F. A., Jr., Knight, B. W., and Toyoda, J., 1968, Voltage noise in Limulus visual cells, *Science* **160:**88–90.

Dreyer, F., Müller, K. D., Peper, K., and Sterz, R., 1976a, The M. Omohyoideus of the mouse as a convenient mammalian muscle preparation, *Pflügers Arch.* **367:**115–122.

Dreyer, F., Walther, G., and Peper, K., 1976b, Junctional and extra junctional acetylcholine receptors in normal and denervated frog muscle fibers, *Pflügers Arch.* **366:**1–9.

Dudel, J., Finger, W., and Stettmeier, H., 1977, GABA-induced membrane current noise and the time course of the inhibitory synaptic current in crayfish muscle, *Neurosci. Lett.* **6:**203–208.

Ehrenstein, G., and Lecar, H., 1977, Electrically gated ionic channels in lipid bilayers, *Q. Rev. Biophys.* **10:**1–34.

Ehrenstein, G., Lecar, H., and Nossal, R., 1970, The nature of the negative resistance in biomolecular lipid membranes containing excitability inducing material, *J. Gen. Physiol.* **55:**119–133.

Eisenberg, M., Hall, J. E., and Mead, C. A., 1973, The nature of the voltage-dependent conductance induced by alamethicin in black lipid membranes, *J. Membr. Biol.* **14:**143–176.

Fischbach, G. D., and Lass, Y., 1978a, Acetylcholine noise in cultured chick myoballs, *J. Physiol. (Lond.)* **280:**515–526.

Fischbach, G. D., and Lass, Y., 1978b, A transition temperature for acetylcholine channel conductance in chick myoballs, *J. Physiol. (Lond.)* **280:**527–536.

Fishman, H. M., 1973, Relaxation spectra of potassium channel noise from squid axon membranes, *Proc. Natl. Acad. Sci. USA* **70:**876–879.

Fishman, H.M., and Moore, L. E., 1977, Ion movements and kinetics in squid axon. II. Spontaneous electrical fluctuations, *Ann. N.Y. Acad. Sci.* **303:**399–423.

Gage, P. W., McBurney, R. N., and Schneider, G. T., 1975, Effects of some aliphatic alcohols on the conductance change caused by a quantum of acetylcholine at the toad end-plate, *J. Physiol. (Lond.)* **244:**409–429.

Gardner, M., 1978, White and brown music, fractal curves and one-over-*f* fluctuations, *Sci. Am.* **239** (April):16–31.

Hagins, W. A., 1965, Electrical signs of information flow in photoreceptors, *Cold Spring Harbor Symp. Quant. Biol.* **30:**403–418.

Hill, T. L., and Chen, Y. D., 1972, On the theory of ion transport across the nerve membrane. IV. Noise from the open-close kinetics of K^+ channels, *Biophys. J.* **12:**948–959.

Jackson, M. B., and Lecar, H., 1979, Single postsynaptic channel currents in tissue cultured muscle, *Nature* **282:**863–864.

Kam, Z., Shore, H. B., and Feher, G., 1975, Simple schemes for measuring autocorrelation functions, *Rev. Sci. Instrum.* **46:**269–277.

Katz, B., and Miledi, R., 1972, The statistical nature of the acetylcholine potential and its molecular components, *J. Physiol. (Lond.)* **224:**665–699.

Katz, B., and Miledi, R., 1973, The characteristics of "endplate noise" produced by different depolarizing drugs, *J. Physiol. (Lond.)* **230:** 707–717.

Katz, B., and Miledi, R., 1975, The effect of procaine on the action of acetylcholine at the neuromuscular junction, *J. Physiol. (Lond.)* **249:** 269–284.

Keynes, R. D., 1975, Organization of the ionic channels in nerve membranes, in: *The Nervous System* (D. R. Tower, ed.), Vol. I: *The Basic Neurosciences,* pp. 165–175, Raven Press, New York.

Kittel, C., 1961, *Elementary Statistical Physics,* John Wiley & Sons, New York.

Krawczyk, S., 1978, Ionic channel formation in a living cell membrane, *Nature* **273:**56–57.

Kuffler, S. W., and Nicholls, J. G., 1976, *From Neuron to Brain,* Sinauer Associates, Sunderland, Massachusetts.

Lamb, T. D., and Simon, E. J., 1977, Analysis of electrical noise in turtle cones, *J. Physiol. (Lond.)* **272:**435–468.

Lass, Y., and Fischbach, G. D., 1976, A discontinuous relationship between the acetylcholine-activated channel conductance and temperature, *Nature* **263**:150–151.

Latorre, R., Alvarez, O., Ehrenstein, G., Espinoza, M., and Reyes, J., 1975, The nature of the voltage-dependent conductance of the hemocyanin channel, *J. Membr. Biol.* **25:**163–182.

Lecar, H., and Nossal, R., 1971, Theory of threshold fluctuations in nerves, *Biophys. J.* **11:**1048–1067, 1068–1083.

Lecar, H., Ehrenstein, G., and Latorre, R., 1975, Mechanism for channel gating in excitable bilayers, *Ann. N.Y. Acad. Sci.* **264:**304–313.

Lee, Y. W., 1960, *Statistical Communication Theory,* John Wiley & Sons, New York.

Lindemann, B., and Van Driessche, W., 1977, Sodium specific membrane channels of frog skin are pores: Current fluctuations reveal high turnover, *Science* **195:**292–294.

Loewenstein, W. R., Kanno, Y., and Socolar, S. J., 1978, Quantum jumps of conductance during formation of membrane channels at cell–cell junction, *Nature* **274:**133–136.

Machlup, S., 1954, Spectrum of a two-parameter random signal, *J. Appl. Phys.* **25:**341–343.

Magleby, K. L., and Stevens, C. F., 1972, A quantitative description of end-plate currents, *J. Physiol. (Lond.)* **223:**173–197.

McBurney, R. N., and Barker, J. L., 1978, Gaba-induced conductance fluctuations in cultured spinal neurones, *Nature* **274:**596–597.

Mueller, P., Rudin, D. O., Tien, H. T., and Wescott, W. C., 1962, Reconstitution of excitable cell membrane structure in vitro, *Circulation* **26:**1167–1177.

Neher, E., and Sakmann, B., 1976a, Noise analysis of drug induced voltage clamp currents in denervated frog muscle fibres, *J. Physiol. (Lond.)* **258:**705–729.

Neher, E., and Sakmann, B., 1976b, Single-channel currents recorded from membrane of denervated frog muscle fibers, *Nature* **260:**799–802.

Neher, E., and Steinbach, H. H., 1978, Local anaesthetics transiently block currents through single acetylcholine-receptor channels, *J. Physiol. (Lond.)* **277:**153–176.

Neher, E., and Stevens, C. F., 1977, Conductance fluctuations and ionic pores in membranes, *Annu. Rev. Biophys. Bioengr.* **6**:345–381.

Neher, E., Sakmann, B., and Steinbach, J. H., 1978, The extracellular patch clamp: A method for resolving currents through individual open channels in biological membranes, *Pflügers Arch.* **375**:219–228.

Nelson, D. J., and Sachs, F., 1979, Single ionic channels observed in tissue cultured muscle, *Nature* **282**:861–863.

Patlak, J., Gration, K. A. F., and Usherwood, P. N. R., 1979, Single glutamate-activated channels in locust muscle, *Nature* **278**:643–645.

Poussart, D. J. M., 1969, Nerve membrane current noise: Direct measurements under voltage clamp, *Proc. Natl. Acad. Sci. USA* **64**:95–99.

Rice, S. O., 1954, Mathematical analysis of noise, in: *Selected Papers on Noise and Stochastic Processes* (N. Wax, ed.), pp. 133–294, Dover Publications, New York.

Ruff, R. L., 1976, Local anesthetic alteration of miniature endplate currents and endplate current fluctuations, *Biophys. J.* **16**:433–439.

Sachs, F., and Lecar, H., 1973, Acetylcholine noise in tissue culture muscle cells, *Nature* [*New Biol.*] **246**:214–216.

Sachs, F., and Lecar, H., 1977, Acetylcholine-induced current fluctuations in tissue-cultured muscle cells under voltage clamp, *Biophys. J.* **17**:129–143.

Sakmann, B., 1975, Noise analysis of acetylcholine-induced currents in normal and denervated rat muscle fibre, *Pflügers Arch.* **359**:R89.

Sakmann, B., and Boheim, G., 1979, Alamethicin-induced single channel conductance fluctuations in biological membranes, *Nature* **282**:336–339.

Schuetze, S. M., Frank, E. F., and Fischbach, G. D., 1978, Channel open time and metabolic stability of synaptic and extrasynaptic acetylcholine receptors on cultured myotubes, *Proc. Natl. Acad. Sci. USA* **75**:520–523.

Schwartz, E. A., 1977, Voltage noise observed in rods of the turtle retina, *J. Physiol. (Lond.)* **272**:217–246.

Siebenga, E., Meyer, A., and Verveen, A. A., 1973, Membrane shot noise in electrically depolarized nodes of Ranvier, *Pflügers Arch.* **341**:87–96.

Sigworth, F. J., 1977, Sodium channels in nerve apparently have two conductance states, *Nature* **270**:215–267.

Stephens, C. L., Jackson, M. B., and Lecar, H., 1980, Single C channel currents in living cells, *J. Immunol.* **124**:1541.

Stevens, C. F., 1976, A comment on Martin's relation, *Biophys. J.* **16**:891–895.

Ten Hoopen, M., and Verveen, A. A., 1962, Nerve-model experiments on fluctuation in excitability, *Proc. Int. Conf. Cybernetics in Medicine* (N. Wiener and J. P. Schade, eds.), pp. 8–21, Elsevier, Amsterdam.

Van Helden, D., Hamill, O. P., and Gage, P. W., 1977, Permeant cations alter endplate channel characteristics, *Nature* **269**:711–713.

Verveen, A. A., and DeFelice, L. J., 1974, Membrane noise, *Prog. Biophys. Mol. Biol.* **28**:189–265.

Verveen, A. A., and Derksen, H. E., 1968, Fluctuation phenomena in nerve membrane, *Proc. IEEE* **56**:906–916.

Verveen, A. A., and Derksen, H. E., 1969, Amplitude distribution of axon membrane noise voltage, *Acta Physiol. Pharmacol. Neerl.* **15**:353–379.

Verveen, A. A., Derksen, H. E., and Schick, K. L., 1967, Voltage fluctuations of neural membrane, *Nature* **216**:588–589.

Wong, F., 1978, Nature of the light-induced conductance changes in neutral photoreceptors of Limulus, *Nature* **276**:76–78.

6

Nerve Cells in Clonal Systems

YOSEF KIMHI

1. INTRODUCTION

A piece of neuronal tube from a frog embryo successfully grown by Harrison (1907) in a drop of clotted lymph was the first "tissue culture" of a nerve cell *in vitro.* Since then, the art of culturing cells for extended periods of time in defined media has been constantly improved and refined. Consequently, the neurobiologist has been provided with the opportunity to study the development of nervous tissue cells and their interactions with other cells under relatively controlled conditions. Primary cultures of brain cells (amphibian, avian, and mammalian) grown as monolayers or rotating aggregates, explants and single cell cultures from the spinal cord, sympathetic and sensory ganglia, and mixed cultures of nerve and muscle cells have all yielded a large, invaluable volume of information concerning the steps that lead to establishing the neuronal network. However, the complexity of the nervous system and the consequent difficulty of separating homogeneous populations of living neurons from their associated satellite cells make the biochemical studies of differentiation a formidable task. The long quest for a simpler system that exhibits neuronal propperties and could still be manipulated under experimental conditions was partially satisfied with the establishment of continuous cell cultures derived from neuronal tumors.

Long-term culture of such cells originating from a human neuroblastoma was already reported in 1957 (Goldstein and Pinkel, 1957). But only after the establishment of a diversity of clones from the mouse C-1300 neuroblastoma tumor (Augusti-Tocco and Sato, 1969; Schubert *et al.,* 1969; Klebe and Ruddle, 1969) did this field rapidly gain popularity. Since then, numerous clones have been isolated and characterized. Electrophysiological studies have dem-

YOSEF KIMHI • Department of Neurobiology, The Weizmann Institute of Science, Rehovot, Israel.

onstrated the development of an excitable cell membrane; a variety of neurotransmitters and their metabolizing enzymes as well as receptors for these molecules have been identified; and fusion of neuroblastoma cells with other cell types has yielded new classes of hybrids that manifest properties different from the parental cells. Morphological changes similar to those observed when normal embryonic nerve cells are cultured *in vitro* were observed, and established functional synapses with muscle cells were demonstrated. These cells can be propagated easily, so sufficient quantities for most biochemical studies can be readily obtained.

Recently, Schubert *et al.* (1974) isolated various cell lines from tumors that were induced by administration of nitrosourea to pregnant rats, and Greene and Tischler (1976) adapted an NGF-sensitive, rat adrenal medulla tumor (pheochromocytoma) to grow in tissue culture. Thus, a variety of cloned cells differing in their properties is now available as a new tool in neurobiology.

Since these cells possess some morphological, biochemical, and electrical characteristics of neurons, they have been extensively used as a model system not only for the study of distinctive specific neurobiological properties but also as a model for nerve cell differentiation. However, the use of the term differentiation in studies of these tumor cells is frequently challenged. Much of the debate stems from one's definition of the term "differentiation." Some workers view differentiation as an expression of *any* specialized property or function for which the cell has become adapted during the development or maturation of the organism (see Wigley, 1975). Others justify the use of this term only when the cell attains its final functional characteristics, thus culminating an irreversible, unidirectional process composed of many steps (Waymouth, 1977).

The differentiating nerve cell *in vivo,* as any other cell in the organism, responds to internal and external signals regulating its replication, transcription, and translation mechanisms and controls. Its maturation involves development of specific morphological structures such as long fibers—axons and dendrites—which travel in some cases very long distances, recognize their specific partner cells, and establish stabilized connections—synapses. Nerve cell maturation involves the development of an electrically excitable membrane, the synthesis of enzymes to metabolize the various neurotransmitters, and special mechanisms that control their release. Different types of neurons attain their final functional forms at various stages of the embryonic life span, whereas some mature only after birth.

At what stage, then, do we decide that the cell "differentiates"? When it attains its characteristic morphology? When specific antigens show up on the membrane? When the level of a neurotransmitter-synthesizing enzyme is increased, or only when a functional synapse is established? Is there one definite stage at which the neuron becomes committed to a specific program, or can it change one course to another when and if an appropriate positive or negative signal is or is not received? Unless these questions are resolved, it is clear that the phenomenon of the outgrowth of an axonlike process from a nerve tumor cell will be regarded as an expression of a differentiated property by some, but will be termed a phenotypic modulation by others.

In essence the question is: are the changes observed in a tumor-derived cell grown *in vitro* a reflection of processes taking place in a normal differentiating cell *in vivo?* A proper answer requires constant comparative studies in which the results obtained in one system are tested in the other.

Furthermore, an alternative viewpoint can also be advocated: if neurobiologically interesting phenomena (Ca^{2+} currents, process and synapse formation, etc.) are effectively studied in neuroblastoma cells or the like, it is worthwhile to investigate these systems even if exact correlations with particular *in vivo* conditions cannot be demonstrated.

2. THE C-1300 MOUSE NEUROBLASTOMA

2.1. Clone Isolation and Some Characteristics

The C-1300 mouse neuroblastoma originated as a spontaneous tumor in the body cavity of an albino A/J mouse in the region of the spinal cord. It was isolated in 1940 by A. M. Claudman and maintained for many years by subcutaneous transfers. The tumor was rediscovered in 1968, whereupon several laboratories successfully isolated from it many clones of different properties. To select cells that were more adaptable to culture conditions and at the same time to maintain their function, Augusti-Tocco and Sato (1969) applied the alternate animal–culture passages technique (Buonassisi *et al.,* 1962). When placed in culture, the tumor cells underwent striking morphological changes, and many of them developed elongated processes similar to those characteristic of nerve cells. These processes began developing soon after subculture of the primary tumor and developed in time to form a complex network. After a few days in culture, round cells reappeared; they piled up forming loose clumps that were easily detached and floated.

Using the single cell plating technique described by Puck *et al.* (1956), clones NB41, NB41A, NB41B, and NB41C were thus isolated by Augusti-Tocco and Sato (1969). Schubert *et al.* (1969) isolated the cells after directly dispersing the solid tumor cells into bacteriological plastic petri dishes. In suspension, the cells retained the round cell morphology similar to that found in the tumor *in vivo* and divided rapidly. However, when transferred to a surface to which they could adhere (glass, collagen-treated plates, or commercial tissue culture dishes), most attached to the surface within 24 hr and sent out processes within 3 days. Cells were cloned twice by spreading a dilute cell suspension in agar and picking visible colonies.

Seeds *et al.* (1970) and Amano *et al.* (1972) reported a procedure that is a combination of the previsouly described techniques. Initially, an isolated colony of cells in agar was selected. Then these cells were further cloned by inclusion of single cells in stainless steel cylinders. In some cases, cells were added to petri dishes containing broken glass cover slips. Pieces containing a single cell were then transferred to a separate dish; a colony was obtained and then analyzed.

Initial work has already indicated that cultured cells derived from C-1300 mouse neuroblastoma tumor are indeed of neural origin. They have the capacity to synthesize neurotransmitters (Augusti-Tocco and Sato, 1969; Schubert *et al.*, 1969; Seeds *et al.*, 1970; Blume *et al.*, 1970), and their electrophysiological characteristics are essentially comparable to those of normal neurons (Harris and Dennis, 1970; Harris *et al.*, 1971; Nelson *et al.*, 1969, 1971a,b).

Isolation of more clones and subclones readily followed (Amano *et al.*, 1972; Prasad *et al.*, 1973c; Siman Tov and Sachs, 1972, 1975a; Breakefield and Nirenberg, 1974). It became evident that the various clones differ in their morphology, level of biosynthetic enzymes, and electrophysiological properties. Amano *et al.* (1972), for example, classified various clones as adrenergic when they were found to contain tyrosine hydroxylase activity (e.g., clone N1E-115), cholinergic when they showed high choline acetyltransferase activity (e.g., clone NS-20), or inactive when they were devoid of both enzymes (e.g., clone N-18). Although initial studies suggested that the expression of a gene required for the synthesis of one neurotransmitter restricted the expression of genes for synthesis of other neurotransmitters (Amano *et al.*, 1972; Minna *et al.*, 1972), subsequent studies proved that two activities may reside within one cell (Prasad *et al.*, 1973c; Hamprecht *et al.*, 1974; Breakefield *et al.*, 1975), as is the case of primary sympathetic neurons (Landis, 1976; Furshpan *et al.*, 1976).

The fact that pure genetic studies on the neuroblastoma cells are difficult, if not impossible, became clear quite early. A normal mouse cell contains 40 chromosomes. In the mouse neuroblastoma C-1300, which arose spontaneously in 1940, a value of 60 to 63 chromosomes with 3 to 5 marker chromosomes was determined by Warter *et al.* in 1974. Amano *et al.* (1972) isolated from the tumor many clones and subclones and found intraclonal, as well as interclonal, heterogeneity with respect to chromosome numbers. Modal values of 59 to 118 were found, for example, in six "cholinergic" clones, 104 and 192 in two adrenergic clones, and 100 to 120 in inactive clones. Warter *et al.* (1974) found that in 55 karyotypes of clone S-21 cells, chromosome number varied between 17 and 175, and the distribution exhibited two modal values. However, in spite of this anomaly and genetic instability, subsequent work showed that the cells maintain and express specific neuronal properties for extended periods. They thus enable detailed biochemical and electrophysiological studies.

The large volume of work addressing itself to practically every aspect of neurobiology makes the discussion of all of the studies that were carried out with cloned cells impossible. Certain aspects, such as electrophysiological and neurotoxicological studies, are dealt with in other chapters of this book. Other aspects, such as studies with cloned human neuroblastoma cells, aging *in vitro,* lipid metabolism, and ganglioside synthesis and function are not discussed in this chapter.

Several specialized reviews that treat various aspects of the mouse neuroblastoma system have been published in the last few years, and the reader is advised to refer to them: Breakefield (1976); Breakefield and Giller (1976); Fischbach and Nelson (1977); Haffke and Seeds (1975); Nelson (1977); Mandel *et al.* (1976); McMorris *et al.* (1973); Phelps and Pfeiffer (1973); Prasad (1975, 1977); Sato (1973); and Richelson (1975, 1976b).

Hamprecht (1976,1977) has reviewed in depth the studies on hybrid cells in which neuroblastoma cells are one of the parental cells. Many of these experiments are not discussed in this chapter.

2.2. Morphological Differentiation

2.2.1. General Features

One of the outstanding morphological features that distinguish nerve cells—the most polymorphic cells in the body—from other types of cells is the possession of long, fiberlike structures, the axons. Although some neurons have very short neurites, and some none at all, this particular cellular structure has been the focus of interest and intensive studies since the early days of modern neurobiology. During and after migration from the neural tube, the neuroblast and neurons establish connection with target cells by mechanisms of which very little is known.

It is obvious that the conditions and factors that direct and guide the sprouting axon *in vivo* seldom, if ever, exist in the nonphysiological, *in vitro* milieu. Whereas chemotrophic effects can be assumed to exist in a co-culture of sympathetic and muscle cells, none are likely to be found in a culture of cloned neuroblastoma cells. Nevertheless, a few cells in every culture send out long processes. What causes these cells to express this phenotype? How are these processes related to the axons and dendrites of normal cells? Does the dish surface play a role, and can one induce neurite growth under controlled conditions? What are the molecular events that take place when the cells "differentiate" morphologically?

2.2.2. Neurite Extension

Time-lapse cinematography has shown that the growth of the neurite tip from the cell body proceeds at a rate of approximately 75–125 μm/hr in serum-free medium (Seeds *et al.,* 1970). Schubert *et al.* (1973) observed that under these conditions the cells initially extended and retracted numerous filopodia at a rapid rate and moved actively about the surface of the culture dish; however, after several hours, some of the filopodia became stabilized and the cells became less mobile. Booher *et al.* (1973) noticed that the migration apparently proceeds in two stages: during the first stage, which lasts about 6 hr, the cells flatten out on the surface, extending small pseudopodic processes; in the second stage, which occurs between the 10th and 24th hours of development, process formation takes place mainly by migration of the cell body away from the more stationary process. Such a mechanism, the authors noted, differs from the one observed in a culture of dissociated peripheral neurons in which the cell body is relatively stationary, and the processes are elongated by outgrowth of the nerve fiber distally from the cell body. However, the authors did not comment about the mechanism by which a second neurite grows out from a cell that has already extended one neurite. Furthermore, on the basis of their observations, Schubert *et al.* (1973) concluded that the process is analogous to that observed in primary culture of neurons as reported by Nakai (1964).

A regular culture of neuroblastoma cells represents a random distribution of cells in different stages of their life cycle. The fact that 70–85% of the cells are capable of extending processes within 1 hr after the removal of the serum (Seeds *et al.*, 1970) indicates that the cells are capable of extending processes during most of the cell cycle.

2.2.3. Transfer of Suspension Cultures to Monolayers

Neuroblastoma cells can proliferate in suspension cultures. They grow as single cells or clusters of round cells that contain a large centrally placed nucleus with one to three nucleoli. The endoplasmic reticulum is diffuse, and ribosomes are dispersed. Electron micrographs indicate the presence of a mixed mitochondrial population as well as dense-core and clear vesicles. A few processes, 5–10 μm long and 1μm in diameter may protrude from the surface. Typical microtubules and microfilaments are sometimes seen in the cytoplasm (Schubert *et al.*, 1969; Augusti-Tocco *et al.*, 1970; McMorris *et al.*, 1973; Graham *et al.*, 1974; Graham and Gonatas, 1976).

Monolayer cultures are generally typified by many cells bearing long extensions that are designated as "neurites" since they can not be readily defined as either dendrites or axons. The size of the cells and number of neurites vary widely, even within a given culture. Examination of the morphologically mature cells has shown that the perikaryon is rather similar to that of the suspension-grown cells, although variable numbers of nuclei and nucleoli have been observed. The total number of ribosomal particles is higher, and, although polysomes are more frequent, free ribosomes are still abundant. The neurites contain mitochondria (normal as well as swollen), neurofilaments, and microtubules (their relative proportion seems to vary with the age of the cells), as well as many dense-core vesicles. Dense-core vesicles are found in varicosities along the neurites and at their tips in greater abundance than in the perikaryon. The pattern of concanavalin A binding sites, which is continuous in the suspension-grown cells, becomes discontinuous in surface-attached cells (McMorris *et al.*, 1973; Augusti-Tocco *et al.*, 1970; Augusti-Tocco and Chiarug:, 1976; Claude and Augusti-Tocco, 1970; Schubert *et al.*, 1973; Graham *et al.*, 1974, 1976; Ross *et al.*, 1975; Anzil *et al.*, 1977).

2.2.4. Serum Removal

Addition of serum is required for growth and maintenance of cells in culture. It is reasonable to assume that many unknown factors that suppress or induce changes in the phenotype of the cell are present in the serum. Sato (1975) formulated the hypothesis that the main role of serum in media is to provide groups of hormones or other accessory factors obligatory for survival and growth of the cell. Indeed, a serum-free synthetic medium to which only few hormones are added has recently been found to support growth of rat as well as mouse neuroblastoma cells in culture (Bottenstein and Sato, 1979). Similar media have been developed for other cell lines. It is quite possible that the use of such defined media to identify additional factors affecting the mode of differentiation of nerve cells *in vitro* will contribute to the understanding of the mechanisms that govern morphogenesis and synaptogenesis *in vivo.*

It appears that certain factors in the serum prevent the extension of neurites. Such factors have been found to be heat-stable and are not eliminated by conditioning the medium (Seeds *et al.*, 1970). The rate of cell division is the same in media containing 10% or 20% serum, but the neurite outgrowth in 20% serum is delayed by about 1 day compared to that in 10% serum (Schubert *et al.*, 1971a). Furthermore, cells that grow in 0.5% serum extend neurites more slowly than do those grown in serum-free medium.

Noticing that the number of neurite-bearing cells is inversely related to the number of dividing cells in the culture, Seeds *et al.* (1970) induced morphological differentiation by removing the serum from the growth medium. Within 30–60 min most cells possessed processes 25–100 μm in length. Relatively long neurites were found after 24 hr, and many cells with neurites up to 2 mm in length were evident after 4 days in serum-free medium. Once serum was added again, the neurites retracted completely within a few hours, and cell division was reinitiated. The outgrowth of axons in several types of cells has been shown to occur only after cessation of DNA synthesis (Jacobson, 1978). However, neurite outgrowth in serum-free medium is not a direct result of the inhibition of DNA synthesis: little or no differentiation occurred in the presence of DNA-synthesis inhibitors such as 5-FdU, mitomycin C, cytosine arabinoside (Ara-C), or thymidine at concentrations that partially or completely inhibit cell division (Schubert *et al.*, 1971a).

Extension of neurites in low-serum (0.5%) and serum-free conditions is not restricted to the neuroblastoma cells and has also been shown to occur in clones derived from rat neuronal tissue (Schubert, 1974). Some of the latter clones extended neurites very quickly, whereas others did so rather slowly, reflecting heterogeneity in the clonal properties.

Schubert *et al.* (1971a) found that neuroblastoma as well as other cells (glia, myoblasts) release a dialyzable compound that induces neurite outgrowth. They found that lactic acid, one of the possible metabolites, is capable of partially inducing a culture of cells to differentiate. They suggested that other secreted compounds may act synergistically with lactic acid to induce differentiation.

Our poor understanding of the role of serum in controlling cell growth and morphogenesis *in vitro* can be exemplified further by the finding that 60% of NB_2A cells undergo spontaneous morphological differentiation when grown in a medium supplemented with delipidated fetal calf serum but not in one with regular serum (Monard *et al.*, 1977; Lindsay and Monard, 1977). Moreover, when fatty acids, in particular oleic acid, are added to delipidated serum, this spontaneous morphological differentiation can be prevented.

2.2.5. Role of Substratum

Cells grown in suspension do not extend neurites; the transfer to a substratum to which the cells can adhere is a necessary, but not always sufficient, condition to induce neurite outgrowth.

It is obvious that the cell surface plays an important part in the adhesiveness of the cells. Formation of a stable bonding with the substratum may mimic the cellular microenvironment and may initiate changes in the cell's phenotype.

Augusti-Tocco and Chiarugi (1976) presented evidence that cells transferred from suspension to monolayer retain their surface heperan sulfate which is shed in suspension conditions. The sulfated polyanions may bind the cell to the substratum partially covalently and partially by calcium ions, thus forming a structure limiting the mobility of membrane components.

Some studies suggest that certain serum proteins interact with a receptor on the cell surface or mask some of the charges on the culture dish so that extension of the neurite is not possible (Seeds *et al.*, 1970; Schubert *et al.*, 1971a). A puzzling observation was made in this context by Miller and Levine (1972) who found that neuroblastoma cells (clones N-2, N-4, and N-18) were unable to extend neurites when plated on collagen. Native collagen has been employed as a supportive substratum in cell and tissue culture for many years. Augusti-Tocco and Sato used gelatin-coated dishes for isolating the first neuroblastoma clone (1969), and Schubert *et al.* (1969) found that suspension cells, after 3 days on collagen, sent out a neurite network. However, Miller and Levine reported that fluorodeoxyuridine (FdU) and bromodeoxyuridine (BrdU), which stimulate axonation in cultures grown on glass (see below, Section 2.2.10), did not overcome the inhibition of neurite outgrowth in cells adherent to collagen. Neurite sprouting occurred only after collagenase treatment. The reason for this anomaly is not known.

Actinomycin D, by an unexplained mechanism of action, was found to induce attachment and cell flattening in bacterial petri dishes (Schubert *et al.*, 1971a).

Ordered and defined columns of neuroblastoma cells (clone NB41A) connected by bipolar processes stretching away from the cell body were obtained by Cooper *et al.* (1976) who changed the properties of the substratum by making an array of silicon monoxide lines on plastic petri dishes. The cells were less flattened compared to cultures conventionally grown on Falcon tissue culture dishes.

2.2.6. Cyclic AMP and Its Modulators

The cyclic nucleotides cyclic AMP and cyclic GMP play key roles in regulation of many cellular processes in growth and differentiation. They are known to have multiple sites of action in the cell and to modulate various cells differently, and they may exert different effects at various stages in the lifespan of the cell. Addition of cyclic nucleotides, of their synthetic analogues, of prostaglandins, or of cyclic nucleotide phosphodiesterase inhibitors (PDI) affects the phenotype of practically every cell system that has been tested in tissue culture. These changes are explained by elevation of the intracellular level of the nucleotides, activation of the adenylate cyclase system, and/or reduction in the rate of cyclic AMP degradation.

Studies with other experimental systems suggest that cyclic AMP may cause morphological changes by affecting microtubules or microfilaments. Observations along this line in the neuroblastoma system were made quite early by Prasad and Hsie (1971) and Furmanski *et al.* (1971) who noted that the cyclic AMP-induced neurite formation was blocked by vinblastine and cytochalasin B.

Various clones were later shown to differentiate in the presence of cyclic AMP and its synthetic analogues, prostaglandins, inhibitors of cyclic nucleotide phosphodiesterase, or combinations of these drugs. The morphological changes are accompanied by enhancement of electrical excitability, synapse formation with muscle cells, changes in enzyme levels, protein synthesis, transport, phosphorylation, and nucleic acid synthesis [Chalazonitis and Greene, 1974 (clone N1E-115); Chalozonitis *et al.*, 1975, 1977 (hybrid clone NX-31); Daniels and Hamprecht, 1974 (clones NL-1F, NL-308 and NL-303); Furmanski *et al.*, 1971; Furmanski, 1973 (clone NB-60); Glazer and Schneider, 1975 (clone T-59); Kumar *et al.*, 1975 (clones NBP_2, NBA_2, and IMR-32);

Lim and Mitsunobu, 1972 (clone NB41A); Nelson *et al.*, 1976 (clone NG108-15); Prasad, 1972a,b,c, 1974, 1977; Prasad *et al.*, 1973a–c, 1975c,d, 1976a (clones NBA_2, NBE, and others); Puro and Nirenberg, 1976 (clone NG 108-15); Reiser *et al.*, 1977 (clones NG108-5 and NG 108-15); Simantov and Sachs, 1975a–c; Waymire *et al.*, 1978a,b (clone NBD-2); Wexler and Katzman, 1975 (clone N20E); Zwiller *et al.*, 1977 (clone M1 and others)].

The scope of this chapter does not allow a critical and detailed evaluation of the large volume of data concerning the effects of cyclic nucleotides on neuroblastoma cells. Various aspects of this problem are covered in the reviews by Friedman (1976), Prasad (1977), and Nathanson (1977), and in the book by Daly (1977).

Experimental results that clearly show the dramatic effects of cyclic AMP and related compounds should, nevertheless, be treated with caution, since the levels of the substances in the medium are generally much higher than those determined intracellularly (cyclic AMP is added on the order of 1 mM). Impermeability and instability of the compounds may justify these high concentrations, but some adverse effects cannot be ruled out. Further, some degradation products may affect the treated cell. Butyric acid, one of the metabolites of dibutyrl cyclic AMP, was shown [contrary to previous reports by Furmanski *et al.* (1971) and Prasad and Mandal (1972)] to induce neurite formation (Glazer and Schneider, 1975; Schneider, 1976). Butyric acid also caused elevation of AChE activity (Prasad and Vernadakis, 1972; Blume *et al.*, 1970; Kates *et al.*, 1971), an effect not related to neurite formation. How is the effect of butyric acid related to that of dibutyryl cyclic AMP? Are the various properties characterizing the differentiated cells expressed independently, or is one change a prerequisite for another? Is cyclic AMP the only trigger of the differentiation process, or are there alternative mechanisms and pathways? One should bear in mind that other treatments (such as those described in Sections 2.2.7–2.2.10) seem to induce neurite formation via mechanisms that do not involve cAMP. Conclusions about exact mechanisms will probably be reached only when more information has been compiled.

2.2.7. Hormones

Treatment of neuroblastoma cells with cortisol (0.1 μg/ml), growth hormone, or thyroxine (1 μg/ml) had no discernible effect on neurite outgrowth or the rate of protein synthesis. In combinations of two, the hormones had a small effect on protein synthesis rate. When the three were present together, there was an increase of protein synthesis followed by an extensive neurite outgrowth after a 5- to 7-day lag. The effect was irreversible after one week, could be manifested in medium containing either 1% or 10% fetal calf serum, but was not seen in the presence of 15% horse serum (DeVellis *et al.*, 1970, 1971).

Dexamethasone, at 50 μg/ml, induced morphological differentiation in about two-thirds of clone NBP_2 cells. Growth was inhibited before the complete expression of morphological differentiation (Sandquist *et al.*, 1978). Steroids without glucocorticoid activity (testosterone and 17β-estradiol at 10 μg/ml) had no effect on differentiation, although they inhibited growth. The time course of differentiation induced by dexamethasone differed from that observed with cyclic AMP or phosphodiesterase inhibitors and was more gradual. The treated cells had larger perikarya, showed an increased protein content per cell, and an increased catecholamine-specific fluorescence. This effect on fluorescence could not be induced by hydrocortisone, even at 20 μg/ml.

2.2.8. Glial Factor

Glial cells are thought to guide neurons migrating from the germinal zone (Sidman and Rakic, 1973). Nonneuronal cells have been shown in a number of *in vitro* systems to support growth and sustain viability of embryonic nerve cells (Ebendal and Jacobson, 1975, 1977). The finding that glioma C-6 cells release a factor that stimulates rapid neurite formation in neuroblastoma cells (Monard *et al.*, 1973) is, therefore, of great interest. A similar factor appears to be produced by normal brain cells in primary culture (Schurch-Rathgeb and Monard, 1978). Moreover, activity of this factor in adult rat brain is two to three times higher than that detected in conditioned medium from cloned glioma cells (C-6) or in primary brain cultures derived from five-day-old rat. Monard *et al.* (1975) have presented evidence that this factor is different from Nerve Growth Factor (NGF). The mechanism of action of this protease-sensitive factor is not clear, but it is somehow connected to the lipid content of the serum (Monard *et al.*, 1977; Lindsay and Monard, 1977). It should be noted here that not all clones of neuroblastoma cells respond to the glial factor.

2.2.9. X-Ray Irradiation

X-ray irradiation induces the development of neurites (Nelson *et al.*, 1976; Prasad, 1972b). This treatment affects differently the activities of various enzymes that are involved in neurotransmitter metabolism and RNA synthesis (Prasad and Mandal, 1972, 1973; Prasad and Vernadakis, 1972; Prasad *et al.*, 1975b).

2.2.10. Bromodeoxyuridine and Other Nucleotide Analogues

The expression of differentiated functions is inhibited when cartilage cells (Abbott and Holtzer, 1968), skeletal muscle cells (Lasher and Cohn, 1969), or melanoma cells (Silagi and Bruce, 1970) are treated by BrdU. However BrdU (at 1.6×10^{-6} M) induced neurite formation in neuroblastoma cells in a serum-supplemented medium (Schubert and Jacob, 1970). At this concentration, the agent did not inhibit cell division or DNA or RNA synthesis. Cells that were pretreated with BrdU when growing in suspension extended neurites upon their transfer to a surface to which they could attach, even in the absence of BrdU.

The possibility that BrdU exerts its effect as a consequence of its incorporation into DNA was ruled out by the observation that morphological differentiation is induced by the agent 6 or 20 hr after inhibition of DNA synthesis with either γ-β-D-arabinofuranosylcytosine (ara-C) or mitomycin C, which by themselves did not induce neurite formation (Schubert and Jacob, 1970).

Protein synthesis was found to be required in this process of neurite formation, but only small quantitative differences in protein composition were revealed by polyacrylamide gel analysis, and no specific function has been attributed to these proteins. Bromodeoxyuridine was found to induce the synthesis of at least one glycoprotein on the cell surface (Brown, 1971). It is interesting to note that this glycoprotein was also expressed in cells that differentiated in the absence of BrdU.

A progressive increase of oxygen consumption and a shift of carbohydrate metabolism from glycolytic to oxidative take place in the presence of BrdU (Ciesielski-Treska *et al.*, 1976).

Can BrdU induce neurite formation in other neuronal systems? Schubert (1974) examined five rat cell lines derived from CNS tumors and found that, in concentrations greater than 1×10^{-6} M, the cells assumed a flattened morphology typical of the BrdU-treated neuroblastoma cells. However, none of the rat cells extended neurites or in any way assumed a more complex morphology. The lack of response was specific: all lines extended neurites in low-serum (0.5%) and serum-free conditions and also responded to the addition of dibutyryl cyclic AMP. Although BrdU was effective in inducing neurite formation in human neuroblastoma cells (Prasad *et al.*, 1973b), it appears that the specific effect of BrdU is not universally characteristic of nerve cells in culture but probably is peculiar to some neuroblastoma tumors. Schubert *et al.* (1971a) reported that ara-C as well as other nucleotide analogues did not induce differentiation. However, other workers have found that 6-thioguanine (Prasad *et al.*, 1973b), ara-C (Kates *et al.*, 1971), and methotrexate (Byfield and Karlson, 1973) induced neurites in mouse neuroblastoma cells. The effects were not consistent and varied from one clone to another. Indeed, as in many other cases, the specific response to a certain drug may depend not only on the properties of a given clone but also on other, still unknown factors.

2.2.11. Dimethylsulfoxide

Addition of dimethylsulfoxide (DMSO) to the growth medium of Friend Leukemic cells results in erythroid differentiation, i.e., hemoglobin synthesis and other typical responses (Friend *et al.,* 1976; Friend, 1977). The mode of action of this agent is still not clear.

In the neuroblastoma system, DMSO proved to be inhibitory when added to serum-free media: the number of neurite-extending cells decreased from 96% to 26% in the presence of 1% DMSO and to 5% in the presence of 2% DMSO (Furmanski and Lubin, 1972). Surprisingly, the effect was transient. After 3–4 days, cells started extending neurites again, although DMSO was still present. The neurite-extending effect produced by dibutyryl cyclic AMP in complete medium was also inhibited by DMSO; even high concentrations of the inducer failed to overcome this inhibition.

However, morphological as well as electrophysiological maturation of neuroblastoma cells (clone NIE-115) could be induced when DMSO was added in serum-containing medium (Kimhi *et al.,* 1976). The effect of the agent was time-and concentration-dependent, and the best results were obtained by treating the cells with 2% DMSO for periods of 1 week or more. Stable cultures with a high degree of morphological differentiation and with excitable membranes were thus obtained. Another compound that induced differentiation in Friend cells—hexamethylenebisacetamide (HMBA) (Reuben *et al.,* 1976)—was found to be active in the neuroblastoma system as well (Palfrey *et al.,* 1977). However, the morphology of the cells treated with this agent differed considerably from that observed in DMSO-differentiated cells. The neurites in the HMBA-treated cells were branched and very thin, whereas those of the DMSO-treated cells were thick and very long. These findings together with biochemical and ion-flux studies suggest that the two compounds act by different mechanisms or affect different genes.

2.2.12. Liposomes

Neurite formation could be induced by liposomes originally used as artificial carriers to discharge NGF into the NB41A3 cells. Electron microscopic studies have shown an abundance of microtubules as well as microfilaments in the neurites (Chen *et al.,* 1976). The authors calculated that a concentration of 3×10^8 phospholipid molecules per cell is sufficient to induce this effect. If all such molecules were incorporated into the membrane, their contribution to the entire surface could be 5–20%.

2.2.13. Hypertonic Medium

Hypertonic medium was also found to induce very long neurites (Ross *et al.,* 1973, 1975). Observations at the ultrastructural level showed the cells to contain large clusters of clear vesicles 40–60 nm in diameter that were morphologically indistiguishable from synaptic vesicles of normal neurons. These organelles were found in the perikaryon as well as in the neurites but were not found in monolayered cells in control medium. Microtubules and microfilaments were evident in the neurites.

Hypertonicity was shown to inhibit protein synthesis in HeLa cells. The possibility of preferential inhibition of certain proteins or of stimulation of other proteins was not tested in the work of Ross *et al.* (1973, 1975; see Wengler and Wengler, 1972).

2.2.14. Selection by Aminopterin

Aminopterin, an inhibitor of folic acid synthesis that also restricts the synthesis of purines and pyrimidines, was used to select a nondividing cell population (Peacock *et al.,* 1972; Spector *et al.,* 1973). The aminopterin-selected cells were quite large, had extensive processes, and showed well-developed electrical properties. These characteristics remained stable for periods of up to 4 weeks. This aminopterin effect is reversible for at least 2 weeks.

2.2.15. The Role of Polyamines

Ornithine decarboxylase (ODC), the rate-limiting enzyme in polyamine biosynthesis, has been implicated in the first events of the response to environmental stimulation. A rapid induction of ODC activity in N1E-115 cells by treatment with compounds that increase intracellular cyclic AMP levels (PGE, adenosine, or PDE inhibitors) was demonstrated by Bachrach (1975, 1976, 1977). The induction of ODC activity in neuroblastoma cells could also be caused by addition of fresh serum, and was antagonized by carbamylcholine and blocked by actinomycin D or cycloheximide.

Apparently, the composition of the medium is important, since Chen and Canellakis (1977) could induce ODC activity in cells in a salts-and-glucose medium by addition of asparagine and glutamine. Cholera toxin and agents that are known to raise intracellular cyclic AMP concentrations had no effect unless suboptimal concentrations of asparagine were present. This induction was not sensitive to actinomycin D, whereas that of cyclic AMP was. The enzyme was found to be more stable in the presence of the two amino acids when protein synthesis was inhibited. Cytolytic concentrations of either bromoacetylcholine or bromoacetate caused a marked decrease in enzyme activity; replication of neuroblastoma cells was inhibited by 1,3-diaminopropane, an ODC inhibitor (Chapman *et al.,* 1978). These results make any conclusion concerning the role of the enzyme in the cellular regulation and differentiation premature.

2.2.16. Transcriptional and Translational Control

The differentiation of neuroblastoma cells involves changes in the size of the soma and nucleus, in the levels of various enzymes, in the development of neurites, and in the maturation of an excitable cell membrane. Do these phenotypic modulations require activation of previously dormant genes and repression of others? Is there evidence for synthesis of new mRNAs? Is there a change in the pattern of newly synthesized intracellular proteins? Are these changes qualitative or are they only quantitative?

In the majority of the studies that addressed themselves to these questions, the criterion of differentiation was the most visible: the extention of neurites. Experiments with enucleated cells suggested that the initial steps of neurite outgrowth do not require the presence of a nucleus. Enucleated cells, for example, are still able to extend neurites when dibutyryl cyclic AMP is added to the medium, even 1 hr after enucleation (Miller and Ruddle, 1974). In a cyclic AMP-sensitive but noninduced culture, only 5–10% of the cells send out neurites (Lieberman and Sachs, 1978). Addition of cyclic AMP causes 40% of the cells to send out neurites within 1 hr. In a cyclic AMP-insensitive clone, only 8% of the cells bear neurites in normal conditions, and addition of cyclic AMP has no effect on the culture's morphology. In the enucleated state, cells of the two clones extended neurites to the same degree after addition of cyclic AMP, theophylline, or papaverine. In both cases, PGE_1 was a better stimulator than cyclic AMP. The neurites in the enucleated cells were longer and more complex than those induced in normal cells. Cycloheximide and puromycin—protein synthesis inhibitors—prevented neurite outgrowth in nucleated cells, but both were not effective in enucleated cells, even at concentrations as high as 40 μg/ml, at which over 90% of protein synthesis is inhibited. These results suggest that protein synthesis in nucleated cells may be required to block or inactivate a nuclear inhibitor of neurite induction (Lieberman and Sachs, 1978). Concanavalin A (Con A), which inhibits the neurite outgrowth in nucleated cells, was less inhibitory in enucleated cells. This finding suggests that the nucleus may be involved in a modulation of surface–cytoskeleton interactions.

In the initial stages, axonal outgrowth apparently involves a mobilization of preexisting molecules such as actin and tubulin and probably a redistribution of other components as well. A rough approximation of the size of this intracellular pool can be made by blocking RNA and protein synthesis and measuring the degree of neurite formation. Seeds *et al.* (1970) reported that cycloheximide (up to 50 μg/ml) had little effect on neurite outgrowth. Schubert and Jacob (1970) found, however, that the same agent (at 40 μg/ml) or puromycin (at 10 μg/ml) prevented process formation within 2 hr. The discrepancy may stem from differences in experimental procedures: the former group added cycloheximide 24 hr after plating; the latter added it at time zero. The finding that cycloheximide and the vinca alkaloids prevent neurite extension only after 2 hr suggests, moreover, that the mechanism operating during the initial 2-hr period is different from that which operates over extended periods.

Cycloheximide consistently caused only moderate inhibition of neurite outgrowth in serum-free medium, but a total inhibition of dibutyryl cyclic AMP induced differentiation (Furmanski *et al.*, 1971). These findings were corroborated by Lim and Mitsunobu (1972) who found that dibutytyl cyclic AMP inhibits DNA and RNA synthesis but stimulates protein synthesis.

Several reports suggest that the differentiation process is inhibited by cycloheximide but is only partially, or not at all, sensitive to actinomycin D. Thus, only partial inhibition of neurite formation by actinomycin D was reported by Schubert and Jacob (1970) and Seeds *et al.* (1970). Prasad *et al.* (1972) found that actinomycin D did not inhibit the dibutyryl cyclic AMP-induced differentiation, whereas cycloheximide did. DNA content in dibutyryl cyclic AMP-treated cultures decreased, but that of RNA, including poly (A)-containing molecules, increased (Bondy *et al.,* 1974; Prasad *et al.,* 1973a). Augusti-Tocco *et al.* (1974) concluded that the flow of mRNA from the nucleus to the cytoplasm is the same in suspension cultures (nondifferentiated cells) and monolayer cultures (differentiated cells). Analysis of poly(A)-containing RNA suggested that the terminal differentiation of N1E-115 cells (induced by withdrawal of serum or by dibutyryl cyclic AMP) is not accompanied by major alterations in transcription and is paralleled by a marked stabilization of the 16 S species (Croizat *et al.,* 1977).

The *in vitro* DNA-dependent RNA transcription of chromatin from proliferating cells was higher than that from the differentiated cultures (Zornetzer and Stein, 1975). However, heterogeneous nuclear and messenger RNA in the nuclei of monolayer and suspension cultures was found to be similar (Augusti-Tocco *et al.,* 1974). These authors concluded that the transfer to monolayer culture involves regulation at the level of processing and stabilization of newly synthesized molecules rather than regulation of the rate of synthesis. Apparently, both soluble factors and ribosomes are involved in the activation of protein synthesis (Zucco *et al.,* 1975). The lower ribosomal RNA content in suspension cultures probably reflects a low rate of transport of newly synthesized rRNA into the cytoplasm along with degraded 45 S pre-rRNA (Casola *et al.,* 1974).

Some of the pitfalls of such studies were illustrated in the work of Glazer and Schneider (1975) whose experiments indicated that changes in the specific activity of cellular RNA produced by dibutyryl cyclic AMP and papaverine resulted from alterations in uridine uptake and changes in the specific radioactivities in the precursor pools.

Attempts to find major differences between the proteins synthesized by differentiated and nondifferentiated cells were rather disappointing. Only minor differences were observed, and in most cases they were quantitative and not qualitative.

The histone proteins in differentiated and proliferating neuroblastoma cells were found to be similar, but significant differences were demonstrated in the nonhistone chromosomal proteins that are synthesized in a cell-free system. (Zornetzer and Stein, 1975). Actin and histones were found to be the major proteins synthesized by nonadenylated mRNA in a cell-free system (Morrison *et al.,* 1978).

2.3. Synapse Formation

One of the primary objectives in generating and characterizing clonal nerve cells was to obtain cells that would be able to form synapses *in vitro* with high frequency. It was hoped that by using such homogeneous cell populations, one would be able to study the molecular events and the requirements for forming the specific nerve–nerve or nerve–muscle cell connections.

Attempts to form synaptic connections between neuroblastoma cells or between neuroblastoma cells and other cells were unsuccessful (Nelson *et al.*, 1969; Harris and Dennis, 1970; Nelson *et al.*, 1971a), although a trophic effect, suggestive of some interaction between neuroblastoma cells (clone C1A) and a rat skeletal myoblast line (clone L-6) was demonstrated (Harris *et al.*, 1971; Steinbach *et al.*, 1973). Recently, however, the neuroblastoma × glioma hybrid cells (clone NG 108-15), which are able to synthesize, store, and excrete acetylcholine, were shown to form synapses with various embryonic striated muscle cells as well as with myotubes of the clonal myogenic cell line G-8 (Nelson *et al.*, 1976; Puro and Nirenberg, 1976; Christian *et al.*, 1977). Further studies have established that although the synapses are immature by physiological and anatomical criteria (Nelson *et al.*, 1978) their incidence in db-cAMP treated cultures was relatively high. The way to study some aspects of the process of synaptogenesis and both pre- and postsynaptic regulatory events was thus opened. The cyclic nucleotide appeared to play some role in establishing synapses because the incidence and strength of the synaptic connections in hybrid cells (which showed excellent morphology and electrical excitability after treatment with the antimetabolite cytosine arabinoside) were markedly reduced relative to those of their db-cAMP-treated sister cultures (Christian *et al.*, 1978). X-irradiated cultures, however, showed a high incidence of synaptic connections even in the absence of db-cAMP.

Physiologically identified functional junctions that were examined by electron microscopy techniques revealed some morphological features of the neuromuscular junctions *in vivo*, but none of the synapses that were examined showed all of them. For example, in one case, 50-nm clear vesicles, similar to those thought to contain acetylcholine in motor axon terminals, were found presynaptically, but the intercellular gap and the structure of the postsynaptic muscle cell in the area were not suggestive of a neuromuscular junction. In another case, the gap and the basement membrane were similar to those found *in vivo* but the presynaptic element contained only very few small clear vesicles. All of these findings may suggest that the initial steps in the organization of the synapse do not depend on clustering of release sites and that various components of the junction can develop relatively independently (Nelson *et al.*, 1978). In addition, the finding that various striated muscle cells have a similar specificity for synapse formation (Puro and Nirenberg, 1976) suggests that at least in the early stage of neurogenesis there are no species- or organ-specific complementary molecules on the nerve and muscle cells.

Such studies will undoubtedly be followed by the isolation of more clones (see Wilson *et al.*, 1978), which will help in examining the molecular events that lead to the establishment of the primitive connection and ultimately to the stabilization and maturation of the synapse.

2.4. Catecholamine Metabolism

2.4.1. General Considerations

The original paraformaldehyde histofluorescence technique of Falk *et al.* (1962) together with enzymatic studies has been used extensively to detect and demonstrate the presence of catecholamines in the C-1300 tumor and its clones (Hermetet *et al.*, 1972a; DeVellis *et al.*, 1970; Breakefield *et al.*, 1975; Narotzky and Bondareff, 1974; Anagnoste *et al.*, 1972; Mandel *et al.*, 1973a,b). Addition of their respective radioactive precursors has made it possible to demonstrate the presence of norepinephrine, dopamine, epinephrine, serotonin, histamine, tyramine, and octopamine (Mandel *et al.*, 1973a,b; Narotzky and Bondareff, 1974). Ultrastructure studies have revealed many clear vesicles 50–70 nm in diameter as well as very large, 100–400 nm, dense, membrane-bound granules (Augusti-Tocco *et al.*, 1970; Daniels and Hamprecht, 1974; Ross *et al.*, 1975; Breakefield *et al.*, 1975; Schubert *et al.*, 1969; DePotter *et al.*, 1978). The suggestion that these vesicles may serve as storage sites for catecholamines in neuroblastoma cells was further supported by autoradiographic studies that showed localization of grains originating from labeled dopamine in cell processes and their endings (Breakefield *et al.*, 1975). Reserpine, a catecholamine-depleting agent, diminished the yellow–green fluorescence localized to the terminals of nerve processes (DeVellis *et al.*, 1970).

The synthetic pathway of catecholamines is illustrated schematically as:

$$\text{Tyrosine} \xrightarrow[\text{hydroxylase}]{\text{Tyrosine}} \underset{\text{(DOPA)}}{\text{Dihydroxyphenylalanine}} \xrightarrow[\text{decarboxylase}]{\text{DOPA}}$$

$$\text{Dopamine} \xrightarrow[\beta\text{-hydroxylase}]{\text{Dopamine}} \text{Norepinephrine} \xrightarrow{\text{PNMT}} \text{Epinephrine}$$

Neuroblastoma cells are thought to be related to sympathoblasts. These pluripotential cells can give rise to neurons that synthesize, store, and release catecholamines, or to cells that use acetylcholine as their neurotransmitter.

Early observations suggested that the expression of either tyrosine hydroxylase or choline acetyltransferase excluded the expression of the other activity (Amano *et al.*, 1972). Later studies, however, did not fully substantiate these conclusions. A clone containing both tyrosine hydroxylase (TH) and choline acetyltranferase (CAT) activity (although at low levels) was isolated by Prasad *et al.* (1973c). A clone (N-TD6) that showed tyrosine hydroxylase as well as CAT activity was selected (Heldman, unpublished data, quoted by Breakefield *et al.*, 1975), and both CAT and dopamine β-hydroxylase (DBH) activities were demonstrated in a neuroblastoma $\times$ glioma hybrid (Hamprecht *et al.*, 1974).

Recently, embryonic sympathetic neurons grown *in vitro* were shown, under suitable culture conditions, to be able to synthesize transmitters by either an adrenergic or a cholinergic pathway. In some cells, both activities were found to be present simultaneously (Furshpan *et al.*, 1976). However, catecholamine synthesis increases while that of acetylcholine decreases and vice versa; therefore expression of these two phenotypes does not seem, in general, to be independent.

2.4.2. Uptake of Tyrosine and Catecholamines

Transport of L-tyrosine into the adrenergic N1E-115 cells was found to be by facilitated diffusion (K_t = 2–17 $\times$ 10^{-5} M; V_{max}, 0.16–6 nmole/min per mg protein) (Richelson and Thompson, 1973; Richelson, 1974). It is a saturable process, sensitive to temperature, and independent of Na and energy. Sulfhydryl reagents and close structural analogues (L-3-iodotyrosine, L-phenylalanine, L-DOPA) inhibit uptake. Many properties of the system suggest that the substrate is transported by the leucine-preferring transport system described by Christensen *et al.* (1973). Similar results were obtained by Archer *et al.* (1977) who studied clones N-T16, N-TD6, and NS-20 cells.

Wexler and Katzman (1975) reported that extended treatment of cells with dibutyryl cyclic AMP caused an increase in tyrosine uptake that was proportional to the increase in tyrosine hydroxylation. Unlike in monolayer culture, the uptake in suspension culture failed to increase with dibutyryl cyclic AMP treatment. The authors suggested that dibutyryl cyclic AMP treatment induces a tyrosine uptake system. However, the larger surface area of differentiated cells in a monolayer culture can partially explain this increased uptake.

At least two neuroblastoma clones were shown to take up norepinephrine by a mechanism similar to that operating at the synaptic terminal. Cells of clone M1 (Zwiller *et al.*, 1975) have a two-component system: a saturable high-affinity component (K_{m_1} = 1.4 $\times$ 10^{-7} M; V_{max} = 0.07 pmole/min per mg protein) and a nonsaturable low-affinity component (K_{m_2} = 2 $\times$ 10^{-5} M; V_{max} = 5.4 pmole/min per mg protein). The high-affinity accumulation was found to be temperature sensitive, sodium-dependent, and energy-requiring. The second system was less dependent on sodium and not sensitive to ouabain. Only the high-affinity system was inhibited by imipramine, desipramine, or amphetamine. A specific high-affinity uptake system for norepinephrine was also demonstrated in a clone that can synthesize this neurotransmitter (Breakefield, 1976).

Differentiated N1E-115 cultures took up dopamine (10^{-4} M at a rate of 37 pmole/min per mg protein. Reserpine (5×10^{-5} M) did not affect the initial rate of uptake but reduced uptake at saturation by 60% (Breakefield, 1975). The conversion of dopamine to norepinephrine and the ability to store dopamine were inhibited in reserpine-treated cells. About a third of the dopamine that was taken up by untreated cells was associated with a particulate fraction. The capability of neuroblastoma cells to store catecholamines was further demonstrated in clone N-TD6 in a comprehensive study in which both histochemical and radiochemical techniques were used (Breakefield *et al.*, 1975).

2.4.3. Tyrosine Hydroxylase (EC 1.14.3a)

Tyrosine hydroxylase is considered to be the catalyst of the rate-limiting step in catecholamine synthesis. High activity of this enzyme was reported in clone N1E-115 (Amano *et al.*, 1972); numerous other clones generally showed lower activity (Augusti-Tocco and Sato, 1969; Amano *et al.*, 1972; Prasad *et al.*, 1973c; Schubert *et al.*, 1974). The capability of TH to convert pheynylalanine to the essential amino acid tyrosine was used by Breakefield and Nirenberg (1974) as a selection method for isolating lines with high activity of this enzyme. They estimated that only 1 out of 70,000 uncloned C-1300 cells gave rise to a cell line with TH level high enough to be able to multiply without supplemental tyrosine.

Tyrosine hydroxylase activity has been shown to be regulated by culture conditions as well as by several inducers: in complete medium, the specific activity of TH increased by more than 12-fold per cell from the early logarithmic growth phase to late stationary phase (Richelson, 1973a). Removal of serum or arrest of cell division by 5-FdU also resulted in an increase of activity, but at a slower rate. Tyrosine (0.1 mM) increased the activity of TH eightfold within 24 hr in N1E-115 and NTD6 cells (Lloyd and Breakefield, 1974). Whether the observed increase in activity was a result of higher *de novo* synthesis or substrate protection against inactivation or degradation is not clear. TH activity could be elevated 15-fold and more by cyclic AMP and various analogues (Waymire *et al.*, 1972, 1977, 1978a,b; Richelson, 1973b), and a similar effect was demonstrated by the phosphodiesterase inhibitors papaverine and Ro20-1724 but not theophylline (Waymire *et al.*, 1978a). The results obtained by the latter group suggested that the cyclic AMP effect is rather specific to TH and is not a general one. Interestingly, butyric acid, which results from the catabolism of dibutyryl cyclic AMP, was also found to stimulate TH activity (Richelson, 1973b; Waymire *et al.*, 1972), and a survey of various monocarboxylic acids showed that butyric acid was the most potent (Lloyd *et al.*, 1978a). RNA synthesis was shown to be essential in the 8-Br-cyclic AMP induction of TH (Waymire *et al.*, 1978b). Tyrosine hydroxylase but not acetylcholinesterase (AChE) activity was found to decrease when cells of clone N1E-115 were induced to differentiate by 2% DMSO (Kimhi *et al.*, 1976).

Tyrosine hydroxylase in a cell-free system had an absolute requirment for the reduced pteridine cofactor, $DMPH_4$ (Richelson, 1973a). Synthesis of DOPA in intact N1E-115 cells (Richelson, 1976a) indicates that the cells contain sufficient quantitites of the natural cofactor, tetrahydrobiopterin, as do cells of the parental clone N1E (Bluff and Dairman, 1975). Sonication of the cells probably destroys the intracellular compartmentalization that allows the synthesis *in vivo.*

Richelson (1976a) studied various parameters of TH activity in intact cells in culture and found that many properties of the reaction were similar to those determined for purified TH from normal tissue. Two apparent K_m values were obtained: K_{m_1} of 10 ± 2 μM and K_{m_2} of 140 ± 10 μM. Substrate inhibition (at 20–50 μM) as well as competitive inhibition by *l*-3-iodotyrosine (K_i = 0.3 μM) were demonstrated. Dopamine was found to be a noncompetitive inhibitor with K_i = 500 μM while *l*-norepinephrine had no effect. The reaction was twice as fast at pH 5.5 as at pH 7.4.

2.4.4. DOPA Decarboxylase (AADC; Aromatic-L-Amino-Acid Decarboxylase; EC 4.1.1.28)

Histofluorescence of formaldehyde-treated tumor cells as well as the identification of [^{14}C]dopamine as one of the products of [^{14}C] tyrosine metabolism in N2a cells (Narotzky and Bondareff, 1974; Wexler and Katzman, 1975) suggest that at least some cells possess AADC activity. However, the enzyme could not be detected *in vitro* (Mandel *et al.*, 1973b). These negative findings were confirmed by Waymire and Gilmar-Waymire (1978), who tested uncloned cells as well as various cloned cells (including N1E-115 and NBD-2). The limited number of clones tested (which unfortunately did not include N2a) does not allow a final conclusion to be made.

2.4.5. Dopamine β-Hydroxylase (DBH; EC 1.1.4.17)

Dopamine β-hydroxylase was found in the original C-1300 tumor (De Potter *et al.*, 1978). Its activity in the serum of tumor-bearing mice was higher than in the serum of control mice (Anagnoste *et al.*, 1972). Clones N2a, N18G (Anagnoste *et al.*, 1972), N1E-115 (Mandel *et al.*, 1973b) and NBD-2 (Waymire *et al.*, 1978a) all show DBH activity.

In clone NBD-2, cyclic AMP analogues and phosphodiesterase inhibitors induced a concomitant elevation of TH and DBH activity, while activities of catechol-*O*-methyl transferase (COMT), MAO, and AChE were unaffected (Waymire *et al.*, 1978a). Very high activities (450–500 pmole/min per mg protein) which could be further enhanced by cyclic AMP were reported in two neuroblastoma × glioma hybrid clones, NG-108-15 and NG108-5 (Hamprecht *et al.*, 1974). The distribution of DBH and of norepinephrine in C-1300 tumor cells differs from that observed in peripheral noradrenergic neurons: most of the enzyme in the tumor cells was associated with the plasma membrane and not with the heavy vesicles (De Potter *et al.*, 1978). In adrenergic tissue, DBH was shown to be present in dense-core vesicles and to be released upon stimulation. It would be of interest to correlate its presence in neuroblastoma clones with release of "neurotransmitter."

2.4.6. Catechol-*O*-Methyl Transferase (COMT; EC 2.3.1.6)

COMT activity has been demonstrated in many cells. The relationship of its intracellular level to culture conditions is not clear: the specific activity calculated per mg protein in culture seems to increase when cells reach confluency but remains unchanged when the data are expressed as activity per cell [Waymire and Waymire-Gilmer, 1978 (clone NBD-2)]. Skaper *et al.* [1976 (clones N1E-115 and N-4)] found that there was no change in the level of the enzyme during the transition from the logarithmic to the stationary phase of growth. Also, there was no change in enzyme level after treatment with dibutyryl cyclic AMP or PGE_1, although these two agents caused morphological differentiation (Skaper *et al.*, 1976; Prasad and Mandal, 1972). Increased COMT activity after treatment with sodium butyrate, which blocks cell division without causing morphological changes, was reported by Prasad and Mandal (1972).

Both glial and neuroblastoma COMT had a similar K_m (200 μM norepinephrine), with a V_{max} of 9–13 pmole normetanephrine/min per mg protein (Skaper *et al.*, 1976).

Addition of 2-hydroxyestradiol to N1E-115 cells caused a specific, dose-dependent reduction in the formation of methylated products from dopamine, and this suggests that COMT is the inhibited enzyme (Lloyd *et al.*, 1978b). The apparent K_m for methylation of the estrogen was about 10 μM (at least an order of magnitude lower than the K_m value for norepinephrine). These findings are consistent with the hypothesis (Paul and Axelrod, 1977) that estrogen may block catecholamine methylation in neuronal tissue and thus may serve as an endogenous modulator of COMT activity. Inhibition of catecholamine methylation was also demonstrated by Michelot *et al.* (1977) who used S-adenosylhomocysteine and *S*-tubercidinylhomocysteine to block metoxytyramine formation from dopamine.

2.4.7. Monoamine Oxidase (MAO; EC 1.4.3.4)

The physiological role of MAO in the brain (where its level is lower than in the liver, kidney, and intestines) is not fully understood. MAO activity was demonstrated in both glioma and neuroblastoma cell lines (Skaper *et al.*, 1976). The affinity of MAO for its naturally occurring substrates (tyramine, tryptamine, serotonin, and norepinephrine) was independent of the presence of the catecholaminergic pathway enzymes in the cell used. Apparent K_m values of 8–14 μM for tryptamine and 510–580 μM for norepinephrine were determined. MAO activity was substantially higher in glioma cells, but V_{max} values varied little with substrate among different cell lines. MAO activity in NBD-2 cells increased with time, based on cell number as well as on amount of cellular protein (Waymire and Gilmer-Waymire, 1978). However, Skaper *et al.* (1976) did not observe significant changes in the specific activity of MAO when clone N-4 cells reached confluency or after treatment with cyclic AMP or dibutyryl cyclic AMP.

In neuroblastoma cells benzylamine was deaminated only 1% as much as serotonin (5-hydroxytryptamine), and the enzyme was over 1000 times more sensitive to inhibition by clorgyline than by deprenyl. These findings led to the conclusion that mainly MAO type A is present in N1E-115 cells (Donnelly *et al.*, 1976), a finding that was confirmed by Hawkins and Breakefield (1978).

2.4.8. Cytotoxicity of 6-Hydroxydopamine (6-OHDA)

The dopamine analogue 6-OHDA was found to be cytotoxic *in vivo* to sympathetic neuroblasts if administered to newborn animals during a critical period of their development (Angeletti and Levi-Montalcini, 1970a). Only the nerve terminals, but not the soma, were affected in adult animals (Thoenen and Tranzer, 1968). The selective toxicity is thought to be the result of a specific uptake into the sympathetic nerve which is followed by rapid oxidation to peroxides, superoxides, and/or quinones that, in turn, cause cell death.

Neuroblastoma seemed to be a suitable model system to study the molecular aspects of this toxicity, and indeed Angeletti and Levi-Montalcini (1970b) demonstrated the cytotoxicity of 6-OHDA for the mouse tumor cells *in vitro*. HeLa and mouse sarcoma cells, which were used as control cells, were not affected by 25–100 μg 6-OHDA/ml, a concentration that proved toxic to the nerve cells within 25 hr. However, *in vivo*, only 2 out of 30 tumors showed complete remission and, while regression was noticeable during the first week of treatment with 6-OHDA, the tumors took. Similar observations were made by Chelmicka-Szorc and Arnason (1976) who reported that the growth of C-1300 was markedly slowed in 6-OHDA-treated mice as was the A-10 breast adenocarcinoma. However, in mice axotomized by pretreatment with 6-OHDA, the growth of the neuroblastoma was slowed by 6-OHDA, whereas that of A-10 was not affected. Thus, a relationship between sympathetic neurons and neuroblastoma cells was suggested.

Prasad (1971) reported that 5×10^{-5} M 6-OHDA caused 50% inhibition of cell division in neuroblastoma cultures, whereas concentrations twofold higher were required to inhibit BHK-21 and CHO-K cells. Dopamine (at 5×10^{-4} M) caused 50% inhibition of neuroblastoma cells, only 20% inhibition of CHO-K cells, and had no effect on BHK-21 cells. Similarly, DeBault and Millard (1973) found that CHB_4, C-6 (glioma lines), and L-929 cells were also susceptible to 6-ODHA, but that the degree of inhibition was correlated with the permeability of the cells to dopamine or 6-OHDA, the neuroblastoma lines showing a significantly greater uptake. Similar observations were made by Rotman *et al.* (1976a), who also noted that N1E-115 cells accumulate nearly four times more 6-OHDA than dopamine during a 15- to 120-min incubation. While reserpine inhibited uptake of dopamine, it did not affect 6-OHDA transport. Neuroblastoma $\times$ L cell hybrids, which initially showed a resistance level similar to the L-cell parent, became more sensitive after chromosome loss (Cronemeyer *et al.,* 1974). Pretreatment of neuroblastoma cells with BRDU ($1 \mu g/ml$) appeared to protect the cells from the cytotoxic effects of 6-OHDA.

The difference in susceptibility of various cells to a given polyphenol may reflect, then, an increased or reduced rate of uptake of the particular compound.

Two mechanisms were suggested as explanations for the cytotoxicity of polyphenols: (1) production of quinone species which kill cells through inhibition of sulfhydryl enzymes and other nucleophilic groups, and (2) production of H_2O_2, $O_2^{\overline{\cdot}}$, and $OH^{\cdot}$. Rotman *et al.* (1976b) compared the protein profiles of cells incubated with 6-OHDA to those treated with dopamine and found, in the 6-OHDA-treated cells, an increased proportion of high molecular weight proteins. In both cases, the DA and 6-OHDA were bound to various proteinaceous complexes. They suggested that the agent exerts its cytotoxicity by its ability to cross-link proteins, an explanation compatible with the first assumption.

However, Graham *et al.* (1978) concluded that 6-OHDA kills cells through the production of H_2O_2, $O_2^{\overline{\cdot}}$, and $OH^{\cdot}$, while dopamine and DOPA cytotoxicity may also involve reactions with quinone oxidation products.

2.5. Acetylcholine Metabolism

2.5.1. General Considerations

Choline cannot be synthesized by nervous tissue, so its transport across the cell membrane is indispensable for the synthesis of cell components as well as for the synthesis of acetylcholine. Isolated synaptosomes exhibit both sodium-dependent high-affinity, and sodium-independent low-affinity uptake systems (Yamamura and Snyder, 1973; Haga and Noda, 1973). Only the high-affinity sodium-dependent system has been implicated in the formation of acetylcholine, but the nature of this linkage has not been established. Brain synaptosomal preparations are not purely cholinergic, and it is difficult, therefore, to study a selective uptake process in such a system. Isolation of "cholinergic" clones—i.e., cells with high choline acetyltransferase activity—offers a better opportunity to study choline uptake and its relationship to ACh synthesis. However, the results obtained thus far suggest that the transport system in these cells does not display the characteristics associated with brain "cholinergic" synaptosomes. It should be noted, though, that a coupled system for high-affinity choline uptake and acetylcholine synthesis was demonstrated only in nerve terminals but not in the cell bodies of the same neurons (Suszkiw *et al.,* 1976). These findings may be related to observations in the neuroblastoma system.

One of the frequently voiced objections to the validity of studies with cloned tumor cells in neurobiology is that these cells are not capable of carrying out a critical task of a nerve cell, synapse formation. Indeed, initial attempts to find synaptic connections among neuroblastoma cells failed, although some features of synaptic interaction between neuroblastoma cells and a rat myoblast cell line (L-8) were observed by Harris *et al.* (1971) and Steinbach *et al.* (1973). Although electrical coupling was occasionally found at the contact points, attempts to record a response across the "gap" were unsuccessful. Still, neuroblastoma × glioma hybrids were recently shown to form functional nicotinic synapses with primary skeletal muscle in culture (Nelson *et al.,* 1976) and thus provide a reasonably convenient system in which to study synaptogenesis.

2.5.2. Choline Uptake

Uptake systems with K_t values (substrate concentration at half maximum velocity of transport) of $1.0–20.0 \times 10^{-6}$ M were demonstrable in many mouse neuroblastoma clones: cholinergic, adrenergic, and inactive (clones N1E-115 and NS-20, Richelson and Thompson, 1973; clones NS20, N-18, and N1E, Lands *et al.,* 1974b; clones N-18, NS-20 and NS-21, Massarelli and Mandel, 1976 and Massarelli *et al.,* 1974a). Cells that were not previously depleted of choline showed K_t values higher than those measured in choline-deprived cells (Richelson and Thompson, 1973; Lanks *et al.,* 1974b). V_{max} values in different clones ranged from 3–70 pmole/min per mg protein, but these differences may reflect differences in surface area. Choline transport systems of this type are not specific to neuroblastoma cells and have been shown to exist in glia, fibroblasts, and other cells (Richelson and Thompson, 1973; Hutchison *et al.,* 1976).

Lanks *et al.* (1974b) concluded that only one transport system operates in neuroblastoma cells (clones N1E, NS20, and N18) and that its affinity depends upon previous exposure of the cells to choline. Massarelli *et al.*

(1974a) presented evidence for two components in NS21 cells, the high-affinity transport ($K_m = 1 \times 10^{-7}$ M) being revealed only at low substrate concentration. Two transport systems (one with a K_m of 10^{-5} M and the other $K_m > 1$ mM) were determined by Haber and Hutchison (clone NB41, 1976) as well as Matthews and Chiou (clone S20 F_3, 1978). The existence of two transport systems was also suggested by McGee *et al.* (1978) and McGee (1978) in the hybrid clone NG-108-15.

Unlike synaptosomal uptake of choline, the transport in neuroblastoma cells showed only partial Na^+ dependence (Haber and Hutchison, 1976; Massarelli *et al.*, 1973, 1974a,b) or no dependence at all (clones N1E, N520, and N-18, Lanks *et al.*, 1974; clone NG-108-15, McGee *et al.*, 1978). Metabolic inhibitors reduced choline uptake in neuroblastoma cells to a degree similar to their effect in synaptosomes (Massarelli *et al.*, 1974 a,b; Massarelli and Mandel, 1976). Most authors have concluded that the low-affinity transport system is mediated by a facilitated-diffusion-like process.

Hemicholinium-3, which inhibits choline uptake in synaptosomal preparations, did not inhibit the transport even at 100 μM in one study (Lanks *et al.*, 1974b), but Hutchison *et al.* (1976) reported that it affects the high-affinity transport system but not the low-affinity (diffusional?) transport (clones NB41 and N2AG). Cholinesterase inhibitors at different concentrations exhibit a complex range of effects on choline uptake, again depending on the particular clone. Eserine sulfate had no effect on the uptake in NS21 cells but increased it in N-18 cells. Iso-OMPA, BW-284C51, and physostigmine were inhibitory (Massarelli, 1973; Massarelli *et al.*, 1973, 1974a, 1976; Hutchison *et al.*, 1976).

Removal of a cell surface sialic acid markedly reduced choline uptake (Stephanovic *et al.*, 1975b).

Only a small fraction (0.5%) of the transported choline was found to be a precursor for acetylcholine synthesis, even after 12 hr of incorporation. Most of it was present as phosphorylcholine (Lanks *et al.*, 1974b), and thus a relation between uptake and ACh synthesis was not demonstrated. A higher conversion value (of 6%) was determined by Kato *et al.* (1977) in the presence of low concentrations of acetate.

Intracellular levels of ACh and choline increased, and those of phosphorylated choline compounds decreased when NG-108-15 cells were treated with dibutyryl cyclic AMP for long periods. In this case as well, ACh synthesis was not found to be preferentially coupled to the high- or low-affinity choline uptake systems (McGee *et al.*, 1978; McGee, 1978).

Acetate can be used as a precursor to ACh in several systems but not in brain cells. Exogenous acetate (at 1.5 mM) was shown to be a precursor of the acetyl moiety of acetylcholine in neuroblastoma cells, even in the presence of glucose (5 mM). Choline enhanced the uptake of acetate and the synthesis of acetylcholine from glucose, whereas acetate did not affect choline uptake. Kinetics analysis suggested that the cholinergic NS-20 cells contain two saturable acetate uptake systems, both modified by choline.

2.5.3. Choline Acetyltransferase (CAT; EC 2.3.1.6)

Choline acetyltransferase activity was demonstrated in numerous clones derived from the C-1300 tumor (Augusti-Tocco and Sato, 1969; Amano *et al.*, 1972; Siman-Tov and Sachs, 1972; Prasad *et al.*, 1973c; Prasad and Mandal, 1973). Though nonexisting or present in very low levels in parental cells, the enzyme is expressed at high levels in some hybrid cells (Amano *et al.*, 1974; Minna *et al.*, 1975; McMorris and Ruddle, 1974). Activity levels in certain clones are similar to those found in brain and cholinergic nerves, i.e., 400 to 1000 pmole/min per mg protein (Amano *et al.*, 1972, 1974).

The optimum pH for neuroblastoma CAT is between 8.0 and 8.5. The K_m is 1.5×10^{-5} M for acetyl-CoA with 2×10^{-3} M choline iodide, and 9.1×10^{-4} M for choline iodide with 2.1×10^{-4} M acetyl-CoA (Wilson *et al.*, 1972). *N*-(2-hydroxyethyl)-4-(β1-naphthylvinyl) pyridinium bromide (NVP-OH) at 5×10^{-5} M inhibits ACh synthesis in neuroblastoma cells by blocking CAT (Steinbach *et al.*, 1974).

Choline acetyltransferase activity increases about sixfold when the culture becomes confluent (Rosenberg *et al.*, 1971; Amano *et al.*, 1974). The enzyme is expressed independently of AChE: addition of adenine, cyclic AMP, or dibutyryl cyclic AMP causes a decrease in CAT activity and, at the same time, an increase in AChE levels. BrdU increases both activities (Siman-Tov and Sachs, 1972).

2.5.4. Acetylcholinesterase (AChE; Acetylcholine Hydrolase; EC 3.1.1.7)

Acetycholinesterase activity was demonstrated in the original C-1300 tumor and in all clones derived from it. The level of the enzyme varies in different clones.

A K_m value of 9.1×10^{-4} M was determined for ACh (clone N18), and the enzyme was found to be sensitive to substrate inhibition (Wilson *et al.*, 1972). The pH optimum for AChE activity is in the range of 8.0–8.5, and the enzyme is stable to freezing and thawing. Only one-fifth of the activity was observed at 1°C. The compound 1,5-*bis*-(4-allyldimethylammonium-phenyl)pentane-1,3-dibromide (BW-284C51), the inhibitor of "true AChE," inhibits more than 90% of the activity at 10^{-6} M. Fifty percent inhibition was observed with 1×10^{-8} M DFP, 6×10^{-8} M BW284C51, or 7×10^{-7} M neostigimine (Blume *et al.*, 1970).

The level of the enzyme is inversely related to the growth rate and increases 25-fold when cells reach confluency. Sulfur mustard, acidic pH, and serum removal also increase the AChE level (Blume *et al.*, 1970; Lanks *et al.*, 1974a, 1975; Bear and Schnieder, 1977; Schubert *et al.*, 1971b; Kates *et al.*, 1971). Actimonycin D (0.5 μg/m1), cordycepin, or cycloheximide (10 μg/ml) prevent these increases. Addition of acetylcholine to a growing culture of neuroblastoma induced a 37-fold increase in the specific activity of AChE, and upon removal of the acetylcholine from the culture after 48 hr the enzyme level returned to normal levels (Harkins *et al.*, 1972). Sodium acetate and choline chloride, the metabolites of ACh, do not alter the level of the enzyme in the cells. Induction is most pronounced in the early stages when the cells are still dividing actively. Histochemical staining showed the presence of AChE in all cells grown in the presence of acetylcholine, even those that did not bear neurites. Co-culturing neuroblastoma cells with various glioma cells caused an increase of AChE level in one case and a decrease in another (Ciesielski-Treska *et al.*, 1975).

Levorphanol, a morphine congener, was found to inhibit the induction of the enzyme in clone 2A at a concentration (0.54 mM) that did not affect DNA or protein synthesis significantly (Hiller and Simon, 1973). These authors also mentioned that the inactivate enantiomorph dextrorphan showed a similar effect, and etonitazine manifested a tenfold greater inhibitory effect. Both dibutyryl cyclic AMP and BrdU induce AChE activity. These two agents stimulate enzyme synthesis by different mechanisms, and only the BrdU effect is sensitive to actinomycin D (Simantov and Sachs, 1975a). Stefanovic *et al.* (1975) reported that the removal of sialic acid from intact cells in culture by neuraminidase results in higher AChE activity, but these observations were not confirmed by Trams *et al.* (1976) who suggested that the higher activity results from leaky membranes.

AChE activity in neuroblastoma cells appears as several molecular forms. Two peaks of activity could be separated on DEAE columns. Analysis on sucrose gradients revealed that one peak sediments as a 4 S entity, and the other as 4 S and 9 S entities (Blume, 1972). Two molecular forms, sedimenting at 3.6 S and 10.3 S in the presence of the nonionic detergent Triton X-100, were found in clone NS-20 (Rieger *et al.*, 1976). Values of 4.0–4.5 S and 10.5–11 S were determined for the two molecular forms in clones N-18, NS-20, and N1E-115 (Vimard *et al.*, 1976; Kimhi *et al.*, 1976, 1980). Three classes sedimenting as 4 S, 6 S, and 9.6 S entities were found by Chang and Blume (1976) who also determined molecular masses of 64,000, 116,000, and 284,000 daltons for the various forms in their preparations. A common subunit with an apparent molecular mass of 70,000 $\pm$ 5,000 daltons was found by [^{3}H]DFP labeling to be the basic oligomer of neuroblastoma AChE (Kimhi *et al.*, 1980).

The smaller and more slowly sedimenting form appears to be the first to accumulate in the cell after irreversible inhibition of the cellular AChE by DFP (Rieger *et al.*, 1976). The heavier form is slower to appear; its level was still low 24 hr after the treatment, a time when the 4S form showed a full recovery. Cycloheximide (10 μM) completely inhibited any restoration of AChE activity, and 10–50 μM of actinomycin D drastically reduced the recovery of the heavier form. An increase of the cellular 11 S form followed by decrease in the 4 S species was demonstrated when protein synthesis was inhibited by cycloheximide (Vimard *et al.*, 1976). The 4 S form is normally predominant among the cellular AChE molecules. However, the heavier 11 S form predominates when the cells produce neurites (Vimard *et al.*, 1976) or when cell maturation is induced with 2% DMSO (Littauer *et al.*, 1978). A similar change in the molecular forms was observed in murine brain AChE (Rieger *et al.*, 1976). The 11 S form is the predominant form in a plasma-membrane-enriched fraction (Chang and Blume, 1976) and among the AChE molecules that are released into the growth medium (Kimhi *et al.*, 1980).

These data indicate that AChE in the cell undergoes a "maturation" process in which the lighter molecular form is converted to a heavier one. If the latter is the "membrane form," then the enrichment in the plasma membrane fraction can be explained, as can the observation that the heavier form is predominant in the released activity, reflecting turnover of a cell membrane component.

2.5.5. Acetylcholine Release

Neuroblastoma × glioma hybrid cells (clone NG-108-15) were shown to synapse with cultured striated muscle cells (clone L-8) that are highly sensitive to acetylcholine (Puro and Nirenberg, 1976; Christian *et al.*, 1977, 1978; Nelson *et al.*, 1976). McGee *et al.* (1978) have shown that these cells spontaneously release both ACh and choline into the medium. However, after treatment for several days with dibutyryl cyclic AMP, the cells shift to a different state and release ACh in response to high K^+, veratridine, serotonin, and PGF_2. Ca^{2+} is required for this evoked release, and Mg^{2+} is inhibitory. These studies were later extended to additional clones by Wilson *et al.* (1978). They found that most of the synapse-forming cells were able to synthesize ACh and released it in response to high K^+ stimulation. Cells of one clone, which formed few synapses, could synthesize ACh but did not respond to high K^+.

Both 5-hydroxytryptamine (5-HT) or the prostaglandin $PGF_{2\alpha}$ caused the release of acetylcholine at the synapse of NG108-15 cells with mouse myotubes (Christian *et al.*, 1978). Application of these compounds also facilitated the synaptic release of ACh elicited by the action potential.

2.6. Cell Membrane

2.6.1. General Considerations

The components of the nerve cell membrane and their spatial arrangement are important factors in the formation of specific synaptic structures: They encode the capability to respond to environmental stimuli, to control ionic permeabilities, to generate action potentials, and to transmit messages to postsynaptic cells. Elucidation of the plasma membrane structure is therefore critical in understanding the molecular mechanisms by which the nerve cell carries out its function. Study at the biochemical level requires large quantities of pure plasma membranes. Cloned cells, which may be maintained at will in a relatively primitive state and which may be induced to differentiate under controlled conditions, are natural candidates for such studies.

An attempt to detect changes in the structure of the cell's membrane requires a clear definition of what a pure plasma membrane is. Do all of the membrane components stay together during "purification" of the membrane fraction? Does one get the same "membranes" when cells in different phases of growth are subjected to a given treatment? A commonly used method for preparation of plasma membranes involves a stabilization step in which Zn^{2+} ions are used. This treatment certainly inactivates various enzymes, making a detailed activity analysis difficult. How does the treatment affect nonenzymatic proteins? This problem and others (such as differences in clonal properties, growth, medium, etc.) must be kept in mind when analyzing structure–function studies. Several methods, biochemical as well as immunologic, have been used to probe into these problems. Although differences have been detected, and some conclusions are beginning to emerge, much more remains to be discovered. An attempt will be made in the following sections to summarize the available data.

2.6.2. Plasma Membrane Composition

Clone N2a cells grow in suspension as round neuriteless cells. They rapidly extend neurites upon transfer to dishes to which they can adhere. From such cultures of morphologically nondifferentiated and differentiated cells, plasma membrane can be purified by the Zn^{2+} stabilization technique (see Warren and Glick, 1969). Analysis of the membrane components labeled by radioactive leucine has revealed only minor differences in the pattern of the proteins synthesized in these two growth conditions (Truding *et al.,* 1974). These findings were corroborated by Charalampous (1977) who slightly modified the Zn^{2+} treatment to obtain a better preparation of the plasma membrane. Several enzymes did show higher levels of activity in the differentiated plasma membrane preparation, but this was not reflected in the protein patterns. Examination of plasma membranes that were internalized during phagocytosis of polystyrene latex beads was more fruitful. Significant differences were observed both in protein profiles and enzyme activities in the two states of maturation. Charalampous demonstrated that cyclic AMP and butyrate affected suspension and monolayer cultures differently.

More information has been obtained by the use of labeled fucose and glucosamine to detect changes in membrane components. A glycoprotein of molecular mass 105,000 daltons was shown to be preferentailly labeled in differentiated clone N2a cells (Truding *et al.*, 1974). The synthesis of this protein always accompanied morphological differentiation in monolayer cultures. However, dibutyryl cyclic AMP treatment of cells in suspension culture could also induce labeling of such a molecule. Thus, the morphological differentiation is not a prerequisite for synthesis of this protein, the function of which is not known.

Another glycoprotein of 78,000 daltons was preferentially labeled by iodination of surface membrane proteins of the differentiated cells (Truding *et al.*, 1974). Unlike the 105,000-dalton molecule, this protein could not be induced by dibutyryl cyclic AMP in suspension cultures. Induction of a glycoprotein was also observed when cells of another clone were induced to differentiate by BrdU (Brown, 1971).

Garvican and Brown (1977) compared the effect of dibutyryl cyclic AMP and BrdU on fucose and amino acid incorporation. They found that both compounds stimulated an increased incorporation of fucose into two proteins of 60,000 and 70,000 daltons. The same compounds increased incorporation of amino acids only into proteins higher than 120,000 daltons. While dibutytyl cyclic AMP affected glycosylation and protein synthesis, with a preference for synthesis of low-molecular-weight proteins, the BrdU primarily stimulated the glycosylation of the membrane proteins. This finding is in line with the observations of Siman-Tov and Sachs (1975a) who found that the two agents induced acetylcholinesterase by different mechanisms. The molecular weights of the preferentially labeled proteins in the Garvican and Brown (1977) experiment differ from those determined by Truding *et al.* (1977). It is not clear whether these results reflect differences in clonal properties or differences in the culture conditions.

Glick *et al.* (1973, 1976) examined the assumption that changes in glycoproteins may reflect steps in the differentiation process and also the possibility that certain phenotypic differences, such as the inability of certain clones to extend neurites, might be correlated with differences in membrane glycoproteins. It was observed that clones of mouse neuroblastoma that have the ability to form neurites had proportionally more of one particular group of glycopeptides on the cell surface. A positive correlation between the capacity to differentiate morphologically and the proportion of these differentiation glycopeptides present on the cell surface was suggested. Haffke and Seeds (1975), however, detected only minor quantitative differences among glycopeptides removed from clone N-103 cells (unable to develop neurites) and from clone N-18 cells (able to extend neurites). They also showed that neuraminidase did not affect extension or maintenance of neurites and suggested that morphological differentiation may not require major changes in glycoproteins.

Apparent differences between experiments in which proteins were labeled by amino acids and those in which they were labeled by fucose or glucosamine may reflect faster turnover of the carbohydrate region than of the polypeptide backbone of the glycoprotein. Several workers addressed themselves to this problem and found that in neuroblastoma, as in other mammalian cells, glycoproteins turn over at heterogenous rates (Mathews *et al.*, 1976; Hudson and Johnson, 1977a,b). Thus, three fucosylated membrane components showed rapid degradation (150,000-, 130,000-, and 48,000-dalton molecules), and another (68,000 daltons) appeared to be turning over more slowly than the rest. The biological roles of these proteins are not known.

A puzzling observation made by Truding and Morrel (1977) is that cells in which differentiation is induced by dibutyryl cyclic AMP treatment release into the medium glycoproteins that were not detected on the surface membrane.

Glycosaminoglycans (GAGs) are another group of molecules that act as regulatory factors of certain cellular events. Moss (1974) observed that neuroblastoma C-1300 cells contain hyaluronidase-resistant sulfated glycosaminoglycans and suggested that this is one cause for their poor adhesive properties. Augusti-Tocco and Chiarugi (1976) reported that surface heparan sulfate (HS) is released into the growth medium of suspension culture and that the adhesion of the cells to the culture dish is accompanied by an increased ability of the cells to retain the HS on their surface. This supports the results of Stoolmiller *et al.* (1973) who reported that, in monolayer, heparan sulfate remains cell-associated. The heparan sulfate molecule is apparently ubiquitous on the cell surface, and as soon as the cell enters mitosis, it sheds heparan sulfate from its surface (Kramer and Tobey, 1972).

2.6.3. Ectoenzymes

Several membrane-bound enzymes show characteristics that suggest that their active sites are exposed to the external medium rather than the cell interior. Of these, 5′-nucleotidase, inorganic pyrophosphatase, ATPase, and acetylcholinesterase were shown by Stefanovic *et al.* (1974a–c, 1975a,b, 1976, 1977) and Ledig *et al.* (1975) to be present in the membranes of neuroblastoma cells. The neuroblastoma ATPase was stimulated by Ca^{2+} better than by Mg^{2+}. For glial cells, the reverse was true. An interesting functional interaction was observed on co-culturing neuroblastoma with an astroblast line (NN): an increase of ecto-ATPase activity that was dependent on the relative amounts of the two cell types in the culture was evident. A similar increase was observed for ectoacetylcholinesterase activity but not for 5′-nucleotidase. Conditioned medium from the NN clone but not of C-6 glioma cells could mimic this effect. In addition, cyclic AMP, which induced morphological differentiation, caused a reduction of the ecto-ATPase activity (Mandel *et al.*, 1977). Thus, a complex set of changes that depend upon the particular clone was revealed.

A partial digestion of the cell membrane with neuraminidase substantially increased ectopyrophosphatase activity in neuroblastoma (clone N-18) and astrocytoma (clone NN) cells (Stefanovic *et al.*, 1975a). A similar activation was observed for *p*-nitrophenylphosphatase activity as well as for acetyl- and butyrylcholinesterase (Rosenberg *et al.*, 1975; Stefanovic *et al.*, 1975a). However, Trams *et al.* (1976), who attempted to repeat these experiments, could not reconcile their observations with the hypothesis that the stimulation of ectoenzyme activity was caused by removal of sialyl residues from the cell surface. The latter authors suggested that the observed "induction" reflects a change in membrane permeability that accounts for accelerated transfer of substrates and products to and from the cytoplasmic compartment.

2.6.4. Cell Surface Antigens

Biochemical techniques show certain differences in glycoproteins and enzyme activities in plasma membranes of differentiated cells. Is it possible to follow this maturation process by immunologic methods? Since the mouse neuroblastoma C-1300 tumor is of sympathetic neuronal origin, it is not surprising that various antigens demonstrable on the surface of its cells are similar to those of normal mouse brain tissue (Schachner, 1973; Schachner and Worthan, 1975; Joseph and Oldstone, 1974; Martin, 1974). During the selection of clones, some marker antigens were lost, but the remainder were detected on differentiated as well as nondifferentiated cells (Schachner, 1973). When antisera to differentiated N-18 cells (grown in serum-free media) were compared to those raised against nondifferentiated cells (suspension culture) or cells of clone 10-D that are unable to extend neurites, a class of antigens that were present only in the mature cells was revealed (Akeson and Herschman, 1974a,b). The latter studies showed, again, that the antigens specific to the differentiated cells are found abundantly in normal mouse brain. The changes in antigens under serum-free conditions were not attributed to a specific function or a specific protein. In another study (Akeson and Herschman, 1975), it was shown that clones that show weak neurite-forming ability in serum-free conditions ($<20\%$) apparently have lost some surface antigens that are found on differentiated N-18 cells.

2.7. Specific Receptors

2.7.1. Acetylcholine Receptors

Chemosensitivity of neuroblastoma cells to acetylcholine was demonstrated in the early electrophysiological studies of this sytem (Harris and Dennis 1970; Nelson *et al.,* 1971a). Acetylcholine could elicit both hyperpolarization and depolarization responses, in some cases in the same cell. The results reflected more than one receptor mechanism in these cells, and the expression of the capacity to produce these receptors appeared to be variable from cell to cell within the population (Nelson *et al.,* 1971a). Binding of the α-toxin from *Naja naja* was shown to be inhibited by both nicotinic and muscarinic blockers (Siman-Tov and Sachs, 1973), and the existence of these two types of receptors was also reported in hybrid cells of clone NG-108-15 (Klee and Nirenberg, 1974). Traber *et al.* (1975c) carried out some interesting electrophysiological studies on differentiated NG-108-15 cells and demonstrated sensitivity to both atropine and *d*-tubocurarine and a blocking effect by α-bungarotoxin. In undifferentiated cells, on the other hand, the sensitivity was only of the muscarinic type. The details of this specificity change during the development of the cell were not described.

Muscarinic receptors have been reported to exist in cells of clone N-18 (Matusuzawa and Nirenberg, 1975), clone N1E-115 (Matsuzawa and Nirenberg, 1975; Richelson, 1977), clone NS20 (Blume *et al.,* 1977b), and clones NBDB and NBEA (Prasad *et al.,* 1974). Muscarinic receptors have also been reported on the neuroblastoma × glioma hybrid clone NG-108-15 (Traber *et al.,* 1975c; Burgermeister *et al.,* 1978), the sympathetic ganglia × neuroblastoma cell hybrid (Blosser *et al.,* 1978; Myers *et al.,* 1978), and the neuroblastoma × L cell hybrid (Chalazonitis *et al.,* 1977). Receptor densities of 25 fmole/mg protein (N1E-115 cells) and 40 fmole/mg protein (NG-108-15 cells) were calculated (Burgermeister *et al.,* 1978). These values should be compared to 35–478 fmole/mg protein in different regions of rat brain and to 28–208 fmole/mg protein in various peripheral tissues (Yamamura and Snyder, 1974a,b).

Muscarinic receptors frequently mediate slow, prolonged responses that may be excitatory or inhibitory. N1E-115 cells sensitive to acetylcholine or carbamylcholine show an inhibitory response (hyperpolarization) and increased synthesis of cyclic GMP (Matsuzawa and Nirenberg, 1975). NG-108-15 cells, on the other hand, possess excitatory receptors and show membrane depolarization and a decrease in cyclic AMP (Boeynaems and Dumont, 1975). Koike and Miyake (1977) reported that Con A (20–25 μg/ml), which does not affect resting or action potentials, did block the muscarinic response. It is not clear if this agent exerts its action directly or indirectly.

Upon activation of the muscarinic receptor, the intracellular level of cyclic

GMP increases 40-fold in clone N-18 and up to 210-fold (> 600 pmole/mg protein) in clone N1E-115 (Matsuzawa and Nirenberg, 1975). Extracellular Na^+ and Ca^{2+} ions are required for cyclic GMP production (Richelson *et al.*, 1978). The level of cyclic GMP, in the absence of a cyclic nucleotide phosphodiesterase inhibitor falls rapidly to normal level (50% in 45 sec). Addition of such an inhibitor, isobutylmethylxanthine (IBMX), postponed the time at which peak levels of cyclic GMP were attained [from 15 to 30 sec (Matsuzawa and Nirenberg, 1975) or from 30 to 45 sec (Richelson, 1977)]. It also prolonged the decay period of the cyclic nucleotide. Maximum effects were obtained with 10^{-3} M carbamylcholine and about 2×10^{-4} M acetylcholine. Atropine was inhibitory: 50% of the carbamylcholine stimulation was inhibited by 10^{-7} M atropine, and 97% of the response was inhibited by 10^{-6} M of the antagonist (Matsuzawa and Nirenberg, 1975). Ninety-five percent of the acetylcholine stimulation was inhibited by 1×10^{-8} M atropine (Richelson and Divinetz-Romero, 1977).

Side effects of various tricyclic antidepressants and antipsychotic drugs suggest that they block the muscarinic acetylcholine receptor. This hypothesis was confirmed by binding studies in which the drugs displaced ligands bound to brain receptors (Snyder *et al.*, 1974; Miller and Hiley, 1974). These studies were extended to neuroblastoma cells (clone N1E-115) in which a direct correlation between the pharmacological effect (increased cyclic GMP level) and binding affinity could be demonstrated. The dose–response curves of cyclic GMP stimulation by carbamylcholine were shifted in the presence of the tested drugs in a way which allowed the determination of an equilibrium dissociation constant for the antagonist–receptor complex (Richelson, 1977; Richelson and Divinetz-Romero, 1977). The potencies of the drugs could thus be correlated, by a biological assay, with their therapeutic value as well as with their side effects. For example, the optical isomer *l*-benzetimide, which was shown in other systems to be nearly devoid of anticholinergic activity, was 40,000 times less potent than the active *d*-isomer (dexitimide) in blocking the induced cyclic GMP-production.

The introduction of specific muscarinic ligands such as quinuclidinyl benzilate (QNB) (Yamamura and Snyder, 1974b) and benzylcholine mustards (Gill and Rang, 1966; Burgen *et al.*, 1974) has provided powerful tools for the characterization of the muscarinic receptor. QNB as well as [^{3}H]scopolamine were used by Burgermeister *et al.* (1978) to study the properties of these receptors on both neuroblastoma N1E-115 and the hybrid NG-108-15 cells. Binding and release of scopolamine were found to be a kinetically biphasic process. The apparent dissociation constants in the N1E-115 and NG-108-15 cells were 0.4 and 0.5 nM, respectively, for scopolamine and 0.06 and 0.1 nM, respectively, for QNB. Displacement studies were used to evaluate the binding properties of the receptor. The authors concluded that the antagonists' binding can be described as an interaction with a noncooperative class of receptors, whereas the agonists' binding exhibits negative cooperativity or heterogeneity in binding sites (Burgermeister *et al.*, 1978). This is consistant with the observations of Birdsall *et al.* (1978).

Receptor desensitization, as measured by a reduction in cyclic GMP formation, was demonstrated in N1E-115 cells after short treatments with various cholinergic agonists. The process, as in other receptor systems, was time-dependent (half-time about 4 min) and temperature-dependent and reflected the potency of the agonist. There were no alterations in QNB-binding characteristics under conditions that caused a reduction of the cyclic GMP response (Richelson, 1978a).

Sustained activation of the muscarinic ACh receptors in NG108-15 cells by various activators (i.e., acetylcholine, oxotremorine, arecoline, and carbachol) was found to cause loss of binding sites for the specific marker $[^3H]$-QNB; this finding suggested a mode of regulation by activation (Klein *et al.*, 1979). Blockers of the receptors, such as atropine or scopolamine, caused no such loss and atropine at 3×10^{-9} M completely inhibited the loss caused by 10^{-5} M carbachol. The loss response occurred at concentrations of agonist commensurate with those needed to occupy receptors and inhibit adenylate cyclase (see Nathanson *et al.*, 1978), thus suggesting that the receptor loss phenomenon is dependent on activation. Indeed, in aggregate cultures of embryonic chick cerebellum cells, which are known to form synapses and have spontaneous activity, addition of atropine caused an increase in the receptor level. This finding may reflect a blockade of the *in vivo* endogenous regulation of the muscarinic receptors (Siman and Klein, 1979). In N1E-115 cells, which do not form synapses, atropine did not cause increased $[^3H]$-QNB binding.

Growth of NG-108-15 cells in the presence of ACh caused no significant change in the receptor affinity of $[^3H]$-QNB, a finding which indicated that the reduction of binding sites was not due to a simple conformational change. Loss of receptor was not affected by cycloheximide, while increase of receptor sites after withdrawal of carbachol was sensitive to the drug—a finding which suggested that the reappearance of receptor molecules required a *de novo* synthesis (Klein *et al.*, 1979). Loss of receptors was also blocked by cytochalasin B (at 5 μg/ml) (Siman and Klein, 1979), indicating a role of neurofilaments in the receptor regulation process.

The muscarinic receptor was also shown to be involved in preventing the elevation of cyclic AMP levels mediated by either adenosine or PGE (Matsuzawa and Nirenberg, 1975; Traber *et al.*, 1975c; Blume *et al.*, 1977b). Various local anesthetics showed affinity for the muscarinic receptor in intact cells, and they displaced carbamylcholine as competitive inhibitors in the cyclic GMP formation assay (Richelson *et al.*, 1978). Burgermeister *et al.* (1978), however, reported that one of these drugs, tetracaine, showed noncompetitive characteristics in inhibiting $[^3H]$scopolamine binding in cells of the same clone.

Histrionicotoxin, an alkaloid isolated from the skin of the tropical frog

Dendrobates histrionicus, was shown to inhibit the ACh-elicited depolarization of mammalian and amphibian nerve–muscle preparations (Albuquerque *et al.,* 1973). Its binding to the muscarinic receptor in N1E-115 cells exhibited non-competitive characteristics, blocking 50% of scopolamine binding at 70 μM. These studies suggested that its action on this class of cholinergic receptors is different from its interaction with nicotinic receptors (Burgermeister *et al.,* 1978).

2.7.2. Histamine, 5-HT, and Dopamine

Histamine receptors of the H_1 type (defined on the basis of sensitivity to various agonists) have been demonstrated in N1E-115 cells (Richelson, 1978a). Some characteristics of these receptors are similar to those of the muscarinic receptors in these cells: for example, the kinetics of activation of cyclic GMP synthesis (rapid, peaking in 30 sec) and a 7- to 50-fold increase in cyclic GMP levels in the cells. The major difference between the two responses appears to be that "spare" receptors are involved in the muscarine- but not the histamine-mediated response (see Taylor and Richelson, 1979).

Electrophysiological studies by Christian *et al.* (1978) demonstrated that NG108-15 cells are sensitive to acetylcholine, 5-hydroxytryptamine, and dopamine (see also Myers and Livengood, 1975). Continuous application of 5-HT or dopamine desensitized the responses. A desensitizing application of one of these neurotransmitters also desensitized the hybrid cell to the other. 5-HT and actylcholine did not cross-desensitize. Interestingly, the 5-HT response was not attenuated by D-LSD but was blocked by 10^{-5} M morphine, although not via binding to nalozone-sensitive opiate receptors.

2.7.3. NGF Receptors

Since the original tumor C-1300 arose in the spinal cord region, the possibility that it responded to NGF was immediately tested. Hermetet *et al.* (1972b) found that NGF induced a neuronal-like differentiation in some clones similar to that obtained with BrdU and dibutyryl cyclic AMP. The treatment also caused an increase in acetylcholinesterase activity in clones N-1 and N1E-115 but not in N2A. Such effects were also observed in certain clones after treatment with insulin, which has structural similarities to NGF (Hermetet *et al.,* 1973). However, antiserum to NGF did not prevent the growth of processes induced by NGF (Mandel *et al.,* 1976). Furthermore, Brodeur and Goldstein (1976) reported that NGF did not induce neurite formation in neuroblastoma cells. However, treated cells stained more intensely for AChE than control cells. Also, Revoltella *et al.* (1974b) showed that NGF could induce microtubule formation but was incapable of promoting neurofilament formation.

Mouse neuroblastoma C-1300 cells display surface receptors for NGF

(Revoltella *et al.*, 1974b; Bosman *et al.*, 1975), and this property was shown to be correlated with cell growth cycle: receptors appeared mainly in the late G_1 and early S phase (Revoltella *et al.*, 1974a). About 10^6 molecules of NGF could be bound, at saturation, by each cell. Binding of NGF-coated red blood cells to neuroblastoma cells could be prevented by pretreatment of the cells with proteolytic enzymes, and new binding sites reappeared 1–2 hr later (Revoltella, 1974b). This effect was not found after treatment of cells with neuraminidase or phospholipases (Revoltella *et al.*, 1975). Colchicine and vinblastine (two alkaloids that bind tubulin) or cytochalasin B (which is a strong inhibitor of microfilament contraction) did not affect binding of NGF-coated erythrocytes. A protein (molecular weight about 52,000) that could bind with high affinity to labeled NGF was released into the medium when neuroblastoma cells were treated with 3 M KCl. Antibodies to this protein could, in the absence of complement, inhibit the binding of [^{125}I]-NGF to the cells and, in its presence, killed them. The amino acid composition of this protein has a remarkable similarity to tubulin, which is in line with the observation of Callisano and Cozzari (1974) that NGF binds with high affinity to this protein. Since tubulin is found in every cell, whereas the NGF effect is a specific one, the relationship of tubulin to this NGF-binding protein is still puzzling.

Neuroblastoma cells were found, like many other tumor cells, to secrete NGF into the growth medium (Murphy *et al.*, 1975). The significance of this phenomenon is not clear.

2.7.4. Opiate and Enkephalin Receptors

The mechanism by which opium and a large group of related agonists and antagonists exert their effect on the central nervous system was not known until recently, when a membrane preparation of brain cells was shown to bind these narcotic analgesics (Pert and Snyder, 1973; Simon *et al.*, 1973; Goldstein *et al.*, 1971; Hughes *et al.*, 1975). Such receptors, which are functionally coupled to adenylate cyclase, were also demonstrated in NG-108-15 cells by specific binding of [^{3}H]dihydrormorphine (Klee and Nirenberg, 1974) and by inhibition of the PGE-induced increase of cyclic AMP levels in these cells (Traber *et al.*, 1974a, 1975a). An opportunity to study under controlled conditions the molecular aspects of opiate action was therefore opened. Morphine-sensitive and -insensitive cell lines were classified, and the degree of sensitivity was shown to be dependent upon the abundance of narcotic receptors (Sharma *et al.*, 1975a). A density of 300,000 receptor sites per NG-108-15 cell was calculated by Klee and Nirenberg (1974). A lower level of 30,000–80,000/cell was later found in clones isolated after fusion of sympathetic ganglia and neuroblastoma cells (Blosser *et al.*, 1976). A clone in which morphine blocked dopamine-induced depolarization but had no effect on the PGE-stimulated adenylate cyclase was also found (Blosser *et al.*, 1978; Myers *et al.*, 1978).

As in other tissues, nonradioactive morphine did not displace all of the bound morphine. The nondisplaceable component represents the nonspecific binding. The saturable (displaceable) component proved to be stereospecific. The analgestic compound levorphanol, for example, displaced dihydromorphine at concentrations three to four orders of magnitude lower than dextrorphan, the inactive enantiomer. The binding affinity for morphine is about ten times lower than that found in the brain but higher than that of the isolated guinea pig ileum (Klee and Nirenberg, 1974; Blosser *et al.,* 1976). However, the relative binding strengths of the various analgesics were similar in brain and neuroblastoma cells.

The inhibitory effect of morphine on the PGE_1-stimulated cyclic AMP production is noncompetitive and is not reversed by very high concentrations of PGE. Naloxone, on the other hand, displaced morphine and reversed its inhibitory effect on the adenylate cyclase activity (Traber *et al.,* 1974a; Sharma *et al.,* 1975a; Klee *et al.,* 1975). Such studies indicate that drugs like naloxone and morphine compete for a site or sites on the same receptor, whereas PGE_1 binds to a different receptor. The two types of receptors may then compete for the same coupling site on the adenylate cyclase complex or, alternatively, both may be coupled to adenylate cyclase simultaneously at separate sites and exert their effect by an allosteric mechanism (Sharma *et al.,* 1977; Klee *et al.,* 1975).

A naloxone-induced stimulation of the adenylate cyclase system (15–20%) was measured in some experiments (Sharma *et al.,* 1975a), but the mechanism of this effect was not studied.

As indicated above, a positive correlation among the relative affinities of various narcotics was demonstrated by measuring naloxone displacement and determing the concentration required for 50% inhibition of adenylate cyclase activity (Sharma *et al.,* 1975a). Hill plot analysis of the data suggested that the formation of the opiate–receptor complex is not a cooperative process, whereas the interaction of the receptor with the adenylate cyclase is.

Morphine-treated cells reveal a rapid reversible inhibition of cyclic AMP production that is followed by a slow, gradual, and long-lasting increase in cyclic AMP level. Withdrawal of the drug at this stage causes on overshooting of adenylate cyclase activity (Sharma *et al.,* 1975b). However, little difference is found when homogenates of opiate-treated and control cells are tested for adenylate cyclase activity in the presence of NaF or Gpp(NH)p, two agents known to uncouple adenylate cyclase from receptors. These and other observations led to the suggestion that in the presence of an opiate more of the enzyme is converted to a different form with an altered activity. As a result, the temporary depletion of cyclic AMP in the cell is reversed, and the cell attains a normal level of this cyclic nucleotide in spite of the opiate presence (Sharma *et al.,* 1977; Klee *et al.,* 1975).

Tolerance and dependence produced by chronic exposure to opiates can also be explained at the cellular level in terms of this dual action on adenylate cyclase activity: opiate-treated cells are tolerant to opiates because the relatively high adenylate cyclase activity of such cells is inhibited to a greater extent by the opiate. They are dependent upon opiates because withdrawal of the opiate reverses the inhibition of adenylate cyclase and thereby increases the cyclic AMP level well above that of control cells. Finally, opiate-treated cells are supersensitive to the opiate antagonist naloxone because opiate-treated cells have more adenylate cyclase activity and more opiate-inhibited enzyme activity than control cells (Sharma *et al.*, 1975b; Hamprecht, 1977).

Inhibition of neuroblastoma cell multiplication by morphine (at 1–100 μM) and etorphine (at 10^{-8} M) was reported by North-Root (1976a,b). Interestingly, only the nucleus and not the plasma membrane was found to bind the opiates with high affinity: saturation was reached at 2×10^{-9} M, and a dissociation constant of 5×10^{-11} M was determined. The binding took place at 37°C but not at 4°C, was susceptible to proteases, and appeared to be chromatin-associated.

Identification of the enkephalins as the natural morphinelike peptides of the brain (Hughes *et al.*, 1975; Siman-Tov and Snyder, 1976) was quickly followed by studies by Brandt *et al.* (1967a,b) and Klee *et al.* (1976), who were able to demonstrate that the two enkephalins effectively regulate the PGE_1 induction of cyclic AMP levels in NG-108-15 cells.

The enkephalins are extremely potent in inhibiting adenylate cyclase: a 50% inhibition of the basal activity is obtained by 12 nM met-enkephalin or 40 nM leu-enkephalin, whereas 1500 nM of morphine are required to achieve this effect (Klee *et al.*, 1976). The effect is blocked by naloxone and manifests a dual regulation consisting of immediate inhibition and delayed stimulation of adenylate cyclase (Klee and Nirenberg, 1976; Brandt *et al.*, 1976a; Lampert *et al.*, 1976). Specific binding of the enkephalins was also demonstrated in N4TG1 cells (Miller *et al.*, 1977). The binding is sensitive to the ionic composition of the medium, in particular to Na^+ and Ca^{2+} (Miller *et al.* 1977; Brandt *et al.*, 1978; Chang *et al.*, 1978), and was used by Dickmann-Gerber *et al.* (1978) to develop a radioreceptor assay for the opioid peptides.

By measuring the binding of [^{125}I]labeled ligands, Chang *et al.* (1978) determined that each neuroblastoma cell (clone N4TG1) has about 18,000 enkephalin-specific receptors. This value is somewhat lower than the number of morphine receptors per cell.

Membrane-bound enkephalin and opiate receptors were found not to be internalized upon interaction with the effectors, even after 60 min at 37°C and were extremely resistant to proteolytic digestion and to phospholipase treatment (Chang *et al.*, 1978). Binding studies of enkephalins, β-lipotropin peptides, and opiates led to the conclusion that the cloned cells (clones NG-108-15 or N4TG1) possess only a single class of enkephalin-binding sites (Wahlstrom *et al.*, 1977; Chang *et al.*, 1978), whereas at least two classes of opiate receptors were inferred in studies of the rat brain.

Rapid inactivation of enkephalins similar to that observed when brain cells are incubated with the peptides was also evident in neuroblastoma cell cultures, providing a model system for studying not only the molecular mechanism of enkephalin action but also the degradation of these natural "narcotics" and similar peptides (Brandt *et al.*, 1976a; Klee *et al.*, 1976). The enzymatic activity is temperature-dependent, and an apparent K_m of 5×10^{-5} M was determined at 37°C. Inhibition by bacitracin ($K_i = 3.2 \times 10^{-5}$ M) and puromycin ($K_i = 2.3 \times 10^{-7}$ M) was demonstrated. Since cell trypsinization greatly diminished the ability to degrade enkephalin, it was suggested that the enzyme is located at the cell surface (Hazum *et al.*, 1979).

Binding of endogenous nonpeptide, morphinelike compounds of the brain to opiate receptors of clone NG-108-15 cells was reported by Blume *et al.* (1977a).

Brandt *et al.* (1978) observed that chronic exposure of hybrid cells to low concentrations of Ca^{2+} gradually evoked an increase in maximal response to PGE similar to that observed with opioid treatment. It was suggested that the effect is mediated via adenylate cyclase activation.

2.7.5. Prostaglandins

Although prostaglandins have been shown to elicit various responses in brain and other nerve tissue (Daly, 1977), their physiological function in the nervous system is not known. They are rapidly inactivated in extraneuronal tissues (e.g., blood) and may serve as short-term, short-range signals between nearby cells.

Neuroblastoma cells can synthesize prostaglandins, mainly PGE and only minor amounts of PGF and PGA. The rate of production is enhanced in the presence of dibutyryl cyclic AMP (Hamprecht *et al.*, 1973).

Most of the work concerning neuroblastoma cells and prostaglandins involves the events taking place when the cells are challenged with these agents. Prostaglandin E_1 stimulates an increase of intracellular levels of cyclic AMP in practically every clone of neuroblastoma tested (Gilman and Nirenberg, 1971; Minna and Gilman, 1973; Prasad, 1972c; Hamprecht and Schultz, 1973a; Blume *et al.*, 1973; Glazer and Schneider, 1975; Matsuzawa and Nirenberg, 1975; Sahu and Prasad, 1975; Siman-Tov and Sachs, 1975b,c; Blume and Foster, 1975, 1976; Penit *et al.*, 1976, 1977; Green and Stanberry, 1977) and neuroblastoma hybrids (Minna and Gilman, 1973; Hamprecht and Schultz, 1973b; Sharma *et al.*, 1975a; Blosser *et al.*, 1978; Myers *et al.*, 1978). When parental lines with little or no response to PGE_1 were fused to cells having better responses to this agent, the majority of progeny hybrid cells expressed high levels of PGE_1 responsiveness (Minna and Gilman, 1973; Hamprecht and Schultz, 1973b).

In most cases, however, the increase in the cellular cyclic AMP levels (which are several-hundred-fold above control levels) could be demonstrated only in the presence of cyclic nucleotide phosphodiesterase inhibitors.

PGA_2 can also induce morphological differentiation (Adolphe *et al.*, 1974), but other prostaglandins such as PGE_2, PGA, PGB_1, $PGF_{1\alpha}$, or $PGF_{2\alpha}$ are either less active than PGE_1 or inactive altogether.

In synchronized NS20 cells, the activity of adenylate cyclase in the presence of 1 μM PGE_1 decreased during G_1 phase, increased rapidly during the first hours of the S phase, and declined before mitosis (Penit *et al.*, 1977). There was no evidence for changes in the K_m values during the entire cell growth cycle. PGE_1 increased the levels of cyclic AMP as well as cyclic GMP. A variety of responses were observed when different pairs of receptors for PGE_1, adenosine, and carbamylcholine were activated in different combinations simultaneously in clone NIE-115 or NS20 (Matsuzawa and Nirenberg, 1975; Penit *et al.*, 1977).

2.7.6. Adenosine

The stimulation of cyclic AMP-generating systems, the potentiation of biogenic amine responses, and other effects of adenosine on nerve cells are now well described. Several analogues of this purine act as agonists, whereas others are antagonists. A typical antagonist is theophylline (as are other methylxanthines), and a typical agonist is 2-chloroadenosine (see Skolnick and Daly 1977).

Adenosine- and 2-chloroadenosine-mediated elevation of cyclic AMP were demonstrated in several clones (Blume *et al.*, 1973: Schultz and Hamprecht, 1973; Blume and Foster 1975; Matsuzawa and Nirenberg 1975; Penit *et al.*, 1976, 1977; Green and Stanberry, 1977). The effect was stereospecific. In most cases, a cyclic nucleotide phosphodiesterase inhibitor had to be present. Theophylline was reported not to antagonize adenosine in one case (Schultz and Hamprecht, 1973) but was found to be inhibitory by other workers (Blume *et al.*, 1973; Blume and Foster, 1975; Green and Stanberry, 1977). There is also some controversy about the mechanism of action: although Blume and Foster (1975) concluded that theophylline is a competitive inhibitor of 2-chloroadenosine (K_i = 35 μM), a comparison among the structure–activity relationships of adenosine analogues (agonists) and xanthine derivatives (antagonists) indicated that the latter act by an allosteric rather than a competitive mechanism (Green and Stanberry, 1977).

The observation that the adenosine response was not blocked by dipyridamole, which inhibits adenosine uptake, was taken to indicate that adenosine mediates its effect by an extracellular "adenosine receptor" (Green and Stanberry, 1977). However, a direct activation of the adenylate cyclase system was suggested by Penit *et al.* (1976), who carried out experiments with permeabilized cells. A direct, noncooperative, stimulatory effect was also suggested by Blume and Foster (1975).

A synergistic interaction between adenosine and catecholamines was not observed in neuroblastoma cells (Schultz and Hamprecht, 1973; Penit *et al.*, 1976).

2.7.7. Somatostatin

A receptor for the tetradecapeptide, somatostatin (somatotropin release inhibitory factor, SRIF) has been demonstrated in NG-108-15 hybrid cells (Traber *et al.*, 1977). Like the opiates, somatostatin also inhibits the PGE_1-induced elevation of cyclic AMP levels. The specific opioid antagonist, naloxone, did not inhibit the action of somatostatin, and the latter did not block the specific binding of [^{3}H]naloxone to the opioid receptors, thus indicating that they have different receptors (Traber *et al.*, 1977). Release of norepinephrine from a human neuroblastoma line was shown to be blocked by somatostatin (Maruyama and Ishikawa, 1977), indicating the presence of somatostatin receptors in these cells.

2.8. Cyclic Nucleotide Metabolism

2.8.1. General Considerations

It is now well-accepted that cyclic AMP is an intracellular messenger in most cells (for reviews, see Robinson *et al.*, 1971). Upon hormone stimulation (or activation), ATP is converted to cyclic AMP by the membrane-bound enzyme adenylate cyclase; this initiates a cascade of events that ultimately result in physiological expression. Many lines of evidence have led to the hypothesis that norepinephrine- and dopamine-activated adenylate cyclase are part of the postsynaptic response machinery (Greengard, 1976). However, at least some aspects of this scheme have been challanged recently (see Libet, 1979).

Clearly, the levels of these two nucleotides modulate many responses of the cell to external stimuli. As in many other cells, the metabolism of cyclic nucleotides in clonal lines is a subject for very extensive research

2.8.2. Adenylate Cyclase

The participation of adenylate cyclase in the synthesis of cyclic AMP in neuroblastoma cells after stimulation by PGE, adenosine, opioids, and enkephalins has been described in Sections 2.7.4–2.7.7.

The response to catecholamines apparently varies from clone to clone (Gilman and Nirenberg, 1971; Blume *et al.*, 1973; Penit *et al.*, 1977). The response of dividing cells to dopamine and norepinephrine differs from that of differentiated cells see Prasad *et al.*, 1976a,b). The responses to dopamine and norepinephrine in differentiated cells were additive, suggesting the existence of different receptors (Blosser *et al.*, 1978; Myers *et al.*, 1978).

(Traber *et al.*, 1974a) and dopamine (Peacock and Nelson, 1973; Myers and Livengood, 1975) have been shown to affect membrane potentials in neuroblastoma cells, but the site of their interaction was not characterized. The response to norepinephrine (via β-receptors) that exists in glial cells is not expressed in the progeny after fusion with neuroblastoma cells (Hamprecht and Schultz, 1973a). However, evidence of an α-type receptor for norepinephrine (inhibitory, depolarizing) in one hybrid was presented by Traber *et al.* (1975b). Responses to dopamine and norepinephrine in rat CNS-derived clones have also been reported (Schubert *et al.*, 1976).

Some properties of adenylate cyclase in cell-free preparations from clone NS-20 have been determined (Blume and Foster, 1976). The K_m for ATP in the presence of Mg^{2+} was found to be 120 $\pm$ 15 μM. The interaction appeared to be noncooperative. The K_m for 2-chloroadenosine was 6 μM; the V_{max} increased in the presence of 2-chloroadenosine, but the K_m for ATP was unchanged. Calcium ions were inhibitory ($K_i = 5 \times 10^{-4}$ M), and the interaction with the enzyme appeared to be cooperative. Mn^{2+}, which substitutes for Mg^{2+} in cyclic AMP production, antagonized the inhibitory action of Ca^{2+}. Activation of adenylate cyclase in cells permeable to ATP and adenosine showed an apparent K_m for adenosine of 5×10^{-6} M (Penit *et al.*, 1976). The effect was stereospecific, since structural analogues of adenosine were inactive. Half maximal cyclic AMP elevations were obtained by 2 μM adenosine in the presence of 0.7 mM Ro20-1724 (Blume *et al.*, 1973).

The hypothesis that activation of adenylate cyclase may lead to a reduction in its activity was tested by Kenimer (1978). A 60–80% decrease in basal chloroadenosine-, NaF-, and PGE_1-stimulated activity was found after 12 hr incubation of NG-108-15 cells with PGE_1 (24 μM). The decay was exponential with a half-life of 6 hr. The recovery within 12 hr (to 80% of control) was sensitive to cycloheximide (20 μg/ml).

In intact NG108-15 cells, carbachol, the muscarinic ACh receptor agonist, mediated transient as well as long-lived effects on adenylate cyclase activity: it rapidly and reversibly inhibited but also evoked an increase in enzyme activity over a period of 24–30 hr. The carbachol-dependent increase in adenylate cyclase activity, once acquired, could be expressed without carbachol present. Both effects of carbachol were blocked by 1 μM atropine (Nathanson *et al.*, 1978).

2.8.3. Guanylate Cyclase

The activation of the guanylate cyclase system in neuroblastoma cells by receptors to acetylcholine, PGE, and histamine was discussed in a previous section (2.7). Little is known about the enzyme complex, its specific properties, or the mechanism by which it acts and is activated after binding of the appropriate ligand to the receptor.

2.8.4. Cyclic Nucleotide Phosphodiesterases

The action of cyclic nucleotides is terminated by specific enzymes—phosphodiesterases that catalyze the hydrolysis of cyclic AMP and cyclic GMP to the 5′-monophosphonucleotides. Multiple forms of cyclic 3′,5′-AMP phosphodiesterases, separable by gel electrophoresis, were found by Uzunov and Weiss (1972) in the rat cerebellum. They had different stabilities and reacted differently to various activators and inhibitors. Only one of these forms (corresponding to peak III from the rat cerebrum) was found in neuroblastoma cells (clones N-1 and N-18). The enzyme was quite labile but could be stabilized with 1% albumin and dithiothreitol. Kinetic data analysis shows that the neuroblastoma enzyme exhibits linear kinetics with a K_m of about 0.1 mM (Uzunov *et al.*, 1974).

As indicated above, various inhibitors affect the isoenzymes differently. Thus, theophylline (at 1 mM) inhibited only 50% of the phosphodiesterase activity (Uzunov *et al.*, 1974), a finding that explains the observation of Hamprecht and Schultz (1973a) that PGE in presence of papaverine and isobutylmethylxanthine (IBMX) induces cyclic AMP levels 25-fold higher than those that were determined earlier with theophylline as inhibitor (Gilman and Nirenberg, 1971). In addition, Hamprecht and Schultz reported that IBMX was generally more effective than papaverine. These data could be explained by assuming the existence of two or more isoenzymes that are affected differently or to a different extent by a given inhibitor. However, such a suggestion was not substantiated in the work of Uzunov *et al.* (1974)

The levels of phosphodiesterase increase in morphologically differentiated cells that are treated by either cyclic AMP, PGE_1, or Ro20-1724 [4-(3-butoxy-4-methoxybenzyl)-2-imidazolidinone], an inhibitor of phosphodiesterase (Prasad and Sheppard, 1972a,b; Prasad and Kumar, 1973). X-ray irradiation induces morphological differentiation similar to that obtained by cyclic AMP but does not change phosphodiesterase activity. Also, sodium butyrate, AMP, and 5′-AMP, as well as theophylline, which inhibits cell division without causing morphological differentiation, do not significantly affect the basal level of PDE (Prasad and Kumar, 1973). Elevation of cyclic AMP PDE activity by cyclic AMP was inhibited by cycloheximide but not by actinomycin D.

Sinha and Prasad (1977) showed that cyclic AMP phosphodiesterase and cyclic GMP phosphodiesterase activities in NBP_2 cells are independently controlled and affected differently by various treatments. Both cyclic GMP and cyclic AMP PDE activities were found in the particulate as well as in the supernatant fraction of the cell homogenate. Both activities showed low (2–8 μM) and high (44–220 μM) K_m values. The two enzymes responded differently to the inhibitor Ro20-1724 and were affected differently by increasing the protein concentration in the assay, suggesting the existence of specific intrinsic inhibitors (Prasad *et al.*, 1975a).

Differences in various phosphodiesterase properties in dividing and undifferentiated cells led to the suggestion that abnormal regulation of cyclic AMP phosphodiesterase is one of the early lesions in malignant transformation (Prasad, 1974).

2.8.5. Cyclic AMP-Binding Proteins and Protein Kinase

If cyclic AMP participates in the differentiation process, can one find differences in activity or level of cyclic AMP-binding proteins at different stages of growth? Are there changes in protein kinase activity? Is regulation reflected as a quantitative or a qualitative difference in protein phosphorylation?

Prasad *et al.* (1978) reported that dibutyryl cyclic AMP induces a threefold increase in activity of a cyclic AMP-binding protein, an event that required RNA and protein synthesis. The binding protein, which consists of two subunits (mol. wt. = 48,000), was found in the cytoplasm as well as among nuclear nonhistone proteins.

Decreases in histone synthesis and Hl-histone phosphorylation were observed in cells that had been induced to differentiate by prolonged treatment with PGE and Ro20-1724 (Lazo *et al.*, 1976). Only minor differences were noted in the nonhistone chromatin-associated protein of these cells.

2.9. Specific Proteins

2.9.1. Tubulin

Tubulin and actin, the subunit proteins of microtubules and microfilaments, respectively, are major components of the cytoskeleton of axons and dendrites. Tubulin, for example, is a major protein in the soluble fraction of brain homogenate (33% in the developing brain and 20% in the adult animal). The intracellular synthesis and state of polymerization of these proteins are probably closely linked to the stability of these cellular structures as well as to the state of axonal transport mechanisms (Lasek and Hoffman, 1976).

Microtubule protein from neuroblastoma cells was first isolated by vin-

An exceptionally high expression of the 10 nm filamentous structure was reported in one of the clones (NB6R) isolated by Bertolini *et al.* (1977). In this cell line and in others, BrdU treatment causes an increase in the number and length of the filaments. The structure did not disappear when the cell formed mitotic spindles or in presence of colcemid (10^{-6} M), vinblastine (5×10^{-6} M), or cytochalasin B (5 μg/ml), which alter the structure of neurotubules and microfilaments. These results of Jorgensen *et al.* and Bertolini *et al.* suggest that the 10-nm filaments, which are found in axons as well as in many other cells, are assembled from a cellular pool of a simple oligomer.

Reversible inhibition of cell differentiation by halothane, a commonly used volatile anasthetic (0.3–2.1% in the gas phase), appears to be mediated by disruption of 4- to 8-nm-diameter microfilaments. Microtubule structures were found to be less sensitive (Hinkley and Telser, 1974).

2.9.4. Protein 14-3-2

The 14-3-2 protein was found by Moore and Perez (1968) to be specific to the nervous system: mammalian brain was found to contain 200 times more 14-3-2 protein than any other organ tested. In 1975, Bock and Dissing showed that this protein exhibits enolase activity. Although the mystery of the protein was thus unveiled, its specific function and abundance in nervous tissue is still not explained (see Bock, 1978).

While Herschman and Lerner (1973) were unable to demonstrate this protein in clones N18 and NB4, Augusti-Tocco *et al.* (1973) showed its presence in clone NB41A3 in both suspension and monolayer cultures. The protein is accumulated rapidly during the lag phase of cell growth unlike, for example, AChE, which accumulates in confluent cultures. Revoltella *et al.* (1976) measured the protein in various cell extracts and showed that, as for many other components, different clones contain different amounts of this protein.

3. RAT PHEOCHROMOCYTOMA

3.1. General Considerations

A transplantable rat pheochromocytoma—a tumor of the adrenal medulla—was isolated by Warren and Chute (1972) in old, irradiated New England Deaconess Hospital strain animals. The transplant-bearing rats developed clinical symptoms that suggested that the tumor cells were functional and released catecholamines. Indeed, subsequent studies by DeLellis *et al.* (1973) confirmed the presence of high intracellular concentrations of primary catecholamines. Two types of granules were found in the same cell: (1) vesicles with an electron-dense core that were separated from the surrounding membrane by an irregular, wide, electron-lucent space and measured up to 0.3 μm in diameter; and (2) smaller vesicles (up to 0.2 μm) composed of a less electron-dense core and separated from the membrane by a narrow, regular, electron-lucent space. The first type predominated in the cells.

From this tumor, Green and Tischler (1976) established a clone designated PC-12 with a near diploid number of 40 chromosomes. PC-12 cells have a round or polygonal shape and tend to grow in small clumps. They show a pluripotency similar to that observed in cultures of sympathetic neurons in being able to synthesize, store, and release catecholamines as well as acetylcholine (see Furshpan *et al.,* 1976). Unlike neuroblastoma cells, they do not extend processes and could not be induced to do so by serum withdrawal, cyclic AMP treatment, or inhibition of RNA or DNA synthesis. They respond to NGF and, by one week of treatment, cease to multiply and begin to extend branched, varicosed neurites (Green and Tichler, 1976). In addition, the NGF treatment induces the development of electrical excitability and sensitivity to acetylcholine (Dichter *et al.,* 1977).

Nerve growth factor (NGF) is a protein that stimulates the growth of sympathetic neurons *in vivo* and *in vitro.* It causes the cells to increase in volume and extend neurites and also controls the levels of various neurotransmitter-synthesizing enzymes (Levi-Montalcine, 1976). Like many other hormones and growth factors, NGF had been shown to interact with specific receptors on the surface plasma membrane of its target cells. In addition, NGF can be internalized and transported into the cell body. Two receptor types for NGF were described in chick embryonic dorsal root neurons (Andres *et al.,* 1977): one is associated with the plasma membrane (or microsomal fraction), is solubilized by Triton X-100, and displays nonsaturable binding characteristics; the second is located in the nucleus (bound to chromatin), is not solubilized by detergent, and displays saturable binding. These findings suggested two discrete, separate loci for the expression of the different NGF responses.

The availability of a stable clonal cell line that responds to NGF by acquiring properties characteristic of sympathetic neurons offers a powerful tool to study various properties of this important protein and the mechanism of induction of nerve cell differentiation. The effects of NGF on the PC-12 cells may be arbitrarily classified as (1) early events that take place within a few minutes and up to several hours and (2) slower processes (long-term) manifested after several days.

3.2. Morphological Differentiation Induced by NGF

PC-12 cells grown without NGF are round or polygonal and tend to form clumps. They possess numerous membrane-bound cytoplasmic granules, most of which are larger than 120 nm in diameter. These morphological features are similar to those found in normal rat chromaffin cells and human pheochromocytomas and are correlated with formaldehyde-induced catecholamine fluorescence. Addition of NGF is followed by flattening of the cells, formation of short cytoplasmic extensions, development of membrane mounds and microspikes, accumulation of granules near the periphery, and formation of neurite-like processes (Tischler and Greene, 1978).

The NGF-treated cultures share more properties with primary sympathetic ganglion cells than with chromaffin cells. Growth of neurites (which may reach 500–1000 μm in length) is accompanied by loss of intracellular granules larger than 120 nm in diameter with a concomitant marked reduction of fluorescence and appearance of clusters of smaller vesicles (up to 70 nm in diameter). While some of the vesicles retain the dense-core characteristics of catecholamine-containing granules of sympathetic neurons, the majority resemble those found in cholinergic nerve cells. In the early stages of process formation, formaldehyde-induced fluorescence can be demonstrated both in cell bodies and in processes, but in later stages, there is a marked diminution of this fluorescence (Tischler and Greene, 1978).

PC-12 cells are not viable in serum-free medium unless NGF is added within a short time after serum withdrawal. Some experiments suggest that this survival is independent of transcription and is different from the neurite outgrowth effect (Greene, 1978). More work is required to clarify these points.

3.2.1. Early Events after Interaction with NGF

NGF was found to increase the initial rates of cell–substratum cell–cell adhesion of sympathetic nerve cells and PC-12 cells (Schubert and Whitlock, 1977). This response is inhibited at 4°C, does not require protein synthesis, and its dose–response curve is superimposable on that of neurite extension effect. Intracellular levels of cyclic AMP were found to increase transiently, and theophylline potentiated the NGF effect on adhesion—observations that pointed out a role for cyclic AMP in the process. However, other compounds that cause membrane depolarization without affecting cyclic AMP were also effective in promoting adhesion. Further studies led Schubert *et al.* (1978) to propose that the NGF effect is mediated by calcium ions which, in turn, cause changes in plasma membranes, triggering changes in the cells. Detailed molecular studies are still undone at this time.

NGF has also been found to induce ornithine decarboxylase (ODC), the activity peaking within 5–8 hr of treatment (Greene, 1978). Induction of ODC activity, which catalyzes the synthesis of polyamines, has been suggested as playing a part in mediating cellular responses to hormone and growth-promoting agents. This induction showed a dose–response relationship to NGF and was sensitive to RNA synthesis inhibitors. However, specific ornithine decarboxylase inhibitors that abolished the enzymatic activity did not affect either cell viability or the ability of NGF to promote neurite outgrowth. Is ODC induction relevant to the NGF effect? An answer to this question requires further studies.

3.2.2. Long-Term Effects of NGF

Although NGF-treated PC-12 cultures differ considerably from non-treated cultures in their morphology, neurotransmitter-synthesizing enzyme levels, vesicle characteristics, electrical excitability, and acetylcholine sensitivity, only few proteins were found to change when NGF-treated cells are compared to control cultures.

By use of immunologic techniques, Lee *et al.* (1977) were able to show that PC-12 cells share specific components with brain and adrenal medulla cells. However, no differences could be detected in the specificity of antisera raised against either NGF-treated or untreated PC-12 cells. Furthermore, total cellular protein patterns of NGF-treated and control cells showed no consistent qualitative and only minor quantitative changes in a small number of species (McGuire *et al.,* 1978). Similar results were also reported by Schubert (1979). The possibility that NGF affects the synthesis of minor components that escaped detection by the currently used methods should still be considered.

Labeling cells with fucose or glucosamine revealed that NGF stimulated the incorporation of radiolabeled fucose into several components, whereas studies with labeled amino acids did not bring out important differences. For example, a glycoprotein with an apparent molecular mass of 230,000 daltons that is at least partially exposed on the cell surface was first detected after 2 days of exposure to NGF (McGuire *et al.,* 1978). Its level increased along with the NGF-stimulated neurite outgrowth, and its synthesis apparently required new transcription. This glycoprotein was also synthesized in suspension cultures treated with NGF, but to a lesser extent.

Do the NGF-induced processes require RNA synthesis? Levi-Montalcini and Angeletti (1968) proposed that RNA does participate in NGF-induced differentiation, but later experiments by Partlow and Larrabee (1971) showed that NGF-stimulated neurite outgrowth can take place even in the presence of actinomycin D. Burstein and Greene (1978), discussing this problem, suggested that the cells that were used in the latter experiments had already been exposed, *in vivo,* to NGF. Such cells were therefore expressing an NGF-stimulated *regeneration* insensitive to RNA synthesis inhibition instead of an *initiation* of neurite outgrowth. Since PC-12 cells can be grown without NGF, they offer an appropriate model system to probe this problem. By use of several RNA-synthesis inhibitors, Burstein and Green (1978) demonstrated that indeed *de novo* synthesis of RNA was required for NGF to stimulate initiation of neurite outgrowth in untreated cultures but not for the regeneration of neurites from cells which had been pretreated with NGF. Since neurite outgrowth could be blocked at inhibitor levels considerably lower than those required for total RNA inhibition, NGF action may involve the transcription of a particularly drug-sensitive species of RNA.

3.3. Catecholamine Metabolism

PC-12 cells contain high levels of tyrosine hydroxylase, DOPA decarboxylase, and dopamine β-hydroxylase, levels comparable to those of the normal rat adrenal gland. Therefore, they are able to synthesize the neurotransmitters dopamine and norepinephrine (Green and Tischler, 1976). Epinephrine and phenylethanolamine-*N*-methyltransferase, the enzyme that catalyzes the synthesis of epinephrine from norephinephrine, were not detected in the cells under various experimental conditions. Unlike the case of adrenals, dopamine, and not norepinephrine, was predominant in cultured cells. However, it was found that the ratio of norepinephrine to dopamine was increased tenfold in the tumors derived from these cells and that the rate of conversion of tyrosine to norepinephrine was enhanced five- to tenfold when reduced ascorbate was added to the culture medium (Greene and Rein, 1978). Ascorbate has been shown to be concentrated by a stereospecific, saturable, energy-dependent transport system. The K_t and V_{max} of this system were similar in NGF-treated and untreated cells (30 μM and 0.3 nmole/min per mg protein, respectively (Spector and Greene, 1977). As in other systems, elevated levels of K^+ (50 mM) as well as dibutyryl cyclic AMP enhanced the synthesis of catecholamines (two- to threefold by K^+, twofold by dibutyryl cyclic AMP) and led to a net accumulation of the newly synthesized material. The stimulating effects of K^+ and dibutyryl cyclic AMP were additive when both agents were applied simultaneously (Greene and Rein, 1978).

Extended NGF treatment does not cause changes in the intracellular levels of catecholamines or their synthetic enzymes measured on a per-cell basis. A four- to sixfold decrease is evident when the levels are expressed on a per-mg-protein basis (Greene and Tischler, 1976). This indicates that other proteins are preferentially synthesized under these conditions.

An energy- and sodium-dependent uptake system for norepinephrine that is blocked by cocaine and desmethylimipramine has been demonstrated in PC-12 cells. Similar K_ms (1.7 μM) were determined for both treated and untreated cells. An apparent change in the V_{max} in NGF-treated cells was shown to be a consequence of NGF-induced fiber outgrowth which increased the cell surface area (Greene and Rein, 1977).

PC-12 cells contain storage vesicles for endogenous norephinephrine and dopamine. These can be released by elevated K^+ levels (Greene and Rein, 1977; Schubert and Klier, 1977), by nicotinic cholinergic stimulation, and by veratridine (Greene and Rein, 1977). Dopamine β-hydroxylase and chromogranin are released concomitantly with the various neurotransmitters in a K^+ and Ca^{2+}-dependent process.

While elevated K^+ causes PC-12 cells to release (within 5 min) a substantial amount of their endogenous norepinephrine and dopamine (30% and 12% for control and NGF-treated cells), the nicotinic cholinergic stimulation was terminated within 1 min and depleted only 1–2% of the [^{3}H]norepinephrine content in control cells, and 5–6% in NGF-treated cells. Ca^{2+} ions are required for the nicotinically stimulated release, but tetrodotoxin has no effect, suggesting that the effect is Na^+-independent.

3.4. Acetylcholine Metabolism

Like monolayer cultures of rat sympathetic neurons, PC-12 cells show a developmental plasticity with respect to neurotransmitter synthesis—i.e., an ability to synthesize acetylcholine as well as catecholamines—that can be regulated and modified by changing the cell's environment. CAT activity was found to increase in these cells as a function of cell density and in response to various effectors including NGF, cyclic AMP and phosphodiesterase inhibitors, and co-culture with various nonneuronal cells, as well as conditioned medium from those cells (Greene and Rein, 1977; Schubert *et al.,* 1977). The augmentation in CAT activity by NGF treatment was followed by higher levels of acetylcholine in the cells.

Cellular acetylcholine can be released in a depolarizing medium (high K^+, in the presence of Ca^{2+}), conditions similar to those which cause release of stored catecholamines. Are the two neurotransmitters stored in the same granules? Several experiments indicate that they are not. Dense-core granules of the type considered to contain catecholamines as well as small agranular vesicles similar to those found in cholinergic neurons were found in the same cells (Greene and Rein, 1977; Schubert *et al.,* 1977). There are also small differences in the density of the vesicles that store the two types of compounds (Schubert and Klier, 1977). Reserpine treatment reduces the number of the larger diameter vesicles. This treatment, which also depletes most of the catecholamines, has a minor effect on acetylcholine synthesis, storage, and release.

PC-12 cells form cholinergic synapses with a clone of skeletal muscle cells (L6) even in the absence of NGF. These synapses were detected within 1 hr after plating (Schubert *et al.,* 1977).

Some information about the nature of the acetylcholine receptor in these cells is available: antibodies against eel acetylcholine receptors block the carbamylcholine-induced Na^+ influx but fail to recognize the α-bungarotoxin-binding component. The toxin, at binding saturation levels, had no effect on Na^+ uptake. These results suggest a difference between the nicotinic receptor and the toxin binding site (Patrick and Stallcup, 1977).

4. Concluding Remarks

The numerous findings described in this review not only are of interest in their own right, but also serve a purpose in pointing out the vast possibilities and problems yet unsolved in the field of neurobiology. Obviously the fact that many of these tumor-cell clones reach only a certain step in their development and are arrested is of advantage in allowing the recording of their history of specialization. Today, the cells in common use are those originating in two tumors of the nervous tissue. More complete documentation requires the selection of additional clones derived from other neuronal sources. A whole array of comparative studies with normal nerve cells as well as the further pursuit of morphological and electrophysiological differentiation, the mode of action of various trophic factors (many still to be discovered), and properties of specific receptors and enzymes await their isolation.

ACKNOWLEDGMENTS. I wish to express my gratitude to Drs. L. Fechter, J. Chang, and W. Ross for critical reading of the manuscript at various stages and for making valuable comments and suggestions. Special thanks are due to Dr. A. M. Goldberg for his help and encouragement and for the facilities provided throughout preparation of the review during my stay in his laboratory. Last but not least, I would like to thank Dr. A. Mahler and D. Saya, whose patience and cooperation in the final stage are greatly appreciated.

This work was supported in part by a United States Public Health Service grant (EHS-00454) and by the United States–Israel Binational Science Foundation.

REFERENCES

Abbott, J., and Holtzer, H., 1968, The loss of phenotypic traits by differentiated cells, V: The effect of 5-bromodeoxyuridine on cloned chondrocytes, *Proc. Natl. Acad. Sci. USA* **59:**1144–1151.

Adolphe, M., Giroud, J. P., Fontagne, J., Lechat P., and Timsit, J., 1974, Action of prostaglandin A_2 on the proliferation and morphologic differentiation of murine neuroblastoma cell line, *C. R. Soc. Biol. (Paris).* **168:**694–698.

Akeson, R., and Herschman, H., 1974a, Neural antigens of morphologically differentiated neuroblastoma cells, *Nature* **249:**620–623.

Akeson, R., and Herschman, H. R., 1974b, Modulation of cell-surface antigens of a murine neuroblastoma, *Proc. Natl. Acad. Sci. USA* **71:**187–191.

Akeson, R., and Herschman, H. R., 1975, Clonal variations in murine neuroblastoma. Morphologic and antigenic differentiation, *Exp. Cell Res.* **93:**492–495.

Albuquerque, E. X., Barnard, E. A., Chiu, T. H., Lapa, A. J., Dolly, J. O., Jansson, S. E., Daly, J., and Witkop, B., 1973, Acetylcholine receptor and ion conductance modulator sites at the murine neuromuscular junction: Evidence from specific toxin reactions, *Proc. Natl. Acad. Sci. USA* **70:**949–953.

Amano, T., Richelson, E., and Nirenberg, M., 1972, Neurotransmitter synthesis of neuroblastoma clones, *Proc. Natl. Acad. Sci. USA* **69:**258–263.

Amano, T., Hamprecht, B., and Kemper, W., 1974, High activity of choline-acetyltransferase induced in neuroblastoma–glia hybrid cells, *Exp. Cell Res.* **85:**399–408.

Anagnoste, B., Freedman, L. S., Goldstein, M., Broome, J., and Fuxe, K., 1972, Dopamine-hydroxylase activity in mouse neuroblastoma tumors and in cell cultures, *Proc. Natl. Acad. Sci. USA* **69:**1883–1886.

Andres, R. Y., Jeug, I., and Bradshaw, R. A., 1977, NGF receptors: Identification of distinct classes in plasma membranes and nuclei of embryonic dorsal root neurons, *Proc. Natl. Acad. Sci. USA* **74:**2785–2789.

Angeletti, P. U., and Levi-Montalcini, R., 1970a, Sympathetic nerve cell destruction in newborn mammals by 6-hydroxydopamine, *Proc. Natl. Acad. Sci. USA* **65:**114–121.

Angeletti, P. U., and Levi-Montalcini, R., 1970b, Cytolytic effect of 6-hydroxydopamine on neuroblastoma cells, *Cancer Res.* **30:**2863–2869.

Anzil, A. P., Stavrou, D., Blinzinger, K., Herrlinger, H., and Dahme, E., 1977, Ultrastructural comparison between the parenchymal cells of tumors derived from parent and hybrid lines of C1300 mouse neuroblastoma and C6 rat glioma, *Cancer Res.* **37:**2236–2245.

Archer, E. G., Breakefield, W. O., and Sharata, M. N., 1977, Transport of tyrosine, phenylalanine, tryptophan and glycine in neuroblastoma clones, *J. Neurochem.* **28:**127–135.

Augusti-Tocco, G., and Chiarugi, V. P., 1976, Surface glycosaminoglycans as a differentiation cofactor in neuroblastoma cell-cultures, *Cell. Differ.* **5:**161–170.

Augusti-Tocco, G., and Sato, G., 1969, Establishment of functional clonal lines of neurons from mouse neuroblastoma, *Proc. Natl. Acad. USA* **64:**311–315.

Augusti-Tocco, G., Sato, G. H., Claude, P., and Potter, D. D., 1970, Clonal cell lines of neurons. Control mechanisms in the expression of cellular phenotypes, *Int. Soc. Cell Biol. Symp.* **9:**109–120.

Augusti-Tocco, G., Casola, L., and Grasso, A., 1973, Neuroblastoma cells and 14-3-2, a brain-specific protein, *Cell. Diff.* **2:**157–161.

Augusti-Tocco, G., Casola, L., and Romano, M., 1974, RNA metabolism in neuroblastoma cultures 2. Synthesis of non-ribosomal RNA, *Cell. Differ.* **3:**313–320.

Bachrach, U., 1975, Cyclic AMP mediated induction of ornithine decarboxylase of glioma and neuroblastoma cells, *Proc. Natl. Acad. Sci. USA* **72:**3087–3091.

Bachrach, U., 1976, Induction of ornithine decarboxylase in glioma and neuroblastoma cells, *FEBS Lett.* **68:**63–67.

Bachrach, U., 1977, Induction of S-adenosyl-L-methionine decarboxylase in glioma and neuroblastoma cells, *FEBS Lett.* **75:**210–204.

Bear, M. P., and Schneider, F. H., 1977, The effect of medium pH on rate of growth, neurite formation and acetylcholinesterase activity in mouse neuroblastoma cells in culture, *J. Cell Physiol.* **91:**63–68.

Bertolini, L., Amini, M., Vigneti, E., Bosman, C., and Revoltella, R., 1977, Intermediate (10mn) filaments in undifferentiated cells of mouse neuroblastoma clones, *Differentiation* **8:**175–181.

Birdsall, N. J. M., Burgen, A. S. V., and Hulme, E. C., 1978, The binding of agonists to brain muscarinic receptors, *Mol. Pharmocol.* **141:**723–736.

Blosser, J., Abbot, J., and Schain, W., 1976, Sympathetic ganglion cell × neuroblastoma hybrids with opiate receptors, *Biochem. Pharmacol.* **25:**2395–2399.

Blosser, J. C., Myers, P. R., and Shain, W., 1978, Neurotransmitter modulation of prostaglandin E_1-stimulated increases in cyclic AMP. I. Characterization of a cultured neuronal cell line in exponential growth phase, *Biochem. Pharmacol.* **27:**1167–1172.

Bluff, K., and Dairman, W., 1975, Biosynthesis of biopterin by two clones of mouse neuroblastoma, *Mol. Pharmacol.* **11**:87–93.

Blume, A. J., 1972, Mouse neuroblastoma AChE: Identification of the active forms, *Fed. Proc.* **31**:841.

Blume, A. J., and Foster, C. J., 1975, Mouse neuroblastoma adenylate cyclase: Adenosine and adenosine analogues as potent effectors of adenylate cyclase activity, *J. Biol. Chem.* **250**:5003–5008.

Blume, A. J., and Foster, C. J., 1976, Mouse neuroblastoma adenylate cyclase: Regulation by 2-chloroadenosine, prostaglandin E_1 and the cations Mg^{2+}, Ca^{2+} and Mn^{2+}, *J. Neurochem.* **26**:305–311.

Blume, A. J., Gilbert, F., Wilson, S., Farber, J., Rosenberg, R., and Nirenberg, M., 1970, Regulation of acetylcholinesterase in neuroblastoma cells, *Proc. Natl. Acad. Sci. USA* **67**:786–792.

Blume, A. J., Dalton, C., and Sheppard, H., 1973, Adenosine-mediated elevation of cyclic 3′:5′-adenosine monophosphate concentrations in cultured mouse neuroblastoma cells, *Proc. Natl. Acad. Sci. USA* **70**:3099–3102.

Blume, A. J., Shorr, J., Finberg, J. P. M., and Spector, S., 1977a, Binding of endogenous non-peptide morphine-like compound to opiate receptors, *Proc. Natl. Acad. Sci. USA* **74**:4972–4981.

Blume, A. J., Chen, C., and Foster, C. J., 1977b, Muscarinic regulation of cAMP in mouse neuroblastoma, *J. Neurochem.* **29**:625–632.

Bock, E., 1978, Nervous system specific proteins, *J. Neurochem.* **30**:7–14.

Boeynaems, J. M., and Dumont, J. E., 1975, Quantitative analysis of the binding of ligands to their receptors, *J. Cyc. Nucl. Res.* **1**:123–142.

Bondy, S. C., Prasad, K. N., and Purdy, J. C., 1974, Neuroblastoma: Drug induced differentiation increases proportion of cytoplasmic RNA that contain polyadenylic acid, *Science* **186**:359–361.

Booher, J., Sensenbrenner, M., and Mandel. P., 1973, Neuroblastoma cell differentiation: A tissue culture study using time-lapse cinematography, *Neurobiology* **3**:335–338.

Bosman, C., Revoltella, R., and Bertolini, L., 1975, Phagocytosis of nerve growth factor-coated erythrocytes in neuroblastoma rosette-forming cells, *Cancer Res.* **35**:896–905.

Bottenstein, J. E., and Sato, G. H., 1979, Growth of a rat neuroblastoma cell line in serum free supplemented medium. *Proc. Natl. Acad. Sci. USA* **76**:514–517.

Brandt, M., Fischer, K., Moroder, L., Wunsch, E., and Hamprecht, B., 1976a, Enkephalin evokes biochemical correlates of opiate tolerance and dependance in neuroblastoma × glioma hybrid cells, *FEBS Lett.* **68**:38–40.

Brandt, M., Gullis, R. J., Fischer, K., Buchen, C., Hamprecht, B., Moroder, H., and Wünsch, E., 1976b, Enkephalin regulates the levels of cyclic nucleotides in neuroblastoma × glioma hybrid cells, *Nature* **262**:311–312.

Brandt, M., Buchen, C., and Hamprecht, B., 1977, Endorphins exert opiate like action on neuroblastoma × glioma hybrid cells, *FEBS Lett.* **80**:251–254.

Brandt, M., Buchen, C., and Hamprecht, B., 1978, Neuroblastoma × glioma hybrid cells as a model system for studying opioid action, in: *Characteristics and Function of Opioids* (Van Ree and Terenius, eds.), pp. 299–310, North Holland Biomedical Press, Amsterdam.

Breakefield, X. O., 1975, Reserpine sensitivity of catecholamine metabolism in murine neuroblastoma clone N1E-115, *J. Neurochem.* **25**:877–882.

Breakefield, X. O., 1976, Neurotransmitter metabolism in murine neuroblastoma cells, *Life Sci.* **18**:267–278.

Breakefield, X. O., and Giller, E. L., 1976, Neurotransmitter metabolism in cell culture, *Biochem. Pharmacol.* **25**:2337–2342.

Breakefield, X. O., and Nirenberg, M., 1974, Selection for neuroblastoma cells that synthesize certain transmitters, *Proc. Natl. Acad. Sci USA* **71**:2530–2533.

Breakefield, X. O., Neale, E. A., Neale, J. H., and Jacobowitz, D. M., 1975, Localized catecholamine storage associated with granules in murine neuroblastoma cells, *Brain Res.* **92:**237–256.

Brodeur, G. M., and Goldstein, M. N., 1976, Histochemical demonstration of an increase in acetylcholinesterase in established lines of human and mouse neuroblastomas by nerve growth factor, *Cytobios* **16:**133–138.

Brown, J. C., 1971, Surface glycoprotein characteristic of the differentiated state of neuroblastoma C-1300 cells, *Exp. Cell Res.* **69:**440–442.

Buonossisi, V., Sato, G., Cohen, A. I., 1962, Hormone producing cultures of adrenal and pituitary tumor origin, *Proc. Natl. Acad. Sci. USA* **48:**1184–1190.

Burgen, A. S. V., Hiley, C. R., and Young, J. M., 1974, The binding of [^{3}H]propylbenzilylcholine mustard by longitudinal muscle strips from guinea-pig small intestine. *Br. J. Pharmacol.* **50:**145–152.

Burgermeister, W., Klein, W. L., Nirenberg, M., and Witkop, B., 1978, Comparative binding studies with cholinergic ligands and histrionicotoxin at muscarinic receptors of neural cell lines, *Mol. Pharmacol.* **14:**751–767.

Burstein, D. E., and Greene, L. A., 1978, Evidence for RNA synthesis-dependent and independent pathways in stimulation of neurite outgrowth by NGF, *Proc. Natl. Acad. Sci. USA* **75:**6059–6063.

Burton, P. R., and Kirkland, W. L., 1972, Actin detected in mouse neuroblastoma cells by binding of heavy meromyosin, *Nature* [*New Biol*]. **239:**244–246.

Byfield, J. E., and Karlsson, U., 1973, Inhibition of replication and differentiation in malignant mouse neuroblasts, *Cell Differ.* **2**:55–65.

Callissano, P., and Cozzari, C., 1974, Interaction of NGF with mouse brain neurotubule protein(s), *Proc. Natl. Acad. Sci. USA* **71:**2131–2135.

Casola, L., Romano, M., DiMatteo, G., Augusti-Tocco, G., and Estenoz, M., 1974, RNA-metabolism in neuroblastoma cultures. 1. Ribosomal-RNA, *Dev. Biol.* **41:**371–379.

Chalazonitis, A., and Greene, L. A., 1974, Enhancement in excitability properties of mouse neuroblastoma cells cultured in the presence of dibutyryl cyclic AMP, *Brain Res.* **72:**340–345.

Chalazonitis, A., Greene, L. A., and Shain, W., 1975, Excitability and chemosensitivity properties of a somatic-cell hybrid between mouse neuroblastoma and sympathetic-ganglion cells, *Exp. Cell Res.* **96:**225–238.

Chalazonitis, A., Minna, J. D., and Nirenberg, M., 1977, Expression and properties of acetylcholine receptors in several clones of mouse neuroblastoma × L cell somatic hybrids, *Exp. Cell Res.* **105:**269–280.

Chang, C. H., and Blume, A. J., 1976, Heterogeneity of acetylcholinesterase in neuroblastoma, *J. Neurochem.* **27:**1427–1435.

Chang, C. M., and Goldman, R. D., 1973, The localization of actin-like fibers in cultured neuroblastoma cells as revealed by heavy meromyosin binding, *J. Cell Biol.* **57:**867–874.

Chang, K. J., Miller, R. J., and Cuatrecasas, P., 1978, Interaction of enkephalin with opiate receptors in intact cultured cells, *Mol. Pharmacol.* **14:**951–970.

Chapman, S. K., Martin, M., Hoover, M. S., and Chiou, C. Y., 1978, Ornithine decarboxylase activity and the growth of neuroblastoma cells. The effects of bromoacetylcholine, bromoacetate and 1,3-diaminopropane, *Biochem. Pharmacol.* **27:**717–721.

Charalampous, F. C., 1977, Differences in plasma-membrane organization of neuroblastoma cells grown in the differentiated and undifferentiated states, *Arch. Biochem. Biophys.* **181:**103–116.

Chelmicka-Szorc, E., and Arnason, B. G. W., 1976, Effect of 6-hydroxydopamine on tumor growth, *Cancer Res.* **36:**2382–2384.

Chen, K. Y., and Canellakis, E. S., 1977, Enzyme regulation in neuroblastoma cells in a salts/glucose medium: Induction of ornithine decarboxylase by asparagine and glutamine, *Proc. Natl. Acad. Sci. USA* **74:**3791–3795.

Chen, J. S., Del Fa, A., Di Luzio, A., and Calissano, P., 1976, Liposome-induced morphological differentiation of murine neuroblastoma, *Nature* **263:**604–606.

Christensen, H. N., De Cespedes, C., Handlogten, M. E., and Ronquist, G., 1973, Energization of amino acid transport, studied in the Ehrlich ascites tumor cell, *Biochim. Biophys. Acta* **300:**487–522.

Christian, C. N., Nelson, P. G., Peacock, J., and Nirenberg, M., 1977, Synapse formation between two clonal cell lines, *Science* **196:**995–998.

Christian, C. N., Nelson, P. G., Bullock, P., Mullinax, D., and Nirenberg, M., 1978, Pharmacologic responses of cells of a neuroblastoma × glioma hybrid clone and modulation of synapses between hybrid cells and mouse myotubes, *Brain Res.* **147:**261–276.

Ciesielski-Treska, J., Stefanov, V., and Mandel, P., 1975, Acetylcholinesterase activity of neuroblastoma and different glial cells in co-culture, *C. R. Acad. Sci.* [*D*] *(Paris)* **281:**1261–1264.

Ciesielski-Treska, J., Tholey, G., Wurtz, B., and Mandel, P., 1976, Enzymic modifications in a cultivated neuroblastoma clone after bromodeoxyuridine treatment, *J. Neurochem.* **26:**465–469.

Claude, P., and Augusti-Tocco, G., 1970, Ultrastructural and cytochemical studies of mouse neuroblastoma cells in tissue culture, *J. Cell Biol.* **47:**88.

Cooper, A., Munden, H. R., and Brown, G. L., 1976, The growth of mouse neuroblastoma cells in controlled orientations on thin films of silicon monoxide, *Exp. Cell Res.* **103:**435–439.

Croizat, B., Berthelot, F., Felsani, A., and Gros, F., 1977, Poly(A)-containing RNA in neuroblastoma: Immature and differentiated cells in culture, *Eur. J. Biochem.* **74:**405–412.

Cronemeyer, R. L., Thuillez, P. E., Shows, T. B., and Morrow, J., 1974, 6-Hydroxydopamine sensitivity in mouse neuroblastoma and neuroblastoma × L-cell hybrids, *Cancer Res.* **34:**1652–1657.

Daley, J., 1977, *Cyclic Nucleotides in the Nervous System,* Plenum Press, New York.

Daniels, M. P., and Hamprecht, B., 1974, The ultrastructure of neuroblastoma × glioma somatic-cell hybrids—expression of neuronal characteristics stimulated by dibutryl adenosine 3′, 5′ cyclic monophosphate, *J. Cell Biol,* **63:**691–699.

Debault, E., and Millard, S. A., 1973, Inhibition of growth by 6-hydroxydopamine in cultured cells of neuronal and nonneuronal origin, *Cancer Res.* **33:**745–749.

DeLellis, R. A., Rabson, A. S., and Albert, D., 1970, The cytochemical distribution of catecholamines in the C-1300 murine neuroblastoma, *J. Histochem. Cytochem.* **18:**913–914.

DeLellis, R. A., Merk, F. B., Deckers, P., Warren, S., and Balogh, K., 1973, Ultrastructure and *in vitro* growth characteristics of a transplantable rat pheochromocytoma, *Cancer* **32:**227–235.

De Potter, W. P., Fraeyman, N. H., Palm, J. W., and De Schaepdryver, A. F., 1978, Localization of noradrenaline and dopamine-β-hydroxylase in C1300 mouse neuroblastoma: A biochemical and electron microscope study, *Life Sci.* **23:**2665–2674.

DeVellis, J., Inglish D., Cole, R., and Molson, J., 1970, Effects of hormones on the differentiation of cloned lines of neurons and glial cells, in: *Influence of Hormones on the Nervous System,* pp. 25–39, S. Karger, Basel.

DeVellis, J., Inglish, D., and Augusti-Tocco, G., 1971, The influence of hormones on neurite formation in a neuroblastoma cell line, in: *Third Internation Meeting of the International Society for Neurochemistry,* Budapest, July 5–9, 1971, p. 191.

Dichter, M. A., Tischler, A. S., and Greene, L. A., 1977, NGF-induced increase in electrical excitability and acetylcholine sensitivity of a rat pheochromocytoma cell line, *Nature* **268:**501–504.

Dickmann Gerber, L., 1978, Binding assay for opioid peptides with neuroblastoma × glioma hybrid cells: Specificity of the receptor site, *Brain Res.* **151:**117–196.

Donnelly, C. H., Richelson, E., and Murphy, D. L., 1976, Properties of monoamine oxidase in mouse neuroblastoma NIE-115 cells, *Biochem. Pharmacol.* **25:**1639–1643.

Ebendal, T., and Jacobson, C. O., 1975, Human glial cells stimulating outgrowth of axons in cultured chick embryo ganglia, *Zoon* **3:**169–172.

Ebendal, T., and Jacobson, C. O., 1977, Tissue explants affecting extension and orientation of axons in cultured chick embryo ganglia, *Exp. Cell Res.* **105**:379–387.

Eiper, B. A., 1972, Rat brain microtubule protein purification and determination of covalently bound phosphate carbohydrate, *Proc. Natl. Acad. Sci. USA* **69**:2283–2287.

Falck, B., Hillarp, N. A., Thieme, G., and Torp, A., 1962, Fluorescence of catecholamines and related compounds condensed with formaldehyde, *J. Histochem. Cytochem.* **10**:348–354.

Fischbach, G. D., and Nelson, P. G., 1977, Cell cultures in neurobiology, in: *Handbook of Physiology,* Chapter 20, American Physiological Society, Bethesda.

Friedman, D. L., 1976, Role of cyclic nucleotides in cell growth and differentiation, *Physiol. Rev.* **56**:652–708.

Friend, C., 1977, The phenomenon of differentiation in murine erythroleukemic cells, *Harvey Lect.* **72**:253–281.

Friend, C., Preisler, H. D., and Scher, W., 1976, Studies on the control of differentiation of murine virus-induced erythroleukemic cells, *Curr. Top. Dev. Biol.,* **8**:81–101.

Furmanski, P., 1973, Neurite extension by mouse neuroblastoma: Evidence for two modes of induction, *Differentiation* **1**:319–322.

Furmanski, P., and Lubin, M., 1972, Effects of dimethylsulfoxide on expression of differentiated functions in mouse neuroblastoma, *J. Natl. Cancer Inst.* **48**:1355–1361.

Furmanski, P., Silverman, D. J., and Lubin, M., 1971, Expression of differentiated functions in mouse neuroblastoma mediated by dibutyryl-cyclic adenosine monophosphate, *Nature* **233**:413–415.

Furshpan, E. J., Macleish, P. R., O'Lague, P. H., and Potter, D. D., 1976, Chemical transmission between rat sympathetic neurons and cardiac myocytes developing in micro-cultures: Evidence for cholinergic, adrenergic and dual-function neurons, *Proc. Natl. Acad. Sci. USA* **73**:4225–4229.

Garvican, J. H., and Brown, G. L., 1977, A comparative analysis of the protein components of plasma membranes isolated from differentiated and undifferentiated mouse neuroblastoma cells in tissue culture, *Eur. J. Biochem.* **76**:251–261.

Gill, E. W., and Rang, H. P., 1966, An alkylating derivative of benzylcholine with specific and long lasting parasympatholytic activity, *Mol. Pharmacol.* **2**:284–297.

Gilman, A. G., and Nirenberg, M., 1971, Regulation of adenosine 3′,5′ -cyclic monophosphate metabolism in cultured neuroblastoma cells, *Nature* **234**:356–358.

Glazer, R. I., and Schneider, F. H., 1975, Effects of adenosine 3′:5′-monophosphate and related agents on ribonucleic acid synthesis and morphological differentiation in mouse neuroblastoma cells in culture, *J. Biol. Chem.* **250**:2745–2749.

Glick, M. C., Kimhi, Y., and Littauer, U. Z., 1973, Glycopeptides from surface membranes of neuroblastoma cells, *Proc. Natl. Acad. Sci. USA* **70**:1682–1687.

Glick, M. C., Kimhi, Y., and Littauer, U. Z., 1976, Surface membrane alterations in somatic cell hybrids of neuroblastoma and glioma cells, *Nature* **259**:230–232.

Goldstein, L. I., Lowney, B., and Pal, K., 1971, Stereospecific and nonspecific interactions of the morphine congener levorphanol in subcellular fractions of mouse brain, *Proc. Natl. Acad. Sci. USA* **68**:1742.

Goldstein, M. N., and Pinkel, D., 1957, Long-term tissue culture of neuroblastoma, *J. Natl. Cancer Inst.* **20**:675–689.

Graham, D. C., Tiffany, S. M., Bell, W. R., and Gutknecht, W. F., 1978, Auto-oxidation versus covalent binding of quinones as the mechanism of toxicity of dopamine, 6-hydroxydopamine, and related compounds toward C1300 neuroblastoma cells *in vitro, Mol. Pharmacol.* **14**:644–653.

Graham, D. I., and Gonatas, N. K., 1976, Subcutaneous C-1300 murine neuroblastoma—Light and ultrastructural study, *Neuropathol. Appl. Neurobiol.* **2**:451–458.

Graham, D. I., Gonatas, N. K., and Charalampous, F. C., 1974, The undifferentiated and extended forms of C1300 murine neuroblastoma. An ultrastructural study and detection of concanavalin A binding sites on the plasma membrane, *Am. J. Pathol.* **76**:285–312.

Green, R. D., and Stanberry, L. R., 1977, Elevation of cyclic-AMP in C-1300 murine neuroblastoma by adenosine and related compounds and the antagonism of this response by methylxanthines, *Biochem. Pharmacol.* **26**:37–43.

Greene, L. A., 1978, NGF prevents the death and stimulates the neuronal differentiation of clonal PC-12 pheochromocytoma cells in serum-free medium, *J. Cell Biol.* **78**:747–755.

Greene, L. A., and Rein, G., 1977, Dopaminergic properties of a somatic cell hybrid line of mouse neuroblastoma × sympathetic ganglion cells, *J. Neurochem.* **29**:141–150.

Greene, L. A., and Rein, G., 1978, Short term regulation of catecholamine biosynthesis in an NGF responsive clonal line of rat pheochromocytoma cells, *J. Neurochem.* **30**:549–555.

Greene, L. A., and Tischler, A. S., 1976, Establishment of a noradrenergic clonal line of rat adrenal pheochromocytoma cells which respond to nerve growth factor, *Proc. Natl. Acad. Sci. USA* **73**:242–2428.

Greengard, P., 1976, Possible role for cyclic nucleotide and phosphorylated membrane proteins in postsynaptic actions of neurotransmitters, *Nature* **260**:101–108.

Haber, B., and Hutchison, H. T., 1976, Uptake of neurotransmitters and precursors by clonal cell lines of neural origin, *Adv. Exp. Med. Biol.* **69**:179–198.

Haffke, S. C., and Seeds, N. W., 1975, Neuroblastoma: The *E. coli* of neurobiology? *Life Sci.* **16**:1649–1658.

Haga, T., and Noda, I. T., 1973, Choline uptake systems of rat brain synaptosomes, *Biochim. Biophys. Acta* **291**:564–575.

Hamprecht, B., 1976, Neuron models, *Angew. Chem.* [*Engl.*] **15**:194–206.

Hamprecht, B., 1977, Structural, electrophysiological, biochemical and pharmacological properties of neuroblastoma–glioma cell hybrids in cell culture, *Int. Rev. Cytol.* **49**:99–170.

Hamprecht, B., and Schultz, J., 1973a, Influence of noradrenaline, prostaglandin E_1 and inhibitors of phosphodiesterase activity on levels of cAMP in somatic cell hybrids, *Hoppe Seylers Z. Physiol. Chem.* **354**:1633–1641.

Hamprecht, B., and Schultz, J., 1973b, Stimulation by prostaglandin E_1 of adenosine 3′:5′-cyclic monophosphate formation in neuroblastoma cells in the presence of phosphodieterase inhibitors, *FEBS Lett.* **34**:85–89.

Hamprecht, B., Jaffee, B. N., and Philpott, G. W., 1973, Prostaglandin production by neuroblastoma, glioma and fibroblast cell lines; stimulation by $N^6,0^{2\prime}$-dibutyryl adenosine 3′:5′-cyclic monophosphate, *FEBS Lett.* **36**:193–198.

Hamprecht, B., Traber, J., and Lamprecht, F., 1974, Dopamine-beta-hydroxylase activity in cholinergic neuroblastoma × glioma hybrid cells; increase of activity by $N^60^{2\prime}$-dibutyryl-adenosine 3′:5′-cyclic monophosphate, *FEBS Lett.* **42**:221–226.

Harkins, J., Arsenault, M., Schlesinger, K., and Kates, J., 1972, Induction of neuronal functions—acetylcholine-induced acetylcholinesterase activity in mouse neuroblastoma cells, *Proc. Natl. Acad. Sci USA* **69**:3161–3164.

Harris, A. J., and Dennis, M. J., 1970, Acetylcholine sensitivity and distribution in mouse neuroblastoma cells, *Science* **167**:1253–1255.

Harris, A. J., Heinemann, S., Schubert, D., and Tarakis, H., 1971, Trophic interaction between cloned tissue culture lines of nerve and muscle, *Nature* **231**:296–301.

Harrison, R. G., 1907, Observations on the living developing nerve fiber, *Anat. Rec.* **1**:116–118.

Hawkins, M., and Breakefield, X. O., 1978, Monoamine oxidase A and B in cultured cells, *J. Neurochem.* **30**:1391–1397.

Hazum, E., Chang, K. J., and Cuatracasas, P., 1979, Rapid degradation of [^{3}H]leucine-enkephalin by intact neuroblastoma cells, *Life Sci.* **24**:137–144.

Hermetet, J. C., Ciesielski-Treska, J., and Mandel, P., 1972a, Cytochemical demonstration of catecholamines and acetylcholinesterase activity in neuroblastoma cells in culture, *J. Histochem. Cytochem.*, **20**:136–138.

Hermetet, J. C., Ciesielski-Treska, J., and Mandel, P., 1972b, Effets du NGF sur les cultures de neuroblastes du neuroblastoma C1300, *C. R. Soc. Biol. (Paris)* **166**:1120–1125.

Hermetet, J. C., Ciesilski-Treska, J., Warter, S., and Mandel, P., 1973, Differential effects of insulin on different clones of neuroblastoma C1300 of mouse, *J. Physiol. (Paris)* **67**:280–80.

Herschman, H. R., and Lerner, M. P., 1973, Production of a nervous-system specific protein (14.3.2) by human neuroblastoma cells in culture, *Nature* [*New Biol.*] **241**:242–244.

Hiller, J. M., and Simon, E. J., 1973, Inhibition by levorphanol of the induction of acetylcholinesterase in a mouse neuroblastoma cell line, *J. Neurochem.* **20**:1789–1792.

Hiller, G., and Weber, K., 1978, Radioimmunoassay for tubulin: A quantitative comparison of the tubulin content of different established tissue culture cells and tissues, *Cell* **14**:795–804.

Hinkley, R. E., and Telser, A. G., 1974, The effects of halothane on cultured mouse neuroblastoma cells. I. Inhibition of morphological differentiation, *J. Cell Biol.* **63**:531–540.

Hudson, J. E., and Johnson, T. C., 1977a, Rapidly metabolized glycoproteins in a neuroblastoma cell line, *Biochim. Biophys. Acta* **497**:567–577.

Hudson, J. E., and Johnson, T. C., 1977b, Degradation and turnover of fucosylated glycoproteins in the plasma-membrane of a neuroblastoma-cell line, *Biochem. J.* **166**:217–223.

Hughes, J., Smith, T. W., Kosterlitz, H. W., Forthergill, L. A., Morgan, B. A., and Morris, H. R., 1975, Identification of 2 related pentapeptides from the brain with potent opiate agonist activity, *Nature* **258**:577–579.

Hutchinson, H. T., Suddith, R. L., Risk, M., and Haber, B., 1976, Uptake of neurotransmitters and precursors by clonal lines of astocytoma and neuroblastoma. 3. Transport of choline, *Neurochem. Res.* **1**:201–215.

Isenberg, G., Rieske, E., and Kreutzberg, G. W., 1978, Distribution of actin and tubulin in neuroblastoma cells, *Cytobiology* **15**:382–389.

Jacobson, M., 1978, *Developmental Neurobiology*, Plenum Press, New York and London.

Jorgensen, A. O., Subrahmanyan, L., Turnbull, C., and Kalnins, V. I. 1976, Localization of the neurofilament protein in neuroblastoma cells by immunofluorescent staining, *Proc. Natl. Acad. Sci. USA* **73**:3192–3196.

Joseph, B. S., and Oldstone, M. B. A., 1974, Expression of selected antigens on the surface of cultured neural cells, *Brain Res.* **80**:421–434.

Kates, J. R., Winterton, R., and Schlessinger, K., 1971, Induction of acetylcholinesterase activity in mouse neuroblastoma tissue culture cells, *Nature* **229**:345–347.

Kato, A. C., Lefresne, P., Berwald-Netter, Y., Beaujouan, J. C., Glowinski, J., and Gross, F., 1977, Choline stimulates the synthesis and accumulation of acetete in a cholinergic neuroblastoma clone, *Biochem. Biophys. Res. Commun.* **78**:350–356.

Kenimer, J. G., 1978, Desensitization of adenylate cyclase by PGE_1 in neuroblastoma × glioma hybrid cells, *Fed. Proc.* **37**:1359.

Kimhi, Y., Palfrey, C., Spector, I., Barak, Y., and Littauer, U. Z., 1976, Maturation of neuroblastoma cells in the presence of dimethylsulfoxide, *Proc. Natl. Acad. Sci. USA* **73**:462–466.

Kimhi, Y., Mahler, A., and Saya, D., 1980, Acetylcholinesterase from mouse neuroblastoma cells, isoenzymes and their properties, *J. Neurochem.* **34**:554–559.

Klebe, R. J., and Ruddle, F. H., 1969, Neuroblastoma: Cell culture analysis of a differentiating stem cell system, *J. Cell Biol.* **43**:69a.

Klee, W. A., and Nirenberg, M., 1974, A neuroblastoma × glioma hybrid cell line with morphine receptors, *Proc. Natl. Acad. Sci. USA* **71**:3474–3477.

Klee, W. A., and Nirenberg, M., 1976, Mode of action of endogenous opiate peptides, *Nature* **263**:609–612.

Klee, W. A., Sharma, S. K., and Nirenberg, M., 1975, Opiate receptors as regulators of adenylate cyclase, *Life Sci.* **16**:1869–1874.

Klee, W. A., Lampert, A., and Nirenberg, M., 1976, Dual regulation of adenylate cyclase by endogenous opiate peptides, in: *Opiates and Endogenous Opioid Peptides* (H. W. Kosterlitz, ed.), pp. 153–159, Elsvier–North Holland, Amsterdam.

Klein, W. L., Nathanson, N., Nirenberg, M., 1979, Muscarinic acetylcholine receptor regulation by accelerated rate of receptor loss, *Biochem. Biophys. Res. Commun.* **90**:506–512.

Koike, T., and Miyake, M., 1977, Effect of concanavalin A on the cholinergic responses of mouse neuroblastoma cells, *Neurosci. Lett.* **5**:209–213.

Kramer, P. M., and Tobey, R. A., 1972, Cell cycle dependent desquamation of heparin sulfate from cell surface, *J. Cell Biol.* **55**:713–717.

Kumar, S., Becker, G., and Prasad, K. N., 1975, Cyclic adenosine 3′-5′-monophosphate phosphodiesterase activity in malignant and cyclic adenosine 3′-5′-monophosphate-induced "differentiated" neuroblastoma cells, *Cancer Res.* **35**:82–87.

Lampert, A., Nirenberg, M., and Klees, W. A., 1976, Tolerance and dependence evoked by an endogenous opiate peptide, *Proc. Natl. Acad. Sci. USA* **73**:3165–3167.

Landis, S. C., 1976, Rat sympathetic neurons and the cardiac myocytes developing in microcultures: Correlation of the fine structure of endings with neurotransmitter function in single neurons, *Proc. Natl. Acad. Sci. USA* **73**:4220–4224.

Lanks, K. W., Dorwin, J. M., and Papirmeister, B., 1974a, Increased rate of acetylcholinesterase synthesis in differentiating neuroblastoma cells, *J. Cell Biol.* **63**:824–830.

Lanks, K., Somers, L., Papirmeister, B., and Yamamura, H., 1974b, Choline transport by neuroblastoma cells in tissue culture, *Nature* **252**:476–478.

Lanks, K. W., Turnbull, J. D., Aloyo, V. J., Dorwin, J., and Papirmeister, B., 1975, Sulfur mustards induce neurite extension and acetylcholinesterase synthesis in cultured neuroblastoma cells, *Exp. Cell Res.* **93**:355–362.

Lasek, R. J., and Hoffman, P. N., 1976, The neuronal cytoskeleton, axonal transport and axonal growth, in: *Cell Motility Book C: Microtubules and Related Proteins* (R. Goldman, J. Pollarad, and J. Rosenbaum, eds), pp. 1021–1051, Cold Spring Harbor Laboratory, Cold Spring Harbor.

Lasher, R., and Cohn. R. D., 1969, The effect of 5-bromodeoxyuridine on the differentiation of chondrocytes *in vitro, Dev. Biol.* **19**:415–435.

Lazo, J. S., and Ruddon, R. W., 1977, Neurite extension and maligancy of neuroblastoma cells after treatment with prostaglandin E_1 and papaverine, *J. Natl. Cancer Inst.* **59**:137–143.

Lazo J. S., Prasad, K. N., and Ruddon, R. W., 1976, Synthesis and phosphorylation of chromatin-associated proteins in cyclic AMP-induced "differentiated" neuroblastoma cells in culture, *Exp. Cell Res.* **100**:41–46.

Ledig, M., Ciesielski-Treska, J., Cam, Y., Montagnon, D., and Mandel, P., 1975, ATPase activity of neuroblastoma cells in culture, *J. Neurochem.* **25**:635–640.

Lee, V., Shelanski, M. L., and Greene, L. A., 1977, Specific neural and adrenal medullary antigens detected by antisera to clonal PC12 pheochromocytoma cells, *Proc. Natl. Acad. Sci. USA* **74**:5021–5025.

Levi-Montalcini, R., 1976, The NGF: Its role in growth, differentiation and function of the sympathetic adrenergic neuron, *Prog. Brain Res.* **45**:235–258.

Levi-Montalcini, R., and Angeletti, R., 1968, Nerve growth factor, *Physiol. Rev.* **48**:534–569.

Libet, B., 1979, Which postsynaptic action of dopamine is mediated by cAMP? *Life Sci.* **24**:1043–1058.

Liebermann, D., and Sachs, L., 1978, Nuclear control of neurite induction in neuroblastoma cells, *Cell Res.* **113**:383–390.

Lim, R., and Mitsunobu, K., 1972, Effect of db-cAMP on nucleic acid synthesis and protein synthesis in neuronal and glial tumor cells, *Life Sci.* **11**:1063–1070.

Lindsay, R. M., and Monard, D., 1977, Influence of serum lipids on morphology of neuroblastoma cells, *Experientia* **33**:823–827.

Littauer, U. Z., Palfrey, C., Kimhi, Y., and Spector, I., 1976, Induction of differentiation in mouse neuroblastoma cells. National Cancer Institute Monograph No. 48, Third Decennial Review Conference, p. 333–337.

Lloyd, T., and Breakefield, X. O., 1974, Tyrosine-dependent increase of tyrosine hydroxylase in neuroblastoma cells, *Nature* **252**:719–720.

Lloyd, T., Jones-Ebersole, B., and Schneider, F. H., 1978a, Simulation of tyrosine hydroxylase activity in cultured mouse neuroblastoma cells by monocarboxylic acids, *J. Neurochem.* **30**:1641–1643.

Lloyd, T., Weizs, J., and Breakefield, X. O., 1978b, The catechol estrogen, 2-hydroxyestradiol inhibits catechol-0-methyltransferase activity in neuroblastoma cells, *J. Neurochem.* **31**:245–250.

Mandel, P., Ciesielski-Treska, J., Hermetet, J. C., Zwiller, J., Mack, G., and Goridis, 1973a, Catecholamines in neuroblastoma cells, *Life Sci.* **13**:113–115.

Mandel, P., Ciesielski-Treska, J., Hermetet, J. C., Zwiller, J., Mack, G., and Goridis, C., 1973b, Neuroblastoma cells as a tool for neuronal molecular biology, in: *Frontiers in Catecholamine Research* (E. Usdin and S. Snyder, eds.), pp. 227–283, Pergamon Press, Oxford.

Mandel, P., Ciesielski-Treska, J., and Sensenbrenner, M., 1976, Neurons *in vitro*, in: *Molecular and Functional Neurobiology* (W. H. Gispen, ed.), pp. 111–157, Elsevier, Amsterdam.

Mandel, P., Ciesielski-Treska, J., and Stefanovic, V., 1977, Neuroblast–glioblast interactions: Ecto enzymes, in: *Cell, Tissue and Organ Culture in Neurobiology* (S. Fedoroff and L. Hertz, eds.), pp. 593–615, Academic Press, New York.

Marchisio, P. C., Osborn, M., and Weber, K., 1978, The intracellular organization of actin and tubulin in cultured C1300 mouse neuroblastoma cells (clone NB41A3), *J. Neurocytol.* **7**:571–582.

Martin, S. E., 1974, Mouse brain antigen detected by rat anti-C-1300 antiserum, *Nature* **249**:71–73.

Maruyama, T., and Ishikawa, H., 1977, Somatostatin: Its inhibiting effect on the release of hormones and IgG from clonal cell strains: Its Ca-influx dependence, *Biochem. Biophys. Res. Commun.* **74**:1083–1088.

Massarelli, R., 1973, Effect of acetylcholinesterase inhibitors on the kinetics of choline incorporation in a clone culture of mouse neuroblastoma, *J. Physiol. (Lond.)* **67**:P351A.

Massarelli, R., and Mandel, P., 1976, On the uptake mechanism of choline in nerve cell cultures, *Adv. Exp. Med. Biol.* **69**:199–209.

Massarelli, R., Ciesielski-Treska, J., Ebel, A., and Mandel, P., 1973, Choline uptake in neuroblastoma cell cultures: Influence of ionic environment, *Pharmacol. Res. Commun.* **5**:397–406.

Massarelli, R., Ciesielski-Treska, J., Ebel, A., and Mandel, P., 1974a, Kinetics of choline uptake in neuroblastoma clones, *Biochem. Pharmacol.* **23**:2857–2865.

Massarelli, R., Sensenbrenner, M., Ebel, A., and Mandel, P., 1974b, Kinetics of choline uptake in mixed neuronal–glial and exclusively glial cultures, *Neurobiology* **4**:414–418.

Mathews, R. A., Johnson, T. C., and Hudson, J. E., 1976, Synthesis and turnover of plasma membrane proteins and glycoproteins in a neuroblastoma cell line, *Biochem. J.* **154**:57–64.

Matsuzawa, H., and Nirenberg, M., 1975, Receptor-mediated shifts in cGMP and cAMP levels in neuroblastoma cells, *Proc. Natl. Acad. Sci. USA* **72**:3472–3476.

Matthews, R. T., and Chiou, C. Y., 1978, Choline and diethylcholine transport into a cholinergic clone of neuroblastoma cells, *Biochem. Pharmacol.* **28**:405–409.

McGee, R., 1978, Choline uptake in neuroblastoma × glioma hybrid NG108-15, *Abstracts of the Society of Neuroscience Meeting, Vol. 4*, p. 320, Abstract no. 1018.

McGee, R., Simpson, P., Christian, C., Mata, M., Nelson, P., and Nirenberg, M., 1978, Regulation of acetylcholine release from neuroblastoma × glioma hybrid cells, *Proc. Natl. Acad. Sci. USA* **75**:1314–1318.

McGuire, J. C., Greene, L. A., and Furano, A. V., 1978, NGF stimulates incorporation of fucose or glucosamine into an external glycoprotein in cultured rat PC12 pheochromocytoma cells, *Cell* **15**:357–365.

McMorris, F. A., and Ruddle, F. H., 1974, Expression of neuronal phenotypes in neuroblastoma cell hybrids, *Dev. Biol.* **39**:226–246.

McMorris, F. A., Nelson, P. G., and Ruddle, F. H., 1973, Contributions of clonal systems to neurobiology—a report based on an NRP work session, *Neurosci. Res. Prog. Bull.* **11**:412–536.

McMorris, F. A., Kolberg, A. R., Moore, B. W., and Perumal, A. S., 1974, Expression of the neuron-specific protein, 14-3-2, and steroid sulfatase in neuroblastoma cell hybrids, *J. Cell. Physiol.* **84:**473–480.

Michelot, R. J., Lesko, N., Stout, R. W., and Coward, J. K., 1977, Effect of S-adenosylhomocysteine and S-tubercidinylhomocysteine on catecholamine methylation in neuroblastoma cells, *Mol. Pharmacol.* **13:**368–373.

Miller, C., and Kuehl, W. M., 1976, Isolation and characterization of myosin from cloned rat glioma and mouse neuroblastoma cells, *Brain Res.* **108:**115–124.

Miller, C. A., and Levine, E. M., 1972, Neuroblastoma: Synchronization of neurite outgrowth in cultures grown on collagen, *Science* **177:**799–802.

Miller, R. A., and Ruddle, F. H., 1974, Enucleated neuroblastoma cells form neurites when treated with dibutyryl cyclic-AMP, *J. Cell Biol.* **63:**295–299.

Miller, R. J., and Hiley, C. R., 1974, Anti-muscarinic properties of neuroleptics and drug induced parkinsonism, *Nature* **248:**596–597.

Miller, R. J., Chang, K. J., Leighton, J., and Cuatrecasas, J., 1977, Interaction of iodinated enkephalin analogues with opiate receptors, *Life Sci.* **22:**379–388.

Minna, J. D., and Gilman, A. G., 1973, Expression of genes for metabolism of cAMP in somatic cells. II. Effects of prostaglandin E_1 and theophylline on parental and hybrid cells, *J. Biol. Chem.* **248:**6618–6625.

Minna, J., Glazer, D., and Nirenberg, M., 1972, Genetic dissection of neuronal properties using somatic cell hybrids, *Nature* **235:**225–231.

Minna, J. D., Yavelow, J., and Coon, H. G., 1975, Expression of phenotypes in hybrid somatic cells derived from the nervous system, *Genetics* **79:**373–383.

Monard, D., Solomon, F., Rentsch, M., and Gysin, R., 1973, Glia-induced morphological differentiation in neuroblastoma cells, *Proc. Natl. Acad. Sci. USA* **70:**1894–1897.

Monard, D., Stockel, K., Goodman, R., and Thoenen, H., 1975, Distinction between nerve growth factor and glial factor, *Nature* **258:**444–445.

Monard, D., Rentsch, M., Schurch-Rathgeb, Y., and Lindsay, R. M., 1977, Morphological differentiation of neuroblastoma cells in medium supplemented with delipidated serum, *Proc. Natl. Acad. Sci. USA* **74:**3893–3897.

Moore, B. W., and Perez, V. J., 1968, Specific acidic proteins of the nervous system, in: *Physiological and Biochemical Aspect of Nervous Integration* (F. D. Carson, ed.), pp. 343–360 Prentice-Hall, Englewood Cliffs, N. J.

Morgan, J. L., and Seeds, N. W., 1975. Tubulin constancy during morphological differentiation of mouse neuroblastoma cells, *J. Cell Biol.* **67:**136–145.

Morrison, M. R., Pardue, S., Brodeur, R., and Rosenberg, R. N., 1978, Actin and histones are the major proteins synthesized by neuroblastoma non-adenylated messenger RNAs, *Fed. Proc.* **37:**1504.

Moss, C. A., 1974, Acid glycosaminoglycans of mouse neuroblastoma C1300 cells, *Histochem. J.* **6:**1–5.

Murphy, R. A., Pantazis, N. J., Arnason, B. G. W., and Young, M., 1975, Secretion of a nerve growth factor by mouse neuroblastoma cells in culture. *Proc. Natl. Acad. Sci. USA* **72:**1895–1898.

Myers, P. R., and Livengood, D. R., 1975, Dopamine depolarising response in a vertebrate neuronal somatic cell hybrid, *Nature* **255:**235–236.

Myers, P. R., Blosser, J., and Shain, W., 1978, Neurotransmitter modulation of prostaglandin E_1-stimulated increases in cAMP. II. Characterization of a cultured neuronal cell line treated with dibutyryl cyclic AMP, *Biochem. Pharmacol.* **27:**1173–1177.

Nagle, B. W., Doenges, K. H., and Bryan, J., 1977, Assembly of tubulin from cultured cells and comparison with the neurotubulin model, *Cell* **12:**573–586.

Nakai, J., 1964, The movements of neurons in tissue-culture, in: *Primitive Motile Systems in Cell Biology* (R. D. Allan and N. Kamiya, eds.), pp. 337–385, Academic Press, New York.

Narotzky, R., and Bondareff, W., 1974, Biogenic amines in cultured neuroblastoma and astrocytoma cells, *J. Cell Biol.* **63:**64–70.

Nathanson, J. A., 1977, Cyclic nucleotides and nervous system function, *Physiol. Rev.* **57:**157–256.

Nathanson, N. M., Klein, W. L., and Nirenberg, M., 1978, Regulation of adenylate cyclase activity mediated by muscarinic acetylcholine receptors, *Proc. Natl. Acad. Sci. USA* **75:**1788–1791.

Nelson, P. G., 1977, Neuronal cell lines, in: *Cell, Tissue and Organ Culture in Neurobiology* (S. Fedoroff and L. Hertz, eds.), pp. 348–365, Academic Press, New York.

Nelson, P. G., Ruffner, W., and Nirenberg, M., 1969, Neuronal tumor cells with excitable membranes grown *in vitro, Proc. Natl. Acad. Sci. USA* **64:**1004–1010.

Nelson, P. G., Peacock, J., and Amano, T., 1971a, Responses of neuroblastoma cells to iontophoretically applied acetylcholine, *J. Cell. Physiol.* **77:**353–362.

Nelson, P. G., Peacock, J. H., Amano, T., and Minna, J., 1971b, Electrogenesis in mouse neuroblastoma cells *in vitro, J. Cell. Physiol.* **77:**337–352.

Nelson, P., Christian, C., and Nirenberg, M., 1976, Synapse formation between clonal neuroblastoma × glioma hybrid cells and striated muscle cells, *Proc. Natl. Acad. Sci. USA* **73:**123–127.

Nelson, P. G., Christian, C. N., Daniels, M. P., Henkart, M., Bullock, P., Mullinax, D., and Niremberg, M., 1978, Formation of synapses between cells of a neuroblastoma × glioma hybrid clone and mouse myotubes, *Brain Res.* **147:**245–259.

North-Root, H., Martin, D. W., and Toliver, A. P., 1976a, Binding of an opiate, levorphanol, to intact neuroblastoma cells in continuous culture, *Physiol. Chem. Phys.* **8:**221–228.

North-Root, H., Martin, D. W., Jr., and Toliver, A. P., 1976b, Evidence for nuclear sites of stereospecific opiate binding in neuroblastoma cells in continuous culture, *Physiol. Chem. Phys.* **8:**437–446.

Olmsted, J. B., Carlson, K., Klebe, R., Ruddle, F., and Rosenbaum, J., 1970, Isolation of microtubule protein from cultured mouse neuroblastoma cells. *Proc. Natl. Acad. Sci. USA* **65:**129–136.

Palfrey, C., Kimhi, Y., and Lettauer, U. Z., 1977, Induction of differentiation in mouse neuroblastoma cells by hexamethylene bisacetamide, *Biochem. Biophys. Res. Commun.* **76:**937–942.

Partlow, L. M., and Larrabee, M. E., 1971, Effects of nerve growth factor, embryo age and metabolic inhibitors on growth of fibers and on synthesis of RNA and protein in embryonic sympathetic cells, *J. Neurochem.* **18:**2101–2118.

Patrick, J., and Stallcup, W. B., 1977, Immunological distinction between acetylcholine receptor and the α-bungarotoxin-binding component on sympathetic neurons, *Proc. Natl. Acad. Sci. USA* **74:**4689–4692.

Paul, S. M., and Axelrod, J., 1977, Catechol estrogens: Presence in brain and endocrine tissues, *Science* **197:**657–659.

Peacock, J. M., and Nelson, P. G., 1973, Chemosensitivity of mouse neuroblastoma cells in vitro, *J. Neurobiol.* **4:**363–374.

Peacock, J. H., Minna, J., Nelson, P. G., and Nirenberg, M., 1972, Use of aminopterin in selecting electrically active neuroblastoma cells, *Exp. Cell Res.* **73:**367–377.

Penit, J., Huot, J., and Jard, S., 1976, Neuroblastoma cell adenylate-cyclase: Direct activation by adenosine and prostaglandins, *J. Neurochem.* **26:**265–273.

Penit, J., Cantau, B., Huot, J., and Jard, S., 1977, Adenylate cyclase from synchronized neuroblastoma cells: Responsiveness to prostaglandin E_1, adenosine, and dopamine during the cell cycle, *Proc. Natl. Acad. Sci. USA* **74:**1575–1579.

Pert, C. B., and Snyder, S. H., 1973, Opiate receptor: Demonstration in nervous tissue, *Science* **179:**1011–1014.

Phelps, C. H., and Pfeiffer, S. E., 1973, Neurogenesis and the cell cycle, in: *Results and Problems in Cell Differentiation, Vol. 7,* pp. 62–83, Springer Verlag, Berlin.

Prasad, K. N., 1971, Effect of dopamine and 6-hydroxydopamine on mouse neuroblastoma cells *in vitro, Cancer Res.* **31:**1457–1460.

Prasad, K. N., 1972a, Morphological differentiation induced by prostaglandin in mouse neuroblastoma cells in culture, *Nature* [*New Biol.*] **236**:49–52.
Prasad, K. N., 1972b, Neuroblastoma clones: Prostaglandin versus dibutyryl cyclic AMP, 8-benzylthio-cyclic AMP, phosphodiesterase inhibitors and X-rays, *Proc. Soc. Exp. Biol. Med.* **140**:126–129.
Prasad, K. N., 1972c, Inhibitors of cyclic nucleotide phosphodiesterase induce morphological differentiation of mouse neuroblastoma cell culture, *Exp. Cell Res.* **73**:436–440.
Prasad, K. N., 1974, Manganese inhibits adenylate cyclase activity and stimulates phosphodiesterase activity in neuroblastoma cells: Its possible implication in manganese-poisoning, *Exp. Neurol.* **45**:554–557.
Prasad, K. N., 1975, Differentiation of neuroblastoma cells in culture, *Biol. Rev.* **2**:129–165.
Prasad, K. N., 1977, Role of cyclic nucleotide in the differentiation of nerve cells, in: *Cell, Organ and Tissue Culture in Neurobiology* (S. Fedorof and L. Hertz, eds.), pp. 448–483, Academic Press, New York.
Prasad, K. N., and Hsie, A. W., 1971, Morphological differentiation of mouse neuroblastoma cells induced in-vitro by db-cAMP, *Nature* [*New Biol.*] **233**:141.-142.
Prasad, K. N., and Kumar, S., 1973, Cyclic 3′, 5′-AMP phosphodiesterase activity during cyclic AMP-induced differentiation of neuroblastoma cells in culture, *Proc. Soc. Exp. Biol. Med.* **142**:406–409.
Prasad, K. N., and Mandal, B., 1972, Catechol-O-methyl-transferase activity in dibutyryl cyclic AMP, prostaglandin and X-ray-induced differentiated neuroblastoma cell culture, *Exp. Cell Res.* **74**:532–534.
Prasad, K. N., and Mandal, B., 1973, Choline acetyltransferase level in cyclic AMP and X-ray induced morphologically differentiated neuroblastoma cells in culture, *Cytobios* **8**:75–80.
Prasad, K. N., and Sheppard, J. R., 1972a, Inhibitors of cyclic-nucleotide phosphodiesterase induced morphological differentiation of mouse neuroblastoma cell culture, *Exp. Cell Res* **73**:436–440.
Prasad, K. N., and Sheppard, J. R., 1972b, Neuroblastoma cell cultures: Membrane changes during cyclic AMP-induced morphological differentiation, *Proc. Soc. Exp. Biol. Med.* **141**:240–243.
Prasad, K. N., and Vernadakis, A., 1972, Morphological and biochemical study in X-ray and dibutyryl cyclic AMP-induced differentiated neuroblastoma cells, *Exp. Cell Res.* **70**:27–32.
Prasad, K. N., Waymire, J. C., and Weiner, N., 1972, A further study on the morphology and biochemistry of X-ray and dibutyryl cyclic AMP-induced differentiated neuroblastoma cells in culture, *Exp. Cell Res.* **74**:100–114.
Prasad, K. N., Kumar, S., Gilmer, K., and Vernadakis, A., 1973a, cAMP induced differentiated neuroblastoma cells: Changes in total nucleic acid and protein contents, *Biochem. Biophys. Res. Commun.* **50**:973–977.
Prasad, K. N., Mandal, B., and Kumar, S., 1973b, Human neuroblastoma cell culture: Effect of 5-BrdU on morphological differentiation and levels of neural enzymes, *Proc. Soc. Exp. Biol. Med.* **144**:38–42.
Prasad, K. N., Mandal, B., Waymire, J. C., Lees, G. J., Vernadakis, A., and Weiner, N., 1973c, Basal level of neurotransmitter synthesizing enzymes and effect of cyclic AMP agents on the morphological differentiation of isolated neuroblastoma clones, *Nature* [*New Biol.*] **241**:117–119.
Prasad, K. N., Gilmer, K. N., and Sahu, S. K., 1974, Demonstration of acetylcholine-sensitive adenyl cyclase in malignant neuroblastoma cells in culture, *Nature* **249**:765–767.
Prasad, K. N., Becker, G., and Tripathy, K., 1975a, Differences and similarities between guanosine 3′, 5′-cyclic monophosphate phosphodiesterase and adenosine 3′, 5′-cyclic monophosphate phosphodiesterase activities in neuroblastoma cells in culture, *Proc. Soc. Exp. Biol. Med.* **149**:757–762.
Prasad, K. N., Bondy, S. C., and Purdy, J. L., 1975b, Polyadenylic acid-containing cytoplasmic RNA increases in X-irradiated neuroblastoma cells in culture, *Radiat. Res.* **62**:585.

Prasad, K. N., Bondy, S. C., and Purdy, J. L., 1975c, Polyadenylic acid-containing cytoplasmic RNA increases in adenosine 3′, 5′-cyclic monophosphate induced differentiation of neuroblastoma cells in culture, *Exp. Cell Res.* **94:**88–94.

Prasad, K. N., Gilmer, K. N., Sahu, S. K., and Becker, G., 1975d, Regulation of adenylate cyclase activity in malignant and cyclic AMP-induced differentiated neuroblastoma cells: Effect of neurotransmitters, GTP and divalent ions, *Cancer Res.* **35:**77–88.

Prasad, K. N., Fogleman, D., Gaschler, M., Sinha, P. K., and Brown, J. L., 1976a, Cyclic nucleotide-dependent protein kinase activity in malignant and cyclic AMP-induced "differentiated" neuroblastoma cells in culture, *Biochem. Biophys. Res. Commun.* **68:**1248–1255.

Prasad, K. N., Sahu, S. K., and Sinha, P. K., 1976b, Cyclic nucleotides in regulation of expression of differentiated functions in neuroblastoma cells, *J. Natl. Cancer Inst.* **57:**619–631.

Prasad, N., Rosenberg, R. N., Ulrich, C., Wischmeyer, B., and Sparkman, D., 1978, Induction of cytoplasmic cAMP receptor proteins and changes in the nuclear non-histone proteins by db-cAMP in differentiated mouse neuroblastoma cells, *Fed. Proc.* **37:**1829.

Puck, T. T., Marcus, P. I., and Cieciura, J., 1956, Clonal growth of mammalian cells *in vitro, J. Exp. Med.* **103:**273–284.

Puro, D. G., and Nirenberg, M., 1976, On the specificity of synapse formation, *Proc. Natl. Acad. Sci. USA* **73:**3544–3548.

Reiser, G., Heumann, R., Kemper, W., Lautenschlager, E., Hamprecht, B., 1977, Influence of cations on the electrical activity of neuroblastoma × glioma hybrid cells, *Brain Res.* **130:**495–504.

Reuben, R. C., Wife, R. L., Breslow, R., Rifkind, R. A., and Marks, P., 1976, A new group of potent inducers of differentiation in murine erythroleukemia cells, *Proc. Natl. Acad. Sci. USA* **78:**862–866.

Revoltella, R., Bertolini, L., and Pediconi, M., 1974a, Unmasking of nerve growth factor membrane-specific binding sites in synchronized murine C-1300 neuroblastoma cells, *Exp. Cell Res.* **85:**89–94.

Revoltella, R., Bertolini, L., Pediconi, M., and Vigneti, E., 1974b, Specific binding of nerve growth factor (NGF) by murine C-1300 neuroblastoma cells, *J. Exp. Med.* **140:**437–451.

Revoltella, R., Bosman, C., and Bertolini, L., 1975, Detection of nerve growth factor binding sites on neuroblastoma cells by rosette formation, *Cancer Res.* **35:**890–985.

Revoltella, R., Bertolini, L., Diamond. L., Vigneti, E., and Grasso, A., 1976, A radio-immunoassay for measuring 14-3-2 protein in cell extracts, *J. Neurochem.* **26:**831–834.

Richelson, E., 1973a, Regulation of tyrosine hydroxylase activity in mouse neuroblastoma clone N1E-115, *J. Neurochem.* **21:**1139–1145.

Richelson, E., 1973b, Stimulation of tyrosine hydroxylase activity in an adrenergic clone of mouse neuroblastoma by dibutyryl cyclic AMP, *Nature* [*New Biol.*] **242:**175–177.

Richelson, E., 1974, Studies on the transport of L-tyrosine into an adrenergic clone of mouse neuroblastoma, *J. Biol. Chem.* **249:**6128–6224.

Richelson, E., 1975, The culture of established clones for neurobiologic investigation, in: *Metabolic Compartmentation and Neurotransmission* (S. Berl, D. D. Clarke, and D. Schneider, eds.), pp. 305–326, Plenum Press, New York.

Richelson, E., 1976a, Properties of tyrosine hydroxylase in living mouse neuroblastoma clone N1E-115, *J. Neurochem.* **27:**1113–1118.

Richelson, E., 1976b, Tissue culture of the nervous system: Applications in neurochemistry and psychopharmacology, in: *Handbook of Psychopharmacology, Vol. 1* (L. L. Iverson, S. D. Iverson, and S. H. Snyder, eds.), pp. 101–135, Plenum Press, New York.

Richelson, E., 1977, Antipsychotics block muscarinic acetylcholine receptor-mediated cyclic GMP formation in cultured mouse neuroblastoma cells, *Nature* **266:**371–373.

Richelson, E., 1978a, Histamine, H_1 receptor-mediated c-GMP formation by cultured mouse neuroblastoma cells, *Science* **201:**69–71.

Richelson, E., 1978b, Desensitization of muscarinic receptor-mediated c-GMP formation by cultured nerve cells, *Nature* **272**:366–368.

Richelson, E., and Divinetz-Romero, S., 1977, Blockade by psychotropic drugs of the muscarinic acetylcholine receptor in cultured nerve cells, *Biol. Psychiatry* **12**:771–785.

Richelson, E., and Thompson, E. J., 1973, Transport of neurotransmitter precursors into cultured cells, *Nature* [*New Biol.*] **241**:201–204.

Richelson, E., Prendergase, F. G., and Divinetz-Romero, S., 1978, Muscarinic receptor-mediated c-GMP formation by cultured nerve cells: Ionic dependence and effects of local anesthetics, *Biochem. Pharmacol.* **27**:2039–2048.

Rieger, F., Faivre-Bauman, A., Benda, P., and Vigny, M., 1976, Molecular forms of acetylcholinesterase: Their *de-novo* synthesis in mouse neuroblastoma cells, *J. Neurochem.* **27**:1059–1063.

Robinson, G. A., Butcher, R. W., and Sutherland, E. W., 1971, *Cyclic AMP*, Academic Press, New York.

Roisen, F., Inczedy-Marcsek, M., Hsu, L., and Yorke, W., 1978, Myosin: Immunofluorescent localization in neuronal and glial cultures, *Science* **199**:1445–1448.

Rosenberg, S. B., and Charalampous, F. C., 1977, Interaction of concanavalian A with differentiated and undifferentiated murine neuroblastoma cells, *Arch Biochem. Biophys.* **181**:117–127.

Rosenberg, R. N., Vandeventer, L., DeFrancesco, L., and Friedkin, M. E., 1971, Regulation of the synthesis of choline-0-acetyltransferase and thymidylate synthetase in mouse neuroblastoma in cell culture, *Proc. Natl. Acad. Sci. USA* **68**:1436–1440.

Ross, J., Granett, S., and Rosenbaum, J. L., 1973, Differentiation of neuroblastoma cells in hypertonic medium, *J. Cell Biol.* **59**:291A.

Ross, J., Olmsted, J. B., and Rosenbaum, J. L., 1975, The ultrastructure of mouse neuroblastoma cells in tissue culture, *Tissue Cell* **7**:107–136.

Rotman, A., Daly, J. W., and Creveling, C. R., 1976a, Oxygen-dependent reaction of 6-hydroxydopamine, 5,6-dihydroxytryptamine and related compounds with proteins *in vitro:* A model for cytotoxicity, *Mol. Pharmacol.* **12**:887–899.

Rotman, A., Daly, J. W., Creveling, C. R., and Breakefield, X. O., 1976b, Uptake and binding of dopamine and 6-hydroxydopamine in murine neuroblastoma and fibroblast cells, *Biochem. Pharmacol.* **25**:383–388.

Sahu, S. K., and Prasad, K. N., 1975, Effect of neurotransmitters and prostaglandin E_1 on cyclic AMP levels in various clones of neuroblastoma cells in culture, *J. Neurochem.* **24**:1267–1269.

Sandquist, D., Williams, T. H., Sahu, S. K., and Kataoka, S., 1978, Morphological differentiation of a murine neuroblastoma clone in monolayer culture induced by dexamethasone, *Exp. Cell Res.* **113**:375–381.

Sato, G., 1973, *Tissue Culture of the Nervous System, Current Topics in Neurobiology, Vol. 1*, Plenum Press, New York.

Sato, G. H., 1975, The role of serum in cell culture, in: *Biochemical Actions of Hormones, Vol. 3* (G. Litwack, ed.), pp. 391–395, Academic Press, New York.

Schachner, M., 1973, Serologically demonstrable cell surface specificities on mouse neuroblastoma C-1300, *Nature* [*New Biol.*] **243**:117–119.

Schachner, R. M., and Worthan, K. A., 1975, Nervous system antigen-3 (NS-3) an antigenic cell surface component expressed on neuroblastoma C1300, *Brain Res.* **99**:210–208.

Schmitt, H., 1976, Control of tubulin and actin synthesis and assembly during differentiation of neuroblastoma cells, *Brain Res.* **115**:165–173.

Schneider, F. H., 1976, Effects of sodium butyrate on mouse neuroblastoma cells in culture, *Biochem. Pharmacol.* **25**:2309–2317.

Schubert, D., 1974, Induced differentiation of clonal rat nerve and glial cells, *Neurobiology* **4**:376–387.

Schubert, D., 1979, Early events after the interaction of NGF with sympathetic nerve cells, *Trends Neurosci.* **2:**17–20.

Schubert, D., and Jacob, F., 1970, 5-Bromodeoxyuridine-induced differentiation of a neuroblastoma, *Proc. Natl. Acad. Sci. USA* **67:**247–254.

Schubert, D., and Klier, F. G., 1977, Storage and release of acetylcholine by a clonal line, *Proc. Natl. Acad. Sci. USA* **74:**5184–5188.

Schubert, D., and Whitlock, C., 1977, Alteration of cellular adhesion by NGF, *Proc. Natl. Acad. Sci. USA* **74:**4055–4058.

Schubert, D., Humphreys, S., Baroni, C., and Cohn, M., 1969, *In-vitro* differentiation of a mouse neuroblastoma, *Proc. Natl. Acad. Sci. USA* **64:**316–323.

Schubert, D., Humphreys, S., Vitry, F. D. E., and Jacob, F., 1971a, Induced differentiation of a neuroblastoma, *Dev. Biol.* **25:**514–546.

Schubert, D., Tarikas, H., Harris, A. J., and Heineman, S., 1971b, Induction of acetylcholinesterase activity in mouse neuroblastoma, *Nature* [*New Biol.*] **233:**79–80.

Schubert, D., Harris, A. J., Heinemann, S., Kidokoro, Y., Patrick, J., and Steinbach, J. H., 1973, Differentiation and interaction of clonal lines of nerve and muscle, in: *Tissue Culture of the Nervous System* (G. H. Sato, ed.), pp. 55–86, Plenum Press, New York.

Schubert, D., Heinemann, S., Carlisle, W., Tarikas, H., Kimes, B., Patrick, J., Steinbach, J. H., Culp, W., and Brandt, B. L., 1974, Clonal cell lines from rat central nervous system, *Nature* **249:**224–227.

Schubert, D., Tarikas, H., and Lacorbiere, M., 1976, Neurotransmitter regulations of adenosine 3′, 5′-monophospate in clonal nerve, glia, and muscle cell lines, *Science* **192:**471–473.

Shubert, D., Heinemann, S., and Kidokoro, Y., 1977a, Cholinergic metabolism and synapse formation by a rat nerve cell line. *Proc. Natl. Acad. Sci. USA* **74:**2579–2583.

Schubert, D., Lacorbiere, M., Whitlcok, C., and Stallcup, W., 1978, Alterations in the surface properties of cells responsive to NGF, *Nature* **273:**718–723.

Schultz, J., and Hamprecht, B., 1973, Adenosine 3′, 5′ monophosphate in cultured neuroblastoma cells: Effect of adenosine phosphodiesterase inhibitors and benzazepines, *Arch. Pharmacol.* **278:**215–225.

Schürch-Rathgeb, Y., and Monard, D., 1978, Brain development influences the appearence of glial factor-like activity in rat brain primary culture, *Nature* **273:**308–309.

Seeds, N. W., Gilman, A. G., Amano, T., and Nirenberg, M., 1970, Regulation of axon formation by clonal lines of a neuronal tumor, *Proc. Natl. Acad. Sci. USA* **66:**160–167.

Sharma, S. K., Nirenberg, M., and Klee, W. A., 1975a, Morphine receptors as regulators of adenylate cyclase activity, *Proc. Natl. Acad. Sci. USA* **72:**590–594.

Sharma, S. K., Klee, W. A., and Nirenberg, M., 1975b, Dual regulation of adenylate cyclase accounts for narcotic dependence and tolerance, *Proc. Natl. Acad. Sci. USA* **72:**3092–3096.

Sharma, S. K., Klee, W. A., and Nirenberg, M., 1977, Opiate-dependent modulation of adenylate cyclase, *Proc. Natl. Acad. Sci. USA* **74:**3365–3369.

Sidman, R. L., and Rakic, P., 1973, Neuronal migration with special reference to developing human brain. A review, *Brain Res.* **62:**1–35.

Silagi, S., and Bruce, S. A., 1970, Suppression of malignancy and differentiation in melanotic melanoma cells, *Proc. Natl. Acad. Sci. USA* **66:**72–78.

Siman, R. G., and Klein, W. L., 1979, Cholinergic activity regulates muscarinic receptors in central nervous system cultures, *Proc. Natl. Acad. Sci. USA* **76:**4141–4145.

Siman-Tov, R., and Sachs, L., 1972, Enzyme Regulation in neuroblastoma cells, Selection of clones with low acetylcholinesterase activity and the independent control of acetylcholinesterase and choline-0-acetyltransferase, *Eur. J. Biochem.* **30:**123–129.

Siman-Tov, R., and Sachs, L., 1973, Regulation of acetylcholine receptors in regulation of acetylcholinesterase in neuroblastoma cells, *Proc. Natl. Acad. Sci. USA* **70:**2902–2905.

Siman-Tov, R., and Sachs, L., 1975a, Different mechanisms for induction of acetylcholinesterase in neuroblastoma cells, *Dev. Biol.* **45:**382–385.

Siman-Tov, R., and Sachs, L., 1975b, Induction of polyadenylate polymerase and differentiation in neuroblastoma cells, *Eur. J. Biochem.* **55:**9–14.

Siman-Tov, R., and Sachs, L., 1975c, Temperature sensitivity of cAMP-binding protein activity of protein kinases and the regulation of cell growth, *Eur. J. Biochem.* **59:**89–95.

Siman-Tov, R., and Snyder, S. H., 1976, Isolation and structure identification of a morphinelike peptide "enkephalin" in bovine brain, *Life Sci.* **18:**781–788.

Simon, E. J., Hiller, J. M., and Edelman, I., 1973, Stereospecific binding of the potent narcotic analgesia [^{3}H]etorphine to rat homogenate, *Proc. Natl. Acad. Sci. USA* **70:**1947–1949.

Sinha, P. K., and Prasad, K. N., 1977, A further study on regulation of cyclic nucleotide phosphodiesterase activity in neuroblastoma cells—Effect of growth, *In Vitro* **13:**497–501.

Skaper, S. D., Adelson, G. L., and Seegmiller, J. E., 1976, Metabolism of biogenic amines in neuroblastoma and glioma cells in culture, *J. Neurochem.* **27:**1065–1070.

Skolnick, P., and Daly, J. W., 1977, Regulation of cAMP formation in brain tissue by putative neurotransmitters, in: *Cyclic Nucleotides: Mechanism of Action* (H. Cramer and J. Schultz, eds.), John Wiley and Sons, New York.

Snyder, S. H., and Siman-Tov, P., 1977, The opiate receptor and opioid peptides, *J. Neurochem.* **28:**13–20.

Snyder, S., Greenberg, D., and Yamamura, H. I., 1974, Antischizophrenic drugs and brain cholinergic receptors, *Arch. Gen. Psychiat.* **31:**58–61.

Solomon, F., Monard, D., and Rentsch, M., 1973, Stabilization of colchicine-binding activity of neuroblastoma, *J. Mol. Biol.* **78:**569–575.

Solomon, F., Gysin, R., Rentsch, M., and Monard, D., 1976, Purification of tubulin from neuroblastoma cells: Absence of covalently bound phosphate in tubulin from normal and morphologically differentiated cells, *FEBS Lett.* **63:**316–319.

Spector, R., and Greene, L. A., 1977, Ascorbic acid transport by a clonal line of pheochromocytoma cells, *Brain Res.* **136:**131–140.

Spector, I., Kimhi, Y., and Nelson, P. G., 1973, Tetrodotoxin and cobalt blockade of neuroblastoma action potentials, *Nature* [*New Biol.*] **246:**124–126.

Stefanovic, V., Ciesielski-Treska, J., Ebel, A., and Mandel, P., 1974a, Nucleoside triphosphatase activity at the external surface of neuroblastoma cells, *Brain Res.* **81:**427–441.

Stefanovic, V., Ciesielski-Treska, J., Ebel, A., and Mandel, P., 1974b, Demonstration of an ATPase sensitive to ouabain on the external surface of the cells of neurobalstomas and of glial cells in culture, *C. R. Acad. Sci.* [*D.*] *(Paris)* **278:**2041–2044.

Stefanovic, V., Ciesielski-Treska, J., Ebel, A., and Mandel, P., 1974c, Ca^{2+}-activated ATPase at the external surface of neuroblastoma cells in culture, *FEBS Lett.* **49:**43–46.

Stefanovic, V., Mandel, P., and Rosenberg, A., 1975a, Activation of acetyl- and butyrylcholinesterase by enzymatic removal of sialic acid from intact neuroblastoma and astroblastoma cells in culture, *Biochemistry* **14:**5257–5260.

Stevanovic, V., Massarelli, R., Mandel, P., and Rosenberg, A., 1975b, Effect of cellular desialylation on choline high affinity uptake and ecto-acetylcholinesterase activity of cholinergic neuroblasts, *Biochem Pharmacol.* **24:**1923–1928.

Stefanovic, V., Ledig, M., and Mandel, P., 1976, Divalent cation-activated ecto-nucleoside triphosphatase activity of nervous system cells in tissue culture, *J. Neurochem.* **27:**799–805.

Stefanovic, V., Ciesielski-Treska, J., and Mandel, P., 1977, Neuroblast–glia interaction in tissue culture as evidenced by the study of ectoenzymes. Ecto-ATPase activity of mouse neuroblastoma cells, *Brain Res.* **122:**313–323.

Steinback, J. H., Harris, E. G., Patrick, G., Schubert, D., and Heineman, S., 1973, Nerve–muscle interaction *in vitro:* Role of acetylcholine, *J. Gen. Physiol.* **62:**255–270.

Steinbach, J. H., Schubert, D., and Tarikas, H., 1974, Inhibition of acetylcholine synthesis in neuroblastoma cells by a styrylpyridine analog, *J. Neurochem.* **22:**611–613.

Stoolmiller, A. C., Dawson, G., and Dorfman, A., 1973, The metabolism of glycosphingolipids and glycosaminoglycans, in: *Tissue Culture of the Nervous System* (G. H. Sato, ed.), pp. 247–280, Plenum Press, New York.

Suszkiw, J. B., Beach, R. L., and Pilar, G. R., 1976, Choline uptake by cholinergic neuron cell somas, *J. Neurochem.* **26:**1123–1131.

Taylor, J. E., and Richelson, E., 1979, Desensitization of histamine H_1 receptor-mediated cyclic GMP. Formation in mouse neuroblastoma cells, *Mol. Pharmacol.* **15:**462–471.

Thoenen, H., and Tranzer, J. P., 1968, Chemical sympathectomy by selective destruction of adrenergic nerve endings with 6-OHDA, *Naunyn-Schmiedebergs Arch. Pharmakol. Exp. Pathol.* **261:**271–288.

Tishler, A. S., and Greene, L. A., 1978, Morphologic and cytochemical properties of a clonal line of rat adrenal pheochromocytoma cells which respond to NGF, *Lab. Invest.* **39:**77–89.

Traber, J., Fischer, K., Latzin, S., and Hamprecht, B., 1974a, Morphine antagonizes action of prostaglandin in neuroblastoma cells but not of prostaglandin and noradrenaline in glioma and glioma × fibroblast hybrid cells, *FEBS Lett.* **49:**260–263.

Traber, J., Fischer, K., Latzin, S., and Hamprecht, B., 1974b, Cultures of cells derived from the nervous system: Synthesis and action of prostaglandin E, in: *Proceedings of the IX Congress of the Collegium International Neuropsychopharmacologium* (O. Virai, Z. Votara, and P. B. Bradley, eds.), pp. 956–969, North-Holland, Amsterdam.

Traber, J., Fischer, K., Latzin, S., and Hamprecht, B., 1975a, Morphine antagonizes action of prostaglandin in neuroblastoma and neuroblastoma × glioma hybrid cells, *Nature* **253:**120–122.

Traber, J., Reiser, G., Fischer, K., and Hamprecht, B., 1975b, Measurements of adenosine 3′, 5′ cyclic monophosphate and membrane potential in neuroblastoma × glioma hybrid cells—Opiates and adrenergic agonists cause effects opposite to those of prostaglandin E_1, *FEBS Lett.* **52:**327–331.

Traber, J., Fischer, K., Buchen, C., and Hamprecht, B., 1975c, Muscarinic response to acetylcholine in neuroblastoma × glioma hybrid cells, *Nature* **255:**558–560.

Traber, J., Glaser, T., Brandt, M., Klebensberger, W., and Hamprecht, B., 1977, Different receptors for somatostatin and opioids in neuroblastoma × glioma hybrid cells, *FEBS Lett.* **81:**351–354.

Trams, E. G., Lauter, C. J., and Banfield, W. G., 1976, On the activation of plasma membrane ecto-enzymes by treatment with neuraminidase, *J. Neurochem.* **27:**1035–1042.

Truding, R., and Morell, P., 1977, Effect of N^6, 0^2-dibutyryl adenosine 3′, 5′-monophosphate on the release of surface proteins by murine neuroblastoma cells, *J. Biol. Chem.* **252:**4850–4854.

Truding, R., Shelanski, M. L., Daniels, M. P., and Morell, P., 1974, Comparison of surface membranes isolated from cultured murine neuroblastoma cells in the differentiated or undifferentiated state, *J. Biol. Chem.* **249:**3973–3982.

Uzunov, P., and Weiss, B., 1972, Separation of multiple forms of cyclic adenosine 3′, 5′-monophosphate phosphodiesterase in rat cerebellum by polyacrylamide gel electrophoresis, *Biochim. Biophys. Acta* **284:**220–226.

Uzunov, P., Shein, H. M., and Weiss, B., 1974, Multiple forms of cyclic 3′, 5′ AMP phosphodiesterase of rat cerebrum and cloned astrocytoma and neuroblastoma cells, *Neuropharmacology* **13:**377–391.

Vimard, C., Jeantet, C., Netter, Y., and Gros, F., 1976, Changes in the sedimentation properties of acetylcholinesterase during neuroblastoma differentiation, *Biochimie* **58:**473–478.

Wahlstrom, A., Brandt, M., Moroder, L., Wunsch, E., Lindeberg, G., Ragnarsson, U., Terenius, L., and Hamprecht, B., 1977, Peptides related to beta-lipotropin with opioid activity—effects on levels of adenosine 3′, 5′ cyclic monophosphate in neuroblastoma × glioma hybrid cells, *FEBS Lett.* **77:**28–32.

Warren, L., and Glick, M., 1969, Isolation of surface membranes of tissue culture cells, in: *Fundamental Techniques in Virology* (K. Habad and N. P. Salzman, eds.), pp. 66–71, Academic Press, New York.

Warren, S., and Chute, R. N., 1972, Pheochromocytoma, *Cancer* **29:**327–331.

Warter, S., Hermetet, J. C., and Cieselski-Treska, J., 1974, Cytogenetic characterization of C-1300 neuroblastoma cells, *Experientia* **30:**291–292.

Waymire, J. C., and Gilmer-Waymire, K., 1978, Adrenergic enzymes in cultured mouse neuroblastoma: Absence of detectable aromatic-L-amino-acid decarboxylase, *J. Neurochem.* **31:**693–698.

Waymire, J. C., Weiner, N., and Prasad, K. N., 1972, Regulation of tyrosine hydroxylase activity in cultured mouse neuroblastoma cells: Elevation induced by analogues of adenosine 3′, 5′ cyclic monophosphate, *Proc. Natl. Acad. Sci. USA* **69:**2241–2245.

Waymire, J. C., Waymire, K. G., Boehme, R., Noritake, D., and Wardell, J., 1977, Regulation of tyrosine hydroxylase by cAMP in cultured neuroblastoma and cultured dissociated bovine adrenal chromaffin cells, in: *Modern Pharmacology–Toxicology, Vol. 10, Conference on Structure and Function of Monoamine Enzymes* (Usdin, Weiner, and Yaudim eds.), pp. 327–363, Marcel Dekker, Basel.

Waymire, J. C., Gilmer-Waymire, K., and Boehme, R. E., 1978a, Concomitant elevation of tyrosine hydroxylase and dopamine β-hydroxylase by cAMP in cultured mouse neuroblastoma cells, *J. Neurochem.* **31:**699–705.

Waymire, J. C., Gilmer-Waymire, K., Noritake, D., Kitayama, D., and Haycock, J. W., 1978b, Induction of tyrosine hydroxylase and dopamine β-hydroxylase in cultured mouse neuroblastoma by 8 Br-cAMP. Involvement of RNA and protein synthesis, *Mol. Pharmacol.* **15:**78–85.

Waymouth, C., 1977, Nutritional requirements of cells in culture with special reference to neural cells, in: *Cell, Tissue and Organ Cultures in Neurobiology* (S. Fedoroff and L. Hertz, eds.), pp. 631–648, Academic Press, New York.

Wengler, G., and Wengler, G., 1972, Medium hypertonicity and polyribosome structure in HeLa cells. The influence of hypertonicity on the growth medium on polyribosomes in HeLa cells, *Eur. J. Biochem.* **27:**162–173.

Wexler, B., and Katzmann, R., 1975, Effects of dibutyryl cyclic AMP on tyrosine uptake and metabolism in neuroblastoma cultures, *Exp. Cell Res.* **92:**291–298.

Wigley, C. B., 1975, Differentiated cells *in vitro, Differentiation* **4:**25–55.

Wilson, S. H., Schrier, B. K., Faber, J. L., Thompson, E. J., Rosenberg, R. N., Blume, A. J., and Nirenberg, N. W., 1972, Markers for gene expression in cultured cells from the nervous system, *J. Biol. Chem.* **247:**3159–3169.

Wilson, S., Higashida, H., Minna, J., and Nirenberg, M., 1978, Defects in synaptic formation and acetylcholine release by neuroblastoma and hybrid cell lines, *Fed. Proc.* **37:**1784.

Yamamura, H. I., and Snyder, S. H., 1973, High affinity transport of choline into synaptosomes of rat brain, *J. Neurochem.* **21:**1355–1374.

Yamamura, H. I., and Snyder, S. H., 1974a, Muscarinic cholinergic binding in rat brain, *Proc. Natl. Acad. Sci. USA* **71:**1725.-1729.

Yamamura, H. I., and Snyder, S. H., 1974b, Muscarinic cholinergic receptor binding in the longitudinal muscle of the guinea pig ileum with ^{3}H quinuclidinyl benzilate, *Mol. Pharmacol.* **10:**861–867.

Zornetzer, M. S., and Stein, G., 1975, Gene expression in mouse neuroblastoma cells; Properties of the genome, *Proc. Natl. Acad. Sci. USA* **72:**3119–3123.

Zucco, F., Persico, M., Felsani, A., Metafora, S., and Augusti-Tocco, G., 1975, Regulation of protein synthesis at the translational level in neuroblastoma cells, *Proc. Natl. Acad. Sci. USA* **72:**2289–2293.

Zwiller, J., Ciesielski-Treska, Mack, G., and Mandel, P., 1975, Uptake of noradrenaline by an adrenergic clone of neuroblastoma cells, *Nature* **254:**443–444.

Zwiller, J., Goridis, C., Ciesielski-Treska, J., and Mandel, P., 1977, Cyclic GMP in neuroblastoma clones: Possible involvement in morphological differentiation induced by dibutyryl cyclic AMP, *J. Neurochem.* **29:**273–278.

7

Electrophysiology of Clonal Nerve Cell Lines

ILAN SPECTOR

1. INTRODUCTION

In recent years it has become apparent that excitable cells grown in tissue culture provide some of the most advantageous preparations for investigating fundamental questions in neurobiology. Among the different *in vitro* neuronal systems presently available (for reviews, see Sato, 1973; Nelson, 1975; Fischbach and Nelson, 1977; Patrick *et al.*, 1978; Patterson, 1978; Federoff and Hertz, 1977), clonal cell lines permit important types of experimentation that are impossible in other preparations *in vivo* or *in vitro*. The unique features of cell lines include: (1) the ability to multiply indefinitely with fairly consistent phenotypes from generation to generation; (2) the ability to differentiate in response to changes in culture conditions; (3) the feasibility of genetic manipulations such as somatic cell hybridization; (4) the ease in obtaining large quantities of homogeneous material for biochemical experiments; (5) the unusually large size of individual cells in some neuronal lines which makes them particularly accessible to detailed analysis of electrically and chemically activated ionic conductances; and (6) the possibility of obtaining large cells by artificial fusion.

Despite these apparent *a priori* advantages and the increasing use of clonal cell lines in neurobiological research (for review, see Kimhi, this volume), their potential as experimental models for the study of neurodifferentiation and of excitable membrane physiology and pharmacology has not been fully realized. Since the establishment of clonal lines from the mouse C-1300

ILAN SPECTOR • Laboratory of Biochemical Genetics, National Heart, Lung, and Blood Institute, National Institutes of Health, Bethesda, Maryland 20205.

neuroblastoma a decade ago, one nagging issue has persevered: What is the relationship of neuronal lines to nerve cells in the living organism? This issue is difficult to resolve unequivocally for any *in vitro* neuronal system, and the answer varies with the neuronal characteristics chosen for comparison. It has been further obscured in the case of clonal systems by the common use of rather arbitrary and often ambiguous morphological rather than physiological criteria to assess the state of the cells. In addition, very little detailed electrophysiological research had been carried out until the past few years. It is not surprising, therefore, that even such a fundamental question as the ability of neuronal cell lines to undergo a process of electrophysiological differentiation is still a matter of dispute (Nelson, 1973, 1975; Schubert *et al.*, 1973; Patrick *et al.*, 1978). This chapter summarizes recent electrophysiological studies of clonal nerve cell lines. Its first aim is to clarify to some extent the issues outlined above. Its second aim is to discuss these studies as they relate to some current ideas about the electrical excitability mechanisms in nerve cells and about the development of these mechanisms during differentiation.

2. SOME GENERAL CONSIDERATIONS ABOUT GROWTH AND DIFFERENTIATION OF CLONAL NERVE CELL LINES

Numerous neuronal lines from diverse sources have been established during the past decade (for review, see Kimhi, this volume). However, only a few of the lines have been investigated with electrophysiological techniques in more than a superficial manner. These include several clonal lines derived from the mouse neuroblastoma C-1300 tumor (e.g., N1E-115, N-18; see Amano *et al.*, 1972) and hybrid cell lines formed by somatic cell hybridization techniques from neuroblastoma parent lines and other parent cell types (e.g., NG108-15, the mouse neuroblastoma × rat glioma line; Hamprecht, Amano, and Nirenberg, unpublished data). The small average cell size of lines derived from neuroendocrine tumors [e.g., the rat anterior pituitary cell line GH_3 (Kidokoro, 1975) and the popular rat pheochromocytoma PC-12 line (Greene and Tischler, 1976; Dichter *et al.*, 1977)] and from chemically induced tumors of the central nervous system (Schubert *et al.*, 1974) has, until recently, hampered adequate electrophysiological measurements. This disadvantage has recently been overcome by O'Lague and Huttner (1980), who used polyethylene glycol to fuse individual PC-12 cells and obtained giant multinucleated cells suitable for electrophysiology.

The most extensive electrophysiological work has been done on the neuroblastoma adrenergic clone N1E-115 (Spector *et al.*, 1973, 1975; Chalazonitis and Greene, 1974; Tuttle and Richelson, 1975, 1979; Richelson and Tuttle, 1975; Moolenaar and Spector, 1977, 1978, 1979a,b). This clone is highly suitable for voltage clamp studies because of the large size (up to 150 μm in diameter) and favorable geometry of the cells (Fig. 1). The cells exhibit electrically activated ionic conductance mechanisms characteristic of mature neurons and respond to at least three putative neurotransmitters: acetylcholine, serotonin, and histamine (MacDermont *et al.,* 1979; Higashida, Wilson, and Nirenberg, in preparation; Christian and Nelson, unpublished observations). In addition, this line has several important growth characteristics that make it highly manipulable in developmental studies. When grown in an appropriate medium [Dulbecco's Modified Eagle Medium (DMEM) with 12.2 g/liter $NaHCO_3$, supplemented with 5–10% fetal calf serum, and equilibrated with an atmosphere of 5% CO_2 in air], the cells adhere well to the plate and have less of a tendency to pile up after entering the confluent phase of growth than do many other lines. It is, therefore, much easier to maintain the cultures in the confluent phase of growth and to subject them to different culture regimens. Cellular differentiation can be obtained with nearly all of the procedures that have been employed for other lines (Kimhi, this volume). These features make it possible to study the regulations of electrophysiological, biochemical, and morphological properties in the same culture in different phases of growth and under different treatment regimens.

Although the normal growth medium promotes cell multiplication, in the stationary phase of growth N1E-115 cultures, as well as cultures of other lines, exhibit some degree of morphological differentiation (Fig. 1). Neurites, however, are labile, and cell morphology is extremely heterogeneous. Morphological differentiation (defined as enlargement of cells and outgrowth of neurites) can be induced by a host of interventions and is marked by a transition to a more homogeneous quasi-stable state in which cell division is essentially blocked. The cells at this stage become less susceptible to changes in the culture conditions.

Serum deprivation and treatment with agents as diverse as aminopterin (Spector *et al.,* 1973), dibutyryl cyclic AMP (dbcAMP) (Chalazonitis and Greene, 1974), dimethylsulfoxide (DMSO) (Kimhi *et al.,* 1976), hexamethylene bisacetamide (HMBA) (Palfrey *et al.,* 1977), and prostaglandin E_1 plus theophylline (Spector, unpublished observations) all induce morphological differentiation in N1E-115 cultures. There does not seem to be a common mode of action because the time course of differentiation, the appearance of the cells, and the long-term maintenance of the differentiated state vary greatly from agent to agent. Furthermore, morphological differentiation can take place without an enhancement of electrical excitability or an increase in levels of neurospecific enzymes (Kimhi *et al.,* 1976; Palfrey, 1976). Similarly, overshooting action potentials can be recorded from morphologically undifferentiated round cells of various lines (Schubert *et al.,* 1973; Dichter *et al,* 1977). It is, therefore, more reasonable to use different culture regimens in studying the regulation of a particular function in a given line rather than to equate the physiologically "differentiated" state with the ability of the cells to extend neurites.

For investigating electrical excitability mechanisms and their develop-

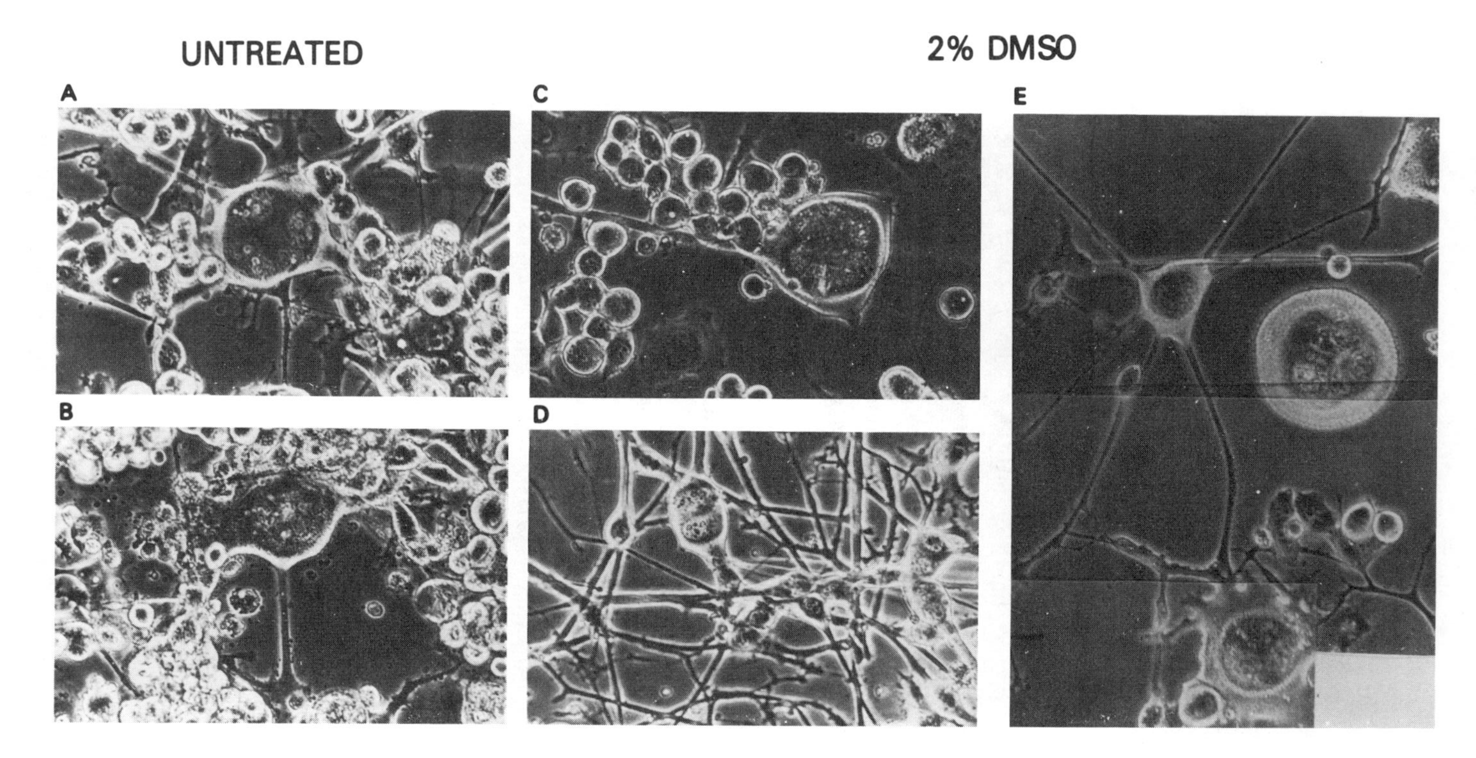

FIGURE 1. Morphological differentiation of neuroblastoma cells induced by DMSO. Confluent N1E-115 cells were replated into medium without (A and B), and with (C, D, and E) 2% DMSO. Phase contrast microphotographs were taken: A and C, 2 days; B and D, 9 days; and E, 22 days after subculture. Note the presence of large morphologically differentiated cells in untreated cultures. The large round cell in E is highly suitable for voltage clamping. Scale bar = 100 μm.

ment, N1E-115 cells cultured in DMSO offer several advantages. These include well-developed excitability properties, large cell soma size, and favorable geometry of the cells, as well as relatively rapid and synchronous kinetics of differentiation (Fig. 1; Kimhi *et al.*, 1976; Moolenaar and Spector, 1978). In addition, DMSO-treated cells maintain their differentiated state for several months and are not susceptible to the deterioration of serum-deprived cultures or to the reinitiation of cell division characteristic of dbcAMP-treated cultures. A major disadvantage of DMSO treatment is that morphological and electrical differentiation are not accompanied by biochemical differentiation. Levels of neurospecific enzymes remain low, and biochemical and electrical responses to neurotransmitters seem to be absent in DMSO-treated cultures. In contrast, such responses are common in confluent cultures (Matsuzawa and Nirenberg, 1975; Study *et al.*, 1978; MacDermont *et al.*, 1979) and in cultures treated with agents such as dbcAMP or prostaglandin E_1 and theophylline that modulate cyclic cAMP levels (Higashida and Nirenberg, in preparation; Morris and Spector, in preparation). It should also be noted that other cloned and hybrid lines do not respond in the same manner to DMSO treatment as does the N1E-115 clone. A different type of treatment is therefore needed in each case to obtain optimal conditions for electrophysiological measurement.

3. RESTING MEMBRANE POTENTIAL

No systematic analysis of resting potential (RP) in neuronal cell lines has been carried out thus far. This is unfortunate because the resting potential regulates membrane excitability and may also play a critical role in the growth and differentiation characteristic of these transformed cells. Reported microelectrode measurements in various neuroblastoma and hybrid lines range from about −10 to −70 mV, depending on the state of growth of the cells, the line examined, and the culture conditions (Nelson *et al.*, 1971b; Peacock *et al.*, 1972, 1973; Spector *et al.*, 1973; Chalazonitis and Greene, 1974; Tuttle and Richelson, 1975; Kimhi *et al.*, 1976; Miyake, 1978). Thus, for any given line, low resting potential values (−10 to −25 mV) characterize exponentially growing cells. The transition from log phase to stationary phase of growth is marked by a shift towards large RP values (−15 to −50 mV). A further hyperpolarization occurs when the cells undergo differentiation, so that differentiated cells characteristically have more homogeneous resting potentials of between −45 and −55 mV.

While similar findings have been reported by various investigators, interpretation of resting potentials based on microeletrode measurements remains a matter of dispute. One point of view is that the observed values are real and that the resting potential is a developmentally regulated property. The other view is that the low RP values are experimental artifacts that reflect a leaky membrane damaged by the microelectrode impalement (Kidokoro, this volume; Jaffe and Robinson, 1978). Implicit in such a position is the assumption that the real RP of cells of all types at all states of development is constant and, in most cases, large. Although this argument is difficult to refute, its predicitive power as a working hypothesis is limited. Perhaps the most direct evidence for this premise are the observations that in the rat skeletal muscle cell line L6, no significant changes in the RP occur during differentiation (Kidokoro, this volume). Both the small mononucleated myoblasts and the large multinucleated muscle fibers exhibit an RP of about −70 mV. Along these lines, a recent study (O'Lague and Huttner, 1980) on artificially fused cells of the rat PC12 line revealed large resting potentials (−50 to −60 mV) that did not change following differentiation induced by nerve growth factor (NGF) treatment. However, high resistance microelectrodes (80–200 MΩ) were used in both of these cases; these, while causing less damage to the membranes of small cells, can introduce artifacts in measuring the membrane potential. This is because such high resistance electrodes have relatively large and unstable tip potentials. It is not clear why such high resistance electrodes have been used to record from large cells, in particular the fused, giant PC12 cells (up to 300 μm in diameter) (O'Lague and Huttner, 1980).

These observations contrast sharply with those obtained in primary muscle cultures by several investigators (for review, see Kidokoro, this volume). In these preparations, not only do the small mononucleated myoblasts have low RPs, but a progressive increase in RP is observed well after fusion has taken place when multinucleated myotubes have reached quite a large size. The difference between the primary muscle cells and the transformed rat skeletal muscle cells has been confirmed by Ritchie and Fambrough (1975) who used identical techniques to measure the RPs of both types of cells. In both cases, external K^+ ion concentration seems to play an important role (Engelhardt *et al.*, 1980). Furthermore, the transformed rat skeletal muscle cells differ considerably from primary muscle cells in important active electrical properties (see Spector and Prives, 1977).

The question has not been so directly approached for neuroblastoma cells. It is true that the low RP values obtained in exponentially dividing neuroblastoma cells may be due to the fact that these cells are small and readily damaged by the impaling microelectrode. However, if this were so, it might be expected that large membrane potentials would be seen transiently during impalement of the cells; this has generally not been the case.

It is quite unlikely that the values obtained in stationary phase and in differentiated cells are due to a damaged membrane. First, these cells are large, and stable recording can be obtained easily. Second, cells with relatively low RPs can have high specific resistance and no significant component of leakage current under voltage clamp. Finally, recordings under identical conditions in co-cultures of hybrid cells and mature skeletal muscle cells show that the hybrid RP is, as a rule, about 20 mV lower than that of the muscle cells. These observations all indicate that factors other than cell injury must account for the low recorded resting potentials of these cells.

If it is accepted that the low RPs are real, the observed fluctuations with changes in the state of the culture and the growth conditions become meaningful. For instance, 2- to 12-week-old large confluent cells have RPs ranging between −16 and −35 mV (Peacock *et al.*, 1972; Spector *et al.*, 1975). When these cells are replated, the culture passes through a lag period before cell division recommences. During this lag period, RPs average about −45 mV but thereafter decline when the culture reattains confluence. This cycle can be repeated with each new passage and appears to be linked to the initiation of cell division. Furthermore, if the replated cells are grown in the presence of agents that promote differentiation, a consistent increase in the RP to values above −45 mV is observed after the lag period.

A rapid increase in the resting potential of stationary phase cells can also be obtained by adding to the medium the potassium ionophore valinomycin (Spector *et al.*, 1975), the sodium ionophore monensin (Lichtshtein *et al.*, 1979a), or by increasing extracellular calcium concentrations (Moolennaar and Spector, 1979a). Interestingly, when the calcium ionophore A23187 is added to differentiated cells bathed in a high calcium medium, the membrane potential hyperpolarizes to values close to the theoretical potassium equilibrium potential (−89 mV) (Spector, unpublished results). All three ionophores appear to act, however, via different mechanisms. The valinomycin-induced hyperpolarization is associated with a marked decrease in the membrane conductance (Spector *et al.*, 1975). The A23187-induced hyperpolarization is associated with a large increase in the membrane conductance, suggesting the activation of a calcium-dependent potassium conductance. The monensin-induced hyperpolarization appears to involve an electroneutral increase in sodium uptake that acts to stimulate the electrogenic activity of the Na^+, K^+–ATPase (Lichtshtein *et al.*, 1979b). These results indicate that the resting potential is not a stable property in the neuroblastoma system but can be modulated by the state of the culture and the growth conditions. In this respect, the changes in the resting potential may be developmentally regulated, since they are analogous to neurite outgrowth and perhaps to changes in biochemical (Schubert *et al.*, 1973; Patrick *et al.*, 1978; Kimhi, this volume) and other electrical properties (see below) of clonal nerve cell systems. All modes of differentiation are highly susceptible to the culture conditions and are reversible but are not always linked to one another. Thus, the most dramatic increase in RP has been obtained in cells treated with 3–4% DMSO. Although these cells are small and round and exhibit poor excitability, they have RPs of −60 to −70 mV (Kimhi *et al.*, 1976).

The resting potential of the differentiated neuroblastoma membrane, like that of other neuronal membranes, is determined primarily by both the selective permeability of the membrane to K^+ ions and the higher concentration of K^+ ions inside the cell (about 150 mM) compared to the normal external K^+ concentration of 5.5 mM (Spector *et al.,* 1975). Increasing the external K^+ concentration depolarizes the membrane of these cells in a manner that approaches, at high external K^+, the behavior of a K^+-permselective membrane as predicted by the Nernst equation (Palfrey, 1976). On the other hand, omitting Na^+ or Cl^- from the medium has little effect on the resting potential, indicating a low permeability of the resting membrane to these ions despite their large concentration gradients across the membrane (Palfrey, 1976). Clearly, a change in any of these factors may account for the modulations of the resting potential by the growth conditions. In addition, the Na^+, K^+–ATPase is apparently functional in neuroblastoma cells and maintains the unequal distribution of Na^+ across the membrane (Lichtshtein *et al.,* 1979b). Although the Na^+, K^+-pump does not contribute a significant electrogenic component to the RP of confluent cultures, DMSO treatment may stimulate its activity or alter the Na^+ efflux/K^+ influx ratio, thereby inducing a change in the resting potential. An understanding of the mechanisms underlying the regulation of the resting potential must, therefore, await a detailed analysis of the relative contribution of each of these factors to the resting potential at various states of growth and differentiation.

4. DEVELOPMENT OF ELECTRICAL EXCITABILITY

The initial electrophysiological investigations of the mouse C-1300 neuroblastoma system had demonstrated that excitability of uncloned and cloned cells is similar to that of other neurons in that the cells can generate a wide spectrum of excitation phenomena, including action potentials, afterpotentials, and different oscillatory firing patterns (Nelson *et al.,* 1969; Harris and Dennis, 1970; Augusti-Tocco *et al.,* 1970). Subsequent examination by Nelson, Nirenberg, and their collaborators of a number of cloned and hybrid lines had revealed marked changes in spike configuration, amplitude, and maximum rates of change, depending on culture conditions and state of growth. This led these authors to postulate that electrical excitability in cells of a given line is subject to regulation and is differentially expressed in different lines (for review, see Nelson, 1973, 1975). An opposite point of view, however, has been advanced by Schubert *et al.* (1973) who have observed that exponentially dividing cells grown in suspension can generate overshooting action potentials. These authors interpreted this observation as indicating that electrophysiological differentiation does not occur in neuroblastoma cells. They questioned as well (Patrick *et al.,* 1978) the reported changes in excitability of the rat pheochromocytoma PC-12 line following NGF treatment (Dichter *et al.,* 1977). The confusion caused by these apparently conflicting observations is to a large extent a result of the use of arbitrary morphological and electrophysiological criteria to define the differentiated state and of the assumption that electrophysiological and morphological differentiation are linked. As discussed in Section 2, at least in the neuroblastoma system, the various modes of differentiation can be expressed independently of each other. Therefore, the mere presence of overshooting action potentials in round cells, whether grown in suspension or not, or of differences in action potential characteristics in morphologically differentiated cells would have only limited meaning. Furthermore, the premise that cells grown in suspension cannot differentiate with respect to membrane properties is probably invalid. For example, myoblasts grown in suspension can fuse to form multinucleated myotubes which, without assuming the characteristic shape of muscle fibers, express differentiated membrane properties similar to those of myotubes grown in monolayers (Fischbach and Lass, 1978; Spector, unpublished data). Two other sources of confusion have been the rather poor definition of the initial and fully developed physiological behavior of the cells, and the lack of appropriate methods to induce rapid and synchronous differentiation that has hampered attempts to establish a sequence of developmental changes in excitability.

Despite reports that undifferentiated neuroblastoma cells are electrically inexcitable (Nelson *et al.,* 1971; Chalazonitis and Greene, 1974), it appears that all cells of an excitable line can generate some form of active response (Schubert *et al.,* 1973). However, the response of exponentially dividing neuroblastoma cells (Peacock *et al.,* 1972; Miyake, 1978) or of PC-12 cells not exposed to NGF (O'Lague and Huttner, 1980) is small and similar to that observed in exponentially dividing L-6 myoblasts (Kidokoro, this volume). The inward current of the undifferentiated neuronal cells is probably carried largely by calcium ions (Miyake, 1978; O'Lague and Huttner, 1980) and not by sodium ions as in the L-6 myoblasts (Kidokoro, this volume).

The transition of an exponentially dividing culture of neuroblastoma cells to the confluent state is accompanied by an increase in the action potential amplitude and rates of rise and fall (Peacock *et al.,* 1972). To achieve a further expression of electrical excitability and a stable population of differentiated cells it is, however, necessary to change the culture conditions. Although confluent cells may exhibit large resting potentials and well-developed excitability properties, cell division of clonal nerve cell lines continues in this state, and the cultures become overgrown, heterogeneous, and, as a rule, poorly excitable (Fig. 2A; Spector *et al.,* 1975). In the PC-12 line, cellular differentiation is achieved after exposure of cultures to NGF and has a well-defined time course (Greene and Tischler, 1976). In this line, the sodium action potential is elaborated only after NGF treatment (O'Lague and Huttner, 1980). The situation is more complicated in the neuroblastoma system, where few of the methods used to induce morphological differentiation also induce electrophysiological differentiation with a well-defined time course.

Kimhi *et al.* (1976) have shown that DMSO-treated N1E-115 cells

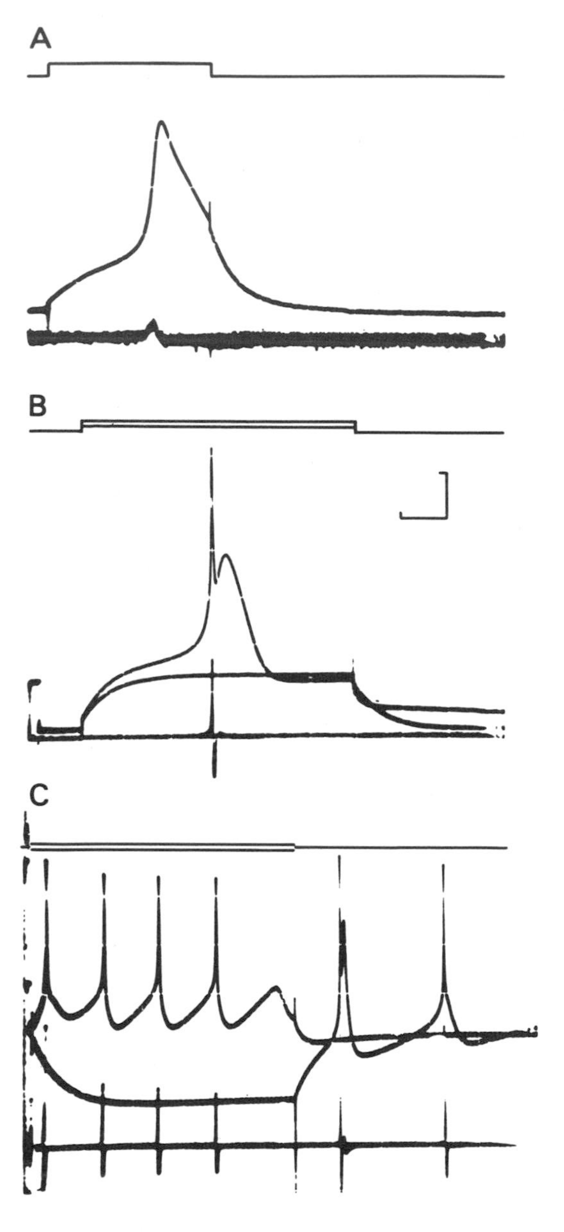

FIGURE 2. Development of electrical excitability during DMSO-induced differentiation of N1E-115 cells. All records are from large, morphologically differentiated cells. A, 2; B, 5; and C, 9 days after subculture into growth medium containing 2% DMSO. The membrane potential was hyperpolarized with steady inward current in A and B to obtain optimal responses. Electrical activity in C is from the resting potential (−55 mV). The regenerative response elicited in cell A is primarily a Ca^{2+} spike. Note the slow rate of rise, small amplitude, and undeveloped falling phase of the spike. This type of response is also seen in old confluent cultures. The action potential in cell B is fully developed, but the cell is not able to fire repetitively. The final stage in the development of electrical excitability is represented in cell C and is marked by the acquisition of the ability to fire repetitively. Note that a depolarizing current pulse applied at the resting potential elicits only fast spikes, whereas anodal break excitation elicits the two-component action potential. This indicates that the threshold for the slow component is more negative than that of the fast spike. These three developmental stages are characteristic of other neuroblastoma and hybrid lines and can be obtained by various differentiation procedures. Upper traces in all records are current (calibration: 5 nA); middle traces are membrane potential (calibration: 20 mV); and lower traces are rate of change of the membrane potential (calibration: A, 10 V/sec; B, 40 V/sec; C, 20 V/sec). Horizontal calibrations: A, 100 msec; B, 50 msec; C, 200 msec. (C is from Moolenaar and Spector, 1978.)

undergo a sequence of developmental changes in electrical excitability. This sequence is similar to that observed in developing skeletal muscle cells in culture (Spector and Prives, 1977). As illustrated in Fig. 2B and C, in both cases the development of electrical excitability includes marked changes in the characteristics of the single action potential followed by the acquisition of the ability to generate repetitive discharges. The increases in the amplitude and rate of change of the depolarizing and repolarizing phases of the action potential have been studied extensively in various developing systems (for review, see Spitzer, 1979). They reflect the elaboration by the membrane of the Na^+ and K^+ conductances that underlie the fast action potential in nerve and muscle cells. One important event in the expression of Na^+ channels in both neuroblastoma and skeletal muscle cells is the acquisition of tetrodotoxin (TTX) sensitivity early in development. This may reflect either structural changes in preexisting Na^+ channels or the elaboration of a different type of channel. A second important consideration is that the development of the Na^+ channels is not coupled to that of the K^+ channels. They proceed at different rates, and the Na^+ conductance system matures before the K^+ system. A third major phenomenon, described by Spector and Prives (1977) in skeletal muscle cells, is that at early states of postfusion development fast action potential generation is restricted to discrete areas of myotube membrane, and only at later states does the entire membrane become capable of fast action potential generation and propagation. These results probably apply to nerve cells and indicate that the maturation of the excitable membrane involves several sequential and possibly independent processes and reflects an increasing degree of organization of the cell membrane.

Although the elaboration of the Na^+ and K^+ conductance systems responsible for the generation of the fast action potential is a major event in the development of the excitable membrane, the mature nerve or muscle cells can exhibit a variety of complex excitation phenomena such as repetitive discharges in which voltage-dependent Ca^{2+} currents play a major role. The onset of this type of behavior occurs only after the maturation of the conductance systems responsible for the fast action potential. It may involve the elaboration of additional ionic conductance systems, a coupling between preexisting systems, or both. As will be considered in more detail in Section 5, differentiated neuroblastoma cells have at least four ionic conductance systems that are involved in electrical excitability: voltage-activated Na^+ and Ca^{2+} systems that carry inward currents, a voltage-activated K^+ system that carries outward current, and a Ca^{2+}-activated K^+ system that carries outward current. In contrast to the marked developmental changes in the voltage-activated Na^+ and K^+ conductances, the Ca^{2+} conductance system that exists in confluent N1E-115 cells is not significantly altered during DMSO-induced differentiation. On the other hand, the Ca^{2+}-activated K^+ system is the last ionic conductance to appear and is associated with the appearance of repetitive firing. Thus, a different type of Ca^{2+} current or a K^+ current or both may be elaborated. Along these lines, an interesting sequence of changes in excitability has recently been observed by Adler and Sabol (in preparation) in a clonal mouse pituitary cell line (AtT-20/D16-16) after exposure to dbcAMP. As shown in Fig. 3, in untreated cells a depolarizing current pulse elicits one or two fast Na^+ spikes followed by a

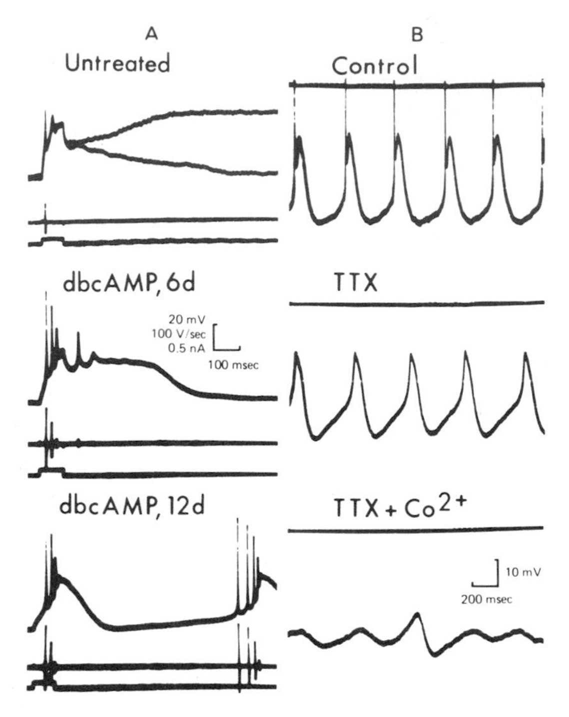

FIGURE 3. Electrical activity of clonal mouse pituitary cell line AtT-20/D16-16. (A) Changes in electrical excitability following treatment with dbcAMP (1 mM). In all cases, recordings are from the resting potential. The untreated cell (upper record) exhibits a fast Na^+ action potential followed by a prolonged Ca^{2+} action potential lasting 23 sec. Note that a slight increase in the stimulus intensity elicits the prolonged action potential (resting potential −63 mV). Six days after exposure to dbcAMP (middle record), multiple fast spikes appear, and the duration of the Ca^{2+} spike is decreased (resting potential −68 mV). Twelve days after exposure to dbcAMP (lower record), there is a further reduction of the Ca^{2+} spike, and rhythmic activity appears (resting potential −65 mV). The upper trace in each record represents the membrane potential; the middle trace is the maximum rate of change of the membrane potential; the lower trace is current. (B) Mechanisms of regenerative activity of dbcAMP-treated AtT-20 cells. Upper record, spontaneous rhythmic discharge of fast and slow action potentials. Middle record, addition of TTX (1 μM) to the bath abolishes the fast spikes, slows the rate of discharge, and increases the threshold of the remaining spike. Lower record shows the initial effects of Co^{2+} addition (5 mM). The residual oscillations disappeared completely 10 min after Co^{2+} was added to the bath. (From Adler and Sabol, in preparation.)

prolonged action potential lasting for tens of seconds and mediated by Ca^{2+}. Exposure of the cultures to 1 mM dbcAMP (9–12 days) produces striking alterations in both Na^+ and Ca^{2+} components of the response. It is seen that the Na^+ spikes increase in overshoot and maximum rate of rise and become repetitive, whereas the Ca^{2+} component is markedly shortened in duration, and a consistent hyperpolarizing afterpotential appears. After 12 days of dbcAMP treatment, the majority of the cells exhibit rhythmic discharges either spontaneously or in response to stimulation. Addition of TTX abolishes the fast spikes but does not eliminate the rhythmic oscillations; on the other hand, Co^{2+}, a Ca^{2+} antagonist, blocks the activity. These observations support the contention that a Ca^{2+}-dependent K^+ conductance is involved in the regulation of slow oscillatory discharge (see Section 5 below). It appears likely that dbcAMP causes the elaboration of such a conductance system in AtT-20 cells, and that the consequent oscillatory electrical activity plays a role in modulating secretion of corticotropin and opioid peptides from these cells (Mains *et al.*, 1977; Giagnino *et al.*, 1977). Any study of the development of the excitable membrane must take into account all of the ionic mechanisms of excitation, their topological organization in the membrane, and the functional interactions among them.

5. ELECTRICAL EXCITABILITY OF DIFFERENTIATED NEUROBLASTOMA CELLS

5.1 Electrical Behavior under Constant-Current Conditions

Typical responses obtained in differentiated cells of the most extensively studied neuroblastoma (N1E-115, N18) and hybrid lines (NG108-15, NBr10A) are very similar. Although different procedures are used to induce maximal differentiation, in all cases an optimum action potential can be obtained when a depolarizing current pulse is applied at an adjusted steady membrane potential of −85 to −100 mV. As illustrated in Fig. 2B, this response consists of a fast action potential followed by a distinct smaller second peak (Spector *et al.*, 1973; Chalazonitis and Greene, 1974; Moolenaar and Spector, 1978; Miyake, 1978; Adler and Nirenberg, in preparation). Maximum rates of rise of the fast spike average about 100 V/sec at room temperature, and peak amplitudes average about +20 mV. The second active component reaches its peak value 10–30 msec after the fast peak and has maximum rates of rise between 2 and 10 V/sec. At membrane potentials more positive than −70 mV, a characteristic afterhyperpolarization (ahp) lasting 50–300 msec is observed in differentiated cells (Moolenaar and Spector, 1978).

In addition to the two-component action potential and the ahp fully differentiated cells usually fire repetitively either spontaneously immediately after penetration or in response to stimulation. Maintained trains of action potentials are, however, rarely observed in normal solution. Optimum repetitive discharge is obtained from a range of membrane potentials between −55 and −70 mV, a range both near the resting potential and near where the ahp can be elicited. At this potential, the discharge usually consists of spikes with only the fast component. However, with steady hyperpolarization or when anodal break stimulation is used, the two-component action potential can be elicited (Fig. 2C).

5.1.1. Ionic Basis of the Action Potential

Spector *et al.* (1973) have shown that the fast action potential is abolished by TTX (10^{-7} g/ml), a specific blocker of voltage-dependent Na^+ channels. Removal of Na^+ from the bathing solution eliminates the fast spike in the same way as does TTX (Moolenaar and Spector, 1977, 1978).

The second, slower component is TTX-resistant, depends on external [Ca^{2+}], can be evoked in Na^+-free solution containing Ca^{2+}, Ba^{2+}, or Sr^{2+}, and is abolished by the Ca^{2+} antagonists La^{3+}, Co^{2+}, or Mn^{2+} (Spector *et al.*, 1973; Moolenaar and Spector, 1979a). These results suggest that the fast component of the action potential is carried by Na^+ and the slow by Ca^{2+}. It is noteworthy that the organic Ca^{2+} antagonist D-600 at concentrations of up to 20 μg/ml does not significantly affect the slow component and that at higher concentrations this drug affects all components of the action potential.

5.1.2. Effects of Potassium Channel Blockers

Moolenaar and Spector (1978) have shown that tetraethylammomium (TEA, 5–10 mM), a blocker of voltage-dependent K^+ channels, causes a marked prolongation of the neuroblastoma action potential without significantly altering the maximum rate of rise. As illustrated in Fig. 4, similar

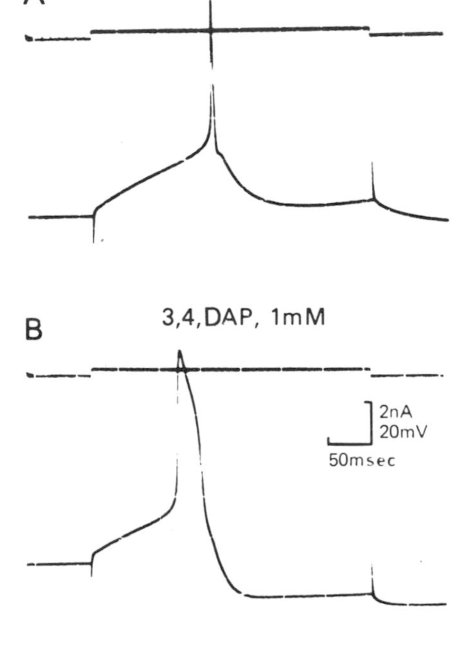

FIGURE 4. Effects of 3,4-diaminopyridine on the action potential of a neuroblastoma × glioma NG108-15 hybrid cell. Recordings are from the resting potential (-60 mV) in a solution containing 5 mM Ca^{2+}. (A) control; (B) after addition of 3,4-diaminopyridine (1 mM). Note in B the increase in amplitude and duration of the action potential and the appearance of a prominent afterhyperpolarization following the action potential. Upper trace in each record is current; lower trace is membrane potential. The NG108-15 cells were co-cultured with normal mouse muscle cells for 9 days in the presence of dbcAMP (1 mM). (Spector, Christian, and Nelson, unpublished data.)

effects can be obtained with lower concentrations (1 mM) of aminopyridines and diaminopyridines, other blockers of K^+ channels (Pelhate and Pichon, 1974; Schauf *et al.,* 1976; Yeh *et al.,* 1976). Both TEA and 3,4-diaminopyridine (3,4-DAP, 1 mM) have, however, additional effects on the neuroblastoma electrical activity: the overshoot of the spike is increased, the amplitude and time course of the ahp are dramatically enhanced, and slow repetitive discharges can be induced in silent cells (Moolenaar and Spector, 1979a).

All of these effects can be obtained in normal solution or in a Na^+-free, high-[Ca^{2+}] solution (Moolenaar and Spector, 1979a; Fig. 5). They can be explained by suppression of a K^+ conductance system that underlies the repolarization phase of the neuroblastoma action potential and counteracts an inward Ca^{2+} current (see Section 5.2 below).

5.1.3. Changing External Calcium Concentration

As illustrated in Fig. 5, when external [Ca^{2+}] is raised to levels between 5 and 20 mM in Na^+-free solution, neuroblastoma and hybrid lines can produce slow all-or-none spikes (Moolenaar and Spector, 1979a). The most prominent features of the electrical activity in elevated [Ca^{2+}] solution with or without Na^+ are, however, a pronounced ahp following the spike and a dramatic

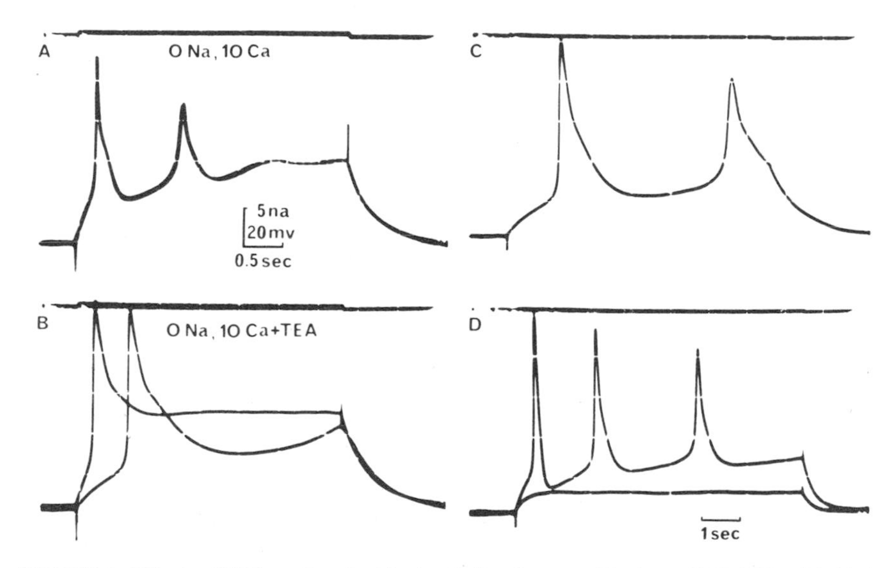

FIGURE 5. Effects of TEA on the electrical activity of a neuroblastoma N1E-115 cell in Na^+-free, high-Ca^{2+} medium. Recordings are from the resting potential (−75 mV). Note in B that addition of TEA prolongs the Ca^{2+} action potential and abolishes the repetitive discharge elicited by a stimulus identical to that in A. Repetitive discharge at a slower rate could be restored in the presence of TEA by decreasing the intensity of the current pulse (C and D). Upper traces in each record are current; lower traces are membrane potential. Calibration bars in A apply to B and C.

increase in the number of cells that fire repetitively (Reiser *et al.*, 1977; Moolenaar and Spector, 1979a). The firing rates in elevated [Ca^{2+}] are slower and can be maintained for a longer time than in normal solution. These features are similar to those obtained after blockade of the voltage-sensitive K^+ conductance.

Evidence that the long-lasting ahp is mediated by an increase in K^+ conductance that is Ca^{2+}-dependent and pharmacologically distinct from the delayed K^+ channel has been presented by Moolenaar and Spector (1979a). It has a reversal potential close to the K^+ equilibrium potential, is enhanced by elevated [Ca^{2+}], is insensitive to TEA and 3,4-DAP, and is inhibited by the same agents that abolish the Ca^{2+} spike or by replacing Ca^{2+} by Ba^{2+} or Sr^{2+}. Since the same manipulations that inhibit the ahp also suppress repetitive activity, these authors suggest that the Ca^{2+}-dependent K^+ conductance underlying the ahp plays an important role in the regulation of repetitive firing in neuroblastoma cells.

Neuroblastoma cells are not unique in manifesting the Ca^{2+}-sensitive K^+ conductance. It appears to underlie the prolonged ahp in many neural cells (for reviews, see Lew and Ferreira, 1978; Meech, 1978). In fact, it seems a ubiquitous conductance mechanism, and calcium regulation of K^+ conductance is demonstrable in red blood cells (Lew and Ferreira, 1978), pancreatic cells (Atwater *et al.*, 1979), liver cells (Burgess *et al.*, 1979), Purkinje fibers (Siegelbaum and Tsien, 1980), and mesenchymal cells (Nelson and Henkart, 1979).

5.1.4. Effect of Quinine

The intimate association between the Ca^{2+} influx and the Ca^{2+}-dependent K^+ conductance underlying the ahp and the inability to separate them pharmacologically had made impossible a direct study of the latter conductance and its physiological consequences. The development of agents to block the Ca^{2+}-dependent K^+ conductance without affecting the Ca^{2+} influx has been noted by several authors (Lux and Heyer, 1979; Meech, 1978) to be the key to any further study. Such agents should ideally not affect calcium action potentials and should act reversibly. Fishman and Spector have recently reported that quinine fulfills such criteria (Fishman and Spector, 1980 and in preparation).

In normal solution, quinine or quinidine (10–20 μM) prolong the action potential as TEA does, indicating an effect on the voltage-sensitive K^+ conductance. Recently, Wong (in preparation) has found that externally applied quinidine (100 μM) causes an inactivation of K^+ currents in *Myxicola* giant axon. This inactivation is similar to that obtained by internal action of TEA derivatives on squid axon (Armstrong, 1971) and frog node of Ranvier (Armstrong and Hille, 1972) and can explain the prolongation of the neuroblastoma action potential. However, as shown in Fig. 6, 10 μM quinine does not affect

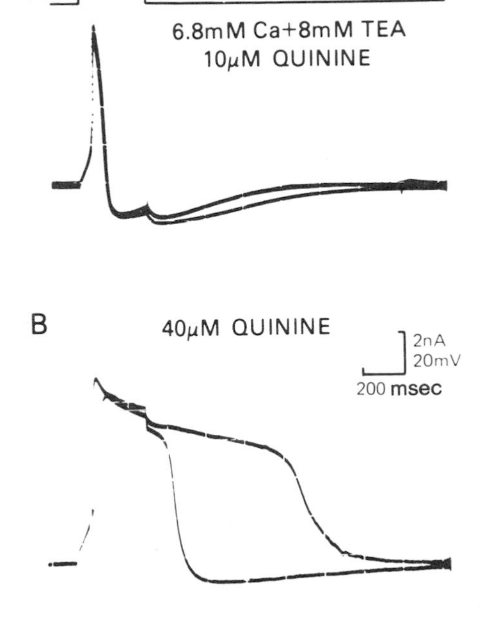

FIGURE 6. Effects of quinine on the electrical activity of a neuroblastoma × glioma NG108-15 hybrid cell. Recordings are from the resting potential (−61 mV) in a normal Na^+ solution containing 6.8 mM Ca^{2+} and 8 mM TEA. (A) Two superimposed traces to show the effects of 10 μM quinine on the prolonged afterhyperpolarization. (B) Two superimposed traces to show the sequential effects of 40 μM quinine on the same cell. Note that a prolongation of the action potential and a reduction in the magnitude of the afterhyperpolarization are followed within 30 sec by a prolonged afterdepolarization. The decrease in the fast spike reflects the effects of quinine on the Na^+ spike. The cell was treated for 5 days with dbcAMP (1 mM). (From Fishman and Spector, in preparation.)

the action potential of NG108-15 cells in a solution containing 6.8 mM Ca^{2+} and 8 mM TEA but causes a diminution of the prolonged ahp. At 40 μM, quinine completely blocks the ahp and induces a prolonged action potential outlasting the duration of the stimulus. These effects are completely reversible, can be demonstrated in Na^+-free medium where no diminution of the calcium action potential occurs, and are similar to those obtained following replacement of Ca^{2+} by Ba^{2+} (Spector, unpublished data).

Dramatic effects of quinine are seen on repetitive activity. In low concentration (12–20 μM) quinine increases the rate of firing and decreases the duration of a repetitive train of impulses. Complete abolition of the ahp with higher concentrations always blocks repetitive firing. This is in contrast to the inhibition of the voltage-sensitive K^+ conductance with TEA which enhances both the ahp and repetitive activity. Furthermore, quinine has little effect on undifferentiated cells. These cells lack the ahp and do not fire repetitively. Thus, the Ca^{2+}-dependent K^+ conductance is a developmental marker acquired to its full extent only in the fully differentiated neuroblastoma cells. Its physiological consequences can be assessed directly using quinine when the voltage-sensitive K^+ currents are blocked.

5.2. Voltage Clamp Experiments

In most investigations concerning the electrical activity of mammalian neurons and the development of excitability, voltage responses to constant cur-

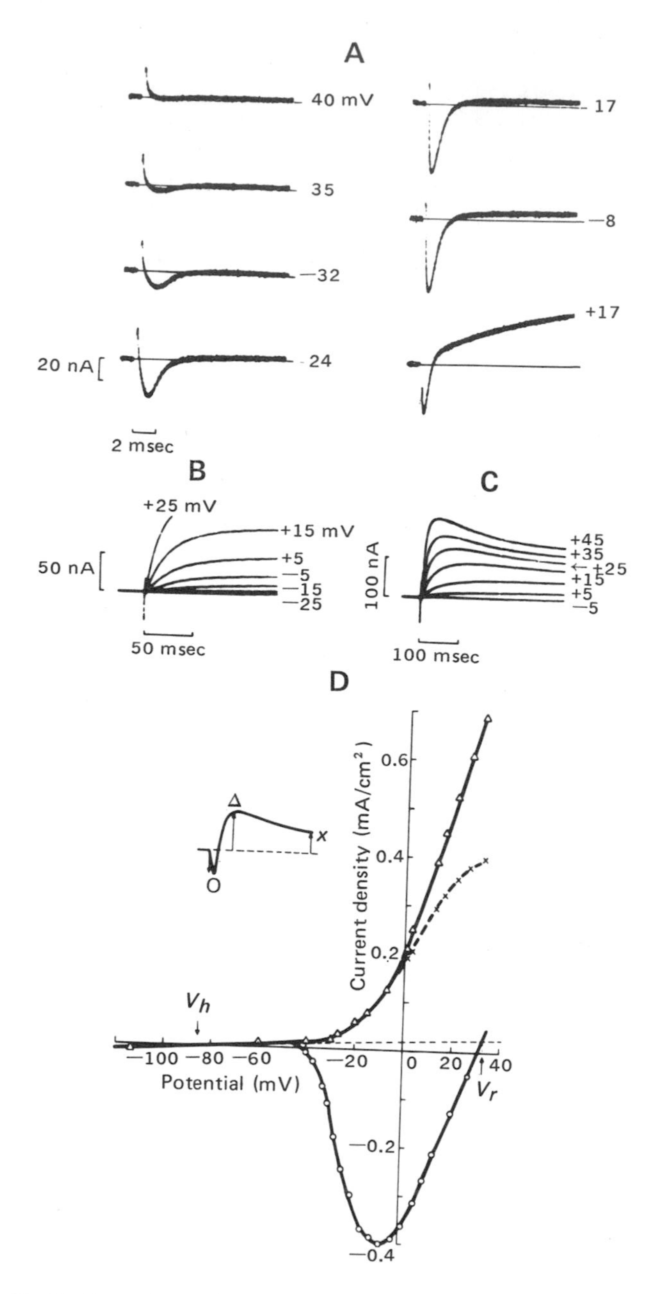

FIGURE 7. (A–C) Current records obtained from neuroblastoma cell under voltage clamp and displayed on fast (A) and slower (B and C) time scales. Command voltages (mV) are indicated next to the current traces. Holding potential −85 mV. Temp. 20°C. (D) Current densities as a function of membrane potential for the same cell. Currents were measured as indicated in inset and relative to a small holding current. O, Peak inward current; Δ, maxi-

rent stimulation have been studied. Interpretations of ionic conductances based on such measurements are, however, indirect and rather limited. In order to understand better the ionic conductance mechanisms that underlie excitability, it is essential to examine directly the ionic currents under voltage clamp conditions. However, in nearly all mammalian neuronal cells *in vivo* and *in vitro,* the prerequisites for successful voltage clamping are difficult to meet (cf. Moolenaar and Spector, 1978; Adams *et al.,* 1980; Smith *et al.,* this volume). It is not surprising, therefore, that voltage clamp studies of the neuronal somata have been restricted primarily to the giant neurons of the gastropod molluscs (Lux and Heyer, 1979; Adams *et al.,* 1980).

The suitability of neuroblastoma cells to voltage clamp was first demonstrated by Moolenaar and Spector (1977). These authors showed that stable impalement with two separate, low-resistance microelectrodes is feasible in large neuroblastoma cells that are round or have few, short processes. Such cells have a spatially uniform membrane potential so that an adequate space clamp and a satisfactory resolution of fast membrane currents can be achieved. In a subsequent series of papers, Moolenaar and Spector (1978, 1979a,b) analyzed the ionic currents in differentiated N1E-115 cells and identified four types of ionic conductances. Two—a fast Na^+ and a slow Ca^{2+}—carry inward currents, and two—a delayed K^+ and a Ca^{2+}-dependent K^+—carry outward currents. These conductances probably represent the major ionic conductances in mammalian neurons but are by no means exhaustive, and others have been, and will continue to be discovered. Veselovskii *et al.* (1977), using the intracellular dialysis method to voltage clamp neuroblastoma cells, observed only weak Na^+ and K^+ currents, probably because of membrane damage.

5.2.1. The Sodium and Potassium Currents

A series of typical membrane currents obtained by applying depolarizing voltage steps from a holding potential of −85 mV to various levels is shown in Fig. 7. The most prominent current patterns in these records are:

1. A transient phase of inward current appearing at −40 mV and showing both voltage-dependent activation and inactivation with fast time courses. This current was identified as Na^+ current; it is blocked by TTX, is absent in Na^+-free solution, and has a reversal potential of about +30 mV, close to the Nernst potential for Na^+ in neuroblastoma cells (Palfrey, 1976).
2. A delayed outward current appearing at −25 mV following the inac-

---

mum value of outward current; *x*, value of outward current measured after 400 msec. Dashed line represents estimate of leakage current. V_h, holding potential (−85 mV); V_r, equilibrium potential for inward current charge carrier. Estimated membrane surface area 2.0×10^{-4} cm^2. (From Moolenaar and Spector, 1978.)

tivation phase of the Na^+ current. It shows a voltage-dependent S-shaped rise to a stationary maximum value and then declines to a lower steady-state level with large and prolonged depolarizations. This current was identified as K^+ current; it is suppressed by TEA and has a mean reversal potential of −71 mV, close to the Nernst potential for K^+ (Palfrey, 1976). These currents underlie the rising and falling phases of the fast action potential, respectively.

Membrane conductance changes derived from voltage clamp experiments can most usefully be described by a system of differential equations such as the one used by Hodgkin and Huxley (1952) for the squid axon. The relatively good time resolution of the neuroblastoma membrane currents and their similarity to those of the squid axon allowed determination of various conductance parameters in the framework of the Hodgkin–Huxley formalism.

The kinetic behavior of the Na^+ conductance (G_{Na}) could be fitted by

$$G_{Na} = \overline{G}_{Na} m^3 h \tag{1}$$

where $\overline{G}_{Na}$ is the maximum Na^+ conductance, and m and h are the parameters for Na^+ activation and inactivation, respectively.

The activation time course of the K^+ conductance (G_K) could be adequately described by

$$G_K = \overline{G}_K n^2 \tag{2}$$

$\overline{G}_K$ is the maximum K^+ conductance, and n is the parameter for K^+ activation. $\overline{G}_{Na}$ and $\overline{G}_K$ in neuroblastoma cells were similar to the values obtained in squid axon.

5.2.2. Effects of Neurotoxins That Modify the Na^+ Permeability

The neuroblastoma system has been used extensively in biochemical assays to probe the molecular properties of voltage-dependent Na^+ channels with neurotoxins that are thought to modify selectively these channels (see Catterall, this volume). Surprisingly, no electrophysiological data concerning the action of these agents on neuroblastoma cells is available.

Some commonly used drugs are batrachotoxin, veratridine, scorpion toxins, and sea anemone toxins. These neurotoxins prolong action potentials in various axonal membranes apparently by slowing down the process of Na^+ channel inactivation, thereby prolonging the open state of these channels (Hille, 1976; Ritchie, 1979).

The effects of two of these neurotoxins, veratridine and the toxin of *Leiurus* scorpion, on neuroblastoma N1E-115 cells are shown in Fig. 8. With crude scorpion venom as well as with the polypeptide toxin purified from the venom, the cells become hyperexcitable and exhibit spontaneous and evoked repetitive discharges. The hyperexcitability occurs within a few minutes of toxin application and is accompanied by an increase in the rate of rise, amplitude, and duration of the fast action potential. With long current pulses, the evoked discharge is occasionally interrupted by a remarkable stepwise depolarization of

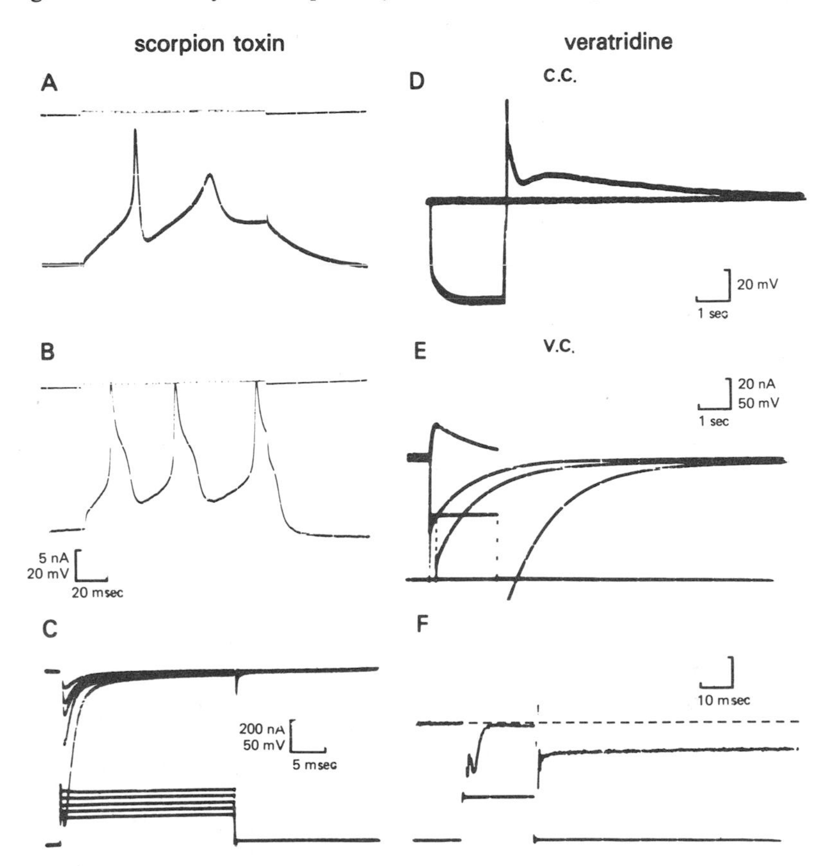

FIGURE 8. Effects of scorpion toxin and veratridine on the electrical activity of neuroblastoma N1E-115 cells. (A, B, and D) Constant current conditions. (C, E, and F) Voltage clamp conditions. (A and B) Recordings from the resting potential (−70 mV) of the same cell before (A) and after (B) application of purified scorpion toxin (1 μg/ml). Note the prolongation of the spike and the rapid repetitive discharge induced by the toxin. Upper traces in A and B are current, lower traces are membrane potential. (C) Current records under voltage clamp in the presence of crude venom (7 μg/ml). Holding potential −130 mV. Upper traces are membrane currents; lower traces are the command voltages. (D, E, and F) Recordings from one cell in the presence of 100 μM veratridine. (D) Voltage response to anodal break stimulation. Current records under voltage clamp are displayed on slow (E) and Fast (F) time scale. Holding potential −90 mV. Upper traces are membrane currents; lower traces are the command voltages. The three slowly decaying inward current traces in E represent tail current to three command voltages of variable durations. (C, Neher and Spector, unpublished data; D–F, Moolenaar and Spector, unpublished data.)

variable duration that terminates with an abrupt repolarization and a continuation of the repetitive discharge. It is as though some channels are immobilized in the open position. In addition, hyperpolarizing the membrane to −90 mV causes marked oscillations in the membrane potential and leads to a spontaneous discharge of giant action potentials reaching 120 mV in amplitude (as compared with the normal amplitude of 60–70 mV).

Figure 8C shows voltage clamp current records elicited in an N1E-115 cell treated with *Leiurus* venom (7 μg/ml) by five voltage steps from a holding potential of −130 mV to levels ranging from −90 to −50 mV. When these records are compared to those obtained in untreated cells (Fig. 7), it is seen that scorpion toxin causes the following modifications of the Na^+ currents: (1) The voltage-dependent activation is shifted by 50 mV from −40 to −90 mV. This effect is similar to that obtained by treating the frog node of Ranvier with *Centruroides* scorpion venom (Cahalan, 1975) or with batrachotoxin (Khodorov and Revenko, 1979). (2) The fast voltage-dependent inactivation phase is only slightly slowed and is followed by a steady-state inward current. (3) The magnitude of the fast current is increased, reaching peak values larger than the corresponding values in untreated cells when the currents are elicited from the same holding potential.

In contrast to scorpion toxin, veratridine (100 μg/ml) has essentially no effect on the fast action potential or on the magnitude and time course of the activation and inactivation of the fast Na^+ current under voltage clamp. The most striking effect of veratridine is to induce, within a few minutes of its application, a Na^+-dependent, TTX-sensitive prolonged depolarization outlasting the current pulse by several seconds (Fig. 8D). Under voltage clamp (Fig. 8E,F), a large Na^+-dependent, TTX-sensitive tail current appears after termination of the depolarizing pulse which declines exponentially with a long time constant. The magnitude and time constant of the veratridine-induced tail current increases dramatically as the depolarizing pulse is lengthened. With long depolarizing pulses, the peak value of the tail current is much larger than the peak of the fast Na^+ current. Taken together, these results clearly indicate that both veratridine and scorpion toxin modify the Na^+ permeability of the neuroblastoma membrane, but by different mechanisms. Thus, scorpion toxin affects both the activation and inactivation phases of the fast Na^+ currents, whereas in the presence of veratridine, a population of Na^+ channels remain open after termination of a depolarizing pulse, although the opening and closing of the Na^+ channels associated with the action potential appear normal.

Ion flux studies have shown a competitive interaction between veratridine and batrachotoxin, indicating that they act at a common receptor site (see Catterall, this volume). While no electrophysiological data is available on the action of batrachotoxin on neuroblastoma cells, the properties of this receptor site have been extrapolated from the effects of batrachotoxin on frog node of Ranvier (Khodorov and Revenko, 1979). The effects of veratridine on the electrophysiological properties of neuroblastoma cells are, however, quite different from those of batrachotoxin on the node of Ranvier. In particular, the batrachotoxin-induced 50 mV shift of Na^+ activation to more negative voltages found in the node of Ranvier has not been seen in veratridine-treated neuroblastoma cells. Similarly, the veratridine-induced large tail currents in neuroblastoma are absent in batrachotoxin-treated axons. Electrophysiological studies of the effects of batrachotoxin on neuroblastoma cells should be useful in understanding the physiological meaning of the competitive interaction between batrachotoxin and veratridine.

Ion-flux and ligand-binding studies have indicated that scorpion toxin and sea anemone toxin II act on a common receptor site of the voltage dependent Na^+ channel but one that is different from the veratridine site (see Catterall, this volume). Both toxins have been assumed, however, to have similar effects: inhibition of the inactivation process. The effects of scorpion toxin on neuroblastoma Na^+ currents are complex; both activation and inactivation processes appear to be modified, and a fraction of Na^+ channels is open at the resting potential so that Na^+ current can flow into the cells. Furthermore, unlike scorpion toxin, the sea anemone toxin appears only to slow Na^+ current inactivation (Neher and Spector, unpublished data) without inducing a steady-state inward current or a voltage shift of the activation process.

The enhancement of the veratridine-induced Na^+ permeability by scorpion toxin or sea anemone toxin can be interpreted as indicating that both toxins act in an allosteric way on the same class of Na^+ channels (see Catterall, this volume). Alternatively, a model in which there are two different classes of Na^+ channels has been proposed by Jacques *et al.* (1978). Clearly, more electrophysiological studies are needed to substantiate molecular interpretations of voltage-dependent Na^+ channels based on ion-flux and ligand-binding studies.

5.2.3. The Ca^{2+} Current

It is now recognized that voltage-sensitive Ca^{2+} currents play an important role in regulating neuronal behavior. Indeed, the existence of Ca^{2+} currents in a variety of excitable tissues has been inferred from voltage responses to current stimulation. However, identification and quantitative study of these currents under voltage clamp have been seriously hampered primarily because of their small size and the slow time course that overlaps opposing outward currents. Thus, even in giant molluscan neurons where these currents have been extensively studied, their complex behavior is not completely understood (Meech, 1978; Lux and Heyer, 1979; Adams *et al.*, 1980).

In neuroblastoma cells bathed in normal medium (containing 1.8 mM Ca^{2+}), the Ca^{2+} current can hardly be detected under voltage clamp, although the action potential exhibits a distinct Ca^{2+} component. Ca^{2+} currents can, however, be seen and separated from the total ionic currents when the external

[Ca^{2+}] is elevated, Na^+ is replaced by Tris, and TEA is added to block the delayed outward K^+ currents (Moolenaar and Spector, 1979b). Records of Ca^{2+} currents obtained under these conditions are shown in Fig. 9.

The Ca^{2+} current in neuroblastoma cells exhibits voltage-dependent activation and inactivation like that of the Na^+ current. However, its magnitude, voltage sensitivity, kinetic and pharmacological properties differ considerably from those of the Na^+ current. The Ca^{2+} current is activated at -55 mV, a far more negative potential than that at which the Na^+ current is activated. This implies that a fraction of Ca^{2+} channels is open at the resting potential so that Ca^{2+} current can flow into the cytoplasm. Its time to peak is slow (about 10 msec), the maximum peak Ca^{2+} current density (reached at -20 mV) is about one order of magnitude smaller than the Na^+ current density, and the inactivation time course follows first-order kinetics with a voltage-dependent time constant ranging from 25 to 100 msec. The pharmacological properties and ion selectivity of the Ca^{2+} channel in neuroblastoma are very similar to those described for molluscan neurons (Lux and Heyer, 1979; Adams *et al.,* 1980): Ba^{2+} and Sr^{2+} appear to carry current through the Ca^{2+} channel, whereas Ni^{2+}, La^{3+}, Co^{2+}, and Mn^{2+} all block Ca^{2+} current. The gating characteristics of the neuroblastoma and molluscan Ca^{2+} currents are, however, quite different. Activation of molluscan Ca^{2+} currents and attainment of maximum Ca^{2+} conductance occur at far more positive potentials than in neuroblastoma. In addition, considerable uncertainty still remains concerning Ca^{2+} current inactivation in molluscan neurons (Adams *et al.,* 1980). Interestingly, the gating characteristics of the neuroblastoma Ca^{2+} are quite similar to those of the mouse oocyte Ca^{2+} currents (see Moolenaar and Spector, 1979b).

5.2.4. The Ca^{2+}-Activated K^+ Current

In addition to the TEA-sensitive K^+ current that underlies the repolarizing phase of the action potential, neuroblastoma cells exhibit TEA-insensitive, slowly rising outward currents that are responsible for the long-lasting hyperpolarization following the action potential. These currents have been identified by Moolenaar and Spector (1979b) as Ca^{2+}-activated K^+ currents ($I_{K(Ca)}$) and can clearly be seen in high-[Ca^{2+}]$_o$ solutions during prolonged depolarizing pulses to values more positive than -20 mV. $I_{K(Ca)}$ has a rise time longer than a second and a bell-shaped voltage dependency that reaches a peak at about $+30$ mV and declines with large depolarizations. The reversal potential of the $I_{K(Ca)}$ is close to the K^+ equilibrium potential, and it is suppressed by procedures that eliminate or steadily inactivate the Ca^{2+} current. The voltage dependency, kinetic, and pharmocological properties of the $I_{K(Ca)}$ are similar to those of the corresponding currents in molluscan neurons (Meech, 1978; Lux and Heyer, 1979; Adams *et al.,* 1980). In the neuroblastoma, however, they cannot be activated by depolarization if Ba^{2+} or Sr^{2+} replace external Ca^{2+}. Furthermore, Moolenaar and Spector (1979b) have shown that Ca^{2+} influx is incap-

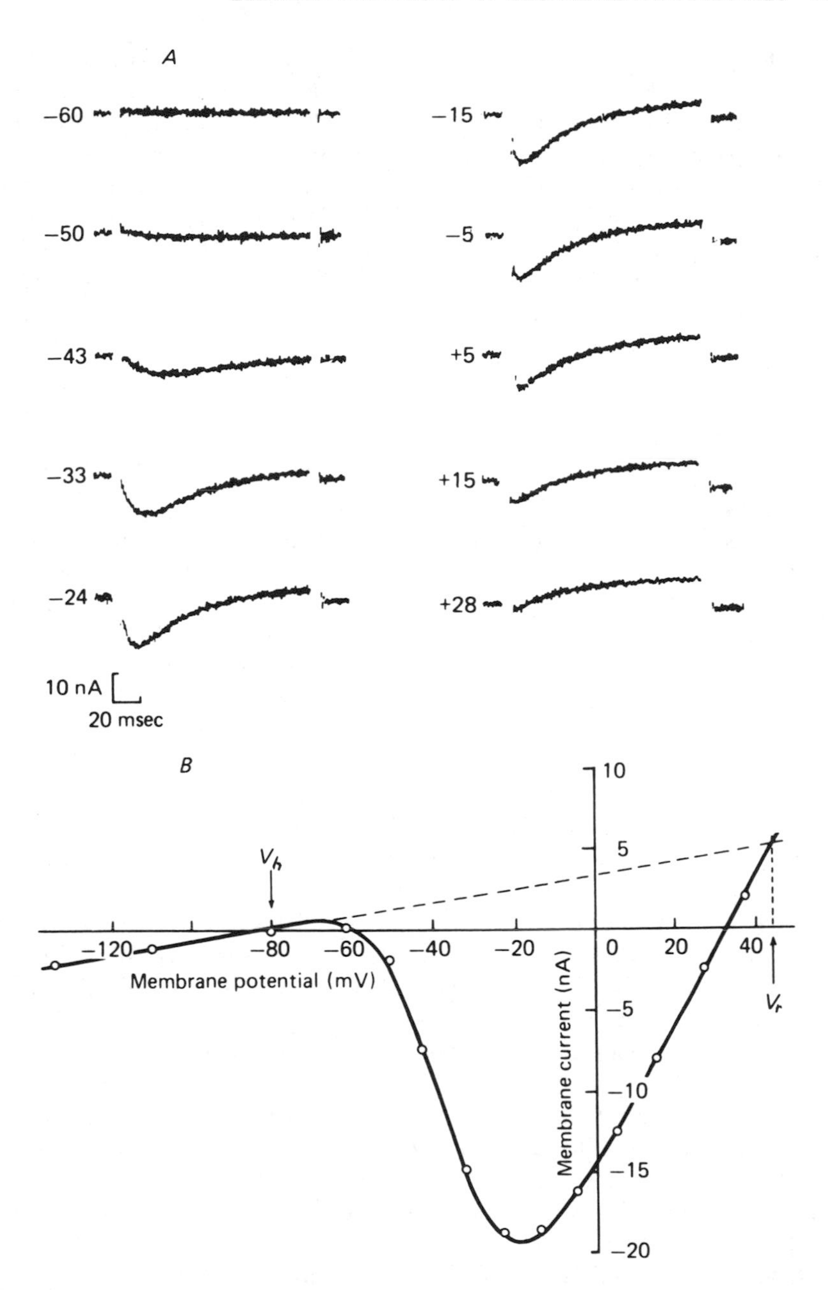

FIGURE 9. (A) Ca^{2+} inward currents (I_{Ca}) in Ca^{2+}-free solution containing 20 mM Ca^{2+} and 25 mM TEA. Command voltages are indicated next to the current traces. Holding potential −80 mV. (B) Peak values of I_{Ca} as a function of membrane potential for the same cell as in A. Dashed line represents linear extrapolation of leakage current as measured by hyperpolarizing voltage steps. V_h, holding potential; V_r reversal potential of I_{Ca}. Estimated membrane surface area 1.7×10^{-4} cm^2. (From Moolenaar and Spector, 1979b.)

able of activating $I_{K(Ca)}$ unless a certain critical voltage level (-20 mV) is reached. Since the Ca^{2+} influx at this potential is already at its maximum, these observations indicate that the relationship between the activation of $I_{K(Ca)}$ and both the Ca^{2+} current and the increase in cytoplasmic free Ca^{2+} is not a direct one and that $I_{K(Ca)}$ activation also depends on membrane potential. It is possible that $I_{K(Ca)}$ is activated by Ca^{2+} influx through a separate population of Ca^{2+} channels that exhibit a voltage dependency similar to that of $I_{K(Ca)}$ rather than to that of I_{Ca}. The properties of $I_{K(Ca)}$ in neuroblastoma are not compatible with a simple model in which an increase in cytoplasmic free Ca^{2+} is sufficient to activate $I_{K(Ca)}$ irrespective of the membrane potential (see Meech, 1978; Lux and Heyer, 1979).

6. CONCLUSIONS

Research of neuronal cell lines has been characterized, thus far, by the abundance of biochemical studies and the paucity of electrophysiological investigations. The predilection of biochemists for this biological system is understandable; it provides a convenient source of large quantities of homogeneous material necessary for biochemical analysis. The successful application of the voltage clamp technique to neuroblastoma cells may perhaps lead to a more frequent use of neuronal lines in electrophysiological studies. The potential of having an *in vitro* differentiating neuronal system in which to quantitatively analyze macrosopic ionic currents, current fluctuations, and currents generated by single ionic channels is immense in addressing questions that relate to electrical excitability, chemosensitivity, and synaptic events of mammalian neurons and to the development of these mechanisms during differentiation.

Some insight into electrical excitability mechanisms has been gained from the voltage clamp studies of neuroblastoma cells. These studies have disclosed voltage-dependent Na^+, Ca^{2+}, and K^+ conductances and a Ca^{2+}-dependent K^+ conductance. Three of these, the Na^+, K^+, and Ca^{2+}-dependent K^+ conductances are developmentally regulated and reflect the sequential acquisition by the neuroblastoma membrane of the ability to generate fast action potentials and to modulate this generation through patterned oscillations of the membrane potential. The developmental regulation of the inactivating Ca^{2+} currents is an intriguing question. These currents appear to underlie a primordial voltage-dependent conductance system that is the dominant one at early developmental stages. Although its dominance is superseded by the other conductances in mature cells, it can still be detected if appropriate experimental conditions are used. Noninactivating Ca^{2+} currents may, however, underlie a second Ca^{2+} system that is subject to developmental regulation. The presence of such a system may be important to the complex oscillatory behavior (Morris and Lecar, submitted for publication) found only in mature cells and may explain the discrepancy in the voltage dependence between the inactivating Ca^{2+} and Ca^{2+}-dependent K^+ currents. Furthermore, the elaboration of a non-inactivating Ca^{2+} conductance may also explain the enhancement of K^+-stimulated Ca^{2+}-uptake during differentiation of several hybrid lines, an enhancement that is critical to the ability of these lines to release acetylcholine and to form functional synapses (Nirenberg *et al.*, 1979; Rotter *et al.*, 1979).

In contrast to electrical excitability mechanisms, electrophysiological research into chemosensitivity is still at a very early stage, although receptor-mediated changes in cyclic nucleotides have been extensively studied (for review, see Kimhi, this volume). Depolarizing and hyperpolarizing responses to acetylcholine application have been demonstrated in early studies on neuroblastoma cells (for reviews, see Nelson, 1973, 1975; Fischbach and Nelson, 1977). Subsequently, serotonin and dopamine were found to produce depolarizing reponses in several lines (Myers and Livengood, 1975; Nelson, 1977; Christian *et al.*, 1978; Higashida *et al.*, 1978). Recently, Higashida and Nirenberg (in preparation) have found multiphasic responses to histamine, neurotensin, somatostatin, and various bradykinins and angiotensins. The presence of such a host of receptor-mediated electrical responses to putative neurotransmitters and peptide hormones is perhaps relevant to the recent observation that addition of serum produces multiphasic ion permeability changes in nondividing neuroblastoma cells (Moolenaar *et al.*, 1979). Biophysical, pharmacological, and molecular characterization of these receptor-mediated electrical responses, their relationship to receptor-mediated biochemical responses, and their regulation during differentiation are major areas for future research.

Functional cholinergic synapses have been found between neuronal cell lines and muscle (Nelson *et al.*, 1976; Christian *et al.*, 1977). The probability of transmitter release is, however, low. It remains to be seen whether this reflects inadequate culture conditions or the inherent inabilities of neuronal lines to form stable, high-efficiency synapses.

As sophistication of biophysical, biochemical and morphological description of the development and function of nerve cells has progressed, it has become more and more difficult to integrate all three aspects into any unified paradigm. Such a paradigm can be advanced through the coordinated application of all three approaches to the same preparation. Clonal nerve cell lines present an ideal system in which to investigate important neurobiological questions using, concomitantly, various techniques in a way that cannot be done in other neuronal systems *in vivo* or *in vitro*. Some of these general issues are (1) the sequence of expression and regulation of membrane properties during maturation of nerve cells, (2) the molecular components and mechanisms underlying membrane permeability and excitability, (3) the prerequisites for the formation and stabilization of synapses, (4) the sequence of developmental events during the interaction between a neuron and a target cell, and (5) the involvement of ionic mechanisms in growth, differentiation, and cellular metabolism.

Ultimately, no matter how ideal the system studied, nor how sophisticated the multidisciplinary approach, answers will remain elusive unless new and probing critical questions are posed.

ACKNOWLEDGMENTS. I am grateful to Dr. P. G. Nelson for his constant encouragement and to Dr. M. C. Fishman for his invaluable help. I would like to thank Drs. M. Nirenberg and M. Adler for making unpublished data available to me and for helpful discussions.

REFERENCES

Adams, D. J., Smith, S. J., and Thompson, S. H., 1980, Ionic currents in molluscan soma, *Annu. Rev. Neurosci.* **3:**141–167.

Amano, T., Richelson, E., and Nirenberg, M., 1972, Neurotransmitter synthesis by neuroblastoma clones, *Proc. Natl. Acad. Sci. USA* **69:**258–263.

Armstrong, C. M., 1971, Interaction of tetraethylammonium ion derivatives with potassium channels of giant axons, *J. Gen. Physiol.* **58:**413–437.

Armstrong, C. M., and Hille, B., 1972, The inner quaternary ammonium ion receptor in potassium channels of the node of Ranvier, *J. Gen. Physiol.* **59:**388–400.

Atwater, I., Dawson, C. M., Ribalet, B., and Rojas, E., 1979, Potassium permeability activated by intracellular calcium ion concentration in the pancreatic β-cell, *J. Physiol. (Lond.)* **288:**575–588.

Augusti-Tocco, B., Sato, G. H., Claude, P., and Potter, D. D., 1970, Clonal cell lines of neurons, *Int. Soc. Cell Biol. Symp.* **9:**109–120.

Burgess, G. M., Claret, M., and Jenkinson, D. H., 1979, Effects of catecholamines, ATP and ionophore A23187 on potassium and calcium movements in isolated hypatocytes, *Nature* **279:**544–546.

Cahalan, M.D., 1975, Modification of sodium channel gating in frog myelinated nerve fibers by *Centruroides sculpturatus* scorpion venom, *J. Physiol. (Lond.)* **244:**511–534.

Chalazonitis, A., and Greene, L. A., 1974, Enhancement in excitability properties of mouse neuroblastoma cells cultured in the presence of dibutyryl cyclic AMP, *Brain Res.* **72:**340–345.

Christian, C. N., Nelson, P. G., Peacock, J., and Nirenberg, M., 1977, Synapse formation between two clonal cell lines, *Science* **196:**995–998.

Christian, C. N., Nelson, P. G., Bullock, P., Mullinax, D., and Nirenberg, M., 1978, Pharmacologic responses of cells of neuroblastoma × glioma hybrid clone and modulation of synapses between hybrid cells and mouse myotubes, *Brain Res.* **147:**261–276.

Dichter, M. A., Tischler, A. S., and Greene, L. A., 1977, Nerve growth factor-induced increase in electrical excitability and acetylcholine sensitivity of a rat pheochromocytoma cell line, *Nature* **268:**501–504.

Engelhardt, J. K., Ishikawa, K., and Katase, D. K., 1980, Low potassium is critical for observing developmental increase in muscle resting potential, *Brain Res.* **190:**564–568.

Fedoroff, S., and Hertz, L. (eds.), 1977, *Cell, Tissue and Organ Cultures in Neurobiology,* Academic Press, New York.

Fischbach, G., and Lass, Y., 1978, Acetylcholine noise in cultured chick myoballs: A voltage clamp analysis, *J. Physiol. (Lond.)* **280:**515–526.

Fischbach, G., and Nelson, P., 1977, Cell culture in neurobiology, in: *Handbook of Physiology—The Nervous System I* (E. R. Kandel, ed.), pp. 719–774, American Physiological Society, Bethesda, Maryland.

Fishman, M. C., and Spector, I., 1980, Blockade of calcium-dependent potassium conductance in neuroblastoma cells by quinine, *J. Supramol. Struct.* [*Suppl.*] **4**:83.
Giagnoni, G., Sabol, S. L., and Nirenberg, M., 1977, Synthesis of opiate peptides by a clonal pituitary tumor cell line, *Proc. Natl. Acad. Sci. USA* **74**:2259–2263.
Greene, L. A., and Tischler, A. S., 1976, Establishment of a noradrenergic clonal line of rat adrenal pheochromocytoma cells which respond to nerve growth factor, *Proc. Natl. Acad. Sci. USA* **73**:2424–2428.
Harris, A. J., and Dennis, M. J., 1970, Acetylocholine sensitivity and distribution on mouse neuroblastoma cells, *Science* **167**:1253–1255.
Higashida, H., Wilson, S. P., Adler, M., and Nirenberg, M., 1975, Synapse formation by neuroblastoma and hybrid cell lines, *Soc. Neurosci. Abst.* **4**:591.
Hille, B., 1976, Gating in sodium channels of nerve, *Annu. Rev. Physiol.* **38**:135–152.
Hille, B., 1978, Ionic channels in excitable membranes, *Biophys. J.* **22**:283–294.
Hodgkin, A. L., and Huxley, A. F., 1952, A quantitative description of membrane current and its application to conduction and excitation in nerve, *J. Physiol. (Lond.)* **117**:500–544.
Jacques, Y., Fosset, M., and Lazdunski, M., 1978, Molecular properties of the action potential Na^+ ionophore in neuroblastoma cells, *J. Biol. Chem.* **253**:7383–7392.
Jaffe, A. L., and Robinson, R. L., 1978, Membrane potential of the unfertilized sea urchin egg, *Dev. Biol.* **62**:215–228.
Khodorov, B. I., and Revenko. S. V., 1979, Further analysis of the mechanisms of action of Batrachotoxin on the membrane of myelinated nerve, *Neuroscience* **4**:1315–1330.
Kidokoro, Y., 1975, Spontaneous Ca action potentials in a clonal pituitary cell line and their relationship to prolactin secretion, *Nature* **258**:741–742.
Kimhi, Y., Palfrey, C., Spector, I., Barak, Y., and Littauer, U. Z., 1976, Maturation of neuroblastoma cells in the presence of dimethylsulfoxide, *Proc. Natl. Acad. Sci. USA* **73**:462–466.
Lew, V. L., and Ferreira, H. G., 1978, Calcium transport and the properties of a calcium-activated potassium channel in red cell membranes, *Curr. Top. Membr. Trans.* **10**:218–278.
Lichtshtein, D., Kaback, H.R., and Blume, A. J., 1979a, Use of lipophilic cation for determination of membrane potential in neuroblastoma–glioma hybrid cell suspension, *Proc. Natl. Acad. Sci. USA* **76**:650–654.
Lichtshtein, D., Dunlop, K., Kaback, H. R., and Blume, A. J., 1979b, Mechanism of monensin-induced hyperpolarization of neuroblastoma–glioma hybrid NG108–15, *Proc. Natl. Acad. Sci. USA* **76**:2580–2584.
Lux, H. D., and Heyer, C. B., 1979, A new electrogenic calcium–potassium system, in: *The Neurosciences Fourth Study Program* (F. O. Schmitt and F. G. Warden, eds.), pp. 601–622, MIT Press, Cambridge, Massachusetts.
Macdermot, J., Higashida, H., Wilson, S., Matsuzawa, H., Minna, J., and Nirenberg, M., 1979, Adenylate cyclase and acetylcholine release regulated by separate serotonin receptors of somatic cell hybrids, *Proc. Natl. Acad. Sci. USA* **76**:1135–1139.
Mains, R. E., Eipper, B. A., and Ling, N., 1977, Common precursor to corticotropins and endorphins, *Proc. Natl. Acad. Sci. USA* **74**:3014–3018.
Matsuzawa, H., and Nirenberg, M., 1975, Receptor-mediated shifts in cGMP and cAMP levels in neuroblastoma cells, *Proc. Natl. Acad. Sci. USA* **72**:3473–3476.
Meech, R. W., 1978, Calcium-dependent potassium activation in nervous tissue, *Annu. Rev. Biophys. Bioeng.* **7**:1–18.
Miyaki, M., 1978, The development of action potential mechanism in a mouse neuronal cell line *in vitro, Brain Res.* **143**:349–354.
Moolennar, W. H., and Spector, I., 1977, Membrane currents examined under voltage clamp in cultured neuroblastoma cells, *Science* **196**:331–333.
Moolenaar, W. H., and Spector, I., 1978, Ionic currents in cultured mouse neuroblastoma cells under voltage-clamp conditions, *J. Physiol. (Lond.)* **278**:265–286.

Moolenaar, W. H., and Spector, I., 1979a, The calcium action potential and a prolonged calcium-dependent after-hyperpolarization in mouse neuroblastoma cells, *J. Physiol. (Lond.)* **292:**297–306.

Moolenaar, W. H., and Spector, I., 1979b, The calcium current and the activation of a slow potassium conductance in voltage-clamped mouse neuroblastoma cells, *J. Physiol. (Lond.)* **292:**307–323.

Moolenaar, W. H., De Laat, S. W., and Van Der Saag, P. T., 1979, Serum triggers a sequence of rapid ionic conductance changes in quiescent neuroblastoma cells, *Nature* **279:**721–723.

Myers, P. R., and Livengood, D. R., 1975, Dopamine depolarizing response in a vertebrate neuronal somatic-cell hybrid, *Nature* **255:**235–237.

Nelson, P. G., 1973, Electrophysiological studies of normal and neoplastic cells in tissue culture, in: *Tissue Culture of the Nervous System* (G. H. Sato, ed.), pp. 135–160, Plenum Press, New York.

Nelson, P. G., 1975, Nerve and muscle cells in culture, *Physiol. Rev.* **55:**1–61.

Nelson, P., 1978, Neuronal cells lines, in: *Cell, Tissue and Organ Cultures in Neurobiology* (S. Federoff and L. Hertz, eds.), pp. 347–365, Academic Press, New York.

Nelson, P. G., and Henkart, M. P., 1979, Oscillatory membrane potential changes in cells of mesenchymal origin: The role of an intracellular calcium regulating system, *J. Exp. Biol.* **81:**49–61.

Nelson, P G., Ruffner, W., and Nirenberg, M., 1969, Neuronal tumor cells with excitable membranes grown *in vitro, Proc. Natl. Acad. Sci. USA* **64:**1004–1010.

Nelson, P. G., Peacock, J., and Amano, T., 1971a, Responses of neuroblastoma cells to iontophoretically applied acetylocholine, *J. Cell. Physiol.* **77:**353–362.

Nelson, P. G., Peacock, J., Amano, T., and Minna, J., 1971b, Electrogenesis in mouse neuroblastoma cells *in vitro, J. Cell Physiol.* **77:**337–352.

Nelson, P., Christian, C., and Nirenberg, M., 1976, Synapse formation between colonal neuroblastoma × glioma hybrid cells and striated muscle cells, *Proc. Natl. Acad. Sci. USA* **73:**123–127.

Nirenberg, M., Wilson, S. P., Higashida, H., Rotter, A., Ray, R., Adler, M., Thompson, J., and Deblas, A., 1979, Synapse plasticity, *Fed. Proc.* **38:**476.

O'Lague, P., and Huttner, S., 1980, Physiological and morphological studies of rat pheochromocytoma cells (PC12) chemically fused and grown in culture, *Proc. Natl. Acad. Sci. USA* **77:**1701–1705.

Palfrey, C., 1976, Development of Membrane Properties and Differentiation in Neuroblastoma Cells, Ph.D. Thesis, The Weizmann Institute of Science, Rehovot, Israel.

Palfrey, C., Kimhi, Y., and Littauer, U.Z., 1977, Induction of differentiation in mouse neuroblastoma cells by hexamethylene bisacetamide, *Biochem. Biophys, Res. Commun* **76:**937–942.

Patrick, J., Heinemann, S., and Schubert, D., 1978, Biology of cultured nerve and muscle, *Annu. Rev. Neurosci.* **1:**417–443.

Patterson, P., 1978, Environmental determination of autonomic neurotransmitter functions, *Annu. Rev. Neurosci.* **1:**1–17.

Peacock, J., Minna, J., Nelson, P., and Nirenberg, M., 1972, Use of aminopterin in selecting electrically active neuroblastoma cells, *Exp. Cell Res.* **73:**367–377.

Peacock, J. H., McMorris, F. A., and Nelson, P. G., 1973, Electrical excitability and chemosensitivity of mouse neuroblastoma × mouse or human fibroblast hybrid, *Exp. Cell Res.* **79:**199–212.

Pelhate, M., and Pichon, Y., 1974, Selective inhibition of potassium current in the giant axon of the cockroach, *J. Physiol. (Lond.)* **242:**90p–91p.

Reiser, G., Neumann, R., Kemper, W., Lautenschlager, E., and Hamprecht, B., 1977, Influence of cations on the electrical activity of neuroblastoma × glioma hybrid cells, *Brain Res.* **130:**497–504.

Richelson, E., and Tuttle, J., 1975, Diphenylhydantoin inhibits ionic excitation of mouse neuroblastoma cells, *Brain Res.* **99:**209–212.

Ritchie, A. K., and Fambrough, D. M., 1975, Electrophysiological properties of the membrane and acetycholine receptor in developing rat and chick myotubes, *J. Gen. Physiol.* **66:**327–355.

Ritchie, J. M., 1979, A pharmacological approach to the structure of sodium channels in myelinated axons, *Annu. Rev. Neurosci.* **2:**341–362.

Rotter, A., Ray, R., and Nirenberg, M., 1979, Regulations of calcium uptake in neuroblastoma or hybrid cells—A possible mechanism for synapse plasticity, *Fed. Proc.* **38:**626.

Sato, G. H. (ed.), 1973, *Tissue Culture of the Nervous System, Current Topics in Neurobiology,* Vol. 1, Plenum Press, New York.

Schauf, C. L., Colton, C. A., Colton, J. S., and Davis, F. A., 1976, Aminopyridine and sparteine as inhibitors of membrane potassium conductance: Effects on *Myxicola* giant axons and the lobster neuromuscular junction, *J. Pharmacol. Exp. Ther.* **197:**414–425.

Schubert, D., Harris, A. J., Heinemann, S., Kidokoro, Y., Patrick, J., and Steinbach, J. H., 1973, Differentiation and interactions of clonal lines of nerve and muscle, in: *Tissue Culture of the Nervous System* (G. H. Sato, ed.), pp. 55–86, Plenum Press, New York.

Schubert, D., Heinemann, S., Carlisle, W., Tarikas, H., Kimes, B., Patrick, J., Steinbach, J. H., Culp, W., and Brandt, B. L., 1974, Clonal cell lines from the rat central nervous system, *Nature* **249:**224–227.

Siegelbaum, S. A., and Tsien, R. W., 1980, Calcium-activated transient outward currents in calf cardiac purkinje fibres, *J. Physiol. (Lond.)* **299:**485–506.

Spector, I., and Prives, J. M., 1977, Development of electrophysiological and biochemical membrane properties during differentiation of embryonic skeletal muscle in culture, *Proc. Natl. Acad. Sci. USA* **74:**5166–5170.

Spector, I., Kimhi, Y., and Nelson, P. G., 1973, Tetrodotoxin and cobalt blockade of neuroblastoma action potentials, *Nature* [*New Biol.*] **246:**124–126.

Spector, I., Palfrey, C., and Littauer, U.Z., 1975, Enhancement of the electrical excitability of neuroblastoma cells by valinomycin, *Nature* **254:**121–124.

Spitzer, N., 1979, Ionic channels in development, *Annu. Rev. Neurosci.* **2:**363–397.

Study, R. E., Breakfield, X. O., Bartfai, T., and Greengard, P., 1978, Voltage-sensitive calcium channels regulate guanosine 3′, 5′-cyclic monophosphate levels in neuroblastoma cells, *Proc. Natl. Acad. Sci. USA* **75:**6295–6299.

Tuttle, J., and Richelson, E., 1975, Ionic excitation of a clone of mouse neuroblastoma, *Brain Res.* **84:**129–135.

Tuttle, J., and Richelson, E., 1979, Phenytoin action on the excitable membrane of mouse neuroblastoma, *J. Pharmcol. Exp. Ther.* **211:**632–637.

Veselovskii, N. S., Kostyuk, P. G., Krishtal, O. A., and Naumov, A. P., 1977, Transmembrane ionic currents in the membrane of neuroblastoma cells, *Neirofiziologiya* **9:**641–643.

Yeh, J. Z., Oxford, G. S., Wu, C. H., and Narahashi, T., 1976, Interactions of aminopyridines with potassium channels of squid axon membranes, *Biophys. J.* **16:**77–80.

8

Studies of Voltage-Sensitive Sodium Channels in Cultured Cells Using Ion-Flux and Ligand-Binding Methods

WILLIAM A. CATTERALL

1. INTRODUCTION

This chapter surveys the development of ion-flux and ligand-binding methods for studies of voltage-sensitive sodium channels and their application to excitable cells in culture. These techniques have proven to be complementary to the electrophysiological methods described in other chapters of this volume in providing new information on the properties of excitable cells in culture and in giving new insight into the fundamental mechanisms of electrical excitability. Electrophysiological methods allow analysis of individual cells; biochemical methods allow analysis of cell populations. Electrophysiological methods give excellent time resolution; biochemical methods are more convenient for pharmacological studies and allow quantitation of ion channel density. One of the principal advantages of cell culture systems for study of electrical excitability is the ability to apply both electrophysiological and biochemical methods in parallel experiments.

The development of ion-flux and ligand-binding methods for studies of voltage-sensitive sodium channels has required the use of neurotoxins as specific probes of ion channel properties. During the last few years, the effects of these neurotoxins have been extensively analyzed in voltage clamp experiments on nerve axons. The electrophysiological effects of these toxins are reviewed briefly here to provide background for discussion of their use in ion-flux and ligand-binding experiments.

WILLIAM A. CATTERALL • Department of Pharmacology, University of Washington, Seattle, Washington 98195.

Neurotoxins that alter the properties of voltage-sensitive sodium channels are of three distinct chemical types. Recent electrophysiological experiments suggest that each of these three classes of neurotoxins has distinct effects on the functional properties of sodium channels. Tetrodotoxin and saxitoxin are water-soluble heterocyclic compounds containing guanidinium moieties. These toxins are specific inhibitors of sodium channels (Narahashi *et al.*, 1964; Hille, 1968; reviewed by Evans, 1972).

Grayanotoxin and the alkaloids batrachotoxin, veratridine, and aconitine are lipid-soluble polycyclic compounds. These toxins depolarize excitable cells by increasing sodium permeability (Albuquerque *et al.*, 1971; Straub, 1956; Ulbricht, 1969; Ohta *et al.*, 1973; Herzog *et al.*, 1964; Seyama and Narahashi, 1973). The increase in sodium permeability is inhibited by tetrodotoxin, indicating that it results from activation of voltage-sensitive sodium channels normally involved in generation of action potentials (Albuquerque *et al.*, 1971; Ulbricht, 1969; Seyama and Narahashi, 1973). Recently, voltage clamp experiments have shown that batrachotoxin and aconitine activate sodium channels at the resting membrane potential by causing a shift in the voltage dependence of activation of sodium channels to more negative membrane potentials and by blocking inactivation of sodium channels (Khodorov, 1978; Schmidt and Schmitt, 1974). It is likely that veratridine and grayanotoxin act by a similar mechanism.

Extracts of sea anemone nematocysts and venoms of North African scorpions cause repetitive action potentials in excitable cells and inhibit inactivation of sodium channels (Koppenhöfer and Schmidt, 1968; Narahashi *et al.*, 1969, 1972). More recent experiments have shown that basic polypeptides with molecular masses of 4000 to 7000 daltons isolated from sea anemones or from North African scorpion venoms also specifically inhibit inactivation of sodium channels (Romey *et al.*, 1975, 1976; Bergman *et al.*, 1976; Conti *et al.*, 1976; Okamoto *et al.*, 1977).

By using these neurotoxins as specific pharmacological probes, it has been possible to establish biochemical methods to study sodium channel properties and to analyze the effects of these toxins on sodium channel function. These results are reviewed in the following sections of this chapter.

2. DEVELOPMENT OF METHODS

2.1. Ion-Flux Procedures for Studies of Sodium Channels

2.1.1. Flux Measurements in Physiological Solution

The initial goal in developing ion-flux methods for studies of voltage-sensitive ion channels was to provide a convenient method for assessing the electrical excitability of populations of cultured cells (Catterall and Nirenberg,

1973). The main difficulty was finding a way to activate a substantial fraction of the ion channels for long enough to carry out an ion-flux measurement. Depolarization of the cell population activates sodium channels for only a few milliseconds before inactivation occurs. We therefore chose to activate sodium channels pharmacologically using veratridine (Catterall and Nirenberg, 1973). Neuroblastoma and muscle cells in monolayer culture were incubated with 5 mM ouabain to inhibit Na^+–K^+ ATPase and up to 200 μM veratridine in a HEPES-buffered physiological solution containing ^{22}NaCl. Uptake of $^{22}Na^+$ was measured after various incubation times by aspirating the radioactive assay medium and washing the cell cultures extensively with nonradioactive medium. Veratridine increased the rate of uptake of $^{22}Na^+$ of electrically excitable neuroblastoma and muscle cells up to sixfold. The increase in $^{22}Na^+$ uptake was inhibited completely by tetrodotoxin, with half-maximal inhibition at approximately 10 nM. These experiments demonstrated the feasibility of using isotopic tracer methods to measure increases in steady-state ion flux caused by persistent pharmacological activation of sodium channels. Inhibition of the veratridine-dependent $^{22}Na^+$ uptake by tetrodotoxin at low concentration suggested that the sodium channels persistently activated by veratridine were the same ion channels normally involved in action potential generation in neuroblastoma and muscle cells. These methods involving measurements of $^{22}Na^+$ uptake in physiological solution provide a useful way to test populations of cultured cells for the presence of voltage-sensitive sodium channels. In general, however, these methods do not allow quantitative assessment of the number of sodium channels in cells or of the fraction of sodium channels activated by neurotoxins because the measured flux is not linearly proportional to the number of active ion channels. The following section considers modifications of these methods that allow flux measurements under conditions where the measured uptake is proportional to the number of active ion channels.

2.1.2. Quantitative Measurements of Sodium Channel Activation by Neurotoxins

In designing procedures to measure activation of ion channels by neurotoxins quantitatively, attention must be given to three phases of the experiment. First, the neurotoxins must be allowed to reach equilibrium with their receptor sites without altering ion gradients. Second, initial rates of sodium flux must be measured under conditions where the sodium pump is effectively inhibited and the membrane potential of the cells remains constant. Third, cells must be washed to remove extracellular $^{22}Na^+$ under conditions where influx is stopped and efflux of $^{22}Na^+$ is prevented. In each of these three phases of the experiment, the technical requirements are most stringent when the neurotoxin-dependent permeability increase is largest. Modifications of our initial experimental technique which overcome these difficulties are discussed below.

Although the action of veratridine is nearly instantaneous on the time scale of the ion-flux experiments (Catterall, 1975a), many of the other neurotoxins that affect sodium channels act slowly and must be incubated with excitable cells for up to 1 hr to equilibrate with their receptor sites (Catterall, 1975a,1976a). During this incubation, Na^+ permeability is dramatically increased. In order to prevent entry of Na^+ into cells, these incubations are carried out in Na^+-free medium. Preincubation in Na^+-free medium also reduces the intracellular Na^+ concentration to less than 0.5 mM (unpublished data), allowing efflux of Na^+ during the subsequent ion-flux assay to be ignored. Toxins have been allowed to equilibrate with their receptor sites in Na^+-free medium in all of our recent studies (Catterall, 1975a,b, 1976a,b, 1977a,b; Catterall *et al.,* 1976; Catterall and Beress, 1978).

Incubation in Na^+-free medium is sufficient to prevent toxin-dependent changes in intracellular Na^+. However, when the sodium permeability of neuroblastoma cells (Catterall, 1976a) or muscle cells (unpublished data) is dramatically increased by activation of all the voltage-sensitive sodium channels, intracellular K^+ is lost during the incubations. The loss results from K^+ exit via voltage-sensitive sodium channels which are approximately 10% as permeable to K^+ as to Na^+ (Catterall, 1976a). In order to prevent this K^+ loss, equilibration with toxins that activate a large fraction of voltage-sensitive sodium channels can be carried out in medium with K^+ replacing Na^+ as the major ion (see Catterall, 1977b). Under these conditions ($[K^+]_{out} = 130$ mM, $[Na^+]_{out} = 0$, $V = 0$), there are no ionic gradients, and therefore the increase in ion permeability does not cause major changes in intracellular ion concentrations.

The second phase of the experiment is measurement of the neurotoxin-dependent ion permeability increase. It is essential in this phase of the experiment to ascertain that the measured ion flux is linearly proportional to the number of active ion channels. The principal difficulty in designing experiments in which the measured $^{22}Na^+$ uptake is linearly proportional to the number of active ion channels is control of the membrane potential. The number of active sodium channels is, by definition, proportional to the sodium permeability (P_{Na}). Goldman (1943) and Hodgkin and Katz (1949) have derived a relationship between ion flux and ion permeability that is obeyed by many cells:

$$J_{Na} = P_{Na}\,[Na^+]_{out}\,(F/RT)[V/(e^{VF/RT} - 1)] \qquad (1)$$

This relationship predicts that $^{22}Na^+$ influx (J_{Na}) is linearly proportional to Na^+ permeability only if the membrane potential (V) is constant. However, V depends on ionic permeabilities and concentrations according to equation (2) (Hodgkin and Katz, 1949):

$$V = (RT/F)\ln\left(\frac{P_K[K^+]_{out} + P_{Na}[Na^+]_{out} + P_{Cl}[Cl^-]_{in}}{P_K[K^+]_{in} + P_{Na}[Na^+]_{in} + P_{Cl}[Cl^-]_{out}}\right) \qquad (2)$$

Thus, in general, when cells are treated with veratridine or other toxins that increase P_{Na}, the relationship between P_{Na} and J_{Na} is nonlinear. This nonlinearity has important consequences for any studies made in medium with physiological ion concentrations. For example, in comparing ion flux of two cell lines having different densities of sodium channels, treatment of the more active line with veratridine will cause a greater depolarization (equation 2), and therefore the driving force for $^{22}Na^+$ influx will be reduced relative to the less active cell line (equation 1). Under this condition, the difference in measured flux must substantially underestimate the true difference in ion channel density. Similarly, in determining concentration–effect relationships for toxins that activate sodium channels, cells in physiological medium will have a different membrane potential at each toxin concentration studied, and both the maximum flux measured and the concentration of toxin required to give half-maximal flux will be seriously underestimated. These values will be more seriously underestimated for cell lines having a high density of sodium channels. Apparent positive cooperativity, unusual allosteric interactions, erroneous ion specificities, and incorrect K_D values can all result from quantitative interpretation of data obtained under these conditions.

One approach to controlling the membrane potential during sodium flux experiments is simply to lower the extracellular sodium concentration so that the increase in P_{Na} has no effect on the membrane potential. This approach has been used in all of our recent ion flux experiments (Catterall, 1975a,b, 1976a,b, 1977a,b; Catterall *et al.*, 1976; Catterall and Beress, 1978). Microelectrode studies indicate that veratridine has no effect on the membrane potential of neuroblastoma clone N18 up to 50 mM Na^+. Veratridine, however, activates only a fraction [approximately 8% (Catterall, 1977b)] of the sodium channels in the cells. If all of the ion channels in N18 cells are activated by batrachotoxin plus scorpion toxin (Catterall, 1977b), depolarization becomes important at sodium concentrations greater than 15 mM. These microelectrode measurements on selected large cells in the cell culture population have been confirmed by measurements of membrane potential from $^{35}SCN^-$ distribution as described in the following section. In addition, equation 1 suggests a simple and direct test to determine the maximum concentration of sodium that can be used without causing depolarization. Equation 1 predicts that $^{22}Na^+$ uptake should vary linearly with sodium concentration at low $[Na^+]_{out}$ where V is constant but should become nonlinear at higher sodium concentrations where depolarization occurs. Figure 1A illustrates an experiment testing this point carried out with mouse neuroblastoma clone N18. After incubation with bactrachotoxin, J_{Na} varies linearly with $[Na^+]_{out}$ up to 10 mM. These data confirm that, in N18 cells, flux experiments carried out in 10 mM Na^+ give rates of $^{22}Na^+$ uptake that are proportional to P_{Na}. The experiment of Fig. 1A provides a simple control to ascertain that the measured ion flux is proportional to P_{Na}. If quantitative conclusions are to be made, it is essential that a concentration of Na^+ be chosen in the range where uptake varies linearly with $[Na^+]_{out}$.

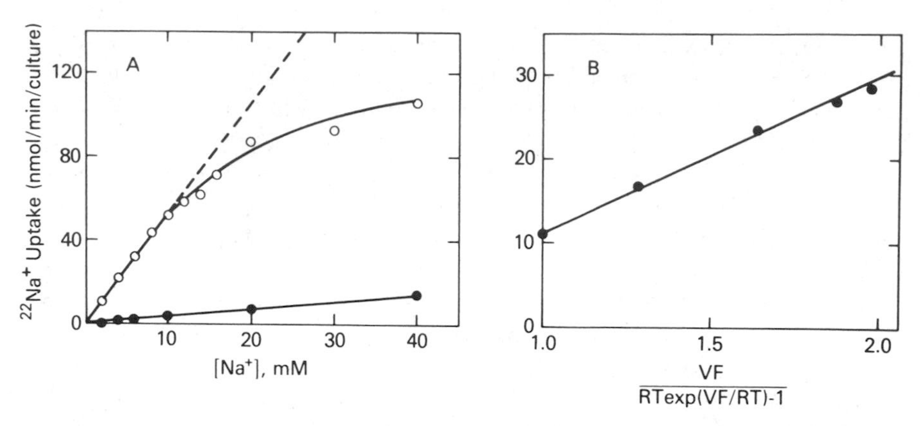

FIGURE 1. Dependence of batrachotoxin-stimulated $^{22}Na^+$ uptake on sodium concentration and membrane potential. (A) N18 cells were incubated for 30 min in preincubation medium I (Table I) either with (○) or without (●) 1 μM batrachotoxin. $^{22}Na^+$ uptake was then measured for 30 sec in assay medium (Table I) with increasing sodium concentrations so that $[Na^+] + [choline^+] = 130$ mM. (B) N18 cells were incubated with or without batrachotoxin as in A. $^{22}Na^+$ uptake was then measured in assay medium containing 5 mM Na^+ plus increasing K^+ concentrations so that $[K^+] + [choline] = 130$ mM. The membrane potentials under these conditions are: $[K^+] = 5.4$ mM, $V = -41$ mV; $[K^+] = 10$ mM, $V = -37$ mV; $[K^+] = 25$ mM, $V = -28$ mV; $[K^+] = 60$ mM, $V = -13$ mV; $[K^+] = 135$ mM, $V = 0$ mV (Catterall *et al.*, 1976). (From Catterall, 1977b, with permission of the American Society of Biological Chemists.)

This discussion has relied heavily upon equation 1 as a description of the relationship between J_{Na} and P_{Na} in excitable cells in culture. In order to test further the validity of equation 1, ion-flux experiments have been carried out at different membrane potentials between -41 mV and 0 mV by varying the extracellular K^+ concentration (Catterall *et al.*, 1976; Catterall, 1977b). These data show a linear relationship between $^{22}Na^+$ uptake and $VF/[RT \exp(VF/RT) - 1]$ as predicted by equation 1 (Fig. 1B). Preliminary experiments with cultured muscle cells and cultured heart cells indicate that similar relationships hold for these cells. The slope of the line in Fig. 1B (1.45) is somewhat greater than predicted by the Goldman–Hodgkin–Katz equation (1.0). This discrepancy is likely to result from competitive inhibition of $^{22}Na^+$ influx by extracellular K^+. The work of Hille (1975) shows that K^+ competitively blocks sodium channels in the node of Ranvier with a K_D of 220 mM. Correcting the data of Fig. 1B for this interaction with K^+ gives a straight line with a slope of 0.92. Therefore, ion flux mediated by sodium channels in N18 cells follows the predictions of the Goldman–Hodgkin–Katz equation.

In the experiments of Fig. 1, and in most of our ion-flux experiments on neuroblastoma clone N18, choline has been used as a sodium replacement. Osmotically equivalent concentrations of Tris or tetraethylammonium are also satisfactory. Choline is not a suitable sodium replacement in muscle cells because it is a weak agonist of the nicotinic acetylcholine receptor. However, if the cells are pretreated with α-bungarotoxin, choline can be used. The compositions of the solutions used at different steps in our assay procedures are presented in Table I.

In principle, the membrane potential during the ion-flux assay can also be controlled by using conditions where there are no ion gradients and $V = 0$. This might be achieved by long-term incubation with ouabain to equilibrate the intracellular and extracellular sodium concentrations or by imposing extracellular ion concentrations (high K^+, low Na^+) during the ion-flux measurement that are equal to the intracellular concentrations. Neither of these approaches is satisfactory in rat or mouse cells. Prolonged incubation with ouabain makes the cells very leaky and difficult to study quantitatively. High extracellular K^+ makes it impossible to inhibit effectively the Na^+–K^+ ATPase with ouabain or other cardiac glycosides because K^+ is a potent inhibitor of glycoside binding. Therefore, in these cell lines, the most useful method of controlling membrane potential during the flux procedure is to replace sodium with impermeant cations and measure flux at a constant membrane potential near the resting potential of the cells.

The final concern in designing the ion-flux phase of the experiment is linearity of influx with time. When all of the sodium channels in N18 cells are activated, uptake remains linear for 2 min. For cells with a higher density of ion channels, the period of linearity would be shorter. Clearly, selection of an incubation time during the linear period of uptake is essential.

The third phase of the ion-flux assay is the washing step. The radioactive assay medium is removed by aspiration, and the cells are washed extensively with ice-cold buffer. These steps effectively stop influx by diluting the $^{22}Na^+$

TABLE I. Compositions of Media Used

	Preincubation media (mM)			
	I	II	Assay medium (mM)	Wash medium (mM)
Choline Cl		130.0	120.0	163.0
NaCl			10.0	
KCl	135.4	5.4	5.4	
$CaCl_2$	1.8	1.8	1.8	1.8
$MgSO_4$	0.8	0.8	0.8	0.8
Glucose	5.5	5.5	5.5	
HEPES[a]	50.0	50.0	50.0	5.0
Ouabain			5.0	

[a]Adjusted to pH 7.4 with Tris base.

manyfold. However, when the sodium permeability of the cells is greatly increased by toxins whose effect is slowly reversible, the sodium permeability remains high during the wash period, and intracellular $^{22}Na^+$ can be lost. Measurements with neuroblastoma clone N18 treated with scorpion toxin plus veratridine give a half-time for sodium loss of 1.5 min at 0°C (Catterall, 1976b). Since loss of intracellular $^{22}Na^+$ must be accompanied by influx of extracellular Na^+ or K^+ to maintain charge neutrality, removal of Na^+ and K^+ from the wash medium should reduce efflux rate. Substitution of Na^+ and K^+ by choline increases the half-time of efflux to 10 min in N18 cells (Catterall, 1976b). This wash medium (Table I) has been used in all of our recent experiments (Catterall, 1976a,b, 1977a,b; Catterall *et al.*, 1976; Catterall and Beress, 1978). Since the washing period is 15 sec long, the loss of $^{22}Na^+$ is negligible.

Our present ion-flux assay conditions can be summarized as follows. In the first phase, cells are allowed to equilibrate with neurotoxins for up to 60 min in sodium-free medium either at $V = 0$ with K^+ as a sodium replacement (preincubation medium I, Table I) or, if no increase in P_{Na} occurs during preincubation, at V near the resting potential with choline as a sodium replacement (preincubation medium II, Table I). Then, the initial rate of $^{22}Na^+$ influx is measured in medium with 5 mM or 10 mM Na^+ and choline as sodium replacement at V near the resting membrane potential. Finally, the cells are washed in Na^+- and K^+-free wash medium containing choline as the major cation at V near the resting membrane potential. A wide variety of alternative approaches was tested in developing these conditions. Most alternatives are inconvenient or unsatisfactory because the linear relationship between flux and permeability is not maintained.

2.2. Ion-Flux Procedures for Studies of Acetylcholine Receptors and Calcium Channels

Ion-flux procedures generally similar to those described for sodium channels have been used to study acetylcholine-activated ion channels and voltage-sensitive Ca^{2+} channels in cultured nerve and muscle cells. Since a detailed discussion of this work is beyond the scope of this chapter, only a brief review is presented. More detailed information can be obtained from the references cited.

Treatment of primary chick muscle cultures or clonal rat muscle cultures with nicotinic cholinergic agonists such as carbamylcholine leads to a large increase in $^{22}Na^+$ uptake (Catterall, 1975c; Stallcup and Cohn, 1976a). The permeability increase is blocked by acetylcholine receptor antagonists such as α-bungarotoxin or tubocurarine but is unaffected by tetrodotoxin. The permeability increase desensitizes on prolonged ($>$ 30 sec) exposure to saturating concentrations of carbamylcholine. These results demonstrate the feasibility of using ion-flux procedures to study acetylcholine receptors in cultured cells.

Several new aspects of receptor function have been studied using these methods. The steady-state ion transport activity of the receptor has been estimated [2×10^8 ions/min (Catterall, 1975c)]. The mechanism of action of a new class of receptor ligands, the histrionicotoxins, has been defined (Burgermeister *et al.*, 1977). The selectivity of the acetylcholine-activated ion channel for cations and nonelectrolytes and some new aspects of ion transport by this ion channel have been described (Catterall, 1975c; Huang *et al.*, 1978). Finally, ion-flux methods designed for studies of nicotinic acetylcholine receptors in cultured muscle cells have been adapted to studies of neuronal nicotinic acetylcholine receptors present in a pheochromocytoma cell line (Patrick and Stallcup, 1977). These studies have led to new insight into the pharmacological and biochemical relationships between nicotinic acetylcholine receptors in muscle and nerve.

A method for studying voltage-sensitive calcium channels in cultured muscle cells has also been reported (Stallcup and Cohn, 1976a). Calcium influx into a clonal line of cultured muscle cells was increased by depolarization with K^+, and the enhanced $^{45}Ca^{2+}$ uptake was blocked with the calcium channel blocker Mn^{2+}. These experiments may provide a new approach to studies of excitation–contraction coupling in cultured muscle cells.

2.3. Measurement of the Membrane Potential of Cultured Cells from the Distribution of Lipid-Soluble Ions

It has proven useful in our ion-flux studies on voltage-sensitive sodium channels to be able to estimate the membrane potential of populations of cultured cells. While this is certainly possible using electrophysiological methods, large numbers of cells must be impaled to generate meaningful data, and the small cells that are the major cell type in the population are often very difficult to impale without injury. Ion-distribution procedures offer an alternative approach which allows determination of the absolute value of the membrane potential of the cell population. The equilibrium distribution of a permeant ion between the intracellular and extracellular compartments should depend on the membrane potential as predicted by the Nernst equation if the ion is neither pumped nor bound by the cells. In this case,

$$V = (RT/F) \ln ([A^-]_{out}/[A^-]_{in}) = (RT/F) \ln ([C^+]_{in}/[C^+]_{out}) \quad (3)$$

This approach to measuring membrane potential has been used in mitochondria (Grinius *et al.*, 1970; Nichols, 1974) and bacteria (Altendorf *et al.*, 1974; Schuldiner and Kaback, 1975).

There are four essential features required of an ion in distribution experiments: (1) The ion must be sufficiently lipid-soluble to penetrate the cell membrane and must have no accessible uncharged forms that can contribute to the measured distribution. (2) The ion must not be bound or actively transported by the cells. (3) The ion must not alter cell function appreciably. (4) The ion must be available in a radiolabeled form. A number of ions have been synthesized for this purpose. The thiocyanate anion, SCN^-, fulfills the criteria outlined above and is particularly convenient since it is available commercially in labeled form (New England Nuclear Corp.). Thiocyanate is a strong acid and is essentially entirely charged at neutral pH. Penetration of uncharged species can therefore be ignored. Thiocyanate is highly lipid-soluble and therefore is freely permeable through bilayer membranes. It has the additional advantage that it is not amphipathic and therefore should not bind tightly to membranes or have strong detergent effects at low concentration. We have used $^{35}SCN^-$ distribution to assess the membrane potential of cultures of N18 and have correlated those results with the results of microelectrode experiments (Catterall *et al.*, 1976).

The rate of equilibration of 5 mM extracellular $^{35}SCN^-$ with clone N18 neuroblastoma cells in culture is illustrated in Fig. 2. Equilibration is complete in 5 to 10 min at 36°C (Fig. 2) and is nearly as fast at 0°C (not illustrated), indicating that SCN^- is freely permeable through the cell membrane. All of the $^{35}SCN^-$ taken up by the cells is readily removed ($t_{1/2}$ = 1 min) by washing in nonradioactive medium, indicating that no slowly reversible binding to cell components occurs. If the membrane potential of the cells is changed by

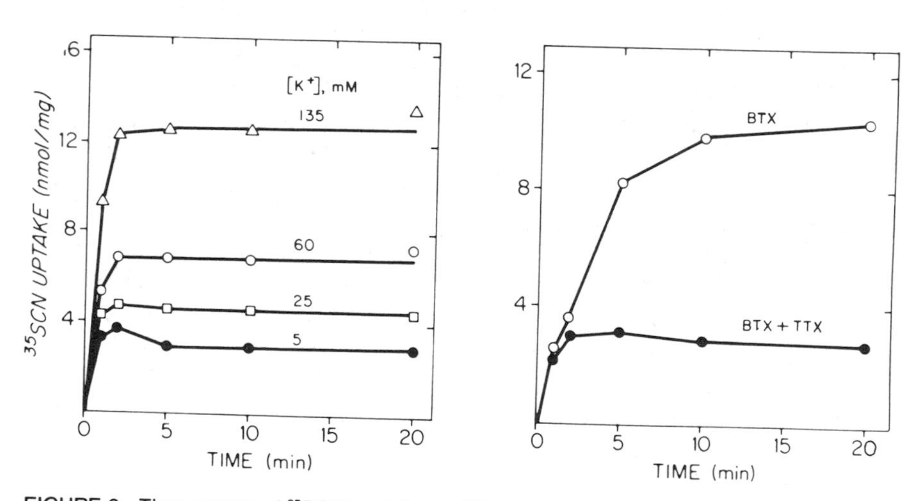

FIGURE 2. Time course of $^{35}SCN^-$ uptake at different membrane potentials. (A) N18 cells were incubated for the indicated times in preincubation medium I containing 5.4 mM $K^{35}SCN$ and increasing concentrations of KCl so that $[K^+] + [choline^+] = 135$ mM. After the incubation, the cells were rapidly washed 3 times with ice-cold wash medium, and $^{35}SCN^-$ uptake was measured by liquid scintillation counting. (B) N18 cells were incubated for 30 min with 1 μM batrachotoxin in preincubation medium I. The time course of $^{35}SCN^-$ uptake was then measured in preincubation medium with 130 mM NaCl and 1 μM batrachotoxin (○) or 1 μM batrachotoxin plus 1 μM tetrodotoxin (●).

increasing the extracellular K^+ concentration, the rate of $^{35}SCN^-$ is not markedly affected, but the equilibrium concentration of $^{35}SCN^-$ is increased by depolarization as predicted by the Nernst equation. The mean values of equilibrium $^{35}SCN^-$ uptake in experiments like those in Fig. 2A are 2.8 nmole $^{35}SCN^-$/mg protein at 5 mM K^+ and 13.9 nmole $^{35}SCN^-$/mg protein at 135 mM K^+. This corresponds to a depolarization of 42 mV on changing from 5 mM K^+ to 135 mM K^+.

Although the $^{35}SCN^-$ uptake data are sufficient to measure relative membrane potentials, the intracellular volume must be known in order to calculate absolute values of membrane potential. For clone N18 neuroblastoma cells, the intracellular volumes have been measured by determining the difference between the volumes available to ^3H_2O or [^{14}C]urea and that available to [^{14}C]inulin. These data give a mean value of 3.3 μl/mg cell protein at both 5 mM K^+ and 135 mM K^+. Using this value, the absolute values of the membrane potential can be calculated from uptake data like that of Fig. 2. Membrane potentials calculated in this way are illustrated as a function of K^+ concentration in Fig. 3 (△). The calculated membrane potential varies from −46 mV to −4 mV as the K^+ concentration is increased from 5 mM to 135 mM. These estimates of membrane potential are compared with direct determinations made by microelectrode impalement of large cells in the population (Fig. 3, ○). While there is a significant difference between the two sets of measurements, the difference is not large. Two main sources of error probably contribute to the observed difference. The microelectrode measurements probably underestimate the true membrane potential slightly because of low-resistance leakage pathways around the pipette. The $^{35}SCN^-$ distribution measurements probably overestimate the membrane potential slightly because of small losses of intracellular $^{35}SCN^-$ during the 10-sec wash period. In view of these two sources of small systematic error, the agreement between the data must be considered good. The slope of the line relating membrane potential to [K^+] and the absolute value of the potential are both smaller than expected if K^+ is the dominant permeant ion. Since the solutions are Na^+-free, the results suggest that ions other than Na^+ and K^+ are significant current carriers in these cells.

In view of the agreement between the measurements of membrane potential by microelectrode impalement and by $^{35}SCN^-$ distribution, we have used the $^{35}SCN^-$-distribution method to monitor membrane potential in many of our ion-flux experiments. One example of such a measurement is illustrated in Fig. 2B. In this experiment, cells in physiological solution (130 mM Na^+) were incubated for 30 min with either 2 μM batrachotoxin or 2 μM batrachotoxin plus 1 μM tetrodotoxin. Then, uptake of $^{35}SCN^-$ was measured. In the cultures treated with batrachotoxin, the average membrane potential was −5 mV. In those treated with both toxins, the average membrane potential was −41 mV, essentially the resting potential. Thus, as expected from our ion-flux studies

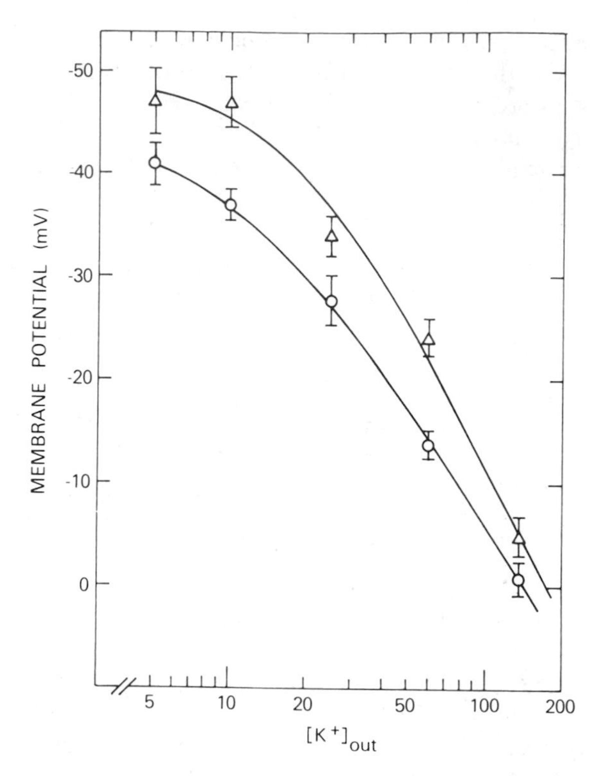

FIGURE 3. Dependence of membrane potential on extracellular K^+. N18 cells were added to 60-mm petri dishes at 10,000 cells per cm^2 and cultured in Dulbecco's modified minimal essential medium + 5% fetal bovine serum for 7–9 days. Membrane potentials of large cells were measured by microelectrode impalement in Na^+-free medium with the indicated K^+ concentrations and choline Cl such that $[K^+] + [\text{choline}^+] = 135$ mM as described in text. Each point represents the mean of 15 to 25 stable impalement ± SEM (○). N18 cells were cultured in multiwell plates as for transport and binding assays. Cells were incubated for 20 min in Na^+-free medium containing the indicated concentrations of K^+ and 5 mM $^{35}SCN^-$ (1 μCi/ml). Cells were then washed and dissolved, and radioactivity was determined. Intracellular SCN^- concentrations were calculated using a cell volume of 3.3 μl/mg of protein. Membrane potentials were calculated from $E_m = 61 \text{ mV} \log([SCN^-]_{in}/[SCN^-]_{out})$ (△).

(Catterall, 1975a), batrachotoxin depolarizes N18 cells nearly to 0 mV, and this depolarization is reversed by tetrodotoxin.

While providing a useful method for measuring membrane potential in cultured cells, the $^{35}SCN^-$ method also has some clear disadvantages. The measurements must be made at equilibrium. Since equilibration requires 5 to 10 min, rapid membrane potential changes cannot be resolved. Figure 2A shows that rates $^{35}SCN^-$ uptake are not markedly potential-dependent and, in any case, there is no theoretical basis for estimating membrane potential from rates of influx of lipid-soluble ions. The main value of the method is therefore measurement of steady-state values of membrane potential in cell culture populations.

2.4. Scorpion Toxin as a Specific Radioligand for Voltage-Sensitive Sodium Channels in Excitable Cells

2.4.1. Purification of Scorpion Toxin

A general method for large-scale purification of toxic polypeptides from a number of North African scorpion venoms has been described by Miranda *et al.* (1970). These studies used mouse lethality as an assay for the active venom components. I developed a small-scale purification procedure using a specific ion-flux assay in order to isolate the venom component responsible for the cooperative enhancement of neurotoxin activation of sodium channels by scorpion venom (Catterall, 1976a). This procedure has been described in detail. I have, therefore, given only a brief description here and indicated minor modifications that improve the method. The procedure involves three steps. The lyophilized crude venom (*Leiurus quinquestriatus,* Sigma Chemical Co.) is first dissolved at low concentration (1 mg/ml) in distilled water. Under these conditions, mucopolysaccharides and other nontoxic venom components are insoluble and are removed by centrifugation. In the second step, the aqueous extract of the venom is fractionated by chromatography on Amberlite CG-50. We currently use a 50 × 0.9 cm column. Careful preparation of the resin (Hirs, 1955) is required to obtain good yields with this longer column. The column is washed with 100 ml of 10 mM ammonium acetate, pH 7.0, followed by 50 ml of 175 mM ammonium acetate, pH 7.0, and finally a 200 ml linear gradient from 175 mM ammonium acetate, pH 7.0, to 275 mM ammonium acetate, pH 7.0. A typical column profile starting from the 175 mM ammonium acetate wash is illustrated in Fig. 4. Most of the protein adsorbed to the column elutes in the 10 mM ammonium acetate wash (Catterall, 1976a). Another large fraction of the input protein elutes in the 175 mM wash (Fig. 4). No biological activity is associated with these fractions. Using the longer column and shallower gradient described above, three protein fractions are resolved by the gradient elution (Fig. 4). Biological activity is associated principally with the last fraction. The third step of the procedure involves electrophoretic analysis of individual fractions which contain biological activity. In the preparation illustrated in Fig. 4, fractions 110 through 160 were analyzed by gel electrophoresis in urea/acetic acid and by isoelectric focusing as described previously (Catterall, 1976a). Fractions in peak 2 contained two or three protein bands. Fractions in peak 3 (138 to 160) contained only a single protein band. These fractions were pooled for use in experiments. The ratios of protein in peaks 1 through 3 vary among different samples of venom. We have

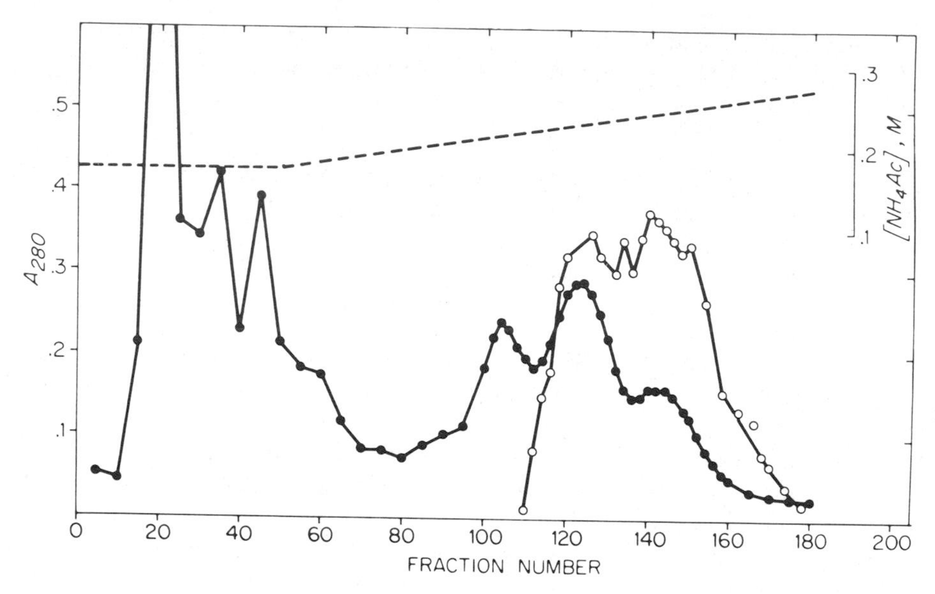

FIGURE 4. Purification of scorpion toxin by column chromatography on Amberlite CG-50. Scorpion venom extract prepared as described previously (Catterall, 1976a) was applied to a column (0.9 × 50 cm) of Amberlite CG-50 (NH_4^+ form) equilibrated with 10 mM ammonium acetate, pH 7.0. The column was eluted with 20 ml of 10 mM ammonium acetate, pH 7.0; 50 ml of 175 mM ammonium acetate, pH 7.0; and then a 200-ml linear gradient from 175 mM ammonium acetate, pH 7.0, to 275 mM ammonium acetate, pH 7.0. Absorbance of representative fractions was monitored at 280 nm (●). Ammonium acetate concentration was determined from conductivity measurements (---). Aliquots (0.1μl) of representative fractions were tested for enhancement of veratridine activation of neuroblastoma cells (Catterall, 1976a). Biological activity is plotted in nmole $^{22}Na^+$ uptake/min per culture. Uptake in the absence of scorpion toxin has been subtracted. Fractions were analyzed by gel electrophoresis.

found it to be essential to carefully measure biological activity by ion-flux methods in individual fractions and to analyze the protein components by gel electrophoresis in individual fractions in each preparation. If these precautions are taken, electrophoretically pure toxin can be reproducibly prepared in 30 to 50% yield.

2.4.2. Preparation of ^{125}I-Labeled Scorpion Toxin

We have iodinated scorpion toxin by two different procedures. The first uses the reagent of Bolton and Hunter (1973), succinimidyl 3-[^{125}]iodo-4-hydroxyphenylpropionate, to derivatize a free primary amino group on the toxin (Catterall *et al.*, 1976; Ray and Catterall, 1978). The monosubstituted derivative retains biological activity and binds specifically to receptor sites associated with voltage-sensitive sodium channels in neuroblastoma cells. The second method uses lactoperoxidase to catalyze iodination of a tyrosine residue in the toxin (Catterall, 1977a). Both monoiodo and diiodo derivatives retain biological activity and bind specifically to sodium channels. The characteristics of the specific binding sites observed with these two toxin derivatives are essentially identical. The derivative prepared by lactoperoxidase labeling is superior in two respects, however. The iodination can be carried out in good yield with relatively inexpensive reagents. The ratio of specific to nonspecific binding is better with the lactoperoxidase-labeled derivative because its K_D is somewhat lower, and nonspecific binding is substantially lower. We now use the lactoperoxidase-labeled derivative exclusively. Since the preparation of the toxin has been described in detail (Catterall, 1977a), and no modifications have been made, no additional description of the method is necessary here. It is worth noting that the separation of the labeled products by ion-exchange chromatography is important to obtain good binding results. The separation of diiodo toxin from unlabeled and monoiodo toxin is complete. The separation of unlabeled toxin from monoiodo toxin is incomplete (Catterall, 1977a) and highly purified monoiodo toxin can only be obtained when the iodination reaction is sufficiently complete that little unlabeled toxin remains. The specific radioactivity of the labeled toxin can then be calculated from the specific radioactivity of the input ^{125}I, since all toxin molecules are labeled.

2.4.3. Scorpion Toxin Binding Measurements

Since the binding of scorpion toxin is highly dependent on membrane potential (Catterall *et al.*, 1976; Catterall, 1977a; Ray and Catterall, 1978), measurements must be made on intact cells under conditions where the membrane potential is controlled. Our binding measurements have been carried out in sodium-free, choline-substituted medium (preincubation medium I, Table I). Cells are incubated with toxin for 1 hr at 36 °C. Unbound toxin is then removed in three washes of 1 min each. The details of this procedure have been described (Catterall, 1977b).

Under these conditions, a saturable component of scorpion toxin binding is observed. At 2 nM labeled toxin, 80% of the binding is saturable (Catterall, 1977a). Several lines of evidence (Catterall, 1977a; see Section 2.5) show that the saturable component of binding represents binding to receptor sites associated with voltage-sensitive sodium channels.

2.5. Saxitoxin as a Specific Radioligand for Voltage-Sensitive Sodium Channels in Cultured Cells.

2.5.1. Preparation and Purification of [^3H]Saxitoxin

Saxitoxin is labeled most conveniently by the specific ^3H_2O exchange procedure described by Ritchie *et al.* (1976). The resulting product is then purified by high-voltage electrophoresis. The product of the electrophoretic separation is not completely pure. Both the concentration and the purity of the labeled toxin must be assessed by bioassay or isotopic dilution methods. Ritchie *et al.* (1976) measured concentration using blockade of compound action potentials in an unmyelinated nerve as a bioassay and estimated purity by measuring the fraction of ^{3}H counts per minute that are specifically bound to nerve membranes. We have used $^{22}Na^{+}$ uptake in neuroblastoma cells as a bioassay and binding to rat brain homogenates as an estimate of purity (Catterall and Morrow, 1978). In either case, the purity of different preparations of [^{3}H]saxitoxin ranges from 55% to 75%. Corrections for this purity must be made in estimating specific radioactivity.

2.5.2. Measurement of Specific [^{3}H] Saxitoxin Binding

The inhibition of sodium channels by saxitoxin is rapidly reversible. The saxitoxin–receptor complex dissociates with $t_{1/2}$ = 75 sec in N18 neuroblastoma cells. Binding assay procedures, therefore, must prevent dissociation of the bound toxin. The low density of receptor sites in cultured cells does not allow equilibrium binding measurements. We have measured [^{3}H]saxitoxin binding by incubating cells with toxin for 20 min at either 36°C or 0°C in preincubation solution I (Table I) and then removing the unbound toxin by washing three times for a total of 6 to 8 sec at 0°C. Less than 10% of the bound saxitoxin is lost in this time. Using this procedure, nearly 90% of the [^{3}H]saxitoxin binding is blocked by unlabeled saxitoxin or tetrodotoxin, with half-maximal inhibition of binding at 8 to 10 nM (Catterall and Morrow, 1978). These results indicate that [^{3}H]saxitoxin binds specifically to receptor sites associated with sodium channels in cultured neuroblastoma cells as has been previously found in nerve and muscle preparations (Ritchie *et al.*, 1976). Additional support for this conclusion is provided by the observation that binding of [^{3}H]saxitoxin is markedly reduced in clone N103 neuroblastoma cells which lack voltage-sensitive sodium channels (Catterall and Morrow, 1978). These results verify that saxitoxin, like scorpion toxin, is a specific radioligand for sodium channels in excitable cells.

2.5.3. Comparison of Saxitoxin and Scorpion Toxin as Radioligands for Voltage-Sensitive Sodium Channels

Both saxitoxin and scorpion toxin provide specific ligands for studies of sodium channels in cultured cells. However, these two toxins bind at separate receptor sites (Catterall and Morrow, 1978; Section 3.1.6), and their binding characteristics are markedly different. Binding of saxitoxin is independent of membrane potential, whereas scorpion toxin binding is inhibited by depolarization. Saxitoxin binding is therefore the method of choice if cells are to be homogenized or if solubilization and purification of receptor sites are planned. On the other hand, binding of saxitoxin to intact cells is rapidly reversible, whereas binding of scorpion toxin is slowly reversible at 36°C and very slowly reversible at 0°C. Diiodo scorpion toxin can be prepared with a specific radioactivity of 4400 Ci/mmole compared to 40 Ci/mmole for saxitoxin. Scorpion toxin binding is therefore the method of choice if intact cells are to be used and the density of binding sites is low or slow reversibility of labeling is important.

3. ANALYSIS OF THE PROPERTIES OF VOLTAGE-SENSITIVE SODIUM CHANNELS IN CULTURED CELLS USING ION-FLUX AND LIGAND-BINDING METHODS

3.1. Neuroblastoma Cells

My objective in this part of the chapter is to review the results that have been obtained using ligand-binding and ion-flux methods to analyze the properties of voltage-sensitive sodium channels in cultured cells. Experiments on the N18 clone of mouse neuroblastoma C1300 are considered first, since this cell line has been studied most extensively. Experiments on other neuronal cell lines, on skeletal muscle cells, and on heart muscle cells are considered at the end of the chapter. The electrophysiological and other differentiated properties of clonal cell lines are considered in the chapters by Drs. Kimhi and Spector in this volume.

3.1.1. Activation of Sodium Channels by Neurotoxins

Veratridine, batrachotoxin, aconitine, and grayanotoxin increase the rate of passive $^{22}Na^+$ influx into N18 cells (Catterall and Nirenberg, 1973; Catterall, 1975a,b, 1977b). This effect of the toxins proceeds without a lag, reaches an equilibrium level of sodium permeability that is concentration-dependent, and is completely reversible when the toxin is washed out of the cells (Catterall, 1975a, unpublished data). Complete equilibrium concentration–effect curves for each of these toxins have been obtained under different conditions using the ion-flux procedures described in this chapter. In our most recent experiments, we found that batrachotoxin activated virtually all of the sodium channels in N18 cells, whereas grayanotoxin activated 51%, veratridine 8%, and aconitine 2% (Table II; Catterall, 1977b). Half-maximal activation was obtained at concentrations ranging from 0.7 μM for batrachotoxin to 1.2 mM for grayanotoxin (Table II). In each case, the data fit a simple hyperbolic concentration curve, indicating that binding of a single toxin molecule is sufficient to activate a sodium channel, and no cooperative interactions are required to activate an ion channel (Catterall, 1975a,1977b).

TABLE II. Activation of Sodium Channels by Neurotoxins[a]

Toxin	Concentration for 50% activation (M)	Fractional activation at saturation
Batrachotoxin	7.0×10^{-7}	0.95
Veratridine	2.9×10^{-5}	0.08
Grayanotoxin	1.2×10^{-3}	0.51
Aconitine	3.6×10^{-6}	0.02

[a]Data are derived from computed fits of titration curves to a simple hyperbolic saturation curve of the form $P(A) = P_{\infty}A/(K_{0.5} + A)$ where P_{∞} is the fraction of ion channels activated at saturation, $K_{0.5}$ is the concentration of toxin required for 50% maximal activation, and A is the concentration of toxin activator. The complete data are published in Catterall (1977b).

Different neurotoxins activate different fractions of sodium channels at saturation (Table II). This result might indicate that these toxins act on different classes of ion channels or that they act on the same ion channels at a common receptor site at which batrachotoxin is a full agonist and the other toxins are partial agonists. In order to resolve this question, concentration–effect curves were determined for each toxin in the presence of fixed concentrations of each of the other toxins (Catterall, 1975a,b, 1977b). In each case, the toxin that was the poorer activator reduced the level of sodium permeability in the presence of the toxin that was the better activator. The concentration-dependence of the inhibition was consistent with a competitive interaction between each pair of toxins. Typical results for aconitine and batrachotoxin

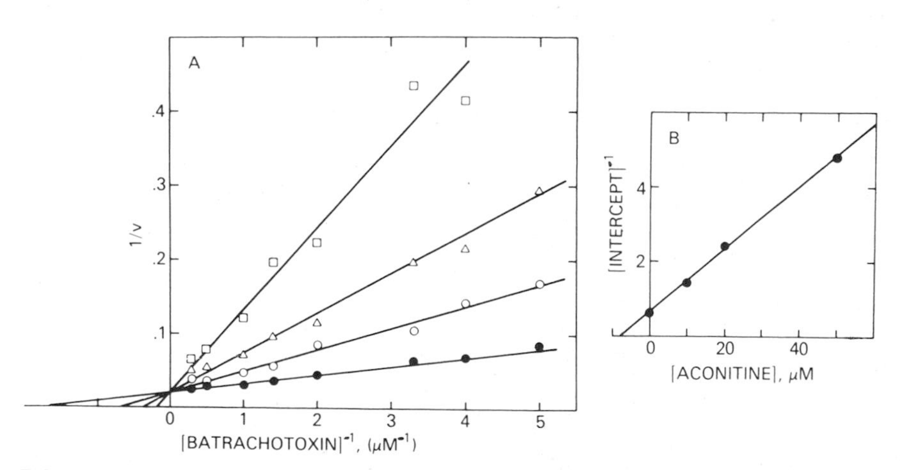

FIGURE 5. Competitive interaction between batrachotoxin and aconitine. N18 cells were incubated for 30 min with the indicated concentrations of batrachotoxin and 0 (●), 10 (○), 20 (△), or 50 (□) μM aconitine. The initial rate of $^{22}Na^{+}$ uptake was then determined. (A) The data are presented as a double-reciprocal plot. (B) Abscissa intercepts are plotted versus aconitine concentration. (From Catterall, 1977b, with permission of the American Society of Biological Chemists.)

are illustrated in Fig. 5. Here batrachotoxin concentration–effect curves at different fixed concentrations of aconitine are plotted as a double reciprocal plot (A), and the intercepts are replotted versus aconitine concentration (B). The data show that aconitine is a linear competitive inhibitor of activation by batrachotoxin. Similar results have been obtained for each pair of neurotoxins, demonstrating that these four toxins act at a common receptor site associated with voltage-sensitive sodium channels. Batrachotoxin is a full agonist at this receptor site, whereas the other toxins are partial agonists.

3.1.3. Specificity of Neurotoxin Activation

The increase in sodium permeability caused by neurotoxins is inhibited by tetrodotoxin in both nerve axons (Albuquerque *et al.,* 1971; Ulbricht, 1969; Seyama and Narahashi, 1973) and in N18 cells. These results suggest that batrachotoxin, veratridine, aconitine, and grayanotoxin specifically activate sodium channels involved in generation of action potentials. Some authors have considered the possibility that these toxins might activate a specific class of tetrodotoxin-insensitive "resting" sodium channels, since the activation occurs in the absence of an electrical stimulus. There are now two other strong lines of evidence in favor of the view that the receptor site for these toxins is associated with voltage-sensitive sodium channels responsible for generation of action potentials.

First, variant mouse neuroblastoma clones that lack sodium-dependent action potentials also lack response to these neurotoxins (Catterall and Nirenberg, 1973; Catterall, 1975a, 1977b). These observations support the conclusion that voltage-sensitive sodium channels must be present for cells to respond to these toxins. Numerous cell lines developed from rat central nervous system tumors have also been studied both electrophysiologically and by ion-flux methods using veratridine (Stallcup and Cohn, 1976b; Stallcup, 1977). These studies also showed almost complete correlation between cell lines that were electrically excitable and those that responded to neurotoxins.

Second, recent voltage clamp studies in frog myelinated nerve show directly that both batrachotoxin and aconitine alter the properties of voltage-sensitive sodium channels (Schmidt and Schmitt, 1974; Khodorov, 1978). Batrachotoxin alters the properties of all of the sodium channels in the node of Ranvier (Khodorov, 1978). Since all four activating neurotoxins act at a common receptor site, all sodium channels in the node of Ranvier must bear a receptor site for grayanotoxin, veratridine, batrachotoxin, and aconitine. Occupancy of this receptor site by neurotoxins leads to persistent activation of a fraction of the sodium channels at the resting membrane potential because of a shift in the voltage dependence of activation and inhibition of inactivation (Khodorov, 1978; Schmidt and Schmitt, 1974). In ion-flux experiments the fraction of sodium channels that is persistently activated is measured as an increase in $^{22}Na^+$ influx.

These results leave little doubt that the sodium channels responsible for action potential generation in nerve and neuroblastoma cells have a specific receptor site for batrachotoxin, veratridine, aconitine, and grayanotoxin and are persistently activated by these toxins. The evidence on this point is less complete in skeletal muscle and in heart muscle (see Sections 3.2 and 3.3). The results available at present also support the conclusion that only electrically excitable cells respond to neurotoxins, since there is a nearly complete correlation between electrically excitable cell lines and those responding to veratridine or other toxins (Catterall and Nirenberg, 1973; Stallcup and Cohn, 1976a,b; Catterall, 1975a, 1977b). Glial cell lines of mammalian origin do not respond to neurotoxins (Catterall and Nirenberg, 1973; Stallcup and Cohn, 1976a,b). However, Schwann cells in the sheath of the squid giant axon are depolarized by veratridine and grayanotoxin (Villegas *et al.*, 1976). Because of the strong correlation between electrical excitability and veratridine response, veratridine stimulation of $^{22}Na^+$ uptake has been used as a marker for neuronal characteristics in studies of cell lines from human tumors and from rat brain (West *et al.*, 1977a,b; Bulloch *et al.*, 1977). While demonstration of a veratridine response is a valuable step in defining cell lines as electrically excitable and therefore neuronal, additional evidence from microelectrode studies and assays of neurotransmitter synthesis greatly strengthens conclusions based on veratridine stimulation alone (cf. West *et al.*, 1977b). It is certainly possible to imagine situations in which veratridine stimulation and electrical excitability would not be correlated. Neuroblastoma cell lines have both voltage-sensitive sodium channels and voltage-sensitive calcium channels (Spector *et al.*, 1973). All lines having only calcium channels would be electrically excitable but would not respond to neurotoxins. On the other hand, electrical excitability requires appropriate passive electrical properties (capacitance, resistance, and membrane potential) in addition to voltage-sensitive ion channels. Cells having sodium channels that are inactivated at accessible membrane potentials would be responsive to neurotoxins but would be electrically inexcitable. Considering the data available at present, it seems correct to conclude that veratridine-stimulated sodium uptake is primarily a neuronal property and therefore that cell lines that are responsive to veratridine and other neurotoxins are likely to be neuronal in origin. Veratridine-stimulated $^{22}Na^+$ uptake should definitely not be considered firm proof of neuronal origin, however. It seems less correct to conclude that cells having neurotoxin-stimulated sodium uptake must necessarily be electrically excitable under normal experimental conditions. Heart cells in monolayer culture provide an example of a cell type having sodium channels that are activated by neurotoxins but not by depolarization of the cells (see Section 3.3).

3.1.3. Enhancement of Neurotoxin Activation of Sodium Channels by Scorpion Toxin and Sea Anemone Toxin

Purified polypeptide toxins from sea anemones and from scorpion venom have little capacity to activate voltage-sensitive sodium channels alone (Catterall, 1976a; Catterall and Beress, 1978). However, these toxins cause a marked enhancement of activation by batrachotoxin, veratridine, aconitine, and grayanotoxin (Catterall, 1976a; Catterall and Beress, 1978). This effect is illustrated in Fig. 6, which compares the enhancement of veratridine activation by scorpion toxin and sea anemone toxin. The presence of either polypeptide toxin causes an increase in sodium permeability at saturation and a decrease in the concentration of veratridine required to give half-maximal activation ($K_{0.5}$). Similar experiments with batrachotoxin, aconitine, and grayanotoxin (Catterall, 1977b; Catterall and Beress, 1978) show that both polypeptide toxins decrease the concentration of batrachotoxin, aconitine, and grayanotoxin required for half-maximal activation, increase the fraction of sodium channels activated by the partial agonists aconitine and grayanotoxin, and have no effect on the fraction of sodium channels activated by the full agonist batrachotoxin.

These results show that scorpion toxin and sea anemone toxin act at sites separate from the receptor site for batrachotoxin, veratridine, aconitine, and grayanotoxin and enhance the action of these toxins at their receptor site. Thus,

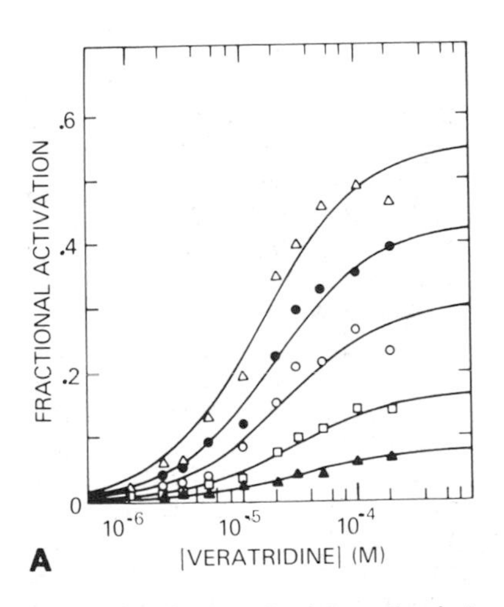

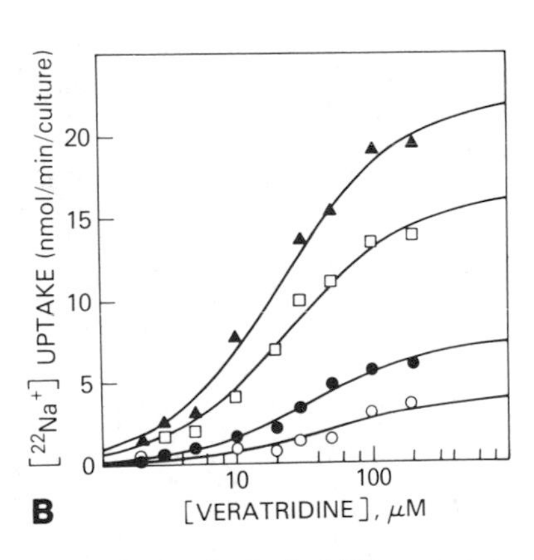

FIGURE 6. Cooperative interaction between veratridine and scorpion toxin (A) or sea anemone toxin (B). N18 cells were incubated with the indicated concentrations of veratridine and 0 (▲), 3 (□), 10 (○), 30 (●), and 100 (△) nM scorpion toxin (A) or 0 (○), 30 (●), 300 (□), or 3000 (▲) nM sea anemone toxin (B). $^{22}Na^+$ uptake was then measured in medium containing the same toxin concentrations.

heterotropic cooperative interactions are observed between two separate receptor sites for neurotoxins. Concentration–effect curves for activation of sodium channels by neurotoxins are hyperbolic in the presence or absence of scorpion toxin or sea anemone toxin, indicating that no homotropic cooperativity is involved in these toxin-dependent processes.

Depolarization of N18 cells by extracellular K^+ causes a marked increase in the concentration of scorpion toxin required for half-maximal enhancement of veratridine activation of sodium channels (Catterall *et al.*, 1976). Concentration–effect curves at five different membrane potentials are illustrated in Fig. 7. Depolarization from −41 mV (○) to 0 mV (▽) causes a 70-fold increase in the concentration of scorpion toxin required for half-maximal activation. Under these conditions, $K_{0.5}$ for sea anemone toxin is increased only fivefold (Catterall and Beress, 1978). In both cases, the value of $K_{0.5}$ is altered by depolarization, but the maximum sodium permeability at saturating toxin concentrations is unaffected (Fig. 7; Catterall *et al.*, 1976). The effect of depolarization is therefore on the affinity of the receptor site for the toxins. These results suggest that scorpion toxin and sea anemone toxin bind to a receptor site located on a region of the sodium channel that undergoes a conformational change to a state with low affinity for toxins on depolarization. This conclusion has been directly demonstrated in studies of scorpion toxin binding described in the following paragraphs.

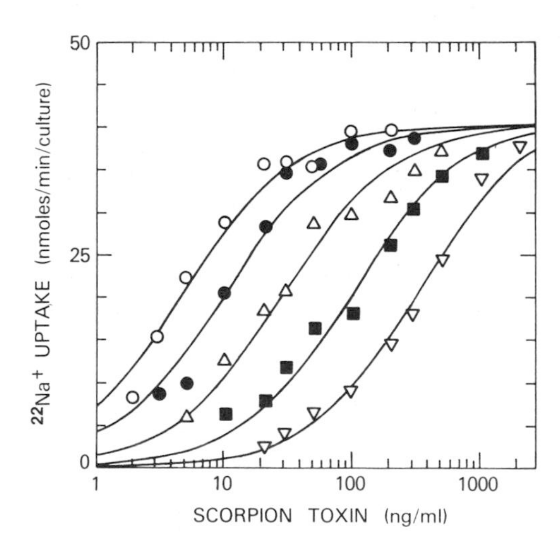

FIGURE 7. Effect of depolarization with K^+ on activation of Na^+ channels by scorpion toxin. N18 cells were incubated for 20 min at 36°C with the indicated concentrations of scorpion toxin in Na^+-free medium containing 5 mM (○), 10 mM (●), 25 mM (△), 60 mM (■), or 135 mM (▽) K^+ and choline Cl such that $[K^+] + [choline^+] = 135$ mM. Cells were rinsed to remove unbound scorpion toxin and excess K^+, and the initial rate of $^{22}Na^+$ uptake was measured in medium containing 5 mM K^+. Curves are least-squares fits to the equation $V = V_\infty A/(K_D + A)$ for $V_\infty = 40.2$ nmole/min per culture and $K_D = 4.8$ ng/m1 at 5 mM K^+, 10.0 ng/ml at 10 mM K^+, 28.6 ng/ml at 25 mM K^+, 98.3 ng/ml at 60 mM K^+, and 331 ng/ml at 135 mM K^+. (From Catterall *et al.*, 1976, with permission of the National Academy of Science.)

3.1.4. Binding of Scorpion Toxin and Sea Anemone Toxin to a Common Receptor Site

Scorpion toxin can be radioactively labeled by reaction of free amino groups with succinimidyl 3-[^{125}I]iodo 4-hydroxyphenyl propionate (Catterall *et al.*, 1976; Ray and Catterall, 1978) or by reaction of tyrosine residues with $^{125}I^-$ catalyzed by lactoperoxidase (Catterall, 1977a). Mono 3-[^{125}I]iodo 4-hydroxyphenyl propionyl scorpion toxin, mono[^{125}I]iodo scorpion toxin, and di[^{125}I]iodo scorpion toxin all retain biological activity. These toxin derivatives all bind to a single class of saturable sites in N18 cells. Three lines of evidence indicate that these sites are associated with sodium channels: (1) The K_D for binding is identical to the concentration of toxin required for half-maximal enhancement of veratridine activation, and the binding and concentration–effect curves are virtually superimposable. (2) Variant neuroblastoma clones that lack sodium-dependent action potentials do not have specific binding sites for scorpion toxin. (3) Depolarization of N18 cells both inhibits specific binding and prevents enhancement of veratridine activation. We have concluded from these results that scorpion toxin binds to a specific receptor site associated with voltage-sensitive sodium channels which undergoes a conformational change on depolarization (Catterall *et al.*, 1976; Catterall, 1977b; Ray and Catterall, 1978).

Scatchard plots of specific scorpion toxin binding at two membrane potentials (-41 mV and -27 mV) are illustrated in Fig. 8. The Scatchard plots are linear at each membrane potential, indicating that there is a single class of sites in each case. Depolarization increases the K_D for scorpion toxin but has no effect on the number of sites observed. K_D is increased tenfold for each 31 mV of depolarization (Figure 8B). These results are consistent with the conclusion that the scorpion toxin receptor site can exist in two conformations with high or low affinity for scorpion toxin and that transitions between these two conformations are rapid, reversible, and voltage-dependent (Catterall, 1977b).

Since scorpion toxin binding affinity is dependent on the conformational state of the sodium channel, the change of state accompanying activation of sodium channels by batrachotoxin and related toxins might also alter scorpion toxin binding. In the presence of tetrodotoxin to prevent depolarization, batrachotoxin reduces the K_D for scorpion toxin binding twofold (Catterall, 1977b). The partial agonists veratridine and aconitine have smaller effects which are not detectable in N18 cells but are observed in synaptosomes (Ray *et al.*, 1978). These results show that the affinity of the scorpion toxin receptor site is affected by two different changes of state: one caused by depolarization which leads to reduced affinity and a second caused by neurotoxin activation which leads to enhanced affinity.

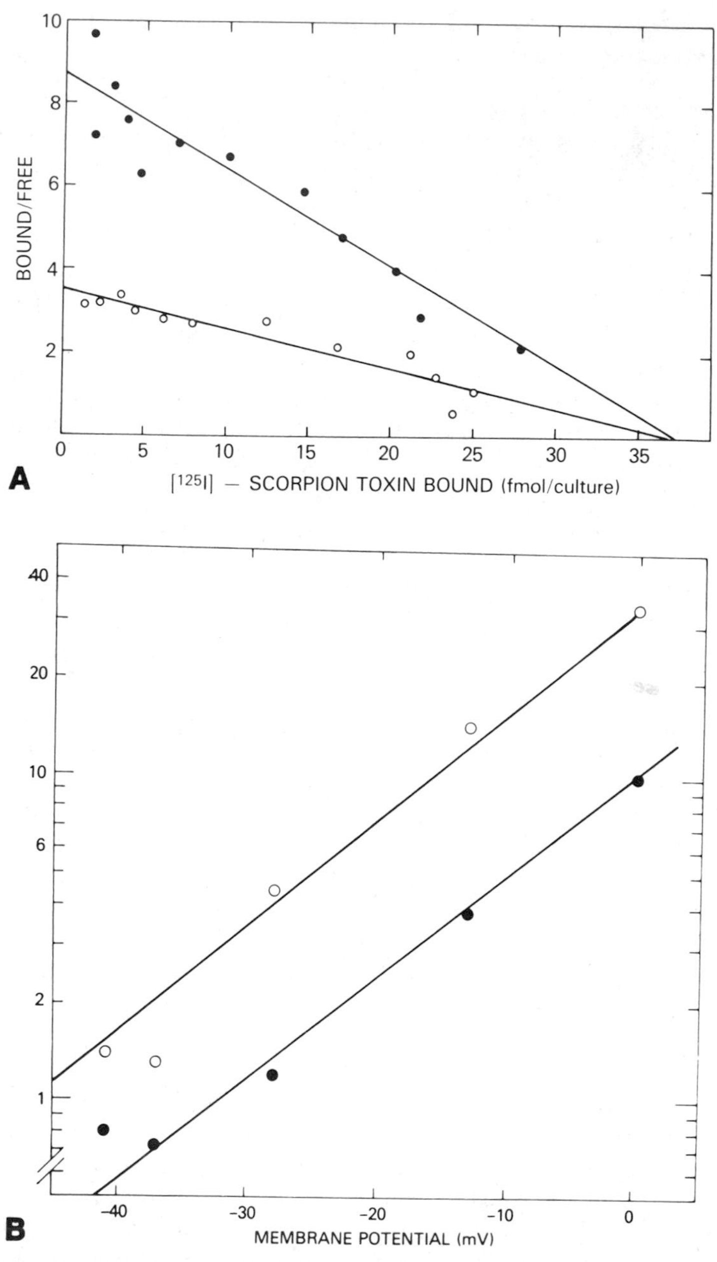

FIGURE 8. Effect of membrane potential on scorpion toxin binding. (A) Specific binding of [^{125}I]diiodo scorpion toxin was measured in the presence of 5 mM K^+ (●) or 25 mM K^+ (○). The data are presented as a Scatchard plot. (B) Specific binding of [^{125}I]monoiodo

Sea anemone toxin and scorpion toxin have similar effects on sodium channels. Both inhibit inactivation specifically in voltage clamp experiments (Romey *et al.*, 1975, 1976; Conti *et al.*, 1976; Bergman *et al.*, 1976; Okamoto *et al.*, 1977). Both enhance activation of sodium channels by neurotoxins (Catterall, 1976a, 1977b; Catterall and Beress, 1978). The enhancement of veratridine stimulation by both is inhibited by depolarization (Catterall *et al.*, 1976; Catterall and Beress, 1978). In order to test whether these two toxins act at a common receptor site, the effect of sea anemone toxin on scorpion toxin binding was measured. We found that sea anemone toxin completely inhibited scorpion toxin binding with a K_D of 90 nM (Catterall and Beress, 1978). Analysis of the binding competition data showed that sea anemone toxin also binds to a single class of sites. These results show that sea anemone toxin and scorpion toxin share a common receptor site associated with sodium channels. This receptor site binds scorpion toxin with 100-fold higher affinity than sea anemone toxin.

Since our binding data allow measurement of the number of scorpion toxin receptor sites in N18 cells, we can estimate both the density and ion transport capacity of the sodium channels in these cells, assuming that each sodium channel is associated with a single scorpion toxin receptor site. We find 22,000 receptor sites/cell or 25 sites/μm^2 of cell surface membrane. Since scorpion toxin shifts the entire concentration–effect curve for batrachotoxin to lower concentrations (Catterall, 1976a, 1977b), all the sodium channels activated by batrachotoxin must contain receptor sites for scorpion toxin. Therefore, we can estimate the ion transport capacity of a batrachotoxin-activated sodium channel by comparing the batrachotoxin-dependent sodium permeability to the number of scorpion toxin receptor sites. We calculate that approximately 1×10^8 ions/min per site would be transported at physiological ion concentrations, corresponding to a conductance of 2.9 pS/site (Catterall, 1977a). This rapid rate of ion movement supports the widely accepted view that ions are transported by a channel mechanism rather than by a mobile carrier mechanism during an action potential.

3.1.5. An Allosteric Model for Activation of Sodium Channels by Neurotoxins

Concentration–effect curves for activation of sodium channels by batrachotoxin, veratridine, aconitine, or grayanotoxin at different fixed scorpion toxin

---

scorpion toxin (●) and [^{125}I]diiodo scorpion toxin (○) was measured at different membrane potentials: 5mM K^+, −41 mV; 10 mM K^+, −37 mV; 25 mM K^+, −28 mV: 60 mM K^+, −13 mV; and 135 mM K^+, 0 mV. Dissociation constants were calculated and plotted as a function of average membrane potential. (From Catterall, 1977a, with permission of the American Society of Biological Chemists.)

or sea anemone toxin concentrations can be fit closely (Catterall, 1977b; Fig. 6) by a two-state allosteric model based on the model of Monod, Wyman, and Changeux (1965). This model makes three assumptions: (1) Sodium channels exist in two states with respect to ion transport activity, active (conducting) and inactive (nonconducting). (2) Alkaloid toxins alter sodium channel properties by binding with greater affinity to active state(s) of sodium channels. These toxins activate sodium channels by shifting a preexisting equilibrium between active and inactive states according to the law of mass action. (3) Scorpion toxin and sea anemone toxin, by binding to their receptor sites, reduce the energy required to activate sodium channels and alter the equilibrium constant between the active and inactive states. These assumptions are identical to those made by Monod *et al.* (1965) in describing heterotropic cooperativity in allosteric enzymes.

In terms of this simple model, the ability of grayanotoxin, veratridine, aconitine, and batrachotoxin to activate sodium channels results from their ability to bind selectively to the active state of sodium channels. This selective binding is most conveniently represented by the ratio of dissocation constants for binding to the inactive and active states (K_T/K_R). These values are presented in Table III. The full agonist batrachotoxin binds 100,000-fold better to the active state of sodium channels, whereas aconitine binds only 140-fold better.

The success of this simple model in fitting our toxin-activation data suggests that there are fundamental similarities between allosteric enzymes and sodium channels. Our current data are most consistent with the view that receptor sites for grayanotoxin and the alkaloid toxins and the receptor site for scorpion toxin and sea anemone toxin reside on separate subunits of a multisubunit complex. Each toxin causes a conformational change in the subunit to which it binds and influences the state of the other subunit through protein–protein interactions.

TABLE III. Selective Binding of Neurotoxins to the Active State of Sodium Channels[a]

Toxin	Binding selectivity (K_T/K_R)
Batrachotoxin	1.4×10^5
Veratridine	6.2×10^2
Grayanotoxin	1.3×10^4
Aconitine	1.4×10^2

[a]The data are derived from computed fits of titration curves to the allosteric model described in the text. The complete data are published in Catterall (1977b).

3.1.6. Binding of Saxitoxin and Tetrodotoxin to Receptor Sites Associated with Sodium Channels

As in nerve axons, sodium channels in neuroblastoma cells are inhibited by tetrodotoxin and saxitoxin (Nelson *et al.*, 1971; Spector *et al.*, 1973). These toxins inhibit sodium channels activated by grayanotoxin and the alkaloid toxins noncompetitively with a K_I of approximately 3 nM (Catterall, 1975a,b; Narahashi and Seyama, 1974). They have no effect on scorpion toxin binding (Catterall, 1977b; Ray and Catterall, 1978). Thus, saxitoxin and tetrodotoxin act at a third receptor site associated with sodium channels. This conclusion is confirmed by direct binding experiments using [^{3}H]saxitoxin prepared as described by Ritchie *et al.* (1976). [^{3}H]saxitoxin binding is inhibited by unlabeled saxitoxin or tetrodotoxin with a K_D of 3 nM (Fig. 9). Scorpion toxin and batrachotoxin at concentrations that saturate their receptor sites have no effect on [^{3}H]saxitoxin binding (Fig. 9).

Since [^{3}H]saxitoxin and [^{125}I]scorpion toxin bind to two separate receptor sites, it is possible to compare the densities of these two receptor sites in neuroblastoma cells. The results of a typical experiment in which binding of saxitoxin and scorpion toxin to companion cultures of N18 cells was measured are

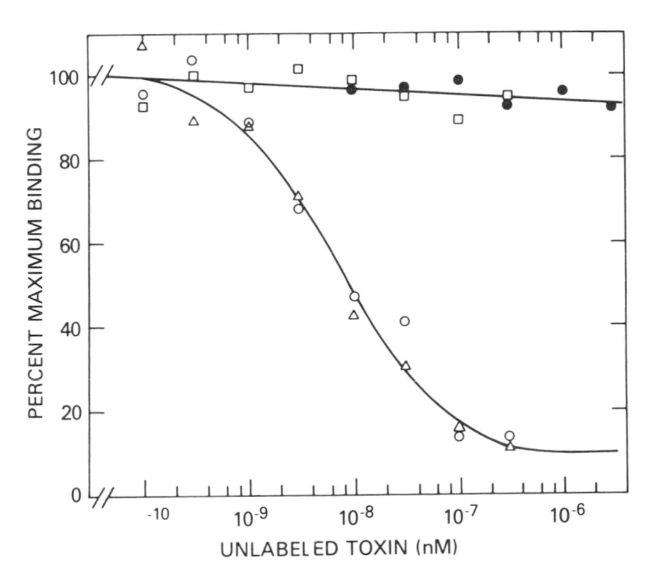

FIGURE 9. Inhibition of [^{3}H] saxitoxin binding by neurotoxins. N18 cells were incubated for 20 min at 36°C with 2 nM [^{3}H] saxitoxin and the indicated concentrations of unlabeled saxitoxin (△), tetrodotoxin (○), batrachotoxin (●), and scorpion toxin (□). Bound saxitoxin was then determined. (From Catterall and Morrow, 1978, with permission of the National Academy of Sciences.)

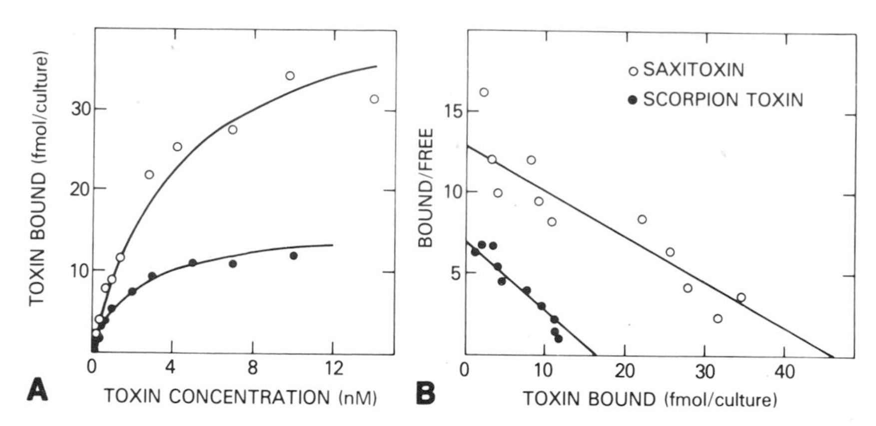

FIGURE 10. Comparison of binding of [^{3}H]saxitoxin and [^{125}I]labeled diiodo scorpion toxin to N18 cells. (A) N18 cells were incubated for 20 min at 36°C with the indicated concentrations of [^{3}H]saxitoxin alone or in the presence of 1 μM tetrodotoxin, and specific binding (○) was determined. Companion cultures of N18 cells were incubated for 60 min at 36°C with the indicated concentrations of [^{125}I]labeled diiodo scorpion toxin alone or in the presence of 200 nM unlabeled scorpion toxin, and specific binding (●) was determined. Only the specific binding component is plotted. (B) Specific binding of saxitoxin (○) and scorpion toxin (●) presented as a Scatchard plot. (From Catterall and Morrow, 1978, with permission of the National Academy of Sciences.)

illustrated in Fig. 10. The results show that there are approximately three times as many saxitoxin receptor sites as scorpion toxin receptor sites in these cells. In several experiments, an average ratio of 2.8 ± 0.2 was observed (Catterall and Morrow, 1978). Thus, despite strong evidence that both of these toxins bind to a single class of sites associated with sodium channels in N18 cells, we find different densities of these two receptor sites. Several possible interpretations of these results have been considered (Catterall and Morrow, 1978). One interesting possibility is the suggestion that this binding site ratio may represent a three-to-one stoichiometry of two separate protein subunits in a multisubunit complex. If this interpretation is correct, these results represent the first direct evidence for a complex subunit structure for the sodium channel.

3.1.7. The Role of the Three Neurotoxin Receptor Sites in Sodium Channel Function

The results summarized in this chapter lead to the conclusion that there are three separate receptor sites for neurotoxins associated with voltage-sensitive sodium channels in excitable cells. The specificity and properties of these sites are outlined in Table IV. Receptor site 1 binds the inhibitors saxitoxin and tetrodotoxin. Occupancy of this site inhibits ion transport by the sodium channel.

Receptor site 2 binds the activators batrachotoxin, aconitine, veratridine, and grayanotoxin. Occupancy of this site causes a shift in the membrane potential dependence of activation and an inhibition of inactivation, resulting in persistent activation of a fraction of sodium channels at the resting membrane potential. Concentration–effect curves for persistent activation are hyperbolic, suggesting that binding of a single toxin molecule is sufficient to activate a sodium channel. The toxins act through an allosteric mechanism by binding selectively to the active state of the sodium channel and causing a shift of ion channels to the active state according to the law of mass action. Since binding of toxins to this site dramatically alters the voltage-dependent regulation of ion channel activity, it is likely that this receptor site resides on an important regulatory component of the sodium channel.

Receptor site 3 binds the polypeptide toxins scorpion toxin and sea anemone toxin. Occupancy of this site inhibits inactivation of sodium channels in voltage clamp experiments and greatly enhances persistent activation of sodium channels by veratridine, batrachotoxin, aconitine, and grayanotoxin. This heterotropic cooperative interaction between the polypeptide toxins acting at receptor site 3 and the activating toxins acting at receptor site 2 indicates that these two receptor sites undergo conformational changes that are strongly coupled. It is likely that these coupled changes of state play an important role in voltage-dependent regulation of sodium channel activity.

The binding of scorpion toxin and sea anemone toxin to receptor site 3 is highly dependent on membrane potential. These results suggest that receptor site 3 is located on a region of the sodium channel that undergoes a voltage-dependent conformational change. Since voltage clamp data on neuroblastoma cells are now available (Moolenaar and Spector, 1977, 1978; Veselovsky *et al.*,

TABLE IV. Properties of the Neurotoxin Receptor Sites

Toxin receptor site	Ligands	Physiological effect
1	Tetrodotoxin Saxitoxin	Inhibition of ion transport
2	Veratridine Batrachotoxin Aconitine Grayanotoxin	Modification of voltage dependence of activation and inactivation resulting in persistent activation
3	Scorpion toxin Sea anemone toxin	Inhibition of inactivation Enhancement of persistent activation by toxins binding to site 2

1977), the voltage dependence of toxin binding can be compared to the voltage dependence of activation and inactivation of the sodium conductance. In Fig. 11, the percent inhibition of toxin binding in N18 is plotted as a function of membrane potential (○). Half-maximal inhibition of binding is obtained at −28 mV. At −2 mV, specific binding is 92% inhibited. The voltage dependence of inactivation of the sodium conductance during prepulses to various membrane potentials is illustrated (●—●) from the data of Moolenaar and Spector (1978) on clone N1E-115 of mouse neuroblastoma C1300. Inactivation of the sodium conductance is 50% complete at −65 mV and nearly 100% complete at −40 mV. Thus, despite the fact that scorpion toxin inhibits inactivation of sodium channels in every system studied, the voltage dependence of

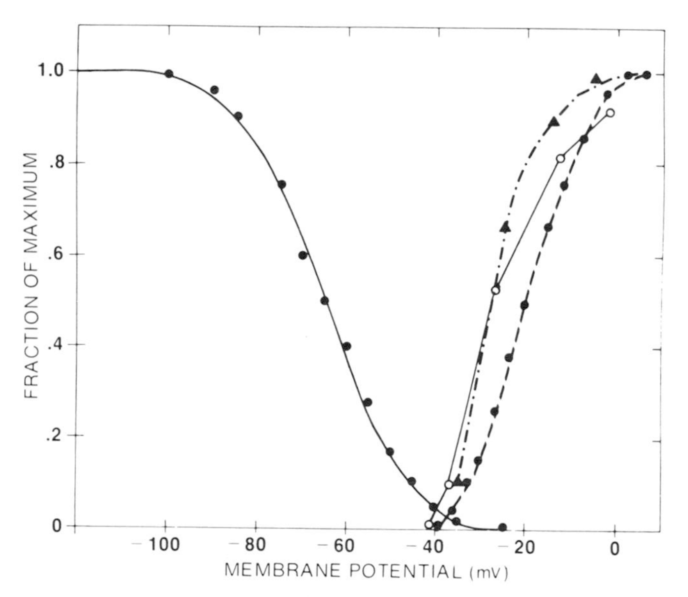

FIGURE 11. Comparison of the voltage dependence of scorpion toxin binding and sodium conductance activation and inactivation. Scorpion toxin binding was measured at five different membrane potentials (see Fig. 8). Fractional inhibition of specific scorpion toxin binding is plotted vs. membrane potential (○). Values of sodium conductance calculated from data in the literature are presented as the fraction of the maximum conductance observed in each experiment. (●——●) Inactivation of the peak sodium conductance during voltage steps from the potentials indicated on the abscissa to −10 mV, from the data of Moolenaar and Spector (1978) on clone N1E-115. (●----●) Activation of the peak sodium conductance during voltage steps from −85 mV to the potentials indicated on the abscissa, from the data of Moolenaar and Spector (1978) on clone N1E-115. (▲—·—▲) Activation of the peak sodium conductance during voltage steps from −100 mV to the potentials indicated on the abscissa, from the data of Veselovsky *et al.* (1977) on clone N18.

toxin binding is completely unrelated to the voltage dependence of inactivation. In contrast, there is a close correlation between the voltage dependence of scorpion toxin binding and the voltage dependence of activation of sodium channels. In Fig. 11, the peak sodium conductance during voltage clamp steps to different membrane potentials is plotted as a function of voltage from the data of Moolenaar and Spector (1978) on clone NlE-115 (●—●) and from the data of Veselovsky *et al.* (1977) on clone N18 (▲—·—▲). In each case, peak currents are essentially zero at −41 mV where scorpion toxin binding is maximum and increase to maximum values at −2 mV where scorpion toxin binding is 92% inhibited. Half-maximal peak currents are obtained at −28 mV for N18 and −21 mV for NlE-115. These values compare favorably to the half-maximal inhibition of scorpion toxin binding at −28 mV. This comparison reveals a strong correlation between the voltage dependence of scorpion toxin binding and the voltage dependence of activation of the peak sodium conductance in neuroblastoma cells. In view of these results, it seems likely that receptor site 3 is located on the primary voltage sensor of the sodium channel which responds to depolarization with a conformational change that leads to activation of the ion channel. Scorpion toxin binding measurements, therefore, may detect the same voltage-dependent change of state that is detected in gating current experiments.

Our current working hypothesis is that each of these three toxin receptor sites resides on a separate subunit of a multisubunit complex comprising the functional sodium channel. Our results suggest that these subunits undergo conformational changes that are dependent on voltage and are coupled through protein–protein interactions such that the conformational state of one subunit influences the conformational states of the others. This hypothesis emphasizes the fundamental similarities between sodium channels and allosteric regulatory enzymes. According to this view, essential steps in understanding the structure and function of the sodium channel are determination of its subunit composition and stoichiometry and delineation of the roles of the individual subunits in the function of the complex. Final resolution of this problem requires purification of sodium channels and reconstitution of function from purified components. However, a first step in this direction can be made by determining the binding stoichiometry of neurotoxins acting at separate receptor sites associated with sodium channels. Our first experiments on this problem have revealed that electrically excitable neuroblastoma cells have 2.8 saxitoxin receptor sites for each scorpion toxin receptor site (Catterall and Morrow, 1978). Although other interpretations have not been ruled out, we are most interested in the possibility that this binding stoichiometry represents the ratio of subunits in a multisubunit sodium channel complex. We are now engaged in improving currently available binding assay methods in order to verify this finding and extend our results to other excitable tissues.

If our interpretation of the 3-to-1 binding stoichiometry in neuroblastoma cells is correct, this finding has important implications for sodium channel function. Hodgkin and Huxley first showed that there was a delay in activation of sodium channels following a depolarization. A similar delay has been found in all excitable tissues studied. This delay has been interpreted in terms of a requirement for multiple changes of state to occur before activation of the ion channel and is represented in the model of Hodgkin and Huxley (1952) by the third-power dependence of activation in the formula m^3h. Our results suggest that these multiple events may be conformational changes in each of three subunits which bind saxitoxin and act together to form an active ion channel. These events would be initiated by protein–protein interactions with the primary voltage-sensitive subunit which binds scorpion toxin.

3.1.8. Sodium Channels in Other Neuronal Cell Lines

A number of other neuronal cell lines from the mouse C1300 neuroblastoma, from tumors of the rat central nervous system, and from human neuroblastomas have been studied by ion-flux procedures (Catterall and Nirenberg, 1973; Stallcup and Cohn, 1976a,b; Stallcup, 1977; West *et al.*, 1977a,b; Palfrey and Littauer, 1976). In each of these cell lines, veratridine increases sodium permeability at concentrations in the range of 50 to 200 μM. A receptor site for veratridine therefore seems to be universal among sodium channels in these different cell lines. Batrachotoxin, aconitine, and grayanotoxin have not been extensively studied with these cell lines.

In each case, the increase in sodium permeability caused by veratridine is inhibited by tetrodotoxin. Therefore, sodium channels from each of these sources have a receptor site for tetrodotoxin. However, the concentration of tetrodotoxin required to give half-maximal inhibition ($K_{0.5}$) varies over a wide range. In all mouse neuroblastoma cells from tumor C1300, the K_I for tetrodotoxin is between 5×10^{-9} M and 2×10^{-8} M (Catterall and Nirenberg, 1973). Neuronal lines from the rat central nervous system have values of K_I from 2×10^{-8} M to 2×10^{-6} M (Stallcup, 1977; West *et al.*, 1977a,b; Huang *et al.*, 1978). Human neuroblastoma lines have K_I values from 2×10^{-8} M to 8×10^{-7} M (Stallcup, 1977; West *et al.*, 1977a,b). Unfortunately, most of these experiments were not performed under conditions where the measured sodium flux was proportional to sodium permeability. Some of these values might therefore substantially overestimate the true K_I for tetrodotoxin. Quantitative studies have been carried out comparing clone N18 of mouse neuroblastoma cells and clone C9 of rat neuroblastoma cells (Huang *et al.*, 1978). These studies confirm that there is a 200-fold difference between the K_I values for tetrodotoxin in these two cell lines (N18, $K_I = 1 \times 10^{-8}$ M; C9, $K_I = 2 \times 10^{-6}$ M). The physiological significance of this wide range of affinities is uncertain. In contrast to the results with rat neuronal cell lines, sodium channels in adult rat brain synaptosomes have receptor sites with high affinity for tetrodotoxin (unpublished results). Therefore, sodium channels may change during development from having low-affinity receptor sites for tetrodotoxin to having high-affinity receptor sites for tetrodotoxin.

The effect of unfractionated scorpion venom on sodium channels in a variety of neuronal cell lines has been studied (Stallcup, 1977). Scorpion venom enhances veratridine activation of sodium channels in human and rat neuronal cell lines as described above for clone N18 of mouse neuroblastoma. Rat cell lines are greater than 100-fold less sensitive to the venom than mouse and human lines, however. This difference in toxin sensitivity may also be modulated during development, since synaptosomes from adult rat brain have 100-fold higher affinity for purified scorpion toxin than do clone C9 rat neuroblastoma cells (unpublished results; Ray *et al.*, 1978).

The experiments reviewed in this section show that neuronal cell lines from rat, mouse, and human origin all have three separate receptor sites for neurotoxins as described for N18 cells. There are differences in the affinities of these receptor sites for their ligands. It is possible that these differences in affinity are related to the developmental status of the different cell lines.

3.2. Skeletal Muscle Cells

Rat skeletal muscle clones L5, L6, and L8 have been studied by ion-flux procedures (Stallcup and Cohn, 1976a; Catterall, 1976b; Stallcup, 1977). These cell lines are electrically excitable and generate sodium-dependent action potentials as both myoblasts and myotubes (Kidokoro, 1973; 1975a,b; Land *et al.*, 1973). These action potentials are relatively insensitive to inhibition by tetrodotoxin. Veratridine, batrachotoxin, and aconitine increase sodium permeability and depolarize both myoblasts and myotubes (Catterall, 1976b; Stallcup and Cohn, 1976a; Sastre and Podleski, 1976; unpublished results). The concentration–effect curves for activation of sodium channels by alkaloid toxins are virtually identical to those for N18 cells (Catterall, 1976b). All three toxins compete for a common receptor site in activating sodium channels (Catterall, 1976b, unpublished results). Thus, sodium channels in rat muscle cells have a receptor site for alkaloid toxins with virtually identical properties to those described for neuroblastoma cells.

As expected from the electrophysiological results (Kidokoro, 1973, 1975a,b; Land *et al.*, 1973), sodium channels activated by alkaloid toxins are relatively insensitive to tetrodotoxin (K_I = 1 to 5 $\times$ 10^{-6} M) and saxitoxin (K_I = 1 to 5 $\times$ 10^{-7} M) (Catterall, 1976b; Stallcup and Cohn, 1976a; Sastre and Podleski, 1976). These observations support the view that the alkaloid toxins specifically activate sodium channels involved in generation of action potentials, since both action potentials and toxin-dependent sodium permeability have similar low sensitivity to tetrodotoxin. In muscle cells, however, it is probable that there are additional toxin-insensitive ion channels that carry inward current including calcium channels and "slow" sodium channels. These ion channels may play a major role in the action potential of cultured muscle cells (Kidokoro, 1973, 1975a).

Adult rat skeletal muscle has high affinity for tetrodotoxin and saxitoxin but, on denervation, action potentials become relatively insensitive to tetrodotoxin (Harris and Thesleff, 1971). Similarly, after denervation, the depolarization caused by batrachotoxin becomes insensitive to tetrodotoxin (Albuquerque and Warnick, 1972). Thus, clonal lines of rat skeletal muscle have sodium channels with pharmacological properties similar to those in denervated adult muscle.

Unfractionated scorpion venom (Stallcup, 1977) and purified scorpion toxin and sea anemone toxin (unpublished results) enhance the activation of sodium channels in rat muscle cells by alkaloid neurotoxins. Much higher concentrations of scorpion toxin are required than with mouse neuroblastoma cells. However, rat muscle cells have higher affinity for sea anemone toxin than do mouse neuroblastoma cells (unpublished results). Thus, sodium channels in rat muscle cells have a receptor site for the polypeptide toxins with altered specificity such that sea anemone toxin is bound substantially more tightly than scorpion toxin.

The results with the clonal lines of rat muscle cells show that these cells have three separate receptor sites for neurotoxins. The properties of these receptor sites are generally similar to those described above for sodium channels in clone N18 neuroblastoma cells. There are allosteric interactions between the receptor site for the alkaloid toxins and the receptor site for the polypeptide toxins. These interactions modulate activation of the ion channels. Tetrodotoxin and saxitoxin noncompetitively block the ion channel. However, the affinities and specificities of these receptor sites differ markedly from those in N18. Comparison of the results on cultured muscle cells with electrophysiological data on normal and denervated adult muscle suggests that the properties of the receptor site for tetrodotoxin and saxitoxin and the receptor site for the polypeptide toxins may be altered by developmental and physiological events.

3.3. Heart Muscle Cells

The electrophysiological properties of embryonic chick heart cells change markedly during development (Sperelakis and Shigenobu, 1972). Before day 5 *in ovo,* beating is insensitive to tetrodotoxin, whereas in older embryonic hearts, beating is inhibited by tetrodotoxin. Embryonic chick heart cells can be cultured in monolayer *in vitro* (Sperelakis, 1972). The beating of these cells is insensitive to tetrodotoxin regardless of the age of the embryo from which they were obtained (Sperelakis, 1972; McDonald *et al.,* 1972). Therefore, tetrodotoxin-sensitive sodium channels are not required for beating in these cultured cells. Despite the insensitivity of beating to tetrodotoxin, veratridine and batrachotoxin increase sodium uptake by these cells in ion-flux assays (Galper and Catterall, 1975, 1978; Fosset *et al.,* 1977), and the increased ion flux is inhibited by low concentrations of tetrodotoxin (K_I = 1 to 6 nM). These results show that embryonic heart cells in culture have sodium channels with the pharmacological properties of sodium channels in adult heart, but the ion channels in cultured heart cells are physiologically inactive. These ion channels may be inactivated at the resting membrane potential of the heart muscle cells and therefore are not activated by depolarization during beating. Since alkaloid neurotoxins block the process of inactivation in nerve (Schmidt and Schmitt, 1974; Khodorov, 1978), these toxins can activate the ion channels pharmacologically.

Both sea anemone toxin and scorpion toxin also increase $^{22}Na^+$ uptake of cultured heart cells (Couraud *et al.,* 1976; DeBarry *et al.,* 1977). These increases in $^{22}Na^+$ uptake are blocked by low tetrodotoxin concentrations (K_I = 10 nM), indicating that they are mediated by sodium channels. These results are in marked contrast to experiments with skeletal muscle cells and neuronal cells where these two polypeptide toxins have no effect on sodium permeability when tested alone but enhance the activation of sodium channels by alkaloid toxins (Catterall, 1976a; Catterall and Beress, 1978; unpublished results). It seems likely that this apparent discrepancy results from the spontaneous electrical activity of the heart muscle cells. Since the studies on heart cells were carried out in physiological solution, the heart cells continued to beat during the flux measurement. The polypeptide toxins therefore need only to allow or enhance activation of sodium channels by the depolarizations during beating in order to increase $^{22}Na^+$ uptake. They do not need to activate sodium channels by a direct allosteric interaction. Ion flux measurements under conditions where beating is inhibited and the measured flux is linearly proportional to permeability are required to analyze the mechanism of toxin activation.

Both veratridine and sea anemone toxin have been shown to increase $^{45}Ca^{2+}$ uptake in heart muscle cells (Fosset *et al.,* 1977; DeBarry *et al.,* 1977). The increase in each case is blocked by tetrodotoxin and is dependent on the presence of Na^+ in the medium. Since Ca^{2+} is not transported by sodium channels, these characteristics indicate that the Ca^{2+} uptake is caused by depolarization and/or Na^+ entry and is mediated by other transport processes specific for Ca^{2+}. Probably both slow Ca^{2+}/Na^+ channels and the Ca^{2+}–Na^+ exchange transport system are involved. The toxins are therefore only indirect effectors of Ca^{2+} permeability of cultured heart cells.

The results with cultured heart cells, although still incomplete, suggest that these cells have sodium channels with pharmacological properties similar to those in nerve and neuroblastoma cells. Veratridine, batrachotoxin, scorpion toxin, and sea anemone toxin enhance activation of the sodium channels. Tetrodotoxin inhibits sodium transport. The affinity of the neurotoxin receptor sites for these toxins is similar to that of receptor sites in clone N18 neuroblastoma cells.

4. CONCLUSION

The emphasis in these experiments has been on understanding the action of neurotoxins that alter sodium channel properties. Detailed understanding of the mechanism of action of these toxins is necessary in order to use them as tools to learn more about the structure and function of sodium channels and to investigate the changes in sodium channel properties occurring during differentiation and physiological function of excitable cells. It now appears that enough is known about the mechanism of toxin action to allow them to be of value in analyzing a number of physiological processes in which sodium channels play an important role. Current work in our laboratory is proceeding in that direction.

REFERENCES

Albuquerque, E. X., and Warnick, J., 1972, The pharmacology of batrachotoxin. IV. Interaction with tetrodotoxin on innervated and chronically denervated rat skeletal muscle, *J. Pharmacol. Exp. Ther.* **180:**683–697.

Albuquerque, E. X., Daly, J. W., and Witkop, B., 1971, Batrachotoxin: Chemistry and pharmacology, *Science* **172:**995–1002.

Altendorf, K., Harold, F. M., Simoni, R. D., 1974, Impairment and restoration of the energized state in membrane vesicles of a mutant of *Escherichia coli* lacking adenosine triphosphate, *J. Biol. Chem.* **249:**4587–4593.

Bergman, C., Dubois, J. M., Rojas, E., and Rathmeyer, W., 1976, Decreased rate of sodium conductance inactivation in the node of Ranvier induced by a polypeptide toxin from *Anemonia sulcata, Biochim. Biophys. Acta* **445:**173–184.

Bolton, A. E., and Hunter, W. M., 1973, The labelling of proteins to high specific radioactivities by conjugation to a ^{125}I-containing acylating agent, *Biochem. J.* **133:**529–539.

Bulloch, K., Stallcup, W. B., and Cohn, M., 1977, The derivation and characterization of neuronal cell lines from rat and mouse brain, *Brain Res.* **135:**25–36.

Burgermeister, W., Catterall, W. A., and Witkot, B., 1977, Histrionicotoxin enhances agonist-induced desensitization of acetylcholine receptor, *Proc. Natl. Acad. Sci. USA* **74:**5754–5758.

Catterall, W. A., 1975a, Activation of the action potential sodium ionophore by veratridine and batrachotoxin, *J. Biol. Chem.* **250:**4053–4059.

Catterall, W. A., 1975b, Cooperative activation of the action potential sodium ionophore by neurotoxins, *Proc. Natl. Acad. Sci. USA* **72:**1782–1786.

Catterall, W. A., 1975c, Sodium transport by the acetylcholine receptor of cultured muscle cells, *J. Biol. Chem.* **250:**1776–1781.

Catterall, W. A., 1976a, Purification of a toxic protein from scorpion venom which activates the action potential Na^+ ionophore, *J. Biol. Chem.* **251:**5528–5536.

Catterall, W. A., 1976b, Activation and inhibition of the action potential Na^+ ionophore of cultured rat muscle cells by neurotoxins, *Biochem. Biophys. Res. Commun.* **68:**136–142.

Catterall, W. A., 1977a, Membrane potential dependent binding of scorpion toxin to the action potential Na^+ ionophore. Studies with a toxin derivative prepared by lactoperoxidase catalyzed iodination, *J. Biol. Chem.* **252**:8660–8668.

Catterall, W. A., 1977b, Activation of the action potential Na^+ ionophore by neurotoxins. An allosteric model, *J. Biol. Chem.* **252**:8669–8676.

Catterall, W. A., and Beress, L., 1978, Sea anemone toxin and scorpion toxin share a common receptor site associated with the action potential sodium ionophore, *J. Biol. Chem.* **253**:7393–7396.

Catterall, W. A., and Morrow, C. S., 1978, Binding of saxitoxin to electrically excitable neuroblastoma cells, *Proc. Natl. Acad. Sci. USA* **75**:218–222.

Catterall, W. A., and Nirenberg, M., 1973, Sodium uptake associated with activation of action potential ionophores of cultured nerve and muscle cells, *Proc. Natl. Acad. Sci. USA* **70**:3759–3763.

Catterall, W. A., Ray, R., and Morrow, C. S., 1976, Membrane potential dependent binding of scorpion toxin to the action potential sodium ionophore, *Proc. Natl. Acad. Sci. USA* **73**:2682–2686.

Conti, F., Hille, B., Neumcke, B., Nonner, W., and Stampfli, R., 1976, Conductance of the sodium channel in myelinated nerve fibers with modified sodium inactivation, *J. Physiol. (Lond.)* **262**:729–742.

Couraud, F., Rochat, H., and Lissitzky, S., 1976, Stimulation of sodium and calcium uptake by scorpion toxin in chick embryo heart cells, *Biochim. Biophys. Acta* **433**:90–100.

DeBarry, J., Fosset, M., and Lazdunski, M., 1977, Molecular mechanism of the cardiotoxic action of a polypeptide neurotoxin from sea anemone on cultured embryonic cardiac cells, *Biochemistry* **16**:3850–3855.

Evans, M. H., 1972, Tetrodotoxin, saxitoxin, and related substances: Their applications in neurobiology, *Int. Rev. Neurobiol.* **15**:83–166.

Fosset, M., DeBarry, J., Lenoir, M.-C., and Lazdunski, M., 1977, Analysis of molecular aspects of Na^+ and Ca^{++} uptakes by embryonic cardiac cells in culture, *J. Biol. Chem.* **252**:6112–6117.

Galper, J., and Catterall, W. A., 1975, Developmental changes in the sensitivity of embryonic heart cells to tetrodotoxin and D600, *J. Cell Biol.* **67**:128a.

Galper, J., and Catterall, W. A., 1978, Developmental changes in the sensitivity of embryonic heart cells to tetrodotoxin and D600, *Dev. Biol.* **65**:216–227.

Goldman, D. E., 1943, Potential, impedance, and rectification in membranes, *J. Gen. Physiol.* **27**:37–60.

Grinius, L. L., Jasaitis, A. A., Kadziauskas, Y. P., Liberman, E. A., Skulachev, V. P., Topali, V. P., Tsofina, L. M., and Vladimirova, M. A., 1970, Conversion of biomembrane-produced energy into electric form, *Biochim. Biophys. Acta* **216**:1–12.

Harris, J. B., and Thesleff, S., 1971, Studies on tetrodotoxin resistant action potentials in denervated skeletal muscle, *Acta Physiol. Scand.* **83**:382–388.

Herzog, W. H., Feibel, R. M., and Bryant, S. H., 1964, The effect of aconitine on the giant axon of the squid, *J. Gen. Physiol.* **47**:719–733.

Hille, B., 1968, Pharmacological modifications of the sodium channels of frog nerve, *J. Gen. Physiol.* **51**:199–219.

Hille, B., 1975, Ionic selectivity, saturation, and block in sodium channels, *J. Gen. Physiol.* **66**:535–560.

Hirs, C. H. W., 1955, Chromatography of enzymes on ion exchange resins, *Meth. Enzymol.* **1**:113–125.

Hodgkin, A. L., and Huxley, A. F., 1952, A quantitative description of membrane current and its application to conduction and excitation in nerve, *J. Physiol. (Lond.)* **117**:500–544.

Hodgkin, A. L., and Katz, B., 1949, The effect of sodium ions on the electrical activity of the giant axon of the squid, *J. Physiol. (Lond.)* **108**:37–77.

Huang, L. M., Catterall, W. A., and Ehrenstein, G., 1978, Selectivity of cations and nonelectrolytes for acetylcholine-activated channels in cultured muscle cells, *J. Gen. Physiol.* **71**:397–410.

Khodorov, B. I., 1978, Chemicals as tools to study nerve fiber sodium channels and effects of batrachotoxin and local anesthetics, in: *Membrane Transport Processes, Vol. II* (D. C. Tosteson, Yu. A. Ovchinnikov, and R. Latorre, eds.), pp. 153–174, Raven Press, New York.

Kidokoro, Y., 1973, Development of action potentials in a clonal rat skeletal muscle cell line, *Nature* **241**:158–159.

Kidokoro, Y., 1975a, Developmental changes of membrane electrical properties in a rat skeletal muscle cell line, *J. Physiol. (Lond.)* **244**:129–143.

Kidokoro, Y., 1975b, Sodium and calcium components of the action potential in a developing skeletal muscle cell line, *J. Physiol. (Lond.)* **244**:145–159.

Koppenhöfer, E., and Schmidt, H., 1968, Die Wirkung von Scorpiongift auf die Ionenströme des Ranvierschen Schnurrings. II. Unvollständige Natrium-Inaktivierung, *Pflügers Arch.* **303**:150–161.

Land, B. R., Sastre, A., and Podleski, T. R., 1973, Tetrodotoxin-sensitive and insensitive action potentials in myotubes, *J. Cell. Physiol.* **82**:497–510.

McDonald, T. F., Sachs, H. G., and DeHaan, R. L., 1972, Development of sensitivity to tetrodotoxin in beating chick embryo hearts, single cells, and aggegrates, *Science* **176**:1248–1250.

Miranda, F., Kupeyan, C., Rochat, H., Rochat, C., and Lissitzky, S., 1970, Purification of animal neurotoxins, *Eur. J. Biochem.* **16**:514–523.

Monod, J., Wyman, J., and Changeux, J.-P., 1965, On the nature of allosteric transitions: A plausible model, *J. Mol. Biol.* **12**:88–118.

Moolenaar, W. H., and Spector, I., 1977, Membrane currents examined under voltage clamp in cultured neuroblastoma cells, *Science* **196**:331–333.

Moolenaar, W. H., and Spector, I., 1978, Ionic currents in cultured mouse neuroblastoma cells under voltage-clamp conditions, *J. Physiol. (Lond.)* **278**:265–286.

Narahashi, T., and Seyama, I., 1974, Mechanism of nerve membrane depolarization caused by grayanotoxin I, *J. Physiol. (Lond.)* **242**:471–487.

Narahashi, T., Moore, J. W., and Scott, W. R., 1964, Tetrodotoxin blockage of sodium conductance increase in lobster giant axons, *J. Gen. Physiol.* **47**:965–974.

Narahashi, T., Moore, J. W., and Shapiro, B. I., 1969, Condylactis toxin: Interaction with nerve membrane ionic conductances, *Science* **163**:680–681.

Narahashi, T., Shapiro, B. I., Deguchi, T., Scuka, M., and Wang, C. M., 1972, Effects of scorpion venom on squid axon membranes, *Am. J. Physiol.* **222**:850–856.

Nelson, P. G., Peacock, J. H., Amano, T., and Minna, J., 1971, Electrogenesis in mouse neuroblastoma cells in vitro, *J. Cell. Physiol.* **77**:337–352.

Nichols, D. G., 1974, Hamster brown adipose tissue mitochondira. The control of respiration and the proton electrochemical potential gradient by possible physiological effectors of the proton conductance of the inner membrane, *Eur. J. Biochem.* **49**:573–583.

Ohta, M., Narahashi, T., and Keeler, R., 1973, Effects of veratrum alkaloids on membrane potential and conductance of squid and crayfish giant axons, *J. Pharmacol. Exp. Ther.* **184**:143–154.

Okamoto, H., Takahashi, K., and Yamashita, N., 1977, One-to-one binding of a purified scorpion toxin to Na^+ channels, *Nature* **266**:465–468.

Palfrey, C., and Littauer, U., 1976, Sodium-dependent efflux of K^+ and Rb^+ through the activated sodium channel of neuroblastoma cells, *Biochem. Biophys. Res. Commun.* **72**:209–215.

Patrick, J., and Stallcup, W., 1977, Bungarotoxin binding and cholinergic receptor function on a rat sympathetic nerve line, *J. Biol. Chem.* **252**:8629–8633.

Ray, R., and Catterall, W. A., 1978, Membrane potential dependent binding of scorpion toxin to the action potential sodium ionophore. Studies with a 3-(4-hydroxy 3-[^{125}I]iodophenyl) propionyl derivative, *J. Neurochem.* **31**:397–407.

Ray, R., Morrow, C. S., and Catterall, W. A., 1978, Binding of scorpion toxin to receptor sites associated with voltage-sensitive sodium channels in synaptic nerve ending particles, *J. Biol. Chem.* **253:**7307–7313.

Ritchie, J. M., Rogart, R. B., and Strichartz, G. R., 1976, A new method for labelling saxitoxin and its binding to non-myelinated fibers of the rabbit vagus, lobster walking leg, and garfish olfactory nerves, *J. Physiol. (Lond.)* **261:**477–494.

Romey, G., Chicheportiche, R., Lazdunski, M., Rochat, H., Miranda, F., and Lissitzky, S., 1975, Scorpion neurotoxin—A presynaptic toxin which affects both Na^+ and K^+ channels in axons, *Biochem. Biophys. Res. Commun.* **64:**115–121.

Romey, G., Abita, J. P., Schweitz, H., Wunderer, G., and Lazdunski, M., 1976, Sea anemone toxin: A tool to study molecular mechanisms of nerve conduction and excitation–secretion coupling, *Proc. Natl. Acad. Sci. USA* **73:**4055–4059.

Sastre, A., and Podleski, T. R., 1976, Pharmacologic characterization of the sodium ionophores in L6 myotubes, *Proc. Natl. Acad. Sci. USA* **73:**1355–1359.

Schmidt, H., and Schmitt, O., 1974, Effect of aconitine on the sodium permeability of the node of Ranvier, *Pflügers Arch.* **394:**133–148.

Schuldiner, S., and Kaback, H. R., 1975, Membrane potential and active transport in membrane vesicles from *Escherichia coli, Biochemistry* **14:**5451–5460.

Seyama, I., and Narahashi, T., 1973, Increase in sodium permeability of squid axon membranes by α-dihydro grayanotoxin II, *J. Pharmacol. Exp. Ther.* **184:**299–307.

Spector, I., Kimhi, Y., and Nelson, P. G., 1973, Tetrodotoxin and cobalt blockade of neuroblastoma action potentials, *Nature* **246:**124–126.

Sperelakis, N., 1972, Electrical properties of embryonic heart cells, in: *Electrical Phenomena in the Heart* (W. De Mello, ed.), pp. 1–56, Academic Press, New York.

Sperelakis, N., and Shigenobu, K., 1972, Changes of membrane properties of chick embryonic hearts during development, *J. Gen. Physiol.* **60:**430–453.

Stallcup, W. B., 1977, Comparative pharmacology of voltage-dependent sodium channels, *Brain Res.* **135:**37–53.

Stallcup, W. B., and Cohn, M., 1976a, Electrical properties of a clonal cell line as determined by measurement of ion fluxes, *Exp. Cell Res.* **98:**277–284.

Stallcup, W. B., and Cohn, M., 1976b, Correlation of surface antigens and cell type in cloned cell lines from the rat central nervous system, *Exp. Cell Res.* **98:**285–297.

Straub, R., 1956, Die Wirkung von Veratridin und Ionen auf das Ruhepotential markhaltiger Nervenfasern des Frosches, *Helv. Physiol. Acta* **14:**1–28.

Ulbricht, W., 1969, The effect of veratridine on excitable membranes of nerve and muscle, *Ergeb. Physiol. Biol. Chem. Exp. Pharmakol.* **61:**18–71.

Veselovsky, N. S., Kostyuk, P. G., Krishtal, D. A., Naumov, A. P., and Pidoplichko, V. I., 1977, Ionic currents in the membrane of neuroblastoma cells, *Neurofysiology* **9:**641–643.

Villegas, J., Sevcik, C., Barnola, F. V., and Villegas, R., 1976, Grayanotoxin, veratridine, and tetrodotoxin-sensitive sodium pathways in the Schwann cell membrane of squid nerve fibers, *J. Gen. Physiol.* **67:**369–380.

West, G. J., Uki, J., Stahn, R., and Herschman, H., 1977a, Neurochemical properties of cell lines from N-ethyl-N-nitroso urea induced rat tumors, *Brain Res.* **130:**387–392.

West, G. J., Uki, J., Herschman, H., and Suger, R. C., 1977b, Adrenergic, cholinergic, and inactive human neuroblastoma cell lines with the action potential sodium ionophore, *Cancer Res.* **37:**1372–1376.

9

Electrophysiological Properties of Developing Skeletal Muscle Cells in Culture

YOSHIAKI KIDOKORO

1. INTRODUCTION

Various electrophysiological properties of skeletal muscles have been extensively studied in adult vertebrates (Hodgkin and Horowicz, 1959; Nakajima *et al.*, 1962; Adrian *et al.*, 1970; Campbell and Hille, 1976; Campbell, 1976; Beaty and Stefani, 1975; Sanchez and Stefani, 1978). Characteristics of ion channels activated by a transmitter, acetylcholine, have also been the subject of exhaustive studies (Katz and Miledi, 1972; Anderson and Stevens, 1973; Neher and Sakmann, 1976; Takeuchi and Takeuchi, 1960; Maeno *et al.*, 1977). All of these distinct membrane properties may be regulated by and expressed through large protein molecules that reside in the lipid bilayer of the sarcolemma (Singer and Nicholson, 1972). These molecules may be inserted into the membrane *de novo* or may be activated during membrane differentiation at certain developmental stages.

Little is known, however, about how these membrane properties are expressed during development. Technical difficulties impede electrophysiological studies of embryonic muscle membranes. In early embryonic stages, muscle cells are difficult to identify *in situ* and are small and too fragile for microelectrode penetration. One way to circumvent these difficulties is to choose an animal that has large skeletal muscle fibers even in early embryonic stages (e.g., tunicates; Takahashi *et al.*, 1971; Miyazaki *et al.*, 1972). Another is to adopt cell-culture techniques in order to visualize and identify early embryonic muscle cells.

YOSHIAKI KIDOKORO • The Salk Institute, San Diego, California 92138.

Mammalian skeletal muscle cells differentiate in a unique sequence. Myogenic precursor cells (myoblasts), which are obtained either from embryonic muscle tissue (primary muscle cultures) or from muscle cell lines, divide exponentially in culture. After cessation of division, myoblasts start to fuse, forming multinucleate myotubes. Subsequently, the myotubes develop myofilaments and cross striations, and the nuclei migrate to the periphery (Fischman, 1972).

In this chapter, various electrophysiological properties studied in cell culture will be compared among myoblasts, multinucleate myotubes, and adult muscle fibers. Several remarkable changes in membrane properties occur during this period of development, and many of these properties are affected by neural factors. Innervation, therefore, seems to play an important part in the development and maturation of muscle membrane properties.

2. METHODS AND MATERIALS

2.1. Cell Cultures

2.1.1. Primary Cultures

The following procedures were used to obtain myoblasts from embryonic rat muscle tissues (Kidokoro, 1980). Other investigators have used different methods to dissociate muscle tissues (Fischbach *et al.,* 1971; Kano and Shimada, 1971a,b; Fambrough and Rash, 1971), but they are basically similar to the methods described below. Muscle tissue was dissected from the legs of 18- to 21-day-old rat fetuses. After the skin, bone, and connective tissue had been removed, the muscle tissue was placed in saline that was free of Ca and Mg and contained NaCl, 137 mM; KCl, 5 mM; Na_2HPO_4, 0.7 mM; HEPES, 25 mM; collagenase, 0.4% w/v (Worthington); and DNase, 10 μg/ml (Sigma) (Vale and Grant, 1975). After approximately 1 hr of enzyme treatment at 37°C, the muscle tissue was returned to the normal culture medium [Dulbecco modified Eagle's medium (DMEM) + 10% fetal calf serum] and triturated several times to dissociate the cells. A total of about 10^5 cells were usually obtained from four legs of one 19-day-old fetus. These cells were suspended in 100 ml of normal culture medium. Cells were then plated in either 35-mm or 60-mm tissue culture dishes with normal culture medium (approximately 1 to 3×10^4 cells per dish) and incubated at 37°C in an atmosphere of 12% CO_2 and 88% air. Myoblasts started to fuse after 2 to 3 days in culture. As soon as small myotubes were formed, cytosine arabinoside (5×10^{-5} M Ara-C, Sigma) was added to the culture medium to kill rapidly dividing cells (fibroblasts and myoblasts) and to prevent further growth of myotubes. Cells were kept in Ara-C for 2 to 3 days and then returned to the normal culture medium.

This procedure results in small myotubes (less than 500 μm in length). Small myotubes are preferable in some experiments since they have high input impedances. Thus, when currents are passed to examine membrane electrical properties, isopotential conditions inside the cell are well satisfied (Kidokoro, 1975a).

2.1.2. Clonal Cell Lines

Various skeletal muscle cell lines are now available. Skeletal muscle cell line L6 is originated from rat thigh muscles (Yaffe, 1968), and H9 (Kimes and Brandt, 1975) and G8 (Christian *et al.*, 1977) are derived from rat and mouse muscles, respectively. The following procedures were used with the L6 cell line for the study of membrane electrical properties (Kidokoro, 1973, 1975a,b). Myoblasts (approximately 5×10^3/ml) were grown in plastic tissue culture dishes in DMEM containing 10% fetal calf serum at 37°C in an atmosphere of 12% CO_2 and 88% air (Vogt and Dulbecco, 1963). Myoblasts started to fuse to form myotubes as soon as they became confluent. In some cases, the Ara-C treatment was used to produce small myotubes (see Section 2.1.1). Some of the myotubes became partially striated and contracted spontaneously.

2.2. Electrophysiology

The following procedures were used during studies of membrane electrical properties of L6 cells (Kidokoro, 1973, 1975a,b). Cells for electrophysiological experiments were grown attached to a round cover glass (diameter 25 mm) at the bottom of a tissue culture dish (diameter 60 mm). After an appropriate time of incubation, the cover glass was removed and placed in a glass-bottomed chamber containing about 3 ml of a saline of the following composition: NaCl, 157.0 mM; $CaCl_2$, 1.8 mM; KCl, 5.6 mM; and Tris-HCl, 2.0 mM (pH 7.4). In some experiments, cells grown in a tissue culture dish were used *in situ*. Myoblasts were examined at least 2–3 days after they attached to the bottom of the culture dish. The cells were observed with a 40 × water-immersion phase contrast objective and 10 × ocular lens (Carl Zeiss, West Germany). Glass microelectrodes were filled with 3 M KCl and had resistances of 80–150 MΩ. Usually two electrodes were placed inside the same cell, one for passing current and the other for recording potential. Each was connected to the input stage of a preamplifier (M-4A, W. P. Instruments). Sometimes a single electrode was used both to record and pass current in order to reduce excess cell damage from penetration by a second electrode. For this purpose, the electrode was connected to the bridge circuit of the amplifier. The current passed through the electrode was measured with a conventional current–voltage converter, having a feedback resistance of either 1 or 10 MΩ, inserted between a Ag–AgCl pellet in the bath and ground. When the external Cl^- concentration was changed, an agar–3 M KCl bridge was used between the bath and the Ag–AgCl pellet.

All experiments were carried out at room temperature (20–24°C).

3. RESULTS

Membrane electrical properties of muscle cells in culture have been studied in various systems. Primary muscle cultures from chick, rat, and mouse embryos have generally been used. Clonal muscle cell lines have also been studied.

It is of interest to compare the membrane electrical properties of adult skeletal muscles with those of myoblasts and myotubes grown *in vitro*. Several differences have been described that may be attributable to developmental processes.

3.1. Resting Membrane Potential

The early developmental changes in the resting potential have been studied in chick skeletal muscle (Fischbach *et al.*, 1971; Dryden *et al.*, 1974; Ritchie and Fambrough, 1975; Spector and Prives, 1977). Fischbach *et al.* (1971) reported a small resting potential (−10 to −15 mV) in mononucleate muscle precursor cells and a larger potential (−60 to −80 mV) in older multinucleate myotubes. Similar results have been reported in rat skeletal muscle in culture where, even after fusion, the resting potential was small, −24 mV, in young myotubes (3 days in culture), whereas in 9-day cultured myotubes it was −51 mV (Ritchie and Fambrough, 1975). Slightly larger resting membrane potentials were reported by Obata (1974): about −40 mV in rat primary myoblasts and young myotubes, and about −60 mV in old myotubes. In contrast to these small resting potentials in chick or rat primary myoblasts, a large value (−71 mV) was reported in myoblasts of an L6 rat skeletal muscle cell line. No significant change in the resting potential was observed as myoblasts fused to form multinucleate myotubes (−69 mV) (Kidokoro, 1975a).

The resting membrane potentials of multinucleate myotubes were examined by various investigators in chick (Fischbach *et al.*, 1971; J. Harris *et al.*, 1973; Dryden *et al.*, 1974; Ritchie and Fambrough, 1975; Kano, 1975a; Spector and Prives, 1977; Engelhardt *et al.*, 1977a), mouse (Christian *et al.*, 1977), and rat primary cultures (Obata, 1974; Ritchie and Fambrough, 1975). The mean resting membrane potential in these reports ranged between −46 mV and −72 mV. Myotubes from mouse skeletal muscle cell line G8 (Christian *et al.*, 1977) were reported to have a resting membrane potential of −49 mV.

While there is a reasonable agreement in the resting membrane potential of multinucleate myotubes, a wide range of values for the resting membrane potentials of myoblasts or young myotubes has been reported. Whether or not there is a progressive change in the resting membrane potential as myoblasts fuse to form myotubes in primary cultures is not yet clear. Alternative explanations for the small resting potential in myoblasts and young myotubes have to be eliminated before we can make this conclusion (see Section 4).

The mean resting membrane potential of primary rat myotubes was reported to be −51 mV (Ritchie and Fambrough, 1975) and −60 mV (Obata, 1974). These values are comparable to that of chronically denervated adult skeletal muscle fibers, −57 mV (Albuquerque and Thesleff, 1968), whereas the mean resting membrane potential of adult innervated rat extensor muscle fibers is −72 mV. Innervation, therefore, seems to affect the muscle resting potential (Albuquerque and Thesleff, 1968; Hasegawa and Kuromi, 1977).

3.2. Ionic Mechanism of the Resting Membrane Potential

Ritchie and Fambrough (1975) examined the ionic mechanism of the resting membrane potential of developing rat and chick myotubes. They changed K concentration in the bathing medium and monitored changes in the resting potential. The data were fitted to the Goldman–Hodgkin–Katz equation (Hodgkin, 1951). They concluded that the P_{Na}/P_K ratio changes progressively during development, from 0.40 in 3-day-old cultures to 0.06 in 9-day-old cultures (P_{Na} and P_K are the permeabilities of the membrane to Na and K ions, respectively). In contrast to this finding, which suggests a high resting Na conductance in young myotubes, Kidokoro (1975a) did not find significant contribution of Na ions to the resting potential of L6 myoblasts or myotubes: the resting membrane potential in Na-free saline was not different from that in normal saline. The resting membrane is predominantly permeable to K ions, both in myoblasts and in myotubes. When the external K ion concentration was varied, the resting membrane potential of myoblasts and myotubes changed with slopes of 50 and 52 mV per decade change of the external K concentration, respectively (Fig. 1). These values are close to the theoretical value (58 mV) for a K electrode.

Interestingly, membrane permeability to Cl ions was found to be small in L6 myotubes. When the bathing solution was changed from normal to Cl-free saline (Cl was replaced by methane sulfonate), the resting potential did not change significantly, and only a slight increase in effective input resistance was noticed (Kidokoro, 1975a). A low Cl permeability was also reported in rat (Ritchie and Fambrough, 1975) and chick myotubes (Engelhardt *et al.*, 1978). This is in contrast to the large Cl permeability found in adult skeletal muscle (Hodgkin and Horowicz, 1959; Hutter and Warner, 1967; Bryant and Morales-Aguilera, 1971; Palade and Barchi, 1977). In the rat diaphragm, for example, 85% of the resting membrane ion conductance has been attributed to Cl (Palade and Barchi, 1977).

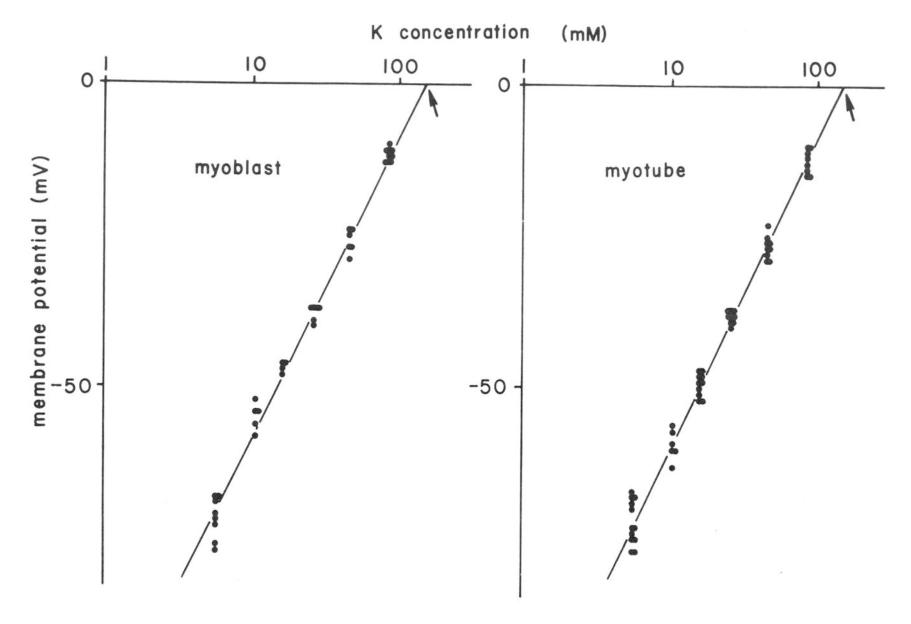

FIGURE 1. Relation between external K concentrations and resting membrane potentials in myoblasts and myotubes. The K concentration is plotted on abscissa in a logarithmic scale. Straight lines were drawn by eye. Arrow indicates an estimate of internal K concentration. (From Kidokoro, 1975a, reprinted with permission.)

3.3. Passive Membrane Electrical Properties

Membrane resistance and capacitance of chick primary myotubes are reported to be 2.5 $k\Omega \cdot cm^2$ and 3.9 $\mu F/cm^2$, respectively (Fischbach *et al.,* 1971). Similar values were reported by others for chick primary myotubes (Engelhardt *et al.,* 1977a). Those for primary mouse myotubes are 0.7 $k\Omega \cdot cm^2$ and 8.4 $\mu F/cm^2$, respectively (Powell and Fambrough, 1973) or 1.9 $k\Omega \cdot cm^2$ and 4.3 $\mu F/cm^2$, respectively (Christian *et al.,* 1977). In a mouse skeletal muscle cell line (G8) the corresponding values are 2.6 $k\Omega \cdot cm^2$ and 4.1 $\mu F/cm^2$ (Christian *et al.,* 1977).

The specific membrane capacitance of the nerve membrane has been found to be about 1 $\mu F/cm^2$ (Cole and Curtis, 1950). This is in contrast to the large values reported for myotubes which may be attributable to the development of the transverse tubular system after fusion (Ezerman and Ishikawa, 1967). The transverse tubular system is known to contribute to a large membrane capacitance in adult frog muscles (Gage and Eisenberg, 1969). In agreement with this idea, Kidokoro (1975a) found in a rat skeletal muscle cell line (L6) a small membrane capacitance (1 $\mu F/cm^2$) in myoblasts and a large value (5 $\mu F/cm^2$) for myotubes. This increase of the membrane capacitance after fusion of myoblasts supports the theory that the large membrane capacitance in myotubes results from the newly developed transverse tubular system.

The large membrane resistance in both myoblasts (8 $k\Omega \cdot cm^2$) and myotubes (12 $k\Omega \cdot cm^2$) of the L6 muscle cell line was attributed to a low Cl permeability (Kidokoro, 1975a). Recently, Engelhardt *et al.* (1976, 1977a,b) showed that chick myotubes have a smaller membrane resistance when cultured with spinal cord explants or when the culture medium contained extracts from spinal cords. This suggests that the smaller membrane resistance observed in adult skeletal muscles may have been induced by the nerve (see Section 4).

3.4. Action Potentials

Small, nonovershooting action potentials were found in exponentially dividing L6 myoblasts (Kidokoro, 1973, 1975a,b; Figs. 2A1 and 3A). From studies of the effects of varying the external ion concentrations, it was concluded that these action potentials result mainly from an inward current carried by Na ions. Because of cell homogeneity in the clonal cell line L6, radioactive Na ion flux studies could be performed (Stallcup and Cohn, 1976). In L6 myoblasts, Na flux was activated by veratridine, an agent which is known to depolarize cells by increasing Na permeability through voltage-dependent channels (Ohta *et al.*, 1973). This veratridine-stimulated Na flux was blocked by high concentrations of tetrodotoxin (TTX). So far, however, no action potentials have been reported to occur in primary myoblasts.

As soon as myoblasts have fused and formed multinucleate myotubes, action potentials become larger and overshooting (Fig. 2B). In agreement with this electrophysiological finding, the veratridine-activated Na flux in these multinucleate myotubes increases dramatically after fusion (Stallcup and Cohn, 1976). Usually action potentials were evoked after the cessation of a hyperpolarizing current. At the level of the resting potential (-50 to -60 mV) Na inward current may be inactivated, and thus depolarization would not generate an action potential. Hyperpolarizing the membrane may reduce or eliminate this inactivation, allowing a regenerative action potential to be evoked upon cessation of the hyperpolarizing pulse. Therefore, the amount of overshoot and the maximum rate of rise of an action potential depend on the level of the preceding membrane potential. When the maximum rate of rise of an action potential was plotted against the membrane potential, a sigmoidal relationship was obtained (Fig. 4) that was analogous to the inactivation curve for Na inward current described in studies of the squid giant axon (Hodgkin and Huxley, 1952). With a higher concentration (10 mM) of Ca ions in the external medium, the curve shifted to a more positive value along the voltage axis (Fig. 4). This is most likely because of a stabilizing action of Ca ions (Frankenheuser and Hodgkin, 1957). If one assumes that the cell surface is negatively charged,

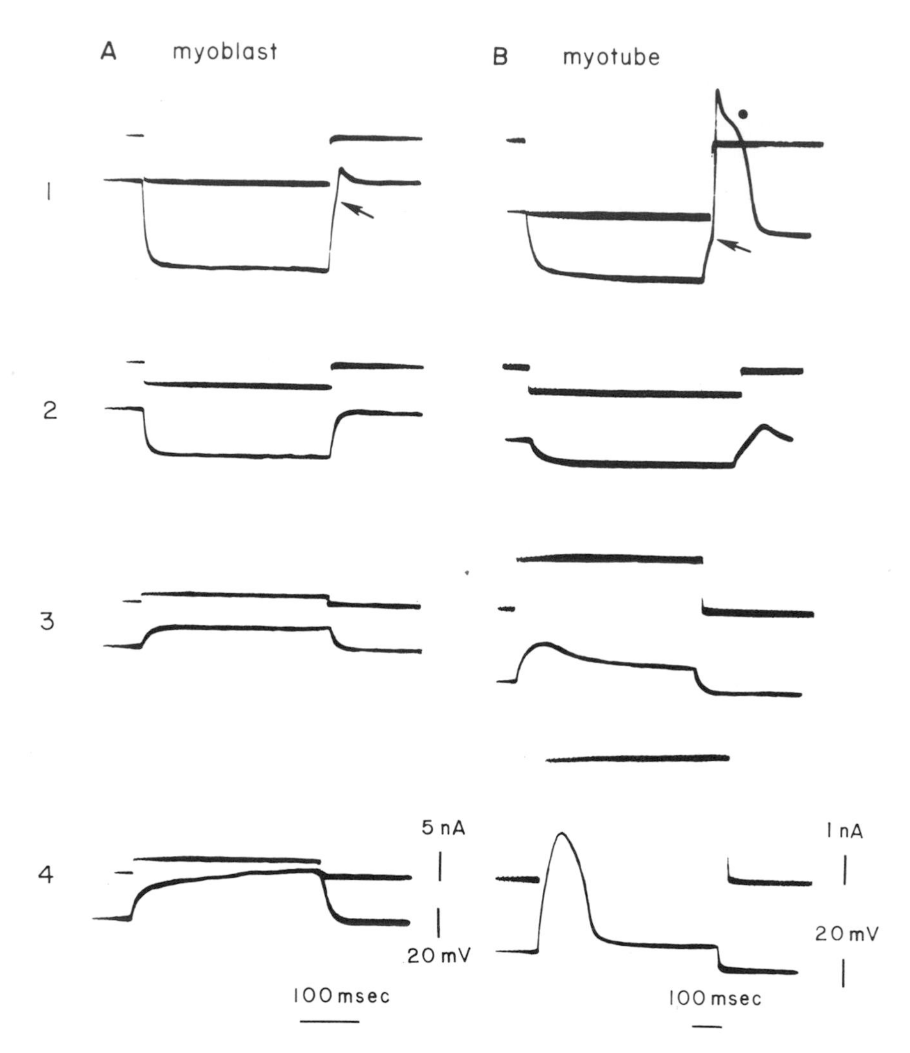

FIGURE 2. Responses to various amounts of applied currents in a myoblast (A) and myotube (B). Upper trace in each pair of records shows current (downward deflection indicates inward current), and zero membrane potential level. Lower trace is the membrane potential record. Calibration shown in A4 applies for all A records; calibration shown in B4 applies for all B records. (From Kidokoro, 1975a, reprinted with permission.)

divalent cations such as Ca ions would reduce the surface negativity by binding to the cell surface. Change in the surface negativity would affect the potential gradient across the membrane and would shift the threshold potential level for a potential-dependent conductance.

The maximum rate of rise of action potentials in chick myotubes was shown to increase as a function of number of days in culture (Spector and

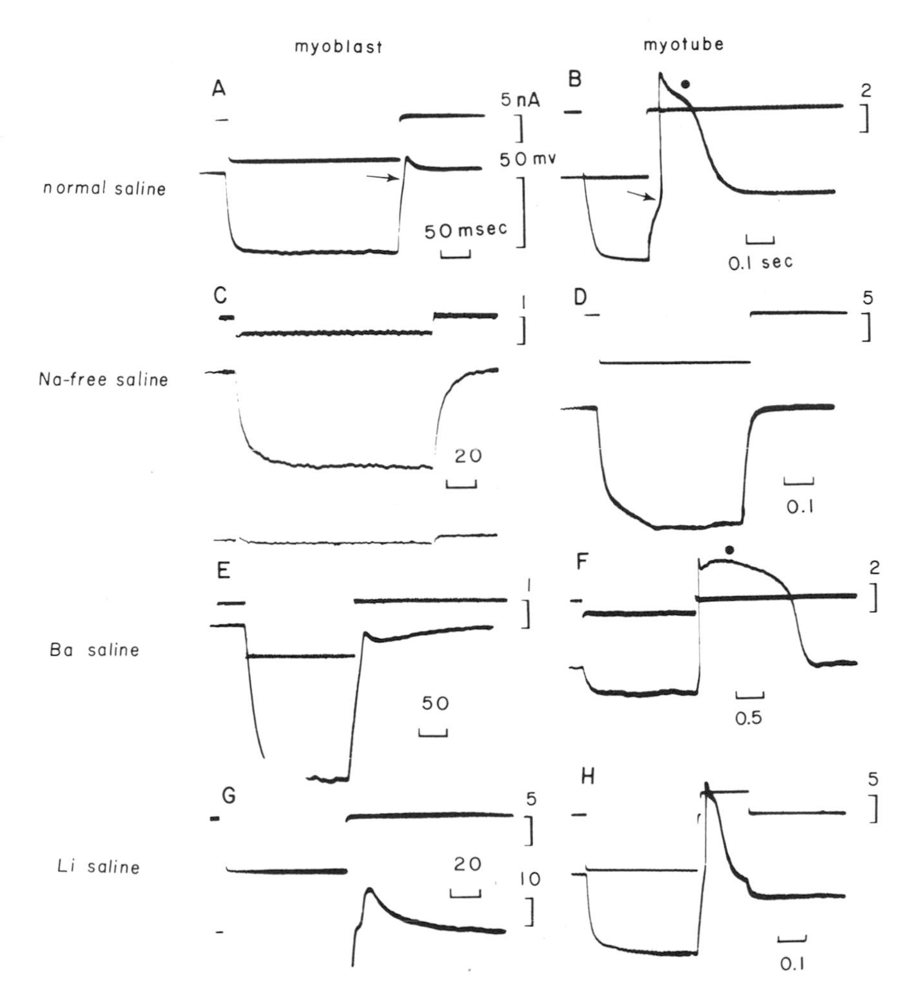

FIGURE 3. Electrical responses to hyperpolarizing current in myoblasts and myotubes under various conditions. Each record except C is composed of two traces, current (upper trace) which also shows the zero potential level, and membrane potential (lower trace). C was obtained with single electrode used for recording and passing current. The third trace is the extracellular potential recorded after the electrode was withdrawn to show the bridge balance. A, C, E, and G are from myoblasts, and B, D, F, and H are from myotubes. A and B were taken in normal saline, C and D in Na-free saline, E and F in Ba saline, and G and H, in Li saline. Voltage scale in A applies to all records except G which has its own. Current and time scales are given in each record. (From Kidokoro, 1975b, reprinted with permission.)

Prives, 1977). This suggests that either the density of ion channels or the properties of each ion channel changes in culture. The maximum rate of rise of L6 myotube action potentials increased approximately twofold when myotubes were cultured with cells from a neuroblastoma cell line (Kidokoro *et al.*, 1975).

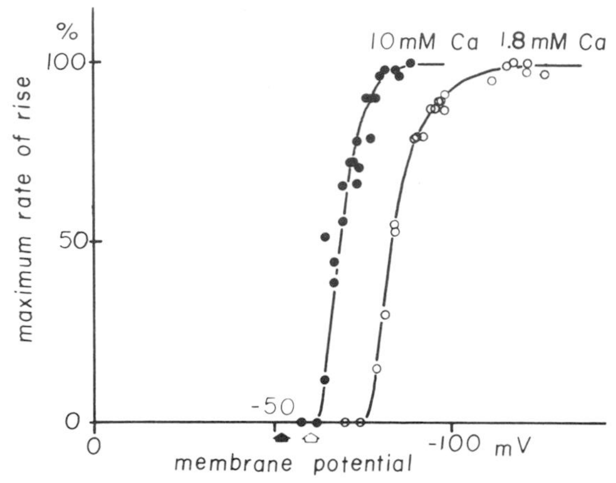

FIGURE 4. Relation between the maximum rate of rise of myotube action potentials and the level of preceding hyperpolarization in normal (○) and 10 mM Ca saline (●). The maximum rate of rise is standardized to the largest value and plotted on the ordinate. The membrane potential level is plotted on the abscissa. Open arrow at the abscissa indicates the resting potential in normal saline (−60 mV), and filled arrow that in 10 mM Ca saline (−57 mV). (From Kidokoro, 1975b, reprinted with permission.)

In the mouse organ culture system, it has also been shown that spinal cord extracts increase the maximum rate of rise of the action potential in previously denervated muscles (Kuromi and Hasegawa, 1975; Hasegawa and Kuromi, 1977). Thus, nerves may exert a trophic influence on membrane excitability by regulating either the density of voltage-dependent ion channels or the properties of each channel. It is known *in vivo* that the maximum rate of rise of the action potential decreases after denervation (Albuquerque and Thesleff, 1968).

The ionic mechanism of the action potential has been studied in chick primary myotubes (Kano *et al.*, 1972; Kano and Shimada, 1973; Fukuda, 1974, 1975; Fukuda *et al.*, 1976a) and in L6 myotubes (Kidokoro, 1973, 1975a,b; Sastre and Podleski, 1976). In both systems Na as well as Ca inward currents contribute to the action potential. The action potential in L6 myotubes was composed of two components, an initial fast spike and a hump on the falling phase (Fig. 2B1, Fig. 3B, Fig. 5A) or in some cases a distinct second peak (Fig. 6A). The overshoot of the initial fast spike decreased when the external

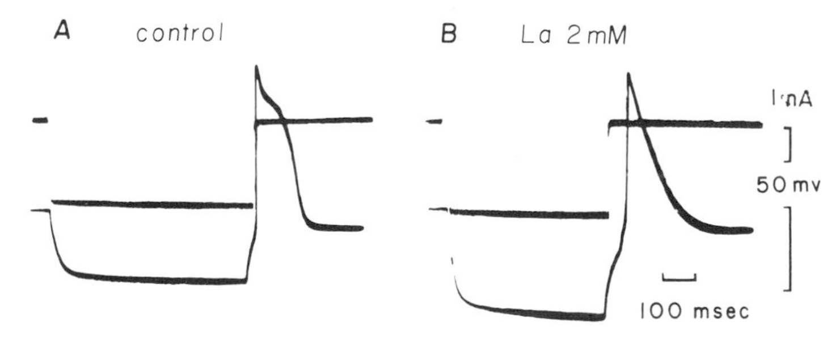

FIGURE 5. Effect of La ions on the shape of a myotube action potential. (A) Control; (B) after application of 2 mM $LaCl_3$. Calibrations at the right of B also apply to A. (From Kidokoro, 1975b, reprinted with permission.)

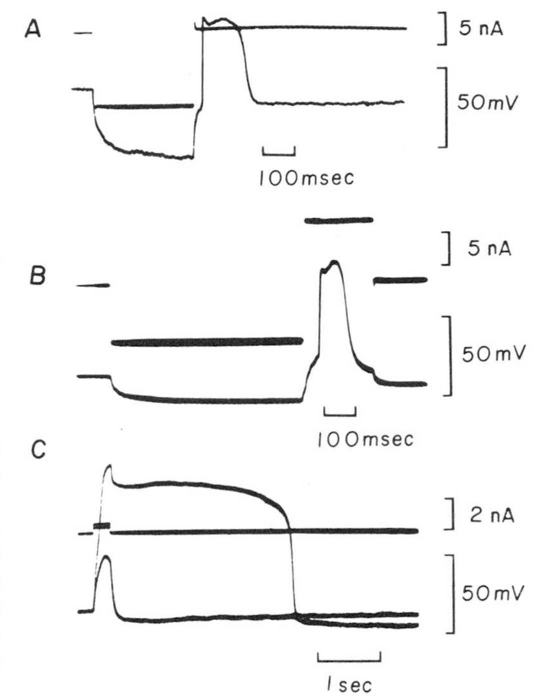

FIGURE 6. Myotube action potentials in various solutions. (A) Normal saline; (B) 10 mM Ca saline; (C) Na-free and 10 mM Ca saline. Upper traces are current recordings and lower traces are membrane potential recordings. Scales are given in each record. (From Kidokoro, 1975b, reprinted with permission.)

Na concentration was decreased. Thus this portion of the action potential was considered to be a Na component (Kidokoro, 1975b).

The Na component in L6 myotubes was relatively insensitive to tetrodotoxin (TTX; half-blocking doses, 4 μM; Kidokoro *et al.*, 1975), whereas that in chick primary myotubes was sensitive to 30 nM TTX (Kano and Shimada, 1973). The relative insensitivity of action potentials to TTX has also been demonstrated in denervated rat muscle cells *in vivo* (J. Harris and Thesleff, 1971; Redfern and Thesleff, 1971). Interestingly, action potentials in innervated newborn rat muscle were also resistant to TTX (J. Harris and Marshall, 1973), suggesting that it requires a certain time for the nerve to affect this property of the muscle action potential. Recently it was reported that spinal cord extracts partially prevented the decrease in TTX sensitivities after denervation (Hasegawa and Kuromi, 1977). Some factor(s) in the spinal cord seems to affect the membrane electrical properties of myotubes.

Li ions could be substituted for Na ions in generating the fast component of action potentials (Fig. 3H), but Cs or Tris ions were ineffective substitutes in L6 myotubes (Kidokoro, 1975b). Thus, the ion selectivity of Na channels so far examined in L6 myotubes was similar to that of nerve (Hille, 1972) and muscle (Campbell, 1976).

A Ca component in the action potentials of L6 myotubes was inferred from the following observations: (1) there is a hump on the falling phase of an action potential which sometimes became a distinct second peak (Fig. 2B1, Fig. 3B, Fig. 5A, Fig. 6A); (2) the second peak or hump is blocked by a small amount of La (Fig. 5B); (3) in Na-free, 10 mM-Ca saline, a prolonged action potential can be evoked (Fig. 6C); (4) in 10 mM-Ba saline, a second slow regenerative potential can be evoked (Fig. 3F), and the peak level of the second action potential is dependent on the external Ba concentration (Kidokoro, 1975b); and (5) in the Na-free Ba saline, a prolonged action potential was evoked, and the membrane conductance at the peak of this action potential was increased (Kidokoro, 1975b).

In chick myotubes, Co (Kano, 1975a,b) and Mn (Kano and Shimada, 1973) ions were shown to block the Ca component of action potentials.

Since Ca action potentials have not been observed in adult rat muscle, it was speculated that a Ca component of the action potential appears only during development and disappears during maturation (Kidokoro, 1975b). A similar disappearance of an ionic component of the action potential during development was observed in tunicate muscle (Takahashi *et al.*, 1971; Miyazaki *et al.*, 1972), in Rohon-Beard neurons (Spitzer and Baccaglini, 1976; Spitzer and Lamborghini, 1976; Spitzer, 1979), and in dorsal root ganglion cells (Baccaglini, 1978) from *Xenopus* embryos.

In addition to the Na and Ca components of the action potential in chick myotubes, Fukuda found slow regenerative potentials that were dependent on external Cl concentrations (Fukuda, 1974, 1975; Fukuda *et al.*, 1976a). It should be noted, however, that these Cl-dependent action potentials were only seen under certain conditions (see Section 4.5).

3.5. Current–Voltage Relationship

The current–voltage (I–V) relationship of cultured muscle membranes has been studied in the chick (J. Harris *et al.*, 1973; Kano and Shimada, 1973; Spector and Prives, 1977; Fukuda *et al.*, 1976b; Engelhardt *et al.*, 1977b) and in the L6 muscle cell line (Kidokoro, 1973, 1975a). The I–V relationship changed drastically when myoblasts fused to form myotubes. In myoblasts, the I–V curve was linear between −90 and 0 mV. The slope resistance increased in the region more positive than 0 mV (Fig. 7A; Kidokoro, 1973, 1975a). In contrast, an I–V curve constructed for myotubes was similar to that found in adult frog skeletal muscle (Nakajima *et al.*, 1962). Outward rectification was prominent in the region more positive than resting potential level (Fig. 7B). This outward rectification was also time-dependent, being more prominent at 500 msec after the onset of current pulse (●) than at 250 msec (○) (Fig. 7B). When a large depolarizing current was passed, the membrane potential depolarized to a large extent at first, but soon the membrane potential displacement resulting from the same constant current became smaller (Fig. 2B3 and 2B4).

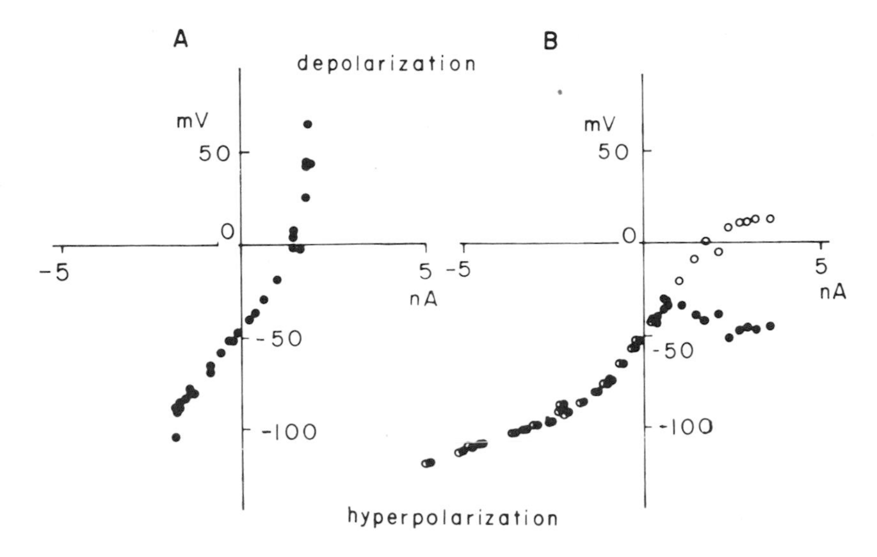

FIGURE 7. Current–voltage relationship in a myoblast (A) and a myotube (B). Current is plotted on the abscissa (outward current is plotted as positive), membrane potential on the ordinate. The potential change was measured 250 msec after the onset of current pulse in the myoblast and 250 msec (○), and 500 msec (●) in the myotube. The intersection of the I–V curve with the ordinate shows the resting potentials (−48 mV in the myoblast and −50 mV in the myotube). (From Kidokoro, 1975b, reprinted with permission.)

Thus, the membrane conductance at the end of the depolarizing current pulse of about 500 msec duration was greatly increased (five- to tenfold). This increase in membrane conductance outlasted both the depolarizing current pulse and the membrane potential after the pulse reached −70 to −77 mV at its peak and lasted 4 to 8 sec (Fig. 2B4). Similar hyperpolarization was observed after an action potential (Figs. 2B1, 3B, and 6A). This time-dependent outward rectification corresponds to delayed rectification (Hodgkin and Horowicz, 1959; Nakajima *et al.*, 1962; Adrian *et al.*, 1970). The hyperpolarization most likely results from activation of K conductance (see Section 4.6).

Inward rectification was found in the region more negative than the resting membrane potential level and was also prominent in isotonic K solution where delayed rectification was completely inactivated (Kidokoro, 1975b). This corresponds to anomalous rectification in adult skeletal muscles (Hodgkin and Horowicz, 1959; Nakajima *et al.*, 1962; Adrian *et al.*, 1970).

3.6. Acetylcholine Sensitivity

Generally neither primary rat myoblasts nor L6 myoblasts respond to iontophoretic application of acetylcholine (ACh) (Fambrough and Rash, 1971; A.

Harris *et al.,* 1971; Steinbach, 1975). Under certain circumstances, however, a transient depolarization was evoked by iontophoretic application of ACh in primary rat (Fambrough and Rash, 1971) and primary chick myoblasts (Obata, 1974). A hyperpolarization with a slow time course was observed in some of L6 myoblasts after application of ACh. This hyperpolarization was not blocked by curare (Steinbach, 1975). As soon as myoblasts fuse to form myotubes, transient membrane depolarizations (ACh potentials) are evoked by iontophoretic application of ACh (Fambrough and Rash, 1971; A. Harris *et al.,* 1971; Steinbach *et al.,* 1973; J. Harris *et al.,* 1973; Obata, 1974; Steinbach, 1975). The ACh potential was blocked by *d*-tubocurarine (*d*TC), cobra neurotoxin, and α-bungarotoxin (J. Harris *et al.,* 1973; Obata, 1974). In agreement with electrophysiological measurements, ion flux studies demonstrated that ACh activates Na flux in myotubes but not in myoblasts (Stallcup and Cohn, 1976). The reversal potential for ACh potentials was reported to be 0 mV for primary rat myotubes (Ritchie and Fambrough, 1975). Values between -3 mV and $+10$ mV have been reported for the reversal potential in chick myotubes (J. Harris *et al.,* 1973; Ritchie and Fambrough, 1975) and in L6 myotubes (Steinbach, 1975; Kidokoro *et al.,* 1975). The ionic mechanism of the ACh potential was studied in rat myotubes (Ritchie and Fambrough, 1975; Steinbach, 1975; Huang *et al.,* 1978). Upon application of ACh the membrane became more permeable to Na, K, and Ca. Cl ions did not contribute to the ACh potential (Ritchie and Fambrough, 1975; Huang *et al.,* 1978). These properties are similar to those found in the adult frog muscle (Takeuchi and Takeuchi, 1960; Maeno *et al.,* 1977).

4. DISCUSSION

4.1. Resting Membrane Potential

The resting membrane potentials in chick myoblasts were reported to be small, -10 to -15 mV (Fischbach *et al.,* 1971; Dryden *et al.,* 1974). Those in rat myoblasts and young myotubes were about -40 mV (Obata, 1974). These values are in contrast to large resting membrane potentials (-71 mV) found in L6 myoblasts (Kidokoro, 1975a). The simplest explanation for this discrepancy is that the rat skeletal muscle cell line (L6) has different properties from primary myogenic cells. However, it is also possible that the small resting membrane potentials found in primary myoblasts may have been artifactual, resulting from damage owing to microelectrode penetration. Generally myoblasts are difficult to penetrate with a microelectrode. This is especially true for the small spindle-shaped cells that are the only cells identified as myoblasts in primary cultures. In the clonal cell line, some myoblasts are large and are easier to penetrate. The difference in cell size could make a difference in the recorded resting membrane potential, since damage from microelectrode penetration is, to a certain extent, inevitable. Electrical leakage through the interface between the microelectrode glass wall and the cell membrane could cause artifactually reduced resting membrane potentials. In addition, even after the seal around the electrode is restored, leakage of K ions from inside the cell could result in small potentials. Both of these artifacts affect smaller cells more severely than larger multinucleate myotubes. Ritchie and Fambrough (1975) attributed small resting potentials in young myotubes to a larger P_{Na}/P_K ratio. If that is the case, replacement of Na ions in the external medium with impermeant molecules should increase the resting potential. But this was not observed: the resting membrane potential in a 350 mM sucrose solution was not different from that in normal saline. Therefore, it is not convincing that the small resting membrane potential found in young myotubes results solely from a relatively high Na permeability. In view of these technical difficulties involved in measuring the resting potentials in small and fragile cells, it is not conclusive whether or not there is a progressive increase of the resting membrane potential in primary muscle cells.

There is a reasonable agreement among the measured resting membrane potentials of myotubes from various sources. The values (−46 to −72 mV) are comparable to those of denervated adult skeletal muscles [−53 mV in the mouse (Hasegawa and Kuromi, 1977); −57 mV in the rat (Albuquerque and Thesleff, 1968)]. Since the resting membrane potential is known to decrease after denervation, neurotrophic influences might be necessary for further increase in the resting potential of myotubes.

4.2. Chloride Permeability

A low membrane permeability to Cl ions was found in chick (Engelhardt *et al.*, 1978), in rat (Ritchie and Fambrough, 1975), and in L6 myotubes (Kidokoro, 1975a). This is in contrast to a large Cl permeability in adult skeletal muscle (Hodgkin and Horowicz, 1959; Hutter and Warner, 1967; Bryant and Morales-Aguilera, 1971; Palade and Barchi, 1977). The membrane resistance has been shown to increase after denervation in the rat (Albuquerque and Thesleff, 1968), and this has been attributed to a decrease in Cl permeability (Camerino and Bryant, 1976). It is thus possible that the nerve is somehow affecting the muscle membrane permeability to Cl ions. The nature of neural influence on muscle membrane properties should be studied in the future.

4.3. Na Action Potential

Small Na action potentials were found in L6 myoblasts (Kidokoro, 1973, 1975a,b). In agreement with this electrophysiological finding, veratridine-stimulated Na flux was observed in L6 myoblasts (Stallcup and Cohn, 1976). So far no action potentials have been reported in primary myoblasts. This, however, could be because of technical difficulties in penetrating a microelectrode into primary myoblasts. Na flux assays on primary myoblasts may be an alternative way to clarify this point.

The action potentials become larger and overshooting soon after the multinucleate myotubes have formed. The action potential in myotubes has two components of inward current carried by Na and Ca ions. Na action potentials in L6 myotubes are resistant to TTX (Kidokoro, 1973; Kidokoro *et al.,* 1975; Land *et al.,* 1973; Sastre and Podleski, 1976). This is comparable to the finding in denervated rat muscle cells (half-blocking doses about 1 μM; J. Harris and Thesleff, 1971). The relative insensitivity of L6 myotube action potentials to TTX did not change even after formation of functional neuromuscular junctions (Kidokoro *et al.,* 1975). Correspondingly, TTX-resistant action potentials were found in neonatal rat muscles that were already functionally innervated (J. Harris and Marshall, 1973). It is also known that spinal cord extract affects TTX sensitivity of organ-cultured mouse skeletal muscle cells (Hasegawa and Kuromi, 1977). It may take some time for the nerve to affect this muscle membrane property even after functional innervation. In contrast to the TTX insensitivity of the rat and mouse muscle membrane, Na action potentials in chick myotubes are highly sensitive to TTX even without innervation (half-blocking dose is less than 30 nM; Kano and Shimada, 1973; J. Harris *et al.,* 1973, Fukuda, 1974).

4.4. Ca Action Potential

A Ca component of action potentials has been found in chick myotubes (Kano and Shimada, 1973; Fukuda *et al.,* 1976a) and in L6 myotubes (Kidokoro, 1973, 1975a,b). Since the Ca component was not detected in adult rat muscle it was speculated that the Ca component found in myotubes disappears during development (Kidokoro, 1975b). S. Bevan (personal communication) has studied the action potential in neonatal rat muscles in the presence of Ba ions which help to reveal the mechanism of voltage-dependent Ca permeability changes by reducing outward K current and carrying inward current through Ca channels. He found a second slow peak in the action potential similar to that found and identified as the Ca component in myotubes in culture. Thus, the composite form of action potentials in myotubes is not unique to cultured cells. Furthermore, Kano (1975b) studied the development of excitability in the embryonic chick muscle and found that at days 19 through hatching the action potential is composed of a spike that is attributable to the Na inward current followed by a plateau that is sensitive to changes in external Ca ion concentration. The significance of the Ca component in the action potential of developing muscle is not known.

4.5. Cl Action Potential

In addition to the Na and Ca components of myotube action potentials, a third component has been found in chick myotubes (Fukuda, 1974, 1975; Fukuda *et al.*, 1976a). The chloride spike could only be evoked under conditions where the Cl equilibrium potential is less negative than the resting potential or holding potential, that is, where a large quantity of Cl ions had been injected into the cell or when the membrane potential was held at a large negative value. Under normal conditions, the Cl equilibrium potential is presumably close to the resting membrane potential, so that a voltage-dependent increase of membrane permeability to Cl ions does not produce regenerative membrane potential changes. Nevertheless, it is intriguing to know that the myotube membrane appears to have a voltage-dependent Cl conductance. The functional significance of this mechanism is not known.

4.6. Delayed Rectification

Delayed rectification (outward rectification) has been found in myotubes but not in myoblasts, although myoblasts already have a mechanism for a voltage-dependent Na inward current (Kidokoro, 1975b). Inward current was also found to develop before delayed rectification in tunicate muscles *in vivo* (Takahashi *et al.*, 1971; Miyazaki *et al.*, 1972). Delayed rectification is most likely the result of a membrane conductance increase to K ions. If so, the level of the membrane potential after a large depolarization or an action potential should approach the K equilibrium potential. From the estimate of internal K concentration obtained in Fig. 1, the K equilibrium potential was calculated to be −83 mV in L6 myotubes. This value is more negative than the level of the membrane potential attained after an action potential (−70 to −77 mV). The explanation for this discrepancy has not been offered.

The time course of delayed rectification is extremely slow in L6 myotubes (the postspike hyperpolarization lasted 4 to 8 sec). This may also contribute to the long duration of an action potential in myotubes (about 100 msec measured at half height; Kidokoro, 1973).

By using cell-culture techniques electrophysiological properties of myogenic precursor cells, myoblasts, and multinucleate myotubes have been examined, and several characteristics of the devleoping muscle membrane have become evident. Since measurements *in vivo* are technically difficult to obtain, and since, by and large, developmental characteristics seem to be reproduced in culture, it may be logical to pursue this research further, taking special interest in neural influences on muscle membrane properties.

ACKNOWLEDGMENTS. I thank Drs. R. Gruener and P. Brehm for criticism of this chapter.

REFERENCES

Adrian, R. H., Chandler, W. K., and Hodgkin, A. L., 1970, Slow changes in potassium permeability in skeletal muscle, *J. Physiol. (Lond.)* **208:**645.

Albuquerque, E. X., and Thesleff, S., 1968, A comparative study of membrane properties of innervated and chronically denervated fast and slow skeletal muscles of the rat, *Acta. Physiol. Scand.* **73:**471.

Anderson, C. R., and Stevens, C. F., 1973, Voltage clamp analysis of acetylcholine produced end-plate current fluctuations at frog neuromuscular junction, *J. Physiol. (Lond.)* **235:**655.

Baccaglini, P., 1978, Action potentials of embryonic dorsal root ganglion neurones in *Xenopus* tadpoles, *J. Physiol. (Lond.)* **283:**585.

Beaty, G. N., and Stefani, E., 1975, Inward calcium current in twitch muscle fibers in the frog, *J. Physiol. (Lond.)* **260:**27P.

Bryant, S. H., and Morales-Aguilera, A., 1971, Chloride conductance in normal and myotonic muscle fibers and the action of monocarboxylic aromatic acids, *J. Physiol. (Lond.)* **219:**367.

Camerino, D., and Bryant, S. H., 1976, Effects of denervation and colchicine treatment on the chloride conductance of rat skeletal muscle fibers, *J. Neurobiol.* **7:**221.

Campbell, D. T., 1976, Ionic selectivity of the sodium channel of frog skeletal muscle, *J. Gen. Physiol.* **67:**295.

Campbell, D. T., and Hille, B., 1976, Kinetic and pharmacological properties of the sodium channel of frog skeletal muscle, *J. Gen. Physiol.* **67:**309.

Christian, C. N., Nelson, P. G., Peacock, J., and Nirenberg, M., 1977, Synapse formation between two clonal cell lines, *Science* **196:**995.

Cole, K. S., and Curtis, J. H., 1950, Bioelectricity: Electric physiology, in: *Medical Physics, Vol. 2* (O. Glaser, ed.), p. 82, Year Book Publishers, Chicago.

Dryden, W. F., Erulkar, S. D., and de la Haba, G., 1974, Properties of the cell membrane of developing skeletal muscle fibers in culture and its sensitivity to acetylcholine, *Clin. Exp. Pharmacol. Physiol.* **1:**369.

Engelhardt, J. K., Ishikawa, K., Lisbin, S. J., and Mori, J., 1976, Neurotrophic effects on passive electrical properties of cultured chick skeletal muscle, *Brain Res.* **110:**170.

Englehardt, J. K., Ishikawa, K., Mori, J., and Schimabukuro, Y., 1977a, Passive electrical properties of cultured chick skeletal muscle: Neurotrophic effect on sample distribution, *Brain Res.* **126:**172.

Engelhardt, J. K., Ishikawa, K., Mori, J., and Shimabukuro, Y., 1977b, Neurotrophic effects on the electrical properties of cultured muscle produced by conditioned medium from spinal cord explants, *Brain Res.* **128:**243.

Engelhardt, J. K., Ishikawa, K., and Shimabukuro, Y., 1978, Neurotrophic regulation of chloride conductance in cultured chick skeletal muscle, *Soc. Neurosci. Abstr.* **4:**603.

Ezerman, E. B., and Ishikawa, H., 1967, Differentiation of the sarcoplasmic reticulum and T system in developing chick skeletal muscle *in vitro, J. Cell Biol.* **25:**405.

Fambrough, D., and Rash, J. E., 1971, Development of acetylcholine sensitivity during myogenesis, *Dev. Biol.* **26:**55.

Fischbach, G. D., Nameroff, M., and Nelson, P. G., 1971, Electrical properties of chick skeletal muscle fibers developing in cell culture, *J. Cell. Physiol.* **78:**289.

Fischman, D. A., 1972, Development of striated muscle, in: *The Structure and Function of Muscle*, Vol. 1 (G. H. Bourne, ed.), 2nd ed., pp. 75–148, Academic Press, New York and London.

Frankenheuser, B., and Hodgkin, A. L., 1957, The action of calcium on the electrical properties of squid axons, *J. Physiol. (Lond.)* **137:**217.

Fukuda, J., 1974, Chloride spike: A third type of action potential in tissue-cultured skeletal muscle cells from the chick, *Science* **185:**76.

Fukuda, J., 1975, Voltage clamp study on inward chloride currents of spherical muscle cells in tissue culture, *Nature* **257**:408.

Fukuda, J., Fischbach, G. D., and Smith, T. G., Jr., 1976a, A voltage clamp study of the sodium, calcium and chloride spikes of chick skeletal muscle cells grown in tissue culture, *Dev. Biol.* **49**:412.

Fukuda, J., Henkart, M. P., Fischbach, G. D., and Smith, T. G., Jr., 1976b, Physiological and structural properties of colchicine-treated chick skeletal muscle cells grown in tissue culture, *Dev. Biol.* **49**:395.

Gage, P. W., and Eisenberg, R. S., 1969, Capacitance of the surface and transverse tubular membrane of frog sartorius muscle fibers, *J. Gen. Physiol.* **53**:265.

Harris, A. J., Heinemann, S., Schubert, D., and Tarikas, H., 1971, Trophic interaction between cloned tissue culture lines of nerve and muscle, *Nature* **231**:296.

Harris, J. B., and Marshall, M. W., 1973, Tetrodotoxin resistant action potentials in newborn rat muscle, *Nature* [*New Biol.*] **243**:191.

Harris, J. B., and Thesleff, S., 1971, Studies on tetrodotoxin resistant action potentials in denervated skeletal muscle, *Acta Physiol. Scand.* **83**:382.

Harris, J. B., Marshall, M. W., and Wilson, P., 1973, A physiological study of chick myotubes grown in tissue culture, *J. Physiol. (Lond.)* **229**:751.

Hasegawa, S., and Kuromi, H., 1977, Effects of spinal cord and other tissue extracts on resting and action potentials of organ-cultured mouse skeletal muscle, *Brain Res.* **119**:133.

Hille, B., 1972, The permeability of the sodium channel to metal cations in myelinated nerve, *J. Gen. Physiol.* **59**:637.

Hodgkin, A. L., and Horowicz, P., 1959, The influence of potassium and chloride ions on the membrane potential of single muscle fibers, *J. Physiol. (Lond.)* **148**:127.

Hodgkin, A. L., and Huxley, A. F., 1952, The dual effect of membrane potential on sodium conductance in the giant axon of *Loligo*, *J. Physiol. (Lond.)* **116**:497.

Huang, L. M., Catterall, W. A., and Ehrenstein, G., 1978, Selectivity of cations and nonelectrolytes for acetylcholine-activated channels in cultured muscle cells, *J. Gen. Physiol.* **71**:397.

Hutter, O. F., and Warner, A. E., 1967, The pH sensitivity of the cloride conductance of frog skeletal muscle, *J. Physiol. (Lond.)* **189**:403.

Kano, M., 1975a, Development of excitability in cultured skeletal muscle cells, *Adv. Neurol. Sci.* **20**:1065.

Kano, M., 1975b, Development of excitability in embryonic chick skeletal muscle cells, *J. Cell. Physiol.* **86**:503.

Kano, M., and Shimada, Y., 1971a, Innervation of skeletal muscle cells differentiated *in vitro* from chick embryo, *Brain Res.* **27**:402.

Kano, M., and Shimada, Y., 1971b, Innervation and acetylcholine sensitivity of skeletal muscle cells differentated *in vitro* from chick embryo, *J. Cell. Physiol.* **78**:233.

Kano, M., and Shimada, Y., 1973, Tetrodotoxin-resistant electric activity in chick skeletal muscle cells differentated *in vitro*, *J. Cell. Physiol.* **81**:85.

Kano, M., Shimada, Y., and Ishikawa, K., 1972, Electrogenesis of embryonic chick skeletal muscle cells differentiated *in vitro*, *J. Cell. Physiol.* **79**:363.

Katz, B., and Miledi, R., 1972, The statistical nature of the acetylcholine potential and its molecular components, *J. Physiol. (Lond.)* **224**:665.

Kidokoro, Y., 1973, Development of action potentials in a clonal rat skeletal muscle cell line, *Nature* [*New Biol.*] **241**:158.

Kidokoro, Y., 1975a, Developmental changes of membrane electrical properties in a rat skeletal muscle cell line, *J. Physiol. (Lond.)* **244**:129.

Kidokoro, Y., 1975b, Sodium and calcium components of the action potential in a developing skeletal muscle cell line, *J. Physiol. (Lond.)* **244**:145.

Kidokoro, Y., 1980, Developmental changes of spontaneous synaptic potential properties in the rat neuromuscular contact formed in culture, *Dev. Biol.* **78:**231.

Kidokoro, Y., Heinemann, S., Schubert, D., Brandt, B. L., and Klier, F. G., 1975, Synapse formation and neurotrophic effect on muscle cell lines, *Cold Spring Harbor Symp. Quant. Biol.* **XL:**373.

Kimes, B. W., and Brandt, B. L., 1975, Properties of a clonal muscle cell line from rat heart, *Exp. Cell Res.* **98:**367.

Kuromi, H., and Hasegawa, S., 1975, Neurotrophic effect of spinal cord extract on membrane potentials of organ-cultured mouse skeletal muscle, *Brain Res.* **100:**178.

Land, B. R., Sastre, A., and Podleski, T. R., 1973, Tetrodotoxin-sensitive and -insensitive action potentials in myotubes, *J. Cell. Physiol.* **82:**497.

Maeno, T., Edwards, C., and Anraku, M., 1977, Permeability of the endplate activated by acetylcholine to some organic cations, *J. Neurobiol.* **8:**173.

Miyazaki, S., Takahashi, K., and Tsuda, K., 1972, Calcium and sodium contributions to regenerative responses in the embryonic excitable cell membrane, *Science* **176:**1441.

Nakajima, S., Iwasaki, S., and Obata, K., 1962, Delayed rectification and anomalous rectification in frog's skeletal muscle membrane, *J. Gen. Physiol.* **46:**97.

Neher, E., and Sakmann, B., 1976, Noise analysis of drug induced voltage clamp currents in denervated frog muscle fibers, *J. Physiol. (Lond.)* **258:**705.

Obata, K., 1974, Transmitter sensitivities of some nerve and muscle cells in culture, *Brain Res.* **73:**71.

Ohta, M., Narahashi, T., and Keeler, R. F., 1973, Effects of veratrum alkaloids on membrane potential and conductance of squid and crayfish giant axons, *J. Pharmacol. Exp. Ther.* **184:**143.

Palade, P. T., and Barchi, R. L., 1977, Characteristics of the chloride conductance in muscle fibers of the rat diaphragm, *J. Gen. Physiol.* **69:**325.

Powell, J. A., and Fambrough, D. M., 1973, Electrical properties of normal and dysgenic mouse skeletal muscle in culture, *J. Cell. Physiol.* **82:**21.

Redfern, P., and Thesleff, S., 1971, Action potential generation in denervated rat skeletal muscle. II. The action of tetrodotoxin, *Acta Physiol. Scand.* **82:**70.

Ritchie, A. K., and Fambrough, D. M., 1975, Electrophysiological properties of the membrane and acetylcholine receptor in developing rat and chick myotubes, *J. Gen. Physiol.* **66:**327.

Sanchez, J. A., and Stefani, D., 1978, Inward calcium current in twitch muscle fibres of the frog, *J. Physiol. (Lond.)* **283:**197.

Sastre, A., and Podleski, T. R., 1976, Pharmacologic characterization of the Na^+ ionophores in L6 myotubes, *Proc. Natl. Acad. Sci. USA* **73:**1355.

Singer, S. J., and Nicolson, G. L., 1972, The fluid mosaic model of the structure of cell membranes, *Science* **175:**720.

Spector, I., and Prives, J. M. 1977, Development of electrophysiological and biochemical membrane properties during differentiation of embryonic skeletal muscle in culture, *Proc. Natl. Acad. Sci. USA* **74:**5166.

Spitzer, N. C., 1979, Ion channels in development, *Annu. Rev. Neurosci.* **2:**363.

Spitzer, N. C., and Baccaglini, P. I., 1976, Development of the action potential in embryonic amphibian neurons *in vivo, Brain Res.* **107:**610.

Spitzer, N. C., and Lamborghini, J. E., 1976, The development of the action potential mechanism of amphibian neurons isolated in culture. *Proc. Natl. Acad. Sci. USA* **73:**1641.

Stallcup, W. B., and Cohn, M., 1976, Electrical properties of a clonal cell line as determined by measurement of ion fluxes, *Exp. Cell Res.* **98:**277.

Steinbach, J. H., 1975, Acetylcholine responses on clonal myogenic cells *in vitro, J. Physiol. (Lond.)* **247:**393.

Steinbach, J. H., Harris, A. J., Patrick, J., Schubert, D., and Heinemann, S., 1973, Nerve–muscle interaction in vitro. Role of acetylcholine, *J. Gen. Physiol.* **62**:255.

Takahashi, K., Miyazaki, S., and Kidokoro, Y., 1971, Development of excitability in embryonic muscle cell membranes in certain tunicates, *Science* **171**:415.

Takeuchi, A., and Takeuchi, N., 1960, On the permeability of end-plate membrane during the action of transmitter, *J. Physiol. (Lond.)* **154**:52.

Vale, W., and Grant, G., 1975, *In vitro* pituitary hormone assay for hypophysiotropic substances, in: *Methods in Enzymology, Vol. 37, Hormone Action, Part B, Peptide Hormones* (B. W. O'Malley and J. G. Hardman, eds.), pp. 80–83, Academic Press, Oxford.

Vogt, M., and Dulbecco, R., 1963, Steps in the neoplastic transformation of hamster embryo cells by polyoma virus, *Proc. Natl. Acad. Sci. USA* **49**:171.

Yaffe, D., 1968, Retention of differentiation of potentialities during prolonged cultivation of myogenic cells, *Proc. Natl. Acad. Sci. USA* **61**:477.

10

The Physiology of Smooth Muscle Cells in Tissue Culture

ROBERT D. PURVES

1. INTRODUCTION

Electrical recordings made from smooth muscles have always posed problems of interpretation. Early attempts at extracellular recording gave little insight into the peculiarities of electrogenesis in smooth muscle except in a few well-defined cases (for example, junctional transmission in the nictitating membrane; Eccles and Magladery, 1937). Even today the simpler forms of extracellular recording do not offer much more than an indication of whether or not electrical activity is present. The introduction of intracellular microelectrode techniques to smooth muscle physiology (Bülbring, 1954; Bülbring *et al.*, 1958) allowed a much more accurate account of the membrane potential changes during activity. Three classes of potential change may be identified, although not all are present in any given smooth muscle: (1) action potentials, which usually provide the signal for contractions and whose upstrokes appear to result from an influx of calcium ions (Tomita, 1975); (2) depolarizing slow waves of several types that are seen in spontaneously active smooth muscle, notably in the intestine; and (3) junction potentials, which are the result of autonomic nerve activity and which may be either depolarizing or hyperpolarizing.

Smooth muscle cells are 2–5 μm in diameter and 100–300 μm in length. The smooth muscle walls of many viscera are several millimeters thick and many centimeters in linear extent and contain immense numbers of smooth muscle cells. A physiological question of great importance is how the activity of individual members is synchronized and coordinated to conform to the

ROBERT D. PURVES • Department of Anatomy and Embryology, University College London, London WC1E 6BT, England.

requirements of the organ. Bozler (1948) in his pioneering studies recognized the importance of this question and argued forcefully that there must be functional coupling among cells. As in the heart, the coupling is electrical via low resistance junctions between apposed cells (Barr *et al.,* 1968; Tomita, 1975). But the electrical activity of smooth muscle, unlike that of the heart, does not usually consist of a regular sequence of propagating action potentials. It is characterized by extreme variability. Smooth muscle from different organs behaves differently. Action potentials in one smooth muscle may vary in frequency, configuration, and amplitude and may change from second to second or minute to minute. Electrical events seen by an intracellular electrode may not have arisen at the recording site; they may have been transmitted from other cells.

Even the passive electrical properties of the three-dimensional coupled network of cells are far from obvious, a fact which impeded progress in smooth muscle physiology for several years. For example, intracellular injection of current through the recording microelectrode appears to give little information about the smooth muscle membrane. This and other puzzling observations are discussed by Tomita (1975) and Purves (1976). Since intracellular polarization is a poor method for studying the membrane properties of intact smooth muscle, other methods of current injection have been developed (partition method; double sucrose gap) in which extracellularly applied current is forced to flow through many cells in parallel, chiefly in one dimension. Under these conditions, the smooth muscle "syncytium" takes on many of the properties of a cable, and semiquantitative estimates of membrane parameters can be made.

Evidently many of the problems of smooth muscle electrophysiology would disappear if it were possible to make intracellular recordings from an electrotonically simple preparation. An ambitious program might consist of the determination of passive and active electrical behavior of single isolated smooth muscle cells, followed by the computation of the expected behavior of coupled assemblies of cells. Single cells and small clusters can be grown in tissue culture which, therefore, seems to offer some promise as a preparative tool. Further reasons for the study of smooth muscle in culture are shared with other excitable and, indeed, inexcitable cells. These include the excellent visibility of cultured preparations, the opportunity of co-culturing with neurons to study synapse formation, and the ease with which neurotransmitters and other drugs may be applied iontophoretically.

This chapter will survey the physiological uses of smooth muscle tissue culture. Studies have been made on primary dissociated cell cultures from taenia coli (Purves *et al.,* 1973; Purves, 1974), vas deferens (Hill-Smith and Purves, 1978), and vascular smooth muscle (Hermsmeyer, 1976; Hermsmeyer *et al.,* 1976; McLean and Sperelakis, 1977), on explant cultures from the sphincter pupillae muscle (Purves *et al.,* 1974; Hill *et al.,* 1976), vas deferens, and taenia coli (Purves *et al.*, 1974), and on cell lines from aorta (Kimes and Brandt, 1976) and human oviduct (Sinback and Shain, 1979). Work on acutely isolated smooth muscle cells will also be discussed, since this experimental approach shares its aims and methods with tissue culture.

2. METHODS

2.1. Tissue Culture

The methods for preparation of dissociated cells are generally similar to those used with other tissues such as heart muscle and involve enzyme treatment together with gentle shaking or stirring. Details of the method of Mark *et al.* (1973) are as follows.

Fragments of smooth muscle tissue dissected with sterile technique are suspended in Hank's balanced salt solution containing up to 0.5% collagenase for 1 hr at 37°C and then transferred to a Ca- and Mg-free solution with trypsin (usually 0.125%). Gentle stirring with changes of the solution rapidly releases single cells and small clusters. The supernatant is added to precooled fetal calf serum (final concentration 10% or, more recently, 50%) to inactivate the trypsin. The cells are centrifuged at 500 *g* for 5 min and resuspended in culture medium before seeding into the growth chambers. The suspension contains fibroblasts as well as smooth muscle cells, but a useful degree of enrichment can be obtained if double-sided (Rose, 1954) chambers are used. After inoculation, the fibroblasts attach rapidly to the bottom coverslip; if the chamber is inverted after 20 min, these fibroblasts remain on the (now) upper coverslip. A further period of 6–12 hr allows the smooth muscle cells to attach firmly (Fig. 1), after which they may be covered with a strip of dialyzing cellophane to help maintain differentiation.

Treatment with trypsin is sometimes felt to be rather severe, objective evidence for this view being provided by Masson-Pévet *et al.* (1976) for heart cultures and by Gröschel-Stewart *et al.* (1976) for smooth muscle. Accordingly, there are variations of the above method that do not use trypsin (see Hermsmeyer *et al.*, 1976). Vascular smooth muscle may be dissociated by collagenase and collagenase/elastase treatment (Chamley *et al.*, 1977), and Travo (personal communication) has on occasion obtained excellent cultures from the vas deferens with the aid of hyaluronidase alone.

Of the numerous culture media that could be used, we have preferred M199, containing fetal calf serum (10%), penicillin (100 U/ml), and sometimes extra calcium (up to 2.5 mM). When autonomic ganglia are to be cocultured, additional supplementation with glucose (5 mg/ml) and insulin (0.05 U/ml) is advisable.

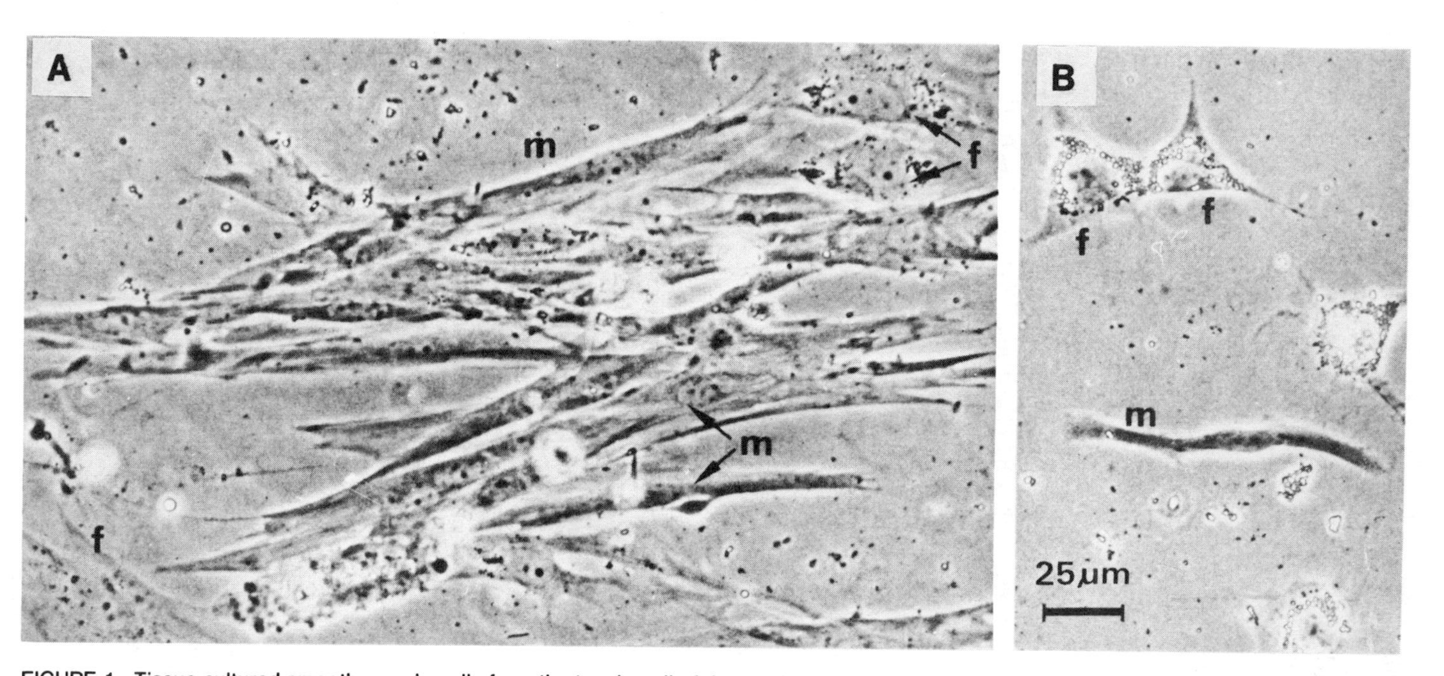

FIGURE 1. Tissue-cultured smooth muscle cells from the taenia coli of the newborn guinea pig; phase contrast microscopy. (A) Cluster of cells in 5-day culture. (B) Single isolated smooth muscle cell in 2-day culture. m, Muscle cells; f, fibroblasts; scale: 25 μm. (From Purves *et al.*, 1973, with permission.)

2.2. Acute Isolation

Although it is possible to isolate single cells from adult mammalian smooth muscle (see Small, 1974), they are rather fragile and have not been used for physiological work. More robust cells can be obtained from the stomach of the toad *Bufo marinus*. Bagby *et al.* (1971) described a procedure using trypsin and collagenase followed by collagenase alone. A refinement of this technique is detailed by Singer and Fay (1977). The cells are up to several hundred μm long and 5–20 μm in diameter. They contract when stimulated electrically or with acetylcholine and remain usable for many hours. A collagenase method was described by Caffrey and Anderson (1978) for salamander stomach.

Single cells have been isolated from the chicken gizzard with collagenase (Kominz and Gröschel-Stewart, 1972; Small, 1974). They do not appear to be viable, although they can be induced to contract irreversibly.

2.3. Recording Methods

The technical difficulties of making intracellular electrical records from smooth muscle in culture are formidable but surmountable. Of the cultured tissues with which this investigator has had experience, skeletal myotubes have proved the easiest to impale and smooth muscle the hardest, with cardiac muscle and small neurons occupying an intermediate position. Microelectrodes made from soda glass of relatively low softening temperature have been used to good effect; the extraordinarily small tips of these are shown in dark-field scanning transmission electron micrographs by Fry (1975). The more convenient borosilicate capillaries with an internal glass fiber are also suitable, provided that electrodes with resistances of at least 80–150 MΩ can be pulled. Brown and Flaming (1977) have designed a new electrode puller which should, if it lives up to its early promise, go a long way toward solving microelectrode problems.

Given suitable microelectrodes, the only obstacle to successful recording will be lack of mechanical stability in the micromanipulator/microscope assembly. A quick check on this can be made by observing the tip of an electrode with an optical magnification of about 500. If the slightest vibration is visible, the mechanical arrangements need to be improved. Commercial anti-vibration tables are extremely effective against vertical components of building vibration, but less so against horizontal components. Smooth muscle cells in culture may be very thin, often 2 μm or less. This poses a special problem of longer term mechanical stability, since a small "creep" will force the microelectrode through the cell onto the supporting coverglass. In the writer's laboratory this problem was traced to thermal expansion in the first few hours of an experiment as the room warmed up. The solution was to switch on the electrical apparatus some hours earlier to allow time for thermal equilibration.

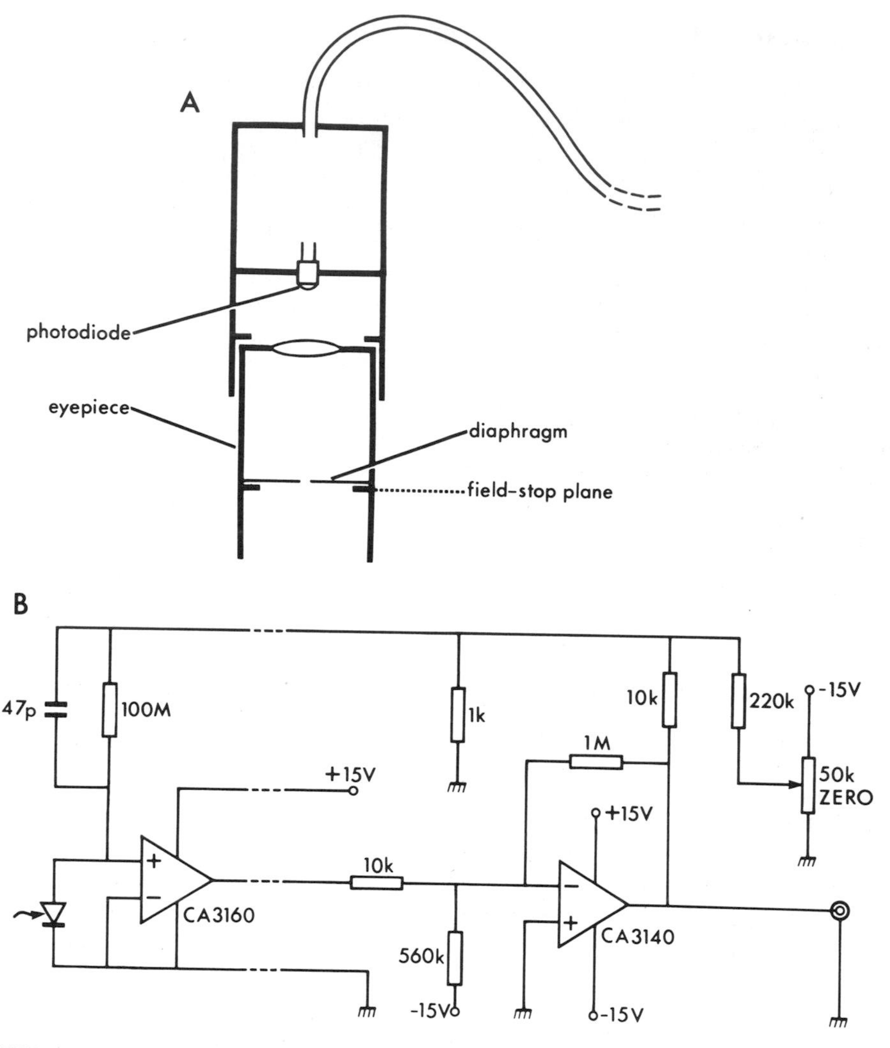

FIGURE 2. Photoelectric recording device. (A) Attachment to one eyepiece of a microscope. The photodiode is mounted at the exit pupil of the eyepiece. Its "receptive field" is controlled by a small hole in a diaphragm placed in the field-stop plane. (B) Electronic circuit. The components to the left are mounted within the eyepiece attachment. Photodiode: Siemens BPX 63.

An elaborate perfusion apparatus is unnecessary in many experiments, and it is often easiest to make recordings with the cells simply sitting in the medium in which they have been grown. A layer of ordinary medicinal paraffin oil run over the surface retards evaporation and loss of CO_2. It does not clog the tips of microelectrodes as they are lowered through it. The culture may be kept warm by direct electrical heating of the medium using a proportional-control feedback circuit like that described by Saltz and Geller (1976). The heating coil of two or three turns is made from a length (10–20 cm) of wire taken from a domestic radiator and inserted into a similar length of polythene (or better, polytetrafluorethylene) insulation. The insulation resistance is initially more than 10^{10} Ω, but after a few months of use cracks appear, and a new coil must be made. Toxins must be leached out of a new coil before it is used by soaking for a week in distilled water.

No doubt almost any compound microscope could be made suitable for observation of impalement, but it is very convenient to have inverted optics with tube focusing. The Zeiss Invertoscope D is moderately priced and can be fitted with long working distance Zeiss phase contrast illuminators. The working distance of several centimeters means that there is plenty of space above the culture for microelectrodes which can be inserted vertically (or nearly so). The stage of this microscope needs to be strengthened by the addition of supporting struts at the front. If a microscope with upright optics is used, the electrodes have to be bent near their tips (Hudspeth and Corey, 1978) so as to clear the objective lens. Sometimes the difficulties of microelectrode methods can be avoided by making photoelectric recordings of the contractions of groups of smooth muscle cells (Purves *et al.*, 1974). A suitable electronic circuit which uses a photodiode as the light sensor is given in Fig. 2. The photodiode and associated circuitry are enclosed in a light-tight box that fits over one eyepiece of the microscope in such a way that the diode is at the exit pupil.

3. APPLICATIONS

There are many descriptive accounts of the microscopic appearance and histochemistry of smooth muscle in culture (for example, Wada and Pollak, 1967; Ross, 1971; Campbell *et al.*, 1971, 1974; Chamley and Campbell, 1975; Gröschel-Stewart *et al.*, 1975; Mauger *et al.*, 1975; Hermsmeyer *et al.*, 1976; Chamley *et al.*, 1974, 1977). Earlier references are given by Murray (1965). Biochemical measurements have been made on norepinephrine metabolism (Yates *et al.*, 1972; Powis, 1973) and cyclic AMP production (Honeyman *et al.*, 1978). An important general review of smooth muscle in culture is given by Chamley-Campbell *et al.* (1979). Some more specifically physiological applications are considered below.

3.1. Electrical Activity

The electrical behavior of some smooth muscles in culture is disappointing. Primary cultures of mammalian vas deferens (Chamley *et al.*, 1974) and ureter (Chamley and Campbell, 1975) undergo a cycle of dedifferentiation, proliferation, and redifferentiation, and spontaneous contractions can be seen in the early and late phases. Nevertheless, intracellular recording (Purves, unpublished observations) has so far failed to reveal any electrical activity.

The taenia coli of the guinea pig adapts more happily to dissociated cell culture. Activity, both contractile and electrical, is evident for 3 weeks or more before the cells lose their differentiated smooth muscle appearance and behavior. Intracellular microelectrode recordings (Purves *et al.*, 1973) from clusters of cells show complicated spontaneous activity consisting of slow waves and action potentials (Figs. 3 and 4) remarkably like those that have been seen in the adult organ. The complexity of these records and the variable height of the action potentials suggest that activity propagates to a variable extent through connected clusters of smooth muscle cells. Evidently in this respect no useful simplification of the electrotonic properties of the adult organ has resulted from tissue culture. Nevertheless, the results of current injection, discussed in Section 3.2, show that the electrotonic properties are somewhat simpler than in whole smooth muscle. Most attempts to impale single isolated cells have been unsuccessful, but records from six cells have been made with a total recording

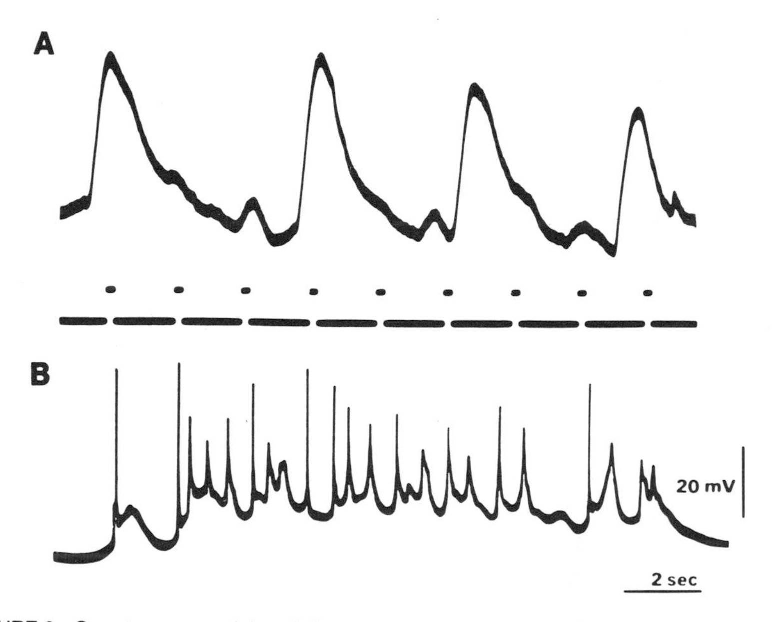

FIGURE 3. Spontaneous activity of clusters of muscle cells in 9-day culture from taenia coli of the guinea pig. (A) Contractions recorded photoelectrically. Time marker: 10-sec intervals. (B) Intracellular electrical recording. The trace shows a complete burst of activity following which electrical silence was maintained for 1 min. Same culture as A but different cluster. (From Purves *et al.*, 1973, with permission.)

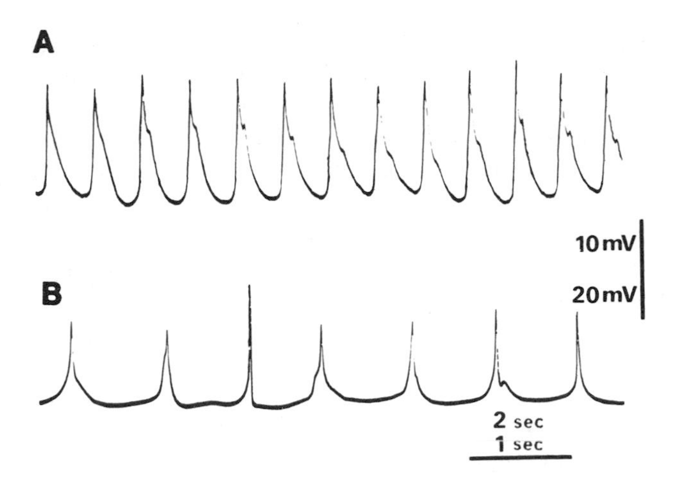

FIGURE 4. Regular spike activity in clusters of taenia coli cells from two different cultures. (A) 3-week culture. (B) 9-day culture. (From Purves *et al.*, 1973, with permission.)

time less than 10 min. These fragmentary results showed a more uniform behavior (Fig. 5), the action potentials being of nearly constant height. The frequency of occurrence of spontaneous action potentials was, however, quite irregular. Thus, some aspects of the variability of smooth muscle arise at the single cell level, and others (such as spike height changes) are presumably the result of interaction among numerous individual cells.

McLean and Sperelakis (1977) recorded from aggregates of muscle cells

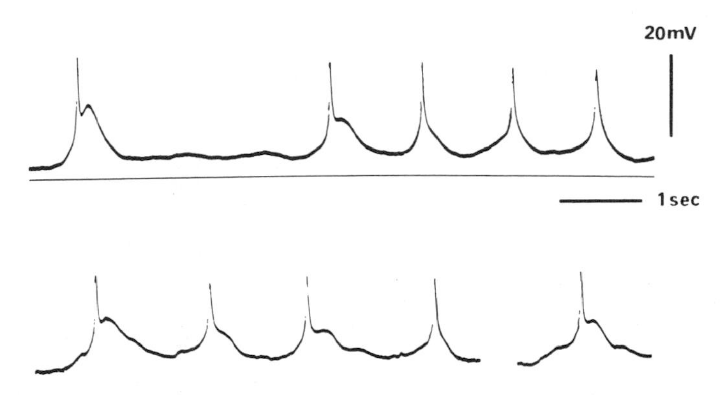

FIGURE 5. Spontaneous activity in single isolated cells from taenia coli. Upper trace: 3-week culture. The cell was electrically silent for 30 sec before and 14 sec after this record was made. (From Purves *et al.*, 1973, with permission.) Lower trace: 8-day culture. The break in the trace indicates a period of electrical silence lasting 8 sec.

cultured from embryonic chick blood vessels. The electrical activity of their cultures did not resemble that of smooth muscle, as ordinarily conceived, but consisted of regularly occurring action potentials of cardiac type. Hermsmeyer (personal communication) has described, in cultures from the aortas of rats and chicks, a population of muscle cells whose contractions are much faster than those of normal smooth muscle. Both of these sets of observations are difficult to fit into the mainstream of smooth muscle physiology. They are doubtless related to the ability of some blood vessels to undergo rhythmic pulsatile contractions (Attardi and Attardi-Gandini, 1955) and to the presence of frank cardiac muscle in the great vessels of some animals (Paes de Almeida *et al.,* 1975).

A continuous cell line derived from a mouse neoplasm was studied by Schubert *et al.* (1974) and showed some properties akin to those of smooth muscle. The tissue of origin of this line is unknown, and the electrical behavior was not especially characteristic of smooth muscle.

Spontaneous action potentials and slow waves (Fig. 6) have been recorded from smooth muscle cells cultured from human uterine tube (Sinback and Shain, 1979). Freshly isolated cells from the stomach of the toad are not spontaneously active, but spikes can be evoked by current injection (Singer and Walsh, 1977). Active responses from isolated cells from salamander stomach have been reported (Caffrey and Anderson, 1979).

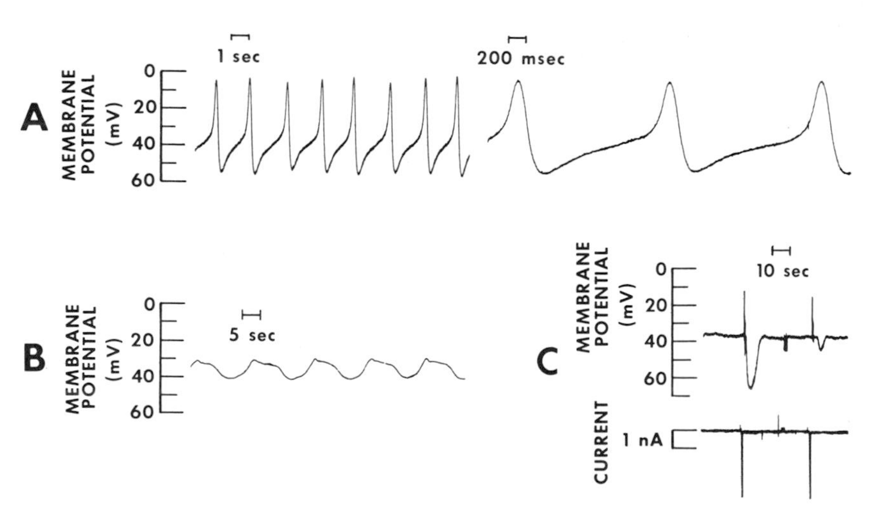

FIGURE 6. Intracellular electrical records from human oviduct smooth muscle cells in culture. (A) Spontaneous action potentials recorded at two different sweep speeds. (B) Slow waves. (C) Hyperpolarizing responses to iontophoretically applied histamine (current shown on bottom trace). (Records kindly supplied by Dr. C. N. Sinback.)

3.2. Passive Electrical Properties

The membrane resistance of smooth muscle, as estimated by extracellular current injection in whole tissue, is extremely high, of the order of 100 kΩ · cm^2 (Tomita, 1975). This value applied to a cell 5 μm in diameter and 150 μm long predicts an input resistance of several GΩ. The technical difficulty of measuring such a high resistance directly by current injection is immense; nevertheless, Singer and Walsh (1977) have succeeded in obtaining values of over 3 GΩ from isolated toad stomach muscle cells. Many of their measured values were much less than this (down to 100 MΩ), the likely cause being leakage through an imperfect seal between microelectrode and membrane. The oviduct cells studied by Sinback and Shain (1979) are much bigger and have a mean input resistance of only 66 MΩ. The calculated membrane resistance is 63 kΩ · cm^2. The older results from whole tissue and these impressive new results from single cells are thus in fair agreement.

Electrical coupling between neighboring muscle cells in culture can be shown convincingly by simultaneous use of two microelectrodes (Schubert *et al.*, 1974; Sinback and Shain, 1976, 1979). For obvious technical reasons few experiments of this type have been done, and it has not been possible to estimate the junctional resistance between cells.

3.3. Response to Neurotransmitters

Tissue-cultured preparations are nearly ideal for the precision high-speed iontophoretic application of transmitters and other drugs. Pipettes can be placed with accuracy, and connective tissue barriers to diffusion are minimal. The favorable geometry of the tissue allows the use of an advanced body of theory concerning the time course of cellular responses to iontophoretically applied substances (Purves, 1977; Hill-Smith and Purves, 1978). In addition, smooth muscle in culture is highly sensitive to bath-applied neurotransmitters (Hermsmeyer, 1976).

The first detailed investigation of the speed of response of muscarinic acetylcholine receptors was made on cultured smooth muscle from the taenia coli of the guinea pig (Purves, 1974). The responses to iontophoretic acetylcholine were considerably slower than those of nicotinic receptors of skeletal muscle, and there was an irreducible lag time of the order of 100 msec (Fig. 7). These findings have now been confirmed in a variety of adult intact tissues (see Purves, 1978, for references), and it is clear that the 1000-fold difference in time course of muscarinic and nicotinic responses is an important physiological correlate of the well-known pharmacological differences. It is possible that the pharmacological distinction between α and β adrenoceptors is also associated with different speeds of response. Iontophoretic application of norepinephrine to cultured vas deferens smooth muscle evokes an excitatory response within

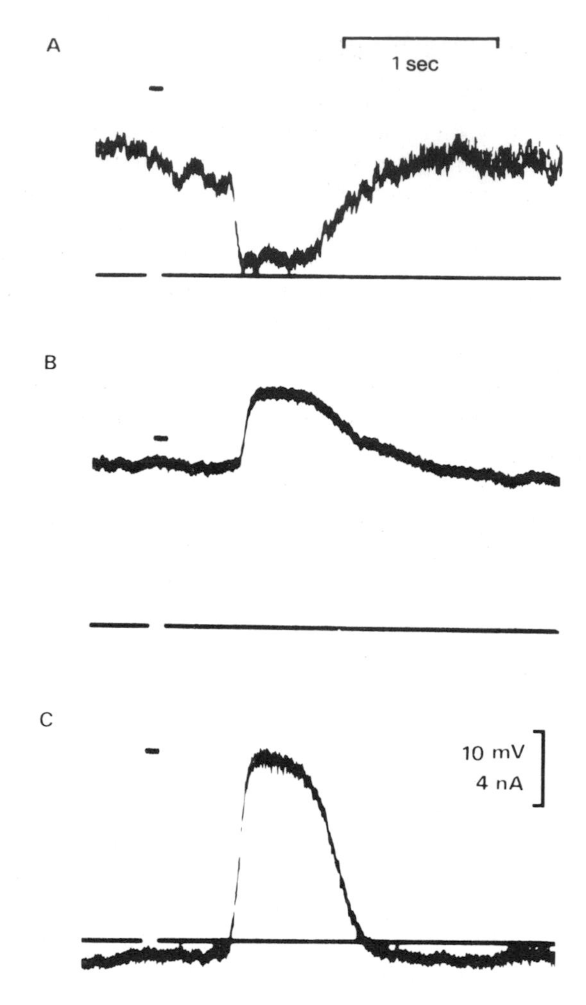

FIGURE 7. Responses of taenia coli muscle cells to iontophoretically applied acetylcholine. In (A), an outward polarizing current of 1.7 nA was passed through the recording electrode. Record (B) was obtained without polarizing current. In record (C), an inward current of 1.6 nA was passed. Note the long latent period of the responses and the reversal by outward current injection. 5-day culture. (From Purves, 1974, with permission.)

a fraction of a second, whereas the beating rate of cultured heart muscle cannot be increased by norepinephrine in less than 3–5 sec (Hill-Smith and Purves, 1978).

A noteworthy feature of the responses shown in Fig. 7 is that the depolarizing response to acetylcholine could be changed in amplitude or even reversed in sign by the injection of steady currents through the recording electrode. This indicates that injected currents reach a large proportion of the membrane affected by the transmitter. In intact smooth muscle, the spatial decay of potential around a current source is abrupt, and this result cannot be obtained. Thus the electrotonic properties of a small cluster of cells in culture are simpler than those of the whole tissue, and changes in membrane conductance can be directly measured with microelectrodes (see Purves, 1976, for further discussion).

No experimental evidence could be obtained in the above studies about the subcellular localization of receptors for neurotransmitters, since the cells were closely entangled in small clusters. The oviduct cell line of Sinback and Shain includes many large single cells whose exposed surface area is great enough to allow accurate application of drugs to different regions (Sinback and Shain, 1977). Responses to histamine were depolarizing, hyperpolarizing (Fig. 6), or biphasic; in some cases, the depolarizing and hyperpolarizing receptors were segregated from each other in two halves of the cell (Sinback and Shain, 1977). This appears to be the only direct evidence to date relating to localization of receptors in smooth muscle.

3.4. Innervation

Interactions between nerves and smooth muscle have been studied in cultures from insects (Aloe and Levi-Montalcini, 1972) and mammals (Mark *et al.*, 1973; Johnson *et al.*, 1973; Cook and Peterson, 1974). Fully functional cholinergic transmission was demonstrated between a variety of autonomic ganglia and smooth muscles by stimulation of the nerves and photoelectric recording of the resulting muscular contractions (Purves *et al.*, 1974; Hill *et al.*, 1976). None of these investigations can be said to have added much to knowledge about the physiology of autonomic neuroeffector transmission.

3.5. Contraction

The microscopic appearance of the contractile apparatus of acutely isolated smooth muscle cells has been studied by Fay and Delise (1973), Small (1974), and Fisher and Bagby (1977). Assemblies of contractile filaments can be recognized in the polarizing microscope. During contraction, these fibrils become angled to the long axis of the cell so as to take on a helical arrangement in contrast to the end-to-end structure of skeletal muscle myofibrils.

Contractions of single cells from the toad's stomach have been measured photographically (Fay and Delise, 1973; Bagby, 1974; Fisher and Bagby, 1977) and with a micro force transducer (Fay, 1977). Interestingly, the cells do not give all-or-none contractions in response to electrical stimulation but respond in a graded fashion. Again, contractions localized to one part of the cell are sometimes seen. It is, unfortunately, not clear whether this behavior results from membrane damage; contraction of the intact cells *in situ* might be more tightly regulated by all-or-none action potentials.

A method for examining the state of contraction of a population of isolated cells was developed by Kominz and Gröschel-Stewart (1972) and further elaborated by Singer and Fay (1977). A suspension of fixed cells is passed through a Coulter counter. Analysis of the electrical signals gives an estimate of the lengths and diameters of the cells. Fay and Singer (1977) used this technique in experiments on the time courses of cholinergic stimulation and antagonism.

4. CONCLUSIONS

It is plain that but little progress has been made in the ambitious program of understanding smooth muscle by literal dissection into single cells followed by conceptual resynthesis. But it is not plain that the quest is hopeless. Some of the passive electrical properties of single cells have been directly measured, and from these the complex electrotonic behavior of an assembly of cells can be computed (Bennett, 1972, 1973; Purves, 1976). In this sense, the passive properties of smooth muscle are understood, although direct measurements of, for instance, junctional resistance are lacking. To carry out the same program for the active electrogenic properties of smooth muscle will be very much harder, since the discomforts of life *in vitro* are more likely to disrupt active membrane properties. The few studies that have been made of neurotransmission and of the effects of artificially applied transmitters have certainly not exhausted the potentialities of the method.

Murray (1965), in describing the tissue culture of smooth muscle, concluded that it had not "contributed importantly to recent advances in the pharmacology or biophysics of this tissue." Work published since 1965 does not compel us to modify Murray's cautious assessment.

REFERENCES

Aloe, L., and Levi-Montalcini, R., 1972, Interrelation and dynamic activity of visceral muscle and nerve cells from insect embryos in long-term cultures, *J. Neurobiol.* **3**:3.

Attardi, G., and Attardi-Gandini, D., 1955, Spontaneous peristaltic activity of veins of chick embryos and newly hatched chickens explanted *in vivo, Experientia* **11**:37.

Bagby, R. M., 1974, Time course of isotonic contractions in single cells and muscle strips from *Bufo marinus* stomach, *Am. J. Physiol.* **227**:789.

Bagby, R. M., Young, A. M., Dotson, R. S., Fisher, B. A., and McKinnon, K., 1971, Contraction of single smooth muscle cells from *Bufo marinus* stomach, *Nature* **234**:351.

Barr, L., Berger, W., and Dewey, M. M., 1968, Electrical transmission at the nexus between smooth muscle cells, *J. Gen. Physiol.* **51**:347.

Bennett, M. R., 1972, *Autonomic Neuromuscular Transmission,* Cambridge University Press, Cambridge.

Bennett, M. R., 1973, Structure and electrical properties of the autonomic neuromuscular junction, *Philos. Trans. R. Soc. London Ser. B* **265**:25.

Bozler, E., 1948, Conduction, automaticity and tonus of visceral muscles, *Experientia* **4**:213.

Brown, K. T., and Flaming, D. G., 1977, New microelectrode techniques for intracellular work in small cells, *Neuroscience* **2**:813.

Bülbring, E., 1954, Membrane potentials of smooth muscle fibres of the taenia coli of the guinea-pig, *J. Physiol. (Lond.)* **125**:302.

Bülbring, E., Burnstock, G., and Holman, M. E., 1958, Excitation and conduction in the smooth muscle of the isolated taenia coli of the guinea-pig, *J. Physiol. (Lond.)* **142**:420.

Caffrey, J. M., and Anderson, N. C., 1978, Isolation of smooth muscle cells from Amphiuma stomach, *Fed. Proc.* **37**:639.

Caffrey, J. M., and Anderson, N. C., 1979, Electrogenesis in isolated smooth muscle cells, *Biophys. J.* **25**:117a.

Campbell, G. R., Uehara, Y., Mark, G., and Burnstock, G., 1971, Fine structure of smooth muscle cells grown in tissue culture, *J. Cell Biol.* **49**:21.

Campbell, G. R., Chamley, J. H., and Burnstock, G., 1974, Development of smooth muscle cells in tissue culture, *J. Anat.* **117**:295.

Chamley, J. H., and Campbell, G. R., 1975, Isolated ureteral smooth muscle cells in culture including their interaction with intrinsic and extrinsic nerves, *Cytobios* **11**:358.

Chamley, J. H., Campbell, G. R., and Burnstock, G., 1974, Dedifferentiation, redifferentiation and bundle formation of smooth muscle cells in tissue culture: The influence of cell number and nerve fibres, *J. Embryol. Exp. Morphol.* **32**:297.

Chamley, J. H., Campbell, G. R., McConnell, J. D., and Gröschel-Stewart, U., 1977, Comparison of vascular smooth muscle cells from adult human, monkey and rabbit in primary culture and in subculture, *Cell Tissue Res.* **177**:503.

Chamley-Campbell, J., Campbell, G. R., and Ross, R., 1979, The smooth muscle cell in culture, *Physiol. Rev.* **59**:1.

Cook, R. D., and Peterson, E. R., 1974, The growth of smooth muscle and sympathetic ganglion in organotypic tissue cultures, *J. Neurol. Sci.* **22**:25.

Eccles, J. C., and Magladery, J. W., 1937, The excitation and response of smooth muscle, *J. Physiol. (Lond.)* **90**:31.

Fay, F. S., 1977, Isometric contractile properties of single isolated smooth muscle cells, *Nature* **265**:553.

Fay, F. S., and Delise, C. M., 1973, Contraction of isolated smooth muscle cells—Structural changes, *Proc. Natl. Acad. Sci. USA* **70**:641.

Fay, F. S., and Singer, J. J., 1977, Characteristics of response of isolated smooth muscle cells to cholinergic drugs, *Am. J. Physiol.* **232**:C144.

Fisher, B. A., and Bagby, R. M., 1977, Reorientation of myofilaments during contraction of a vertebrate smooth muscle, *Am. J. Physiol.* **232**:C5.

Fry, D. M., 1975, A scanning electron microscope method for the examination of glass microelectrode tips either before or after use, *Experientia* **31**:695.

Gröschel-Stewart, U., Chamley, J. H., Campbell, G. R., and Burnstock, G., 1975, Changes in myosin distribution in dedifferentiating smooth muscle cells in tissue culture, *Cell Tissue Res.* **165**:13.

Gröschel-Stewart, U., Chamley, J. H., McConnell, J. D., and Burnstock, G., 1976, Membrane alteration of trypsin-treated smooth muscle cells and penetration by antibodies to myosin, *Histochemistry* **47**:285.

Hermsmeyer, K., 1976, Cellular basis for increased sensitivity of vascular smooth muscle in spontaneously hypertensive rats, *Circ. Res. (Suppl.)* **38**:1153.

Hermsmeyer, K., DeCino, P., and White, R., 1976, Spontaneous contractions of dispersed vascular muscle in cell culture, *In Vitro* **12**:628.

Hill, C. E., Purves, R. D., Watanabe, H., and Burnstock, G., 1976, Specificity of innervation of iris musculature by sympathetic nerve fibres in tissue culture, *Pflügers Arch.* **361**:127.

Hill, Smith, I., and Purves, R. D., 1978, Synaptic delay in the heart: An ionophoretic study, *J. Physiol. (Lond.)* **279:**31.

Honeyman, T., Merriam, P., and Fay, F. S., 1978, The effects of isoproterenol on adenosine cyclic 3′5′ monophosphate and contractility in isolated smooth muscle cells, *Mol. Pharmacol.* **14:**86.

Hudspeth, A. J., and Corey, D. P., 1978, Controlled bending of high resistance glass microelectrodes, *Am. J. Physiol.* **234:**C56.

Johnson, D. G., Weise, V. K., Hanbauer, I., Silberstein, S. D., and Kopin, I. J., 1973, Dopamine-β-hydroxylase activity during sympathetic reinnervation of rat iris in organ culture, *Neurobiology* **3:**88.

Kimes, B. W., and Brandt, B. L., 1976, Characterisation of two putative smooth muscle cell lines from rat thoracic aorta, *Exp. Cell Res.* **98:**349.

Kominz, D. R., and Gröschel-Stewart, U., 1972, Antibody-dependent size changes of myofibrils and isolated smooth muscle cells, *J. Mechanochem. Cell Motil.* **2:**181.

Mark, G. E., Chamley, J. H., and Burnstock, G., 1973, Interactions between autonomic nerves and smooth and cardiac muscle cells in tissue culture, *Dev. Biol.* **32:**194.

Masson-Pévet, M., Jongsma, H. J., and DeBruijne, J., 1976, Collagenase- and trypsin-dissociated heart cells: A comparative ultrastructural study, *J. Mol. Cell Cardiol.* **8:**747.

Mauger, J. P., Worcel, M., Tassin, J., and Courtois, Y., 1975, Contractility of smooth muscle cells of rabbit aorta in tissue culture, *Nature* **255:**337.

McLean, M. J., and Sperelakis, N., 1977, Electrophysiological recordings from spontaneously contracting reaggregates of cultured vascular smooth muscle cells from chick embryos, *Exp. Cell Res.* **104:**309.

Murray, M. R., 1965, Muscle, in: *Cells and Tissues in Culture* (E. N. Willmer, ed.), pp. 311–372, Academic Press, London.

Paes de Almeida, O., Böhm, G. M., de Paula Carvalho, M., and Paes de Carvalho, A., 1975, The cardiac muscle in the pulmonary vein of the rat: A morphological and electrophysiological study, *J. Morphol.* **145:**409.

Powis, F., 1973, The accumulation and metabolism of (−)-noradrenaline by cells in culture, *Br. J. Pharmacol.* **47:**568.

Purves, R. D., 1974, Muscarinic excitation: A microelectrophoretic study on cultured smooth muscle cells, *Br. J. Pharmacol.* **52:**77.

Purves, R. D., 1976, Current flow and potential in a three-dimensional syncytium, *J. Theor. Biol.* **60:**147.

Purves, R. D., 1977, The time course of cellular responses to iontophoretically applied drugs, *J. Theor. Biol.* **65:**327.

Purves, R. D., 1978, The physiology of muscarinic acetylcholine receptors, in: *Cell Membrane Receptors for Drugs and Hormones* (L. Bolis and R. W. Straub, eds.), pp. 69–79, Raven Press, New York.

Purves, R. D., Mark, G. E., and Burnstock, G., 1973, The electrical activity of single isolated smooth muscle cells, *Pflügers Arch.* **341:**325.

Purves, R. D., Hill, C. E., Chamley, J. H., Mark, G. E., Fry, D. M., and Burnstock, G., 1974, Functional autonomic neuromuscular junctions in tissue culture, *Pflügers Arch.* **350:**1.

Rose, G. G., 1954, A separable and multipurpose tissue culture chamber, *Tex. Rep. Biol. Med.* **12:**1074.

Ross, R., 1971, The smooth muscle cell. II. Growth of smooth muscle in culture and formation of elastic fibres, *J. Cell Biol.* **50:**172.

Saltz, E. C., and Geller, H. M., 1976, Simple circuit for laboratory temperature regulation, *Med. Biol. Eng.* **14:**681.

Schubert, D., Harris, A. J., Devine, C. E., and Heinemann, S., 1974, Characterization of a unique cell line, *J. Cell Biol.* **61:**398.

Sinback, C. N., and Shain, W., 1976, Electrophysiological properties of smooth muscle in dissociated cell culture, *Neurosci. Abstr.* **2**:358.

Sinback, C. N., and Shain, W., 1977, Distribution of histamine, acetylcholine, and noradrenaline receptors on single smooth muscle cells in dissociated cell culture, *Neurosci. Abstr.* **3**:529.

Sinback, C. N., and Shain, W., 1979, Electrophysiological properties of human oviduct smooth muscle cells in dissociated cell culture, *J. Cell. Physiol.* **98**:377.

Singer, J. J., and Fay, F. S., 1977, Detection of contraction of isolated smooth muscle cells in suspension, *Am. J. Physiol.* **232**:C138.

Singer, J. J., and Walsh, J. V., 1977, Electrical properties of freshly isolated single smooth muscle cells, in: *Excitation–Contraction Coupling in Smooth Muscle* (R. Casteels, ed.), pp. 53–60, Elsevier/North Holland Biomedical Press, Amsterdam.

Small, J. V., 1974, Contractile units in vertebrate smooth muscle, *Nature* **249**:324.

Tomita, T., 1975, Electrophysiology of mammalian smooth muscle, *Prog. Biophys. Mol. Biol.* **30**:185.

Wada, A., and Pollak, O. J., 1967, Aggregation in vivo of cultured aorta cells of rabbits, *Nature* **214**:1358.

Yates, C. S., Pollock, D., and Muir, T. C., 1972, The uptake of (^{3}H)noradrenaline by human isolated smooth muscle cells, *Life Sci.* **11**:1183.

11

Calcium Exchange in Myocardial Tissue Culture

GLENN A. LANGER

1. INTRODUCTION

Calcium (Ca^{2+}) exchange in whole, functional cardiac tissue is complex (Shine *et al.*, 1971). The exchangeable fraction is characterized by a minimum of five phases. These include vascular, interstitial, and cellular components that are not unequivocally separable. Quantitation of contents ascribable to particular morphological loci and fluxes across identifiable membranes is not possible. A unique model that includes the five phases, describes tissue Ca^{2+} flux, and is capable of being tested experimentally cannot be rigorously defined.

It was obvious that a great deal of information on Ca^{2+} exchange could be derived if an isolated cardiac cell preparation were available. This would eliminate the vascular and interstitial components and make the pattern of cellular exchange much more amenable to direct analysis. Unfortunately, dissection of cells from adult hearts, usually by enzymatic or mechanical means, leaves them in a compromised state (Vahouny *et al.*, 1970; Bloom, 1971; Berry *et al.*, 1970). The cells derived by most techniques rapidly progress into contracture if exposed to solutions in which the Ca^{2+} concentration is more than a few micromolar. This clearly indicates a leaky membrane state—a condition in which definition of transsarcolemmal Ca^{2+} exchange would be meaningless.

The alternative experimental model for cells isolated from other tissue components was cells in culture. These promised to be much less complex from a kinetic point of view and would allow for the application of large molecules, otherwise restricted to the vascular space or surface in whole tissue, directly to

---

GLENN A. LANGER • Departments of Medicine and Physiology and the American Heart Association, Greater Los Angeles Affiliate Cardiovascular Research Laboratories, University of California at Los Angeles, Center for the Health Sciences, Los Angeles, California 90024.

the cellular membrane. The latter would permit histochemical and specific enzymatic techniques to be applied. Such techniques have proved to be essential to the further definition of the factors important in the control of transsarcolemmal Ca^{2+} movements (Frank *et al.*, 1977). Culture techniques have also proven to be valuable in the analysis of kinetics and compartmentalization of Ca^{2+} (Borle, 1968, 1969a,b), since identification of cellular components is clear.

We decided, therefore, to attempt to develop a system that would use cultured myocardial cells and would allow for on-line measurement of ^{45}Ca labeling and washout at high sensitivity. The basis for this system was developed some 10 years ago and has been used with various improvements to the present.

2. THE SCINTILLATION-DISK FLOW CELL TECHNIQUE

The principle of the technique is the growth of the cells directly on the surface of the radioisotopic detector (Langer *et al.*, 1969; Frank *et al.*, 1977). The original detector used was a single slide of glass scintillator with cells grown on both surfaces. The flow cell used to perfuse the slide was adapted to fit the well of a standard scintillation spectrometer. This technique has been greatly modified in order that a greater mass of cells could be used and in order to produce much increased counting efficiency. At present, a cellular monolayer is grown on the surface of each of two disks composed of polystyrene combined with scintillator material (Nuclear Enterprises, San Carlos, California). The disks are 45 mm in diameter and 1 mm thick and are designed to form a portion of each side of the flow cell chamber illustrated in Fig. 1.

The culture technique is a modification of that described by Harary and Farley (1963). Ten to twenty whole hearts from zero- to two-day-old baby rats are washed in ice-cold Saline A* and then minced with scissors. The minced tissue is placed in a spinner flask and, with gentle stirring, subjected to three successive 15-min digestions followed by six 10-min digestions at 37 °C with 0.1% trypsin (Calbiochem, San Diego, California, B grade). The trypsin is diluted with Saline A. The nine digestions are usually sufficient to disintegrate the fragments almost completely.

The supernatants from the first three digestions, which contain fibroblasts and debris, are discarded. Each of the subsequent supernatants is poured into 3 ml of precooled growth medium in a centrifuge tube and placed in ice. The growth medium consists of Ham's F-10 without $NaHCO_3$ added (North American Biologicals, Miami, Florida) plus 10% horse and 10% fetal bovine sera (Microbiological Associates, Walkersville, Maryland) with 100 U penicillin and 0.1 mg streptomycin/ml. Each tube is centrifuged for 5 min at half speed in a clinical centrifuge followed by a wash with growth medium, cooled,

*Composition: NaCl, 137 mM; KCl, 5.4; dextrose, 5.9; $NaHCO_3$, 4.2.

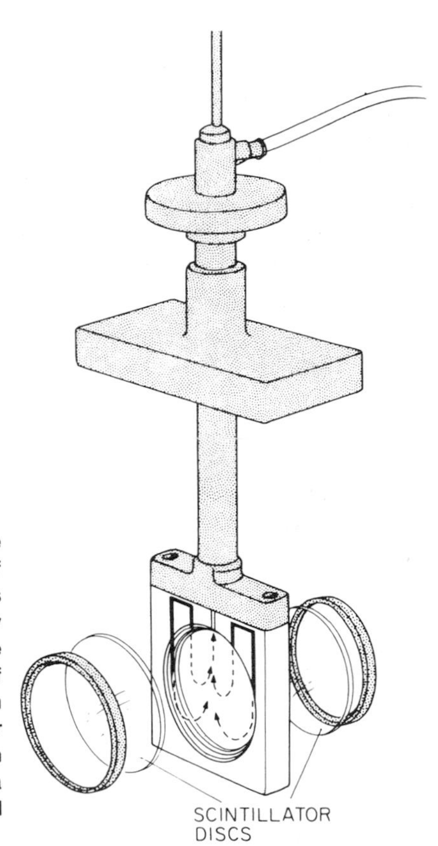

FIGURE 1. The flow cell. Myocardial cells are grown in monolayer on one surface of each of the scintillator disks. The surface with cells attached forms the inner surface of the flow cell wall. With the disks sealed in place, the flow cell is inserted into the lightproof well of the spectrometer so that the disks are 4.5 mm from each of two apposed photomultiplier tubes. Perfusate enters the flow cell through the four ports indicated and exits through a single port at the top of the cell. (Reproduced with permission of *Circulation Research.*)

and centrifuged again for 3 min at half speed. The pellets are then resuspended by repetitive pipetting and distributed into large petri dishes with 10–12 ml of growth medium. The pellets derived from 20 hearts are distributed in six large petri dishes and incubated at 37°C for 2–3 hr (Blondel *et al.*, 1971). During this period the fibroblasts attach to the dishes, while most of the myoblasts remain spherical and suspended. At the end of this preincubation the supernatant of myoblasts from each large dish is pipetted into eight to ten 50-mm diameter dishes, each of which contains a 45-mm polystyrene disk. The disks are pretreated by Falcon Plastics (Oxnard, California) to increase cellular adhesivity and sterilized by exposure to ultraviolet light for 48 hr prior to use. The number of disks finally incubated varies with the number of hearts used and evaluation of the final pellet size. The cells are incubated in an air–CO_2 environment (pH 7.1–7.2) for 24 hr and then given a medium change to remove dead cells and debris. The medium is changed subsequently every 36 hr.

Thin monolayer cultures are desirable, since thicker cultures present problems of diffusion as well as screening of the low energy ^{45}Ca emission from the disks. The latter affects the counting efficiency of the system. The cultures are checked by phase microscopy and are not used if they are not essentially monolayered. The layer of cells on each disk weighs between 0.6 and 1.0 mg dry. In order to determine the percentage of myoblasts, a series of representative cultures were surveyed with electron microscopy and found to be routinely 80–90% myoblastic. In monolayered cultures it is also easy to distinguish the beating myoblasts with their sarcoplasmic inclusions from the flat, nonbeating, more transparent fibroblasts.

After 3 or 4 days, the disks are removed from the culture medium and mounted in the flow cell with the surfaces to which the cells are attached directed inward in contact with the perfusate flowing through the flow cell (see Fig. 1). The flow cell is then inserted into the well of a modified Beta-Mate II spectrometer (Beckman Instruments). The well of the spectrometer includes a Lucite jacket which surrounds the flow cell to contain inadvertent leaks. Any leak of the conductive perfusion solution completes a circuit that sets off an alarm to permit correction of the problem before damage to the spectrometer occurs. The opposed photomultiplier tubes of the spectrometer are mounted flush with the Lucite jacket and aligned with the disk on each side of the flow cell. The disk-to-photomultiplier tube distance is 4.5 mm.

With the disks mounted in each side of the flow cell, the cell is inserted into the well of the spectrometer. The cultured cells are then perfused through four entry ports about the perimeter of the disks (Fig. 1). This creates turbulence and improves mixing. Effluent exits through a single port at the top of the flow cell. ^{45}Ca labeling is usually done at a flow rate of 10 ml/min, and washout at 24 ml/min. The volume of the flow cell is 5 ml, and at a flow rate of 24 ml/min it exchanges with a rate constant of 4.6/min. Rapid components of cellular exchange are perfusion-limited at this rate, but slower components are not (see Section 2.1.2).

The disks, with cells attached, are equilibrated to the nonisotopic perfusion solution for 30 min prior to isotopic labeling. The standard solution is of the following composition (mM): NaCl, 133; KCl, 3.6; $CaCl_2$, 1.0; $MgCl_2$, 0.3; glucose, 16.0; N-2-hydroxyethylpiperazine-N′-2 ethane sulfonic acid (HEPES) buffer, 3.0. Most experiments are done at ambient temperature (23–24°C). The HEPES buffer is used routinely, since much of the experimentation requires the use of lanthanum chloride in the solution. Lanthanum salts, other than the chloride, are very insoluble. ^{45}Ca labeling can be followed continuously, since the efficiency for counting the 0.25 MeV β emission from ^{45}Ca bound to the cellular layer on the disks is 70%, compared with approximately 5% for ^{45}Ca in the perfusion solution flowing past the disks. This results from the high quenching of the weak β emission by the solution. Following a period of labeling, ^{45}Ca washout is commenced by simply switching to a nonisotopic solution.

The counting efficiency is determined in the following manner. At the completion of isotopic labeling or washout the disks are removed, their surfaces briefly rinsed to remove unbound isotope, and the cells are scraped onto a piece of predried, preweighed filter paper. The cells are dried overnight at 100°C, and the dry weight is obtained. The filter paper with the dried cells is then placed in 2 ml Protosol (New England Nuclear Corp.) tissue solubilizer for 10–12 hr, 0.5 ml glacial acetic acid is added to neutralize the alkaline solution (to reduce background luminescence) followed by the addition of 15 ml of Aquasol (New England Nuclear Corp.) scintillator solution. The Protosol dissolves the tissue, leaving the filter paper in the vial. The counting efficiency of this system was determined by placing known activities of ^{45}Ca on identical strips of filter paper and treating these standards in exactly the same manner as described above. The solubilized tissue sample is then counted in the spectrometer (Model LS 200B, Beckman Instruments, Fullerton, California) at known efficiency. The final isotopic activity as recorded from the tissue on the disks in the flow cell is then compared to the corrected isotopic activity from the dissolved tissue to derive the counting efficiency of the scintillator disks. As indicated above, the efficiency for counting ^{45}Ca in the cellular layer is generally 70.0 ± 1.7% (SEM) as compared to 5% for the labeling solution in the flow cell. (The latter value is obtained by simply filling the flow cell with a measured amount of ^{45}Ca solution of known specific activity and recording the counts obtained.)

In each experiment, after removal of the cell layer from the disks, the disks are rinsed with ethylenediaminetetraacetate (EDTA) to remove all traces of radioactivity, returned to the flow cell, and exposed to the same experimental sequence applied to the disks when cells were attached. This procedure establishes the response of the "blank." Subtraction of the "blank disk" sequence from the "disks with cells" defines the isotopic activity of the cellular layer alone.

2.1. Pattern of Calcium Exchange

Experience over the period of the past 6 years indicates that the fundamental pattern of Ca^{2+} exchange and distribution is not invariable in cultured myocardial cells. In our laboratory at UCLA we have documented two distinct patterns despite maintenance of an unchanged culture procedure. It is of course true that successive batches of medium and serum may vary, but one pattern does persist through a number of successive medium and serum shipments. Both patterns will be described.

2.1.1. Pattern I

This pattern is described by Langer and Frank (1972). It has been recorded both from cells grown on glass scintillator slides and on plastic scintillator disks. The pattern of ^{45}Ca uptake is shown in Fig. 2. It is seen that it required 40–50 min for the counts to reach asymptote. In this early 1972 series, 15.0 ± 2.2 mmoles Ca/kg dry wt. was exchangeable. In a later series done in early 1976 (Frank *et al.*, 1977), 16.2 ± 1.35 mmoles Ca/kg dry wt. was exchangeable, with the same pattern of ^{45}Ca uptake. Of the total exchangeable Ca in the first series, 3.0 mmoles, 20% of the exchangeable Ca^{2+} pool, was displaced upon perfusion with 0.5 mM lanthanum ion (La^{3+}). In the later series, 17% was La^{3+}-displaceable. La^{3+} has been shown (Sanborn and Langer, 1970; Langer and Frank, 1972; Frank *et al.*, 1977; Burton *et al.*, 1977) to be restricted in its location to the cellular surface and to be competitive for Ca^{2+}-binding sites. It is, therefore, an effective probe for surface-bound Ca^{2+}. The exchangeable but non-La^{3+}-displaceable Ca^{2+} was, therefore, 12.0 mmoles in the 1972 series and 13.5 mmoles in the later series. Inexchangeable (or very slowly exchangeable) Ca^{2+} was 4.7 mmoles in the 1972 series.

Washout of the ^{45}Ca in these Pattern I cells demonstrated two exponential components that included all of the exchangeable Ca^{2+}. A rapidly exchangeable component, which included all of the La^{2+}-displaceable Ca^{2+}, had a half-time ($t_{1/2}$) of 1.2 min (rate constant = 0.6/min). The remaining Ca^{2+} exchanged with a $t_{1/2}$ of 19.2 min (rate constant = 0.036/min). The fast component accounted for 43%, and the slow component for 57%, of the exchangeable Ca^{2+}. It is notable that McCall (1977) found very similar kinetics in

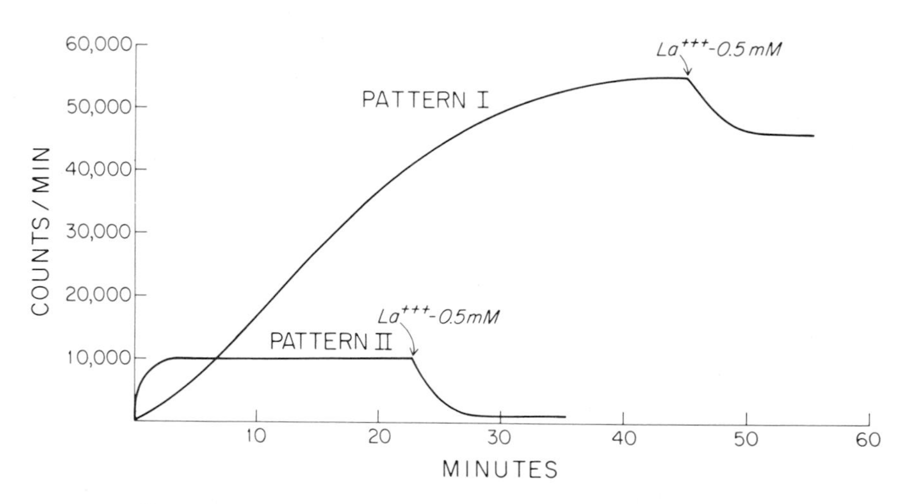

FIGURE 2. ^{45}Ca labeling of cultured cells. Two basic patterns of isotopic uptake are illustrated. Pattern I demonstrates approximately five times more labeling and a significant component of exchange which requires 40–50 min to attain asymptote. La^{3+} displaces less than 20% of the exchangeable Ca^{2+}. Pattern II demonstrates markedly less exchangeable Ca^{2+}, all of which exchanges 15–20 times more rapidly than the major fraction of Ca^{2+} in the Pattern I cells. Of this Ca^{2+}, 90% is La^{3+}-displaceable and, therefore, at the cellular surface.

monolayers of cells also cultured from the neonatal rat. Exchange was separated into two components, a rapid ($t_{1/2}$ = 0.9 min) component accounting for 53% of exchangeable Ca^{2+}, and a slower ($t_{1/2}$ = 24 min) component accounting for the remaining 47%. This was despite the use of different growth medium in the culturing of the cells.

2.1.2. Pattern II

This pattern appeared spontaneously in the fall of 1976 and has been consistently present since that time. The pattern of ^{45}Ca uptake is shown in Fig. 2. Note that only 2–3 min were required for the counts to reach asymptote. This is the time required for complete turnover of the contents of the flow cell and indicates that cellular labeling is very rapid and probably perfusion-limited. In these cells 3.1 ± 0.6 mmoles Ca^{2+}/kg dry wt. was exchangeable. Of this, 2.7 mmoles were La^{3+}-displaceable—almost 90% of the exchangeable pool. The total Ca^{2+} of these cells was 7.4 mmoles, indicating that 4.3 mmoles were inexchangeable.

A summary of the two patterns is presented in Table I. A number of points deserve emphasis: (1) the total Ca^{2+} is almost threefold greater in the Pattern I cells, and the exchangeable Ca^{2+} is fivefold greater; (2) the amount of inexchangeable Ca^{2+} is very similar in the two groups, as is the absolute amount of Ca^{2+} displaced by La^{3+} (3.0 vs. 2.7); and (3) the major difference between the two groups is in the amount of non-La^{3+}-displaceable exchangeable Ca^{2+}, which was 30 times greater in the Pattern I cells. Since the action of La^{3+} is limited to the cellular surface, it is likely that this additional exchangeable pool is located intracellularly.

In an attempt to increase the exchangeable pool in the Pattern II cells, they were warmed from 24 to 35°C and incubated with ^{45}Ca prior to insertion in the flow cell. This increased their spontaneous rate of beating from 85 to 145/min. The exchangeable Ca^{2+} increased from 3.1 to 5.1 mmoles/kg dry wt. over the course of 60 min as measured in the flow cell immediately after the 60-min incubation. The 2.0 mmolar increment in Ca^{2+} was not, however,

TABLE I. Ca^{2+} Exchange in Cultured Myocardium[a]

	Pattern I	Pattern II
Total Ca^{2+}	19.7 ± 3.0[b]	7.4 ± 0.5
Exchangeable Ca^{2+}	15.0 ± 2.2	3.1 ± 0.6
Inexchangeable Ca^{2+}	4.7	4.3
La^{3+} displacement	3.0 (20 ± 1.9%)[c]	2.7 (88.5 ± 6.6%)[c]
Non-La^{3+}-displaceable exchangeable Ca^{2+}	12.0	0.4

[a]Temperature was 24°C. [c]Percent of exchangeable pool.
[b]All values in mmoles/kg dry wt.; wet wt./dry wt. = 5.85.

attributable to intracellular Ca^{2+}, since over 89% of the total 5.1 mmoles was La^{3+}-displaceable and, therefore, localized to the cellular surface.

It is of interest that Borle (1968, 1969a,b), in his studies of Ca kinetics in HeLa cells, found that 90% of the Ca of the cell monolayers was bound to an extracellular coat and could be removed with trypsin–EDTA. The concentration of Ca^{2+} at the cell surface is, then, similar to that of the Pattern II cells. This is, perhaps, not particularly surprising for noncontractile cells but presents a problem for cells in which intracellular contractile components require Ca in order to function. In addition it has been demonstrated that the depolarization of cultured cells in monolayer, including those from the neonatal rat (Schanne *et al.*, 1977), is dependent upon an increased current, probably carried by Ca. In these cells, Schanne *et al.* found an action potential amplitude of 59.8 mV. If we assume a membrane capacitance of 1 $\mu F/cm^2$ for these cells (Jongsma and van Rijn, 1972) and that Ca^{2+} carries the current, the transmembrane Ca flux associated with each depolarization can be calculated according to the relation $M = CV/zF$ (M = flux in moles; C = capacitance in farads; V = depolarization amplitude in volts; z = valence; and F = faraday, 96,500 coulombs/mole). The flux per excitation is 0.3 pmoles/cm^2 of cell membrane. The minimum cell surface area for the cells grown in confluent monolayer (including upper and lower surfaces) is twice the area of the scintillator disk upon which the cells are grown. The wet cellular volume is closely estimated from the dry weight of the cells $\times$ wet weight/dry weight ratio. The mean wet volume of cells distributed over two scintillator disks (minimum surface area = 63.6 cm^2) is 0.0082 cm^3, giving a surface : volume ratio of 7.75×10^3/cm. Using these figures, a minimum transmembrane flux of 2.4×10^{-6} moles Ca^{2+}/kg wet cells per beat is derived. This means that for an increment in excitation rate of 60/min, an additional 8.6 mmoles Ca^{2+}/kg wet cells would cross the sarcolemma in 60 min. This calculation assumes that all of the depolarizing current is attributable to Ca. Though Schanne *et al.* (1977) imply that a large component of the current is attributable to Ca, McCall's (1976) study indicates that a slow Na current dominates. If such is the case, the calculated depolarization-associated Ca flux would be considerably reduced. It should be noted that, whether or not Ca carries the depolarizing current, the Pattern II cells demonstrated an increment of 2 mmoles/kg dry or 0.34 mmoles/kg wet wt. associated with an increase of 60 excitations/min for 60 min. In addition, almost 90% of the increment was La^{3+}-displaceable.

It is clear that the increment in Ca^{2+} flux associated with excitation did not produce a significant increase in non-La^{3+}-displaceable exchangeable Ca^{2+}, which is, presumably, intracellular in location. This means that a large fraction of cellular Ca circulates between the surface of the cell and the cell interior without distributing or mixing intracellularly and provides further evidence for marked compartmentalization of Ca^{2+} in these Pattern II cells. It should also be noted that Pattern II cells cease beating immediately upon removal of Ca from the bathing solution. This is consistent with the concept that a component of the Ca^{2+} involved in excitation–contraction coupling in these cells is located at the cell surface and is very rapidly exchangeable (see Fig. 2). The Pattern I cells were also immediately responsive to changes in Ca^{2+} in the bathing solution. This suggests, along with its virtual absence in Pattern II cells, that the more slowly exchangeable Ca pool has little to do with maintenance of contraction, at least directly.

It is difficult to state with certainty which pattern represents the "normal" pattern of myoblasts in culture. Both groups were functional and reproducible over long periods of time and appeared to be similar ultrastructurally. My opinion is, however, that the Pattern I cells demonstrated an enhanced Ca permeability (for unknown reasons) which allowed Ca to enter the cell in increased quantity where it was adequately buffered or stored. The Pattern II cells, with essentially all of their exchangeable Ca localized to rapidly exchangeable, apparently superficial sites, may provide us with a number of further insights into the excitation–contraction coupling sequence. These myoblasts contain mitochondria and sarcotubular structures, both of which would be expected to be capable of binding and exchanging Ca^{2+}. The kinetic data indicate, however, that the Ca^{2+} that is necessary for contraction enters the cell but does not exchange with these organelles. Though there is an extensive so-called free sarcoplasmic reticulum (SR) surrounding the myofibrils, little is known of its functional state in these cells. The cells also possess an extensive junctional SR located in close apposition to the sarcolemmal membrane (study in preparation). Its precise function in excitation–contraction (EC) coupling is not known, but amphioxus muscle has an extensive junctional SR and is dependent upon an extracellular source of Ca^{2+} for its contraction (Hagiwara *et al.*, 1971) as are the myoblasts. Amphioxus cells demonstrate an increased La^{3+} density overlying the junctional SR, indicative of binding at these sites. As indicated, the myoblasts show an abundant junctional SR within 20–30 nm of the sarcolemmal membrane. The ability of La to displace almost 90% of the Ca^{2+} in these cells (see Table I) raises the possibility that Ca^{2+} localized at the junctional sites is accessible to La^{3+} either directly or by virtue of the fact that Ca rapidly exchanges across the sarcolemma and is then displaced from surface structures by La^{3+}. Therefore, it is possible that Ca^{2+} within the junctional SR plays an important role in EC coupling in these cells. The mitochondrial Ca^{2+}, if present in significant quantity, must either exchange extremely slowly across the sarcolemma or, alternatively, very rapidly via the junctional SR. In either case, its Ca^{2+} might not be recognized kinetically as a separate pool by the use of isotopic techniques.

The kinetic characteristics outlined above focused our attention at the cellular surface, and a series of studies has been directed toward definition of the possible role this region plays in the regulation of Ca^{2+} flux in cultured (as well as adult) cells.

Since this work was submitted for publication it has been found that a slowly exchangeable compartment can be added to the cells that demonstrate only a rapidly exchangeable component. This is done by the substitution of 10 mM NaH_2PO_4 for HEPES in the perfusate (at unchanged pH). Such addition produces an immediate, linear increase in Ca uptake that more than triples exchangeable Ca within 60 min. It was found (Langer and Nudd, 1980) that the introduction of inhibitors of mitochondrial respiration (Warfarin Sodium or Antimycin A, 10^{-5}M) completely prevented the increased uptake induced by phosphate addition. The characteristics of this added Ca compartment strongly indicate that it is localized in the mitochondria. The $H_2PO_4^-$ is visualized as exchanging for matrix hydroxide ions in the mitochondria—a reaction equivalent to the entry of H_3PO_4 with loss of a proton to the alkaline matrix (Lehninger, 1974). The excess anion that results provides the milieu for Ca accumulation as the phosphate salt. On the basis of this recent work it seems possible that the difference between the two types of cells described could be the presence of mitochondrial Ca uptake in the type demonstrating the additional slower component. Though no phosphate was included in the perfusion medium of these cells, they may have retained a higher level of the salt after removal from the incubation medium, which does contain phosphate.

3. THE ROLE OF THE GLYCOCALYX

Bennett (1963) was among the first to draw attention to what he termed the glycocalyx ("sweet husk") on the surface of cells. The glycocalyx denoted the coating external to the unit membrane bimolecular lipid leaflet.

3.1. Histochemistry

This has been investigated recently by Frank *et al.* (1977). The surface of myocardial cells in culture binds positively charged molecules including La^{3+}, ruthenium red, and colloidal iron. These substances are electron dense and, therefore, produce staining of the surface layers.

La^{3+} was chosen as one of the histochemical stains because, aside from being a well-known surface marker (Doggenweiler and Frank, 1965), it has well-defined effects on muscle function (Langer and Frank, 1972; Sanborn and Langer, 1970; Martinez-Palomo *et al.,* 1973). $LaCl_3$ (American Chemicals), in concentrations between 1 and 5 mM, is added to the standard solution. The culture is incubated from 20 to 120 minutes, after which the cells are fixed and processed for microscopy. La^{3+} is not present in any of the fixatives.

Ruthenium red (RR) was chosen as a marker because it is a well-known mucopolysaccharide stain, and its histochemistry and effects on function have been studied in a number of tissues, including muscle (Luft, 1971; Huet and Herzberg, 1973). Staining with RR is performed according to Luft (1971). RR (Sigma) is added to both glutaraldehyde and osmium fixatives in concentrations of 1500 ppm from a stock solution made up in distilled H_2O.

Colloidal iron hydroxide (CIH) was chosen as a histochemical probe because of its documented specificity for sialic acid (Benedetti and Emmelot, 1967; Nicolson, 1973). CIH is prepared according to a modification of the method of Gasic and Berwick (1963). Five milliliters of 0.5 mM $FeCl_3$ (Fisher Chemicals) is added to rapidly boiling distilled water (60 ml). After cooling, 10 ml of glacial acetic acid is added to the CIH solution, and the pH adjusted to below 1.8. At this pH only the strongly acidic sialic acid sites (pK = 2.6) among the membraneous anionic sites remain partially ionized and, therefore, capable of binding the positively charged colloid. The cells are fixed in 2% glutaraldehyde and then rinsed for 30 sec in 12% acetic acid. The culture is exposed to CIH for 40 min, washed for 10 min with several changes of 12% acetic acid, postfixed, and embedded for sectioning.

The pattern with each of these stains was similar in adult and cultured cells and indicates that, qualitatively, there are not major differences between cultured and more mature cells. The cultured cells demonstrate two types of surface layers common to the periphery of most animal cells (Martinez-Palomo, 1970; Parsons and Subjeck, 1972): (1) a surface coat, glycoprotein in nature, that is integrated with the plasma membrane, and (2) a more peripheral external lamina composed of protein and carbohydrate. These two coats together form the glycocalyx. La^{3+} stains the lipid bilayer, the surface coat, and the external lamina. It does not penetrate intracellularly if the cellular surface is intact. Ruthenium red is a general polyanionic stain and adheres to the outer leaflet of the unit membrane, the surface coat, and the external lamina. The total thickness of the glycocalyx is approximately 50 nm in the cultured cells. CIH staining shows particles, indicating the presence of sialic acid, ranging from 3 to 20 nm in size uniformly distributed in two layers within the glycocalyx—one layer on the surface coat and another on the external lamina. Sialic acid is a nine-carbon amino sugar always terminally linked to the side chains of membrane glycoproteins (Lloyd, 1975). It is strongly acidic (pK = 2.6), and its carboxyl group is therefore fully ionized at physiological pH levels. The effect of sialic acid removal on Ca exchange of the cultured cells is marked (Langer *et al.*, 1976; Frank *et al.*, 1977).

3.2. Sialic Acid Removal

Sialic acid residues were specifically removed by exposure of the cultured cells to purified neuraminidase (0.25 U/ml for 15 min). The neuraminidase (Worthington, Freehold, New Jersey) was purified by the method of Hatton and Regoeczi (1973). At concentrations applied to the cells the enzyme was assessed for protease activity using ^{125}I-labeled albumin as substrate. No proteolytic activity was detectable. Phospholipase activity of the enzyme was measured by incubating the neuraminidase for 90 minutes with [^{14}C]phosphatidylcholine as substrate. The extent of hydrolysis was compared to that obtained with phopholipase C, and the neuraminidase was found to contain the equivalent of 0.0025 units of phospholipase activity per ml. The cultured cells were then incubated with 0.0025 units of phospholipase per ml under conditions identical to those used with neuraminidase incubation. No effect on beating or ^{45}Ca exchangeability was seen. The effect of neuraminidase was evaluated by exposure of the disks, with cells attached, to standard solution containing 0.25 u/ml neuraminidase for 15 min at 37°C, pH = 7.1. Following exposure, the disks were mounted in the flow cell, and ^{45}Ca uptake and washout followed in the usual manner.

The technique removed 61% of the total cellular sialic acid, and the effect on Ca exchange was striking. The rate of ^{45}Ca uptake and washout was increased five- to sixfold, and the La^{3+} ion, normally restricted to the lipid bilayer and surface coat–external lamina complex entered the Pattern I cells and displaced more than 80% of the previously La^{3+}-inaccessible Ca^{2+}. The effect of sialic acid removal on Ca^{2+} exchange was in marked contrast to its effect on potassium exchange. K^+ exchange was not significantly affected. Therefore, sialic acid removal did not result in a nonselective disruption of cellular ionic permeability. The Pattern II cells, upon exposure to neuraminidase, increased their exchangeable Ca from 3.07 ± 0.64 to 13.68 ± 2.3 mmoles/kg dry wt. As with the Pattern I cells all of the Ca was very rapidly exchangeable and, in addition, over 80% was displaced by La^{3+}. Therefore, removal of sialic acid from the Pattern II cells produced the same marked increase in Ca^{2+} and La^{3+} permeability that was evident in the Pattern I cells from which sialic acid was removed. Functionally, neuraminidase caused the cells to retract and cease beating. These exchange studies of Langer *et al.* (1976) and Frank *et al.* (1977) indicate that, though the surface coats are altered, the lipid bilayer remained intact after sialic acid removal. Electron microscopy also indicated that neuraminidase treatment markedly reduced surface site binding of the La^{3+} but showed extensive binding to mitochondria. This indicates that the removal of sialic acid permits La, previously excluded from intracellular sites, to enter the cell freely and displace Ca^{2+} from intracellular organelles. This accounts for the fact that 80% of the Ca^{2+} in sialic acid-depleted cells becomes La^{3+}-displaceable. Also, since La^{3+} competes for and displaces Ca^{2+} from rapidly exchangeable sites in intact cells, the results suggest that sialic acid residues account for a significant number of the superficial binding sites on these cells.

It is to be emphasized that for these sialic acid studies to be done it was required that the culture technique be employed. It was necessary that the cellular surface be exposed directly to the large neuraminidase molecule in order that the cellular effect could be unequivocally analyzed. The results indicate that Ca^{2+} binding to sialic acid residues in the surface coat–external lamina complex plays a crucial role in the regulation of transmembrane Ca^{2+} flux in the heart. Removal of this nine-carbon amino sugar essentially destroys the ability of the cell to regulate its Ca^{2+} exchange and, in addition, permits entry into the cell of competitive cations such as La^{3+} which normally do not traverse the sarcolemma. It is clear that an intact lipid bilayer, at least with respect to potassium leakage, is not sufficient to prevent a marked increase of Ca^{2+} (and La^{3+}) permeability. These studies are consistent with the proposal that the glycocalyx may be the rate-limiting site for Ca^{2+} flux in the heart.

The culture technique has also provided an opportunity to explore further the mechanism of excitation–contraction coupling in the heart through analysis of the cell's cation affinity characteristics.

4. CATION AFFINITY SEQUENCE

The flow cell technique was employed in order to compare the ability of a series of cations to uncouple excitation from contraction (in the intact neonatal rat heart) with their ability to displace Ca from the rapidly exchangeable sites of Pattern I cultured cells (Langer *et al.*, 1974). La^{3+}, Cd^{2+}, Zn^{2+}, Mn^{2+}, and Mg^{2+} at equimolar concentrations were compared with respect to their potency as excitation–contraction uncouplers in papillary muscles derived from the same type of neonatal hearts from which the cultures were grown. This sequence was then compared with the Ca^{2+}-displacing ability from the cultured cells of each of the cations. This was accomplished by labeling the cells in the flow cell with essentially carrier-free ^{45}Ca. After 30 min of labeling, 1 mM of the competitive cation was added to the perfusing solution, and the quantity of Ca^{2+} displaced within 2.2 min (time required for mixing within the flow cell) was measured. The affinity for the Ca^{2+} binding sites was in the order $La^{3+} > Cd^{2+} > Zn^{2+} > Mn^{2+} > Mg^{2+}$. This sequence correlated positively with the rates at which the ions diminished contractile tension in the neonatal rat papillary muscle. Both of these properties, i.e., ability to uncouple and ability to displace Ca, depended upon the similarity in size of the ion's nonhydrated radius as compared to that of Ca. The closer this radius was to that of Ca^{2+} at 0.099 nm, the more effectively it uncoupled and the more Ca it displaced from the culture monolayer. For example, Mg^{2+} has a crystal ionic radius of 0.066 nm, is the weakest uncoupler in the series, and displaces 20% of the amount of Ca^{2+} displaced by La^{3+} which has a radius of 0.101 nm and is the strongest uncoupler in the series. The study supports the proposal that Ca^{2+} bound at the surface of the cell at sites of fixed negative charge (e.g., sialic acid) plays a major role in the control of the cell's force development.

The studies summarized in the preceding pages strongly implicate the crucial role of the sarcolemma–glycocalyx complex in the regulation of Ca^{2+} exchange in cultured cardiac cells. It became apparent that further insights could be gained from study of isolated membrane from these cells—especially if the membrane could be obtained without exposure to noxious agents [e.g., mercuric acetate (Barland and Schroeder, 1970)] and could be derived from the same cells in which characteristics had been defined in the whole, intact state. A new, exceedingly simple technique has been developed for preparation of membranes which is, in fact, applicable to any monolayer culture.

5. "GAS-DISSECTED" MEMBRANES

The basis of the procedure is the direction of a stream of nitrogen (N_2) gas at high velocity across the surface of the cellular monolayer (Langer *et al.*, 1978). When the flow is of proper velocity and orientation, the uppermost cell membranes are torn open, and the intracellular contents are blown out and removed. The residual membranes are rolled and flattened on the disk or dish surface. It is desirable to restrict the technique to monolayer cultures. Multilayered cultures produce greater intracellular residue, since the nitrogen stream is not optimally oriented for clean dissection of all layers.

The chamber and valvular arrangement used are illustrated in Fig. 3. The chamber is machined from stainless steel. The walls are at least 1.0 cm in thickness to support pressures of 1800–2200 psi. The purpose of the chamber is to provide a stable support for the culture and to prevent drying of the membranes after their isolation. Compressed N_2 at 1800–2200 psi is released through the valve positioned in the center of the chamber. The valve has a diameter of 8 mm and is brought into firm direct contact with the surface of the disk or dish. This is accomplished by elevation of the chamber floor by means of the turnscrew shown. The disk or dish is held in position and prevented from lateral movement by three pins spaced equidistantly around the outer edge.

The dimensions and configuration of the valve outlet are of critical importance. The internal valve cylinder and valve are beveled (radius of curvature = 1.55 mm or 0.062 in.) as illustrated in order that the N_2 exits from the valve head in a stream which is essentially parallel to the surface of the culture. If the gas exits downward, the cells are swept from the surface, and if it exits even marginally upward, the cells remain intact. The velocity of the N_2 stream is also important. If less than 1800 psi is used to generate the stream, cellular residual is increased. The valve opening is also critical. The opening is set at 0.30 mm (0.012 in.) for optimal results. A larger opening leaves cellular residual, and a smaller opening tends to sweep the cells from the surface.

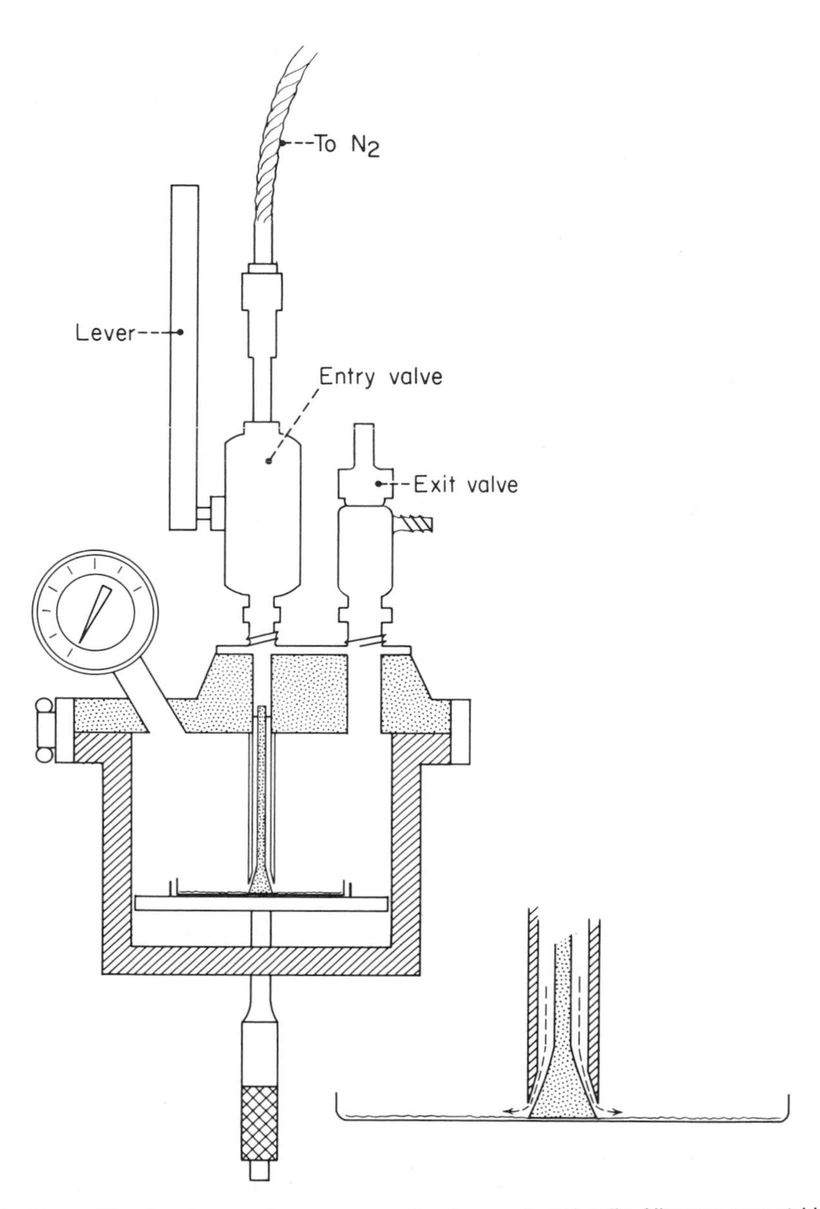

FIGURE 3. Chamber for membrane preparation from cultured cells. Nitrogen gas at high velocity is admitted through the valve (see insert) and directed radially over the culture surface. The upper surface of the cells is torn open, and the cellular contents blown out. The residual membranes remain adherent to the dish surface. See text for details. (Reproduced with permission from *Science.*)

The procedure for membrane preparation is quite simple. The disk or culture dish is mounted on the floor of the chamber with the cellular monolayer directed upward. Prior to mounting, the culture medium is poured off and the disk or dish shaken a few times to remove excess adherent fluid. The top of the chamber is closed and sealed with the two stainless steel semicircular rings as shown. A circular steel band is then fitted and secured over the rings to prevent them from detaching laterally. The floor of the chamber is then elevated by rotation of the turnscrew to bring the center of the culture disk or dish into firm contact with the surface of the valve. The culture is now oriented to receive the N_2 stream. The entry of the gas to the valve applied to the culture is controlled by a valve located in the high-pressure line attached to a large (H size) N_2 tank (Ohio Medical Products, Madison, Wisconsin). A minimum pressure of 1800 psi is used. The valve is opened rapidly (within 1 sec) with the aid of the attached lever. The high-velocity N_2 stream exits through the circumferential 0.30 mm opening and travels radially over the culture surface, tearing open the upper surface of the cells and blowing the cellular material to the sides of the chamber. The pressure in the sealed chamber rapidly equilibrates (<1 sec) to that in the tank, and flow ceases. The entry valve is closed and the chamber is decompressed over the course of 30–60 sec by opening the valve indicated. It should be emphasized that the technique does not depend upon decompression to rupture the cells. The membranes are produced by the shearing force of the N_2 stream tearing the cells open and expelling the intracellular contents. The chamber is opened, and the disk or dish is removed, placed in culture medium or a physiological salt solution, and examined under the phase microscope.

Figure 4 is a scanning electron micrograph of the membranes on the surface of the dish. Enzyme measurements show that Na–K ATPase activity is increased over 15-fold and 5′-AMPase some 12-fold. Succinic dehydrogenase (SDH) (a mitochondrial marker) activity is undetectable in the membrane fraction. In addition, cells labeled with radioactive potassium (^{42}K) prior to membrane preparation show that less than 1% of this intracellular marker is present in the residual membrane layer.

The technique is unique in a number of respects. The cells are never exposed to any chemical agents that might modify the basic membrane characteristics. Exposure to a critically oriented high-velocity N_2 jet disrupts and fractionates the tissue in a single operation which requires less than a second. The membranes remain in sheetlike configuration and adherent to the surface on which the cells were grown. They are amenable to enzymatic, ion-binding, histochemical, and ultrastructural studies. These studies can be preceded by a variety of studies on the same cells prior to isolation of their membranes. The technique should be applicable not only to cardiac cell culture but to any tissue that can be grown in monolayer. This has been shown to be the case in preparation of gas-dissected membranes from rabbit alveolar macrophages (per-

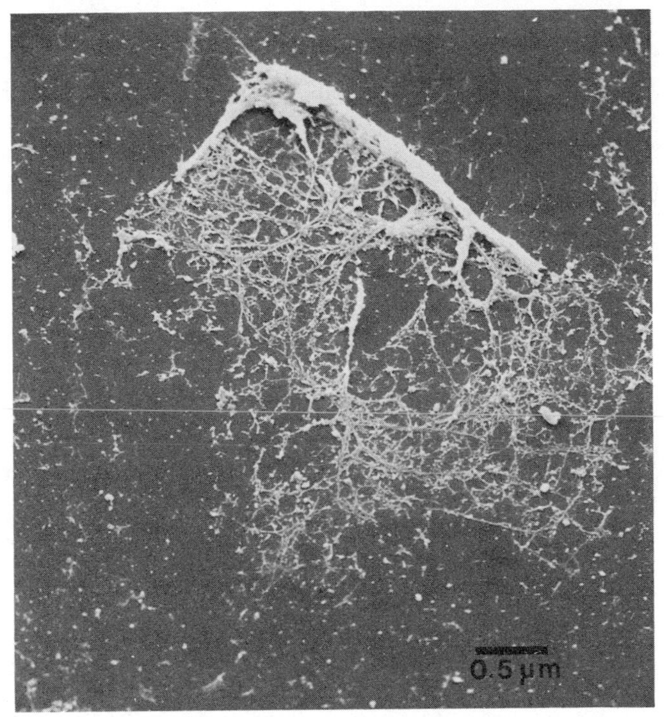

FIGURE 4. Scanning electron micrograph of membranes from myoblasts grown in culture. Membranes cover a large portion of the dish and appear in sheets or, in some cases, as rolled membrane.

sonal communication, Dr. R. Martinez, Department of Bacteriology, UCLA). The preparation shows no detectable lactic dehydrogenase, no lysosomal enzymes, no glutamic dehydrogenase (mitochondrial marker), nor any glucose-6-dehydrogenase—all intracellular enzymes. There is, however, a manyfold increase in Na–K ATPase and in 5′-nucleotidase—both membrane markers.

6. SUMMARY

There is no doubt that cardiac cell culture techniques have provided us with further insights into the characteristics of Ca^{2+} exchange in heart muscle. At present at least two patterns of exchange have been demonstrable: one pattern showing a large component of relatively slowly exchangeable Ca^{2+} ($t_{1/2} \backsim 20$ min) and another pattern which demonstrates virtually no Ca^{2+} in this fraction. The two patterns are similar, however, in the quantity of rapidly exchangeable Ca^{2+} present as well as in the amount of an inexchangeable fraction. Both types of cell contract well and are immediately EC uncoupled by removal of external Ca^{2+} or addition of La^{3+}. Analysis of these patterns indicates that coupling Ca^{2+} circulates between superficial or surface sites and the myofilaments without mixing or distributing intracellularly, as though it moved in and out within a "packet."

The direct availability of the cellular surface in culture allowed the surface structure to be studied. The glycocalyx, with its high concentration of sialic acid residues, binds Ca^{2+} and plays a major role in controlling its movement across the sarcolemma.

The study of the characteristics of cellular Ca exchange has been greatly aided by the use of the scintillator-disk flow cell technique. Further definition of the molecular charactistics of cellular membrane will be facilitated by the new technique of "high-velocity gas dissection" for preparation of a purified membrane fraction from monolayer cultures.

ACKNOWLEDGMENTS. The author expresses his appreciation to Dr. Joy Frank, who supplied the electron micrograph (Fig. 4). The studies reported herein were supported by Grants HL 11351-6-11 from the U.S. Public Health Service and a grant from the Castera Foundation.

REFERENCES

Barland, P., and Schroeder, E. A., 1970, A new rapid method for the isolation of surface membranes from tissue culture cells, *J. Cell Biol.* **45**:3.

Benedetti, E. L., and Emmelot, P., 1967, Studies in plasma membranes. IV. The ultrastructural localization and content of sialic acid in plasma membranes isolated from rat liver and hepatoma, *J. Cell Sci.* **2**:499.

Bennet, H. S., 1963, Morphological aspects of extracellular polysaccharides, *J. Histochem. Cytochem.* **11**:14.

Berry, M. N., Friend, D. S., and Scherer, J., 1970, Morphology and metabolism of intact muscle cells isolated from adult rat heart, *Circ. Res.* **26**:679.

Blondel, B., Roijeir, R., and Cheneval, J. P., 1971, Heart cells in culture: A simple method for increasing the proportion of myoblasts, *Experientia* **27**:356.

Bloom, S., 1971, Requirements for spontaneous contractility in isolated adult mammalian heart muscle cells, *Exp. Cell Res.* **69**:17.

Borle, A. B., 1968, Calcium metabolism in HeLa cells and the effects of parathyroid hormone, *J. Cell Biol.* **36**:567.

Borle, A. B., 1969a, Kinetic analyses of calcium movements in HeLa cell cultures. I. Calcium influx, *J. Gen. Physiol.* **53**:43.

Borle, A. B., 1969b, Kinetic analyses of calcium movements in HeLa cell cultures. II. Calcium efflux, *J. Gen. Physiol.* **53**:57.

Burton, K. P., Hagler, H. K., Templeton, G. H., Willerson, J. T., and Buja, L. M., 1977, Lanthanum probe studies in cellular pathophysiology induced by hypoxia in isolated cardiac muscle, *J. Clin. Invest.* **60**:1289.

Doggenweiler, C., and Frenk, S., 1965, Staining properties of lanthanum on cell membranes, *Proc. Natl. Acad. Sci. USA* **53**:425.

Frank, J. S., Langer, G. A., Nudd, L. M., and Seraydarian, K., 1977, The myocardial cell surface, its histochemistry and the effect of sialic acid and calcium removal on its structure and cellular ionic exchange, *Circ. Res.* **41**:702.

Gasic, G., and Berwick, L., 1963, Hale stain for sialic acid-containing mucins; adaption to electron microscopy, *J. Cell Biol.* **19**:223.

Hagiwara, S., Henkart, M. P., and Kidokoro, Y., 1971, Excitation–contraction coupling in amphioxus muscle cell, *J. Physiol. (Lond.)* **219**:233.

Harary, I., and Farley, B., 1963, *In vitro* studies on single beating rat heart cells. I. Growth and organization, *Exp. Cell Res.* **29**:451.

Hatton, M. W. C., and Regoeczi, E., 1973, A simple method for the purification of commercial neuraminidase preparation-free from proteases, *Biochim. Biophys. Acta.* **327**:114.

Huet, C., and Herzberg, M., 1973, Effects of enzyme and EDTA on ruthenium red and concanavalin A labeling of cell surface, *J. Ultrastruct. Res.* **42**:186.

Jongsma, H. J., and van Rijn, H. E., 1972, Electrotonic spread of current in monolayer cultures of neonatal rat heart cells, *J. Membr. Biol.* **9**:341.

Langer, G. A., and Frank, J. S., 1972, Lanthanum in heart cell culture. Effect on calcium exchange correlated with its localization, *J. Cell Biol.* **54**:441.

Langer, G. A., and Nudd, L. M., 1980, Calcium exchange in myocardial tissue culture: Addition and characterization of a mitochondrial component, *Am. J. Physiol.* in press.

Langer, G. A., Sato, E., and Seraydarian, M., 1969, Calcium exchange in a single layer of rat cardiac cells studied by direct counting of cellular activity of labeled calcium, *Circ. Res.* **24**:589.

Langer, G. A., Serena, S. D., and Nudd, L. M., 1974, Cation exchange in heart cell culture: Correlation with effects on contractile force, *J. Mol. Cell. Cardiol.* **6**:149.

Langer, G. A., Frank, J. S., Nudd, L. M., and Seraydarian, K., 1976, Sialic acid: Effect of removal on calcium exchangeability of cultured heart cells, *Science* **193**:1013.

Langer, G. A., Frank, J. S., and Philipson, K. D., 1978, Preparation of sarcolemmal membrane from myocardial tissue culture monolayer by "high velocity gas dissection," *Science* **200**:1388.

Lehninger, A. L., 1974, Role of phosphate and other proton-donating anions in respiration-coupled transport of Ca^{2+} by mitochondria, *Proc. Natl. Acad. Sci. USA* **71**:1520.

Lloyd, C. W., 1975, Sialic acid and the social behavior of cells, *Biol. Rev.* **50**:325.

Luft, J. H., 1971, Ruthenium red and violet II. Fine structural localization in animal tissues, *Anat. Rec.* **171**:369.

Martinez-Palomo, A., 1970, The surface coats of animal cells, *Int. Rev. Cytol.* **29**:29.

Martinez-Palomo, A., Benitez, D., and Alanis, J., 1973, Selective deposition of lanthanum in mammalian cardiac cell membranes; ultrastructure and electrophysiological evidence, *J. Cell Biol.* **58**:1.

McCall, D., 1976, Effect of verapamil and of extracellular Ca and Na on contraction frequency of cultured heart cells, *J. Gen. Physiol.* **68**:537.

McCall, D., 1977, Na : Ca exchange in cultured heart cells, *Circulation* **56**:207.

Nicolson, G. L., 1973, Anionic sites of human erythrocyte membranes. I. Effects of trypsin, phospholipase C and pH on the topography of bound positively charged colloidal particles, *J. Cell Biol.* **57**:373.

Parsons, D. F., and Subjeck, J. R., 1972, The morphology of the polysaccharide coat of mamalian cells, *Biochim. Biophys. Acta* **265**:85.

Sanborn, W. G., and Langer, G. A., 1970, Specific uncoupling of excitation and contraction in mammalian cardiac tissue by lanthanum, *J. Gen. Physiol.* **56**:191.

Schanne, O. F., Ruiz-Ceretti, E., Rivard, C., and Chartier, D., 1977, Determinants of electrical activity in clusters of cultured cardiac cells from neonatal rats, *J. Mol. Cell. Cardiol.* **9**:269.

Shine, K. I., Serena, S. D., and Langer, G. A., 1971, Kinetic localization of contractile calcium in rabbit myocardium, *Am. J. Physiol.* **221**:1408.

Vahouny, G. V., Wei, R., Starkweather, R., and Davis, C., 1970, Preparation of beating heart cells from adult rats, *Science* **167**:1616.

12

Cardiac Muscle with Controlled Geometry

Application to Electrophysiological and Ion Transport Studies

MELVYN LIEBERMAN, C. RUSSELL HORRES, NORIKAZU SHIGETO, LISA EBIHARA, JAMES F. AITON, and EDWARD A. JOHNSON

1. INTRODUCTION

Heart cells in tissue culture have served as an experimental preparation for nearly 70 years (Burrows, 1912). However, the unique advantages of the preparations for physiological studies became evident after Moscona (1952) succeeded in using the proteolytic enzyme, trypsin, to obtain suspensions of embryonic cells. Isolated heart cells, contained within an appropriate culture medium, could then be introduced to glass or plastic substrates at densities sufficient to promote the formation of either monolayer or relatively thin multilayer preparations that were suitable for electrophysiological (Fänge *et al.*, 1952; Crill *et al.*, 1959) and ion-transport (Burrows and Lamb, 1962) studies. In 1954, a conference was held, jointly sponsored by the New York Academy of Sciences and the Tissue Culture Association, to explore the utilization of

---

MELVYN LIEBERMAN, C. RUSSELL HORRES, NORIKAZU SHIGETO, LISA EBIHARA, JAMES F. AITON, and EDWARD A. JOHNSON • Department of Physiology, Duke University Medical Center, Durham, North Carolina 27710. *Present address for N. S.*: Division of Cardiology, Hiroshima National Chest Hospital, Hiroshima, Japan. *Present address for J. F. A.*: Department of Physiology, University of St. Andrews, St. Andrews, Scotland.

tissue culture in pharmacology (Pomerat, 1954). Although at the time, it was believed that tissue culture had much to offer the study of drug action, recognition was given to the fact that tissue culture was not yet amenable to quantitative determinations.

Numerous reports have since demonstrated the successful adaptation of cultured heart cells as a model system for studying the morphology, biochemistry, pharmacology, pathology, and physiology of cardiac muscle (see chapters in books edited by Lieberman and Sano, 1976; Kobayashi *et al.,* 1978). The growing acceptance of heart cells in tissue culture is borne out by the presence of over 600 citations in a Medline data base for the period 1969–1979. Nevertheless, it has been difficult to obtain accurate quantitative data from preparations grown in the characteristic sheetlike configuration because of uncertainties associated with the state of cardiac cellular differentiation (McLean and Sperelakis, 1974, 1976; Galper and Catterall, 1978), questionable reliability of microelectrode impalements (Sperelakis, 1978), and the extent to which cellular heterogeneity and the inability to control, uniformly, the rate of beating can alter the determination of true transmembrane fluxes (Horres and Lieberman, 1977; McCall, 1979).

In our opinion, the most significant advantage afforded by tissue culture methodology has been the ability to promote the aggregation and orientation of isolated cells to conform to a simple geometric configuration, i.e., a cylindrical bundle or spheroidal aggregate, suitable for experimental analysis and theoretical simulation. In recent years, our laboratory has directed its efforts toward developing preparations of cardiac muscle in the form of a bundle of linearly aligned fibers ("synthetic strand"; Lieberman *et al.,* 1972; Purdy *et al.,* 1972; Lieberman *et al.,* 1975), spheroidal aggregates ("cultured cluster"; Ebihara *et al.,* 1980), and a multistrand preparation of cultured heart cells ("polystrand") grown as thin annuli about nylon monofilament (Horres *et al.,* 1977). This chapter will consider questions that relate directly to the fundamental properties of the cardiac cell membrane and illustrate the feasibility of using growth-oriented cells in tissue culture to resolve them.

2. SYNTHETIC STRAND

The need for a simple preparation of cardiac muscle, free of the complexities associated with naturally occurring cardiac muscle, was deemed essential in order to obtain accurate values for the specific electrical constants of cardiac muscle and to apply the analytic techniques of voltage clamp to determine the nonlinear electrical properties of the cardiac membrane (Johnson and Lieberman, 1971). For these reasons, we turned our attention to tissue culture with the aim of exploring the feasibility of growing a preparation of cardiac muscle with a linear and relatively simple geometry (Lieberman *et al.,* 1972).

Although the traditional monolayer culture of cardiac muscle appeared suitable for quantitative electrophysiological studies (Hyde *et al.*, 1969; Jongsma and van Rijn, 1972), we were concerned about using a point source of current to resolve nonlinear membrane properties in a sheetlike preparation (Noble, 1962). Consequently, trypsin-dissociated embryonic chick heart cells were plated on surfaces and directed to grow in channels cut in an agar gel such that thin bundles of fibers could be grown to any desired length (Fig. 1). Electron micrographs of the synthetic strands (Purdy *et al.*, 1972) clearly demonstrated that the muscle cells differentiated normally, i.e., myofibrils were similar in arrangement and orientation to those in muscle cells of naturally occurring preparations. Junctional specializations (intercalated discs, desmosomes, and nexuses) and the sarcoplasmic reticulum also appeared to conform to the generally accepted scheme for muscle cells.

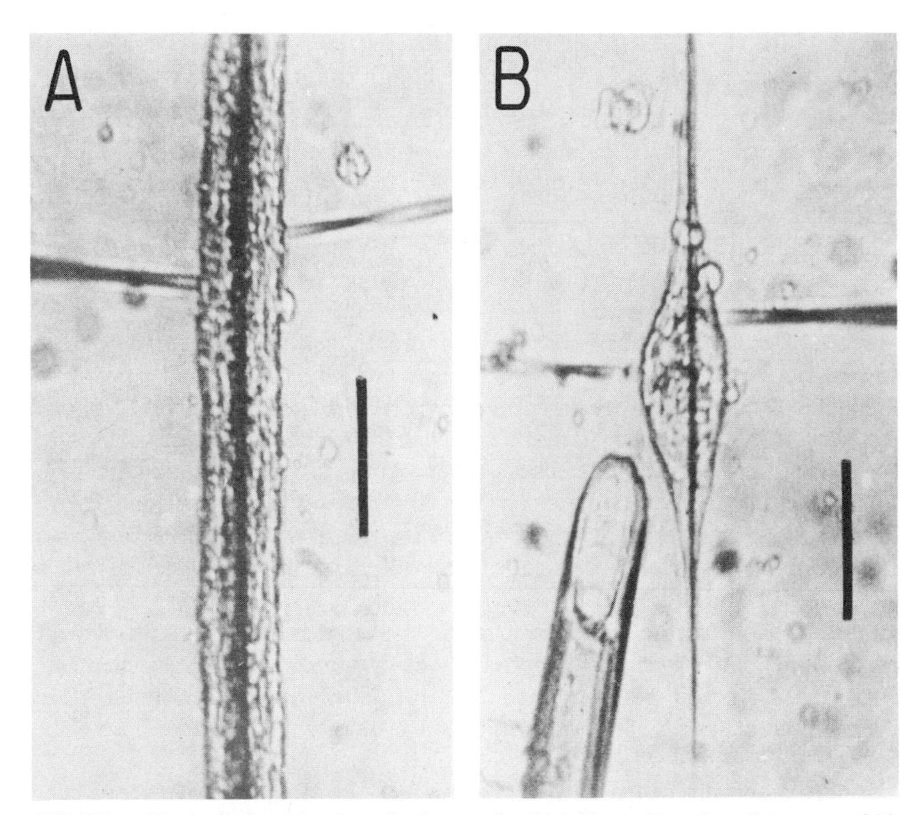

FIGURE 1. Photomicrographs of synthetic strands aligned in position along the groove. (A) A segment of a long preparation showing the position of the current-passing and voltage-recording microelectrodes. (B) A short preparation showing the position of the bevel-edged stimulating electrode, and the current-passing and voltage-recording microelectrodes. Scale: 100 μm.

2.1. Linear Electrical Properties

In a subsequent publication (Lieberman *et al.*, 1975), morphometric measurements of the preparation were combined with the linear electrical properties of the synthetic strand to describe the specific electrical constants of the membrane and cytoplasm. In response to a step of current injected intracellularly, synthetic strands (>10 mm in length) behaved like a single one-dimensional leaky capacitive cable, having a core resistivity of 180 $\Omega \cdot$ cm and specific membrane resistance and capacitance of 20 k$\Omega \cdot$ cm^2 and 1.5 μF $\cdot$ cm^{-2}, respectively. These findings were subsequently corroborated in a similar study by Sachs (1976) and were in general agreement with the values obtained from adult Purkinje fibers when the calculations took into account the true sarcolemmal area per unit length in such preparations (Mobley and Page, 1972).

2.2. Current–Voltage Relationship and Cable Properties

In steady-state conditions, a theoretical comparison of the relationship between current–voltage curves for a point-polarized cable and those for a patch of membrane demonstrated vividly the effect of cable properties in reducing nonlinearities in the membrane current–voltage curve (Noble, 1962; Noble, 1966; Noble and Stein, 1966). Given the implications of the results of an earlier study in which the input current–voltage relationship of a long synthetic strand was linearized either in the prescence of tetrodotoxin or the absence of sodium from the bathing medium (Lieberman, 1973), we decided to test whether our inability to detect nonlinearities in these studies could be attributed to the cable properties of the preparation. The input current–voltage relationships of long (>10 mm) and short (0.2 mm) preparations were obtained using two microelectrodes and conventional voltage-clamp circuitry. The results obtained experimentally (Fig. 2B) show the degree of nonlinearity to be markedly dependent on the length of the preparation and in accord with the I–V relationship for theoretical cables with identical membrane properties (Fig. 2A; Kootsey *et al.*, 1975). From this study, it was demonstrated that steady-state membrane properties can be measured relatively undistorted in preparations of cardiac muscle approximately 0.2 mm in length. A similar conclusion was reached by Noma and Irisawa (1976) from experimental data obtained using specimens of rabbit sinoatrial node varying in length between 0.3 and 1.7 mm.

The distortions of the relationship between the input and membrane current–voltage curves introduced by cable properties need not be entirely of this relatively simple kind, i.e., smearing and obfuscation of nonlinear behavior. Experiments were performed on synthetic strands of various length, in which preparations were clamped momentarily (ca. 20 msec) during the repolariza-

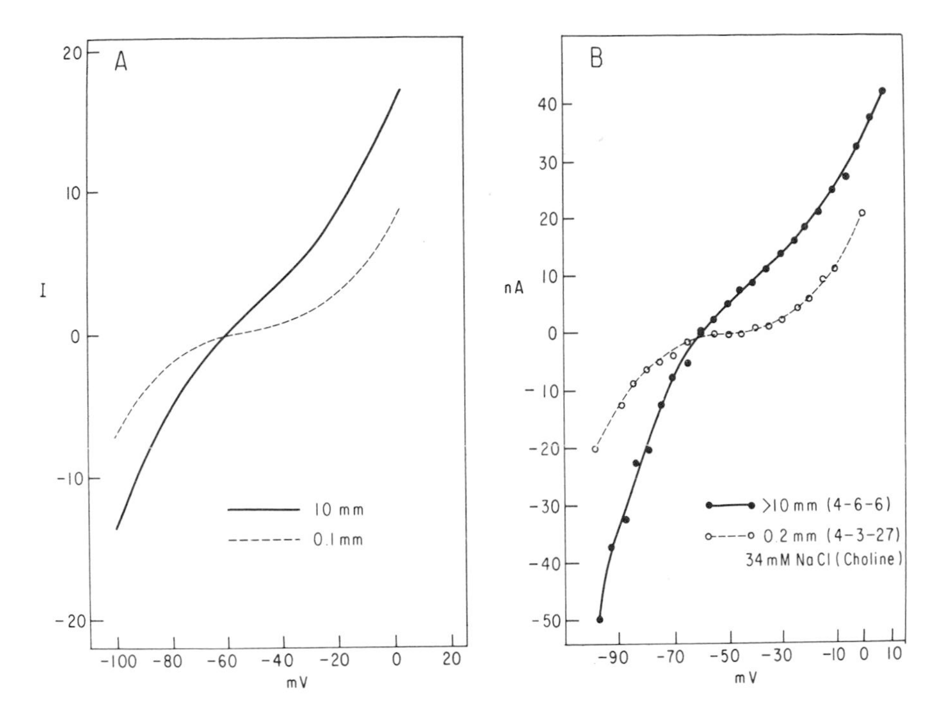

FIGURE 2. Steady-state current–voltage relationships for long and short synthetic strands. (A) Computed input current–voltage relationship for cable lengths of 0.1 mm (---) and 10 mm (—). (B) Experimental input current–voltage relationships for strand lengths of 0.2 mm (○) and 10 mm (●). Measurements were obtained from strands bathed by a solution containing 25% NaCl (choline chloride substitution). Lines drawn by hand. (Unpublished observations, Kootsey *et al.*, 1975.)

tion phase of the action potential to a variety of different (hyperpolarized) values. Figure 3A shows that immediately after the feedback loop was opened, the time course of the voltage was initially positive in slope, showing that the membrane current at the recording site must have been inward. It has been argued that such inward current results from an inherent property of the membrane, i.e., there is a region of negative chord conductance in the membrane current–voltage relationship (Noble and Tsien, 1972; Kass and Tsien, 1976; Goldman and Morad, 1977), as must be the case where the membrane voltage in the rest of the preparation is the same as that at the recording site. There is, however, an alternative explanation, the inward current could arise from membrane that is at a higher voltage, i.e., membrane relatively undisturbed from its normal course of repolarization by the injected clamp current. When the same experiment was performed on a short synthetic strand (Fig. 3B), the phase of inward current at the voltage control site immediately after the momentary clamp was no longer seen. Rather, the repolarization at the end of

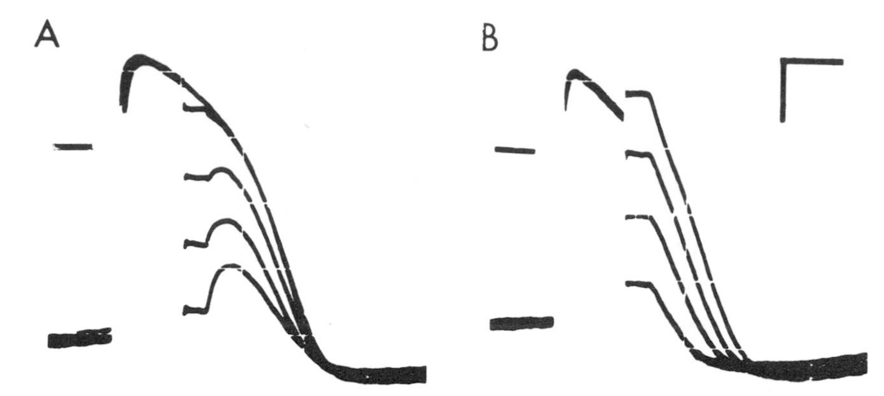

FIGURE 3. Momentary voltage clamp to several (hyperpolarized) values during the repolarization phase of an action potential. (A) Action potential from a long synthetic strand. (B) Action potential from a short (0.15 mm) synthetic strand. Calibration: 20 mV, 50 msec. (Unpublished observations, Kootsey *et al.*, 1975.)

the clamp had a negative slope, i.e., the membrane current was outward. Theoretical stimulations performed on long and short cables in which the action potential was generated by a membrane model, the I–V relationship of which had no inward component of current at any time, confirmed these experimental findings exactly (Kootsey, 1975).

In summary, the purpose of these experiments was to illustrate the complications induced by cable distortions and to show that the time-dependence of the membrane current in voltage clamp of cardiac muscle could, in part, be artifactual. By reducing the length of the growth-orienting channels to approximately 1.0 mm, we were able to obtain short (ca. 100 μm), relatively narrow (50–100 μm) strands that would avoid such distortions (Lieberman *et al.*, 1975, 1976).

2.3. Voltage Clamp of the Synthetic Strand

Using such a preparation, current was injected via an intracellular microelectrode, situated centrally, to control the potential difference between a second intracellular electrode (50–100 μm to one side of the first) and a matching reference electrode in extracellular spcace. Capacitive coupling was minimized by shielding the voltage electrode with stainless steel tubing that was driven by the "guard" of the headstage amplifier. For the first 3 msec following a step change in command potential, the recorded potential deviated noticeably from its final steady value (Fig. 4). Secondary early transient currents (Johnson and Lieberman, 1971) were never observed in uniformly developed preparations of small diameter but were present (Fig. 5) in both irregularly shaped bulbous preparations and those preparations linked by "pseudopods" (<20 μm diameter) in the grooves to an adjacent small preparation. In the latter case, the

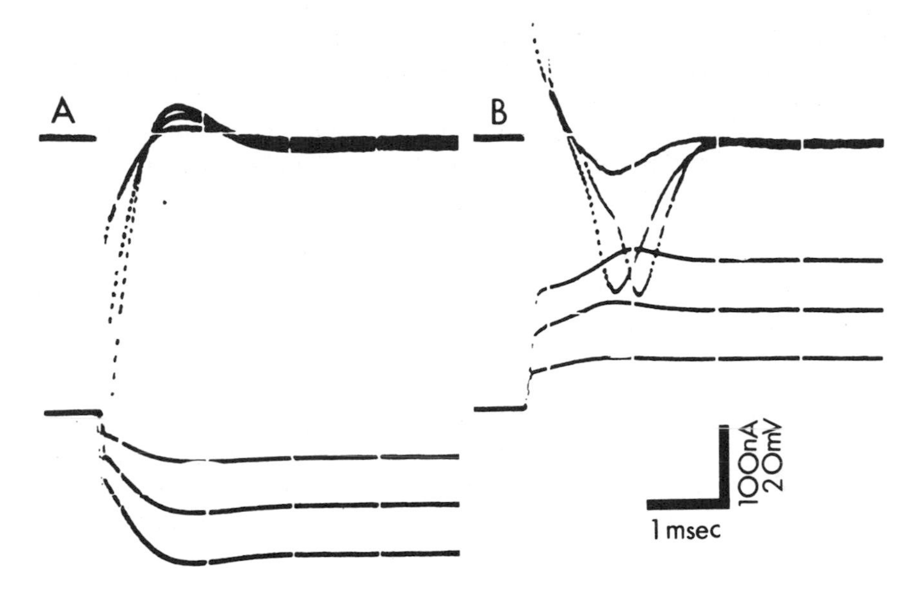

FIGURE 4. Time course of current in response to hyperpolarizing (A) and depolarizing (B) steps in command voltage from a holding potential of −60 mV. Upper trace: current. Lower trace: voltage. Calibration: 20 mV, 100 nA; 1 msec.

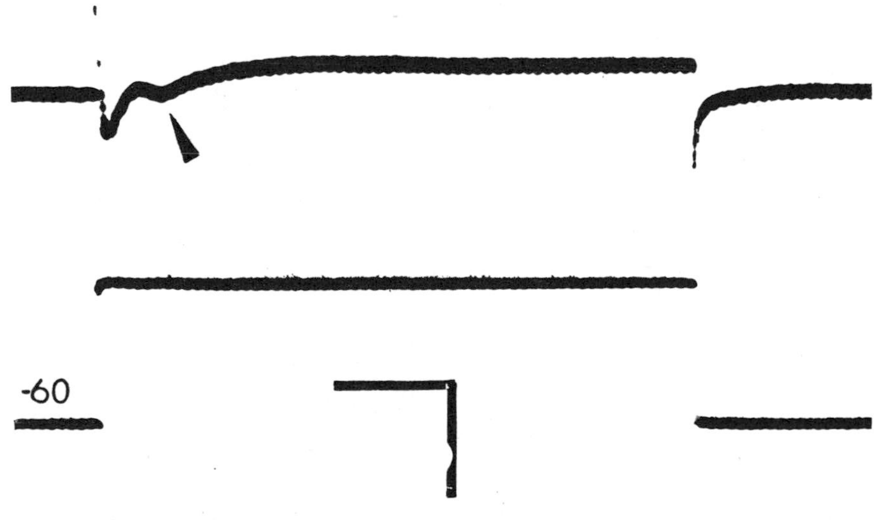

FIGURE 5. Time course of current in response to a depolarizing step in command voltage. Bathing solution contains 25% NaCl (choline chloride substitution). Arrow indicates secondary transient current observed in preparations beyond the limits of size and shape to permit adequate spatial control of membrane potential. Calibration: 20 mV, 50 nA; 200 msec.

small perturbation in holding current disappeared when the groove in which two seemingly separate short strands were located was cut across between them (Chilson, Lieberman, and Johnson, unpublished data).

Plots of the input current as a function of the magnitude of the step, for various times after this initial 3 msec, are displayed in Fig. 6. During depolarizing steps to some voltages, the current remains inward for up to 100 msec, thereafter becoming outward. At these later times, the I–V relationship has a pronounced curvature, the slope resistance increasing markedly only to diminish once again at the extreme of the range of depolarizing steps. Analysis of the time sequence of the current-voltage relationships throughout the clamp steps suggested that the membrane current consisted of two components: a transient current composed largely of sodium ions and a steady-state current dependent on potassium (Lieberman *et al.*, 1975). However, before we could be sure of the meaning of the results of experiments designed to separate the current into its ionic components, it was necessary to know the extent to which an electrogenic Na–K pump could contribute to the electrical properties of the membrane (Horres *et al.*, 1979; Chapman *et al.*, 1979). Theoretical studies showed that during repolarization, the contribution of electrogenic active ion transport to the membrane current–voltage relationship might be comparable

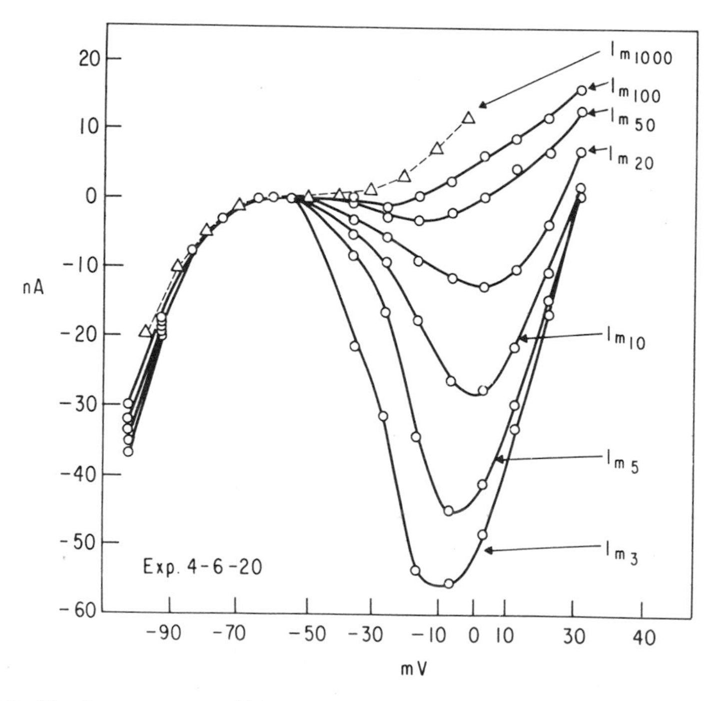

FIGURE 6. Membrane currents (I_m) as a function of membrane potential for various times (msec) after the onset of the voltage clamp.

in magnitude to that of the passive ion movements (Chapman and Johnson, 1976) and, moreover, that a change in external sodium ion concentration could change active transport rates and thereby change the contribution of such transport to the membrane current–voltage relationship.

We therefore decided to analyze the membrane current, using classical voltage clamp methods, at times when the electrogenic pump current is insignificant, i.e., an analysis of the kinetic behavior of the initial inward current. Subsequent efforts were directed to determining which aspects, if any, of the behavior of this early fast sodium current might be studied successfully, without compromises necessitated by deficiencies in the preparation and/or the technique. The approach taken by Ebihara *et al.* (1980) was to reduce the size of the preparation by promoting the growth of spherical aggregates or cultured clusters of heart cells (60–80 μm diameter). The geometry of this preparation comes closest to the ideal (single spherical cell) in minimizing problems associated with using microelectrodes as a point source of current (Eisenberg and Engel, 1970; DeHaan and Fozzard, 1975).

3. CULTURED CLUSTER

Clusters of cardiac muscle cells were obtained by modifying the culture techniques used to prepare synthetic strands (Lieberman *et al.,* 1972). Small openings (20 μm) were formed on the surface of an agar-coated culture dish using a 27 gauge needle. The exposed plastic surface was sufficient in area to provide a point of attachment for preformed small spherical aggregates (50–200 μm diameter). Similar preparations have been developed in other laboratories by gyratory reaggregation (Fischman and Moscona, 1971; Sachs and DeHaan, 1973) or by spontaneous reaggregation on a cellophane sheet (McLean and Sperelakis, 1976). Figure 7 contains a photomicrograph in which cultured clusters are depicted adjacent to a synthetic strand. Typical cardiac action potentials recorded from each spontaneously beating preparation is also shown for comparison. The resting or maximum diastolic potential is about −75 to −80 mV, and the upstrokes of the action potentials have a maximum rate rise of ca. 130 $V \cdot sec^{-1}$. In the presence of tetrodotoxin (30 μg/ml), the initial rate of depolarization is markedly reduced (1 $V \cdot sec^{-1}$) and after 10–15 min, the preparations stopped beating spontaneously. In comparison, the effect of D600 (1 μg/ml) is to accelerate repolarization without affecting the maximum rate of rise of the action potential (Ebihara *et al.,* 1980).

3.1. Voltage Clamp of the Cultured Cluster

For voltage-clamp analysis of the cultured cluster, it was necessary to change the membrane potential very rapidly in order to resolve the capacitive

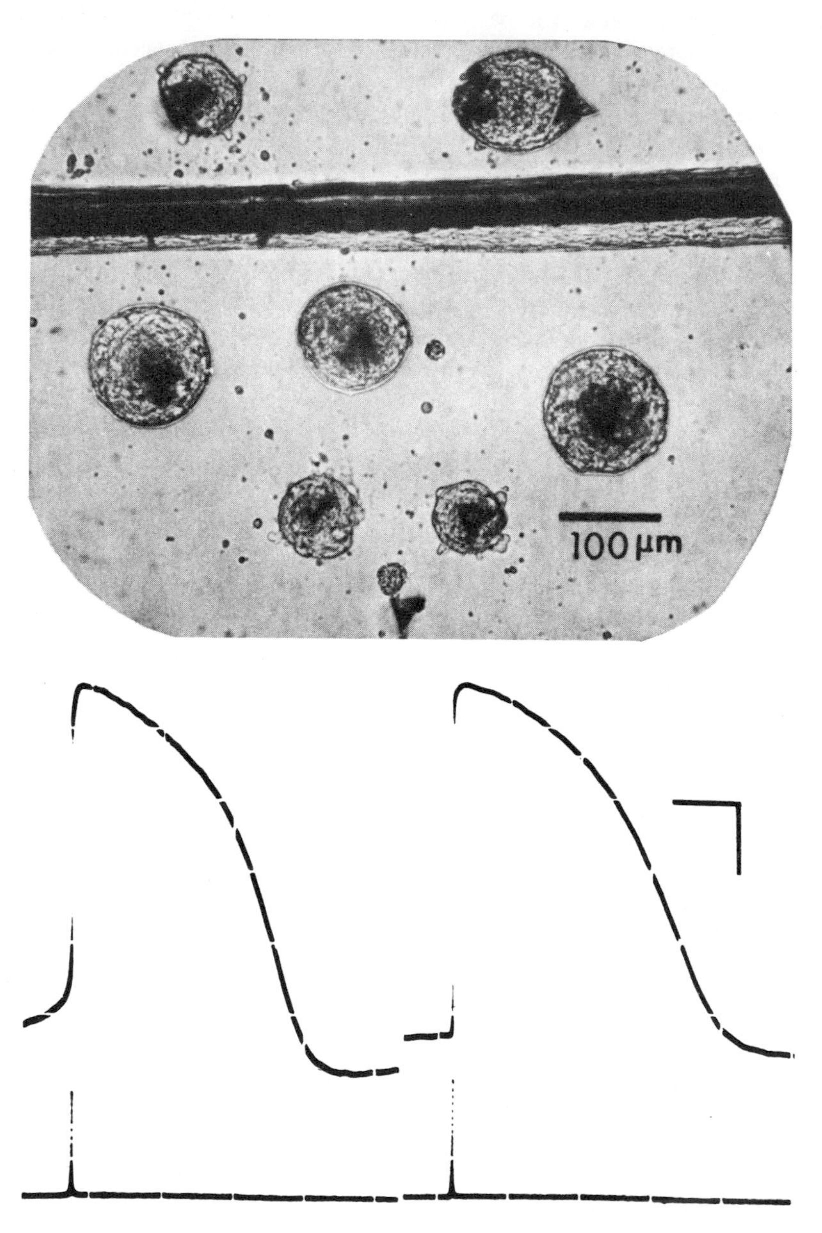

FIGURE 7. Photomicrograph of cultured clusters of embryonic chick heart cells adjacent to a segment of a long synthetic strand (upper panel). Scale: 100 μm. Action potentials recorded from a spontaneously beating cultured cluster (left) and synthetic strand (right). Calibration: vertical bar, 20 mV and 100 V sec^{-1}; horizontal bar, 50 msec.

and ionic currents. Using the two microelectrode technique, the frequency response of the voltage-clamp circuitry is limited by capacitive coupling between the current-passing and voltage electrodes, stray capacitances to ground, and the ability of the microelectrode to pass large currents. The fidelity and speed of clamping have been shown to be dependent on the membrane time constant and the resistance of the current-passing electrode (Katz and Schwartz, 1974; Smith *et al.*, this volume). A schematic drawing of the recording system is shown in Fig. 8. An aluminum shield, driven by the unity gain output of the headstage amplifier, was used to eliminate the coupling capacitance between the current and voltage electrodes. Stray capacitance was reduced by shielding the electrodes with conductive silver paint covered by a lacquer to insulate them from the extracellular solution or with stainless steel tubing either driven by a guard of the headstage amplifier (voltage electrode) or grounded (current electrode). With this system (see Ebihara *et al.*, 1980, for complete details), it was possible to achieve voltage control in less than 50 μsec.

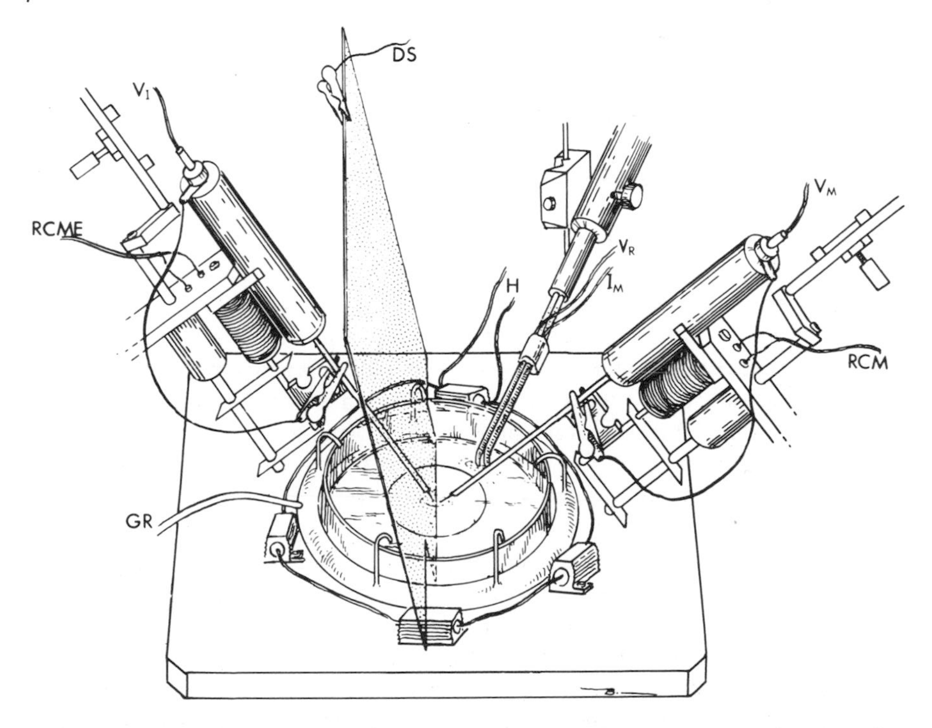

FIGURE 8. Experimental apparatus for voltage-clamp studies. H, heat source (power resistors) for warming microscope stage; GR, gassing ring for air/CO_2 mixture; DS, driven aluminum shield between current (V_i) and voltage (V_m) electrodes; V_R, wire from reference electrode; I_M, wire to current amplifier. RCM and RCME, remote control electromagnetic transducers for positioning microelectrodes.

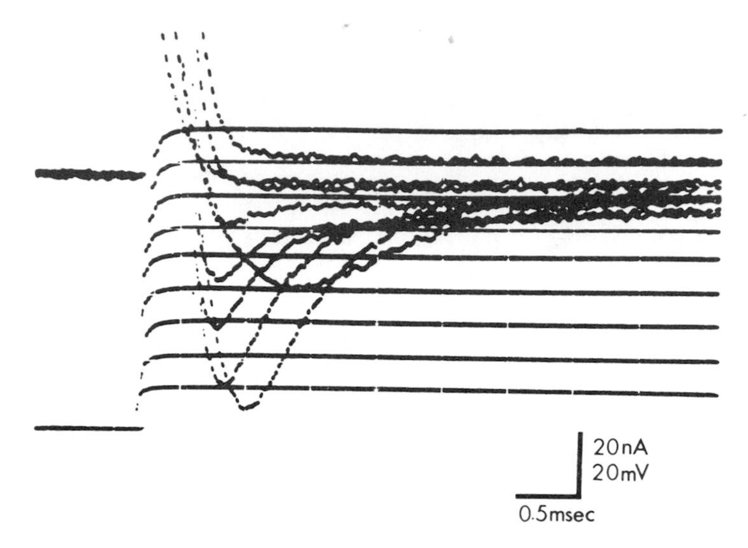

FIGURE 9. Current and voltage traces recorded from a cultured cluster during a series of depolarizing voltage clamp steps from a holding potential of −60 mV.

Figure 9 shows a family of membrane currents recorded at 37°C during a sequence of depolarizing voltage-clamp steps from a holding potential of −60 mV. The transient inward current is initiated at potential steps more positive than −50 mV and reached a maximum value at a potential of about −30 mV (Fig. 10). Detailed analysis of the time course of inactivation and reactivation of the early inward current was implemented by using a PDP 11/40 computer as an on-line experimental control and data system (Ebihara *et al.*, 1980). In summary, no significant differences were noted in the time constants of inactivation as determined by single and double voltage-step protocols. Furthermore, the time constants of inactivation and reactivation at the same potential in the same preparation were similar, showing that the sodium current of heart cells recorded at 37°C can be described by Hodgkin–Huxley kinetics. Studies are currently in progress to analyze the kinetics of the secondary slow inward current* (see inset, Fig. 10). Future attempts to separate the membrane current into its individual ionic components will, of necessity, take into account both the results of experimental and theoretical studies of the electrophysiological consequences of electrogenic transport. Additional methods involving ion-flux measurements with the polystrand preparation (see Section 4) are being explored to elucidate the passive permeability mechanisms and aid in distinguishing between simple electrochemical effects of a given ion substitution from the possible complex and unpredictable pharmacological effects of the substitution itself on the permeability of the membrane.

*A preliminary report by Ebihara *et al.* appears in *Fed. Proc.* **39**(6):2065, 1980.

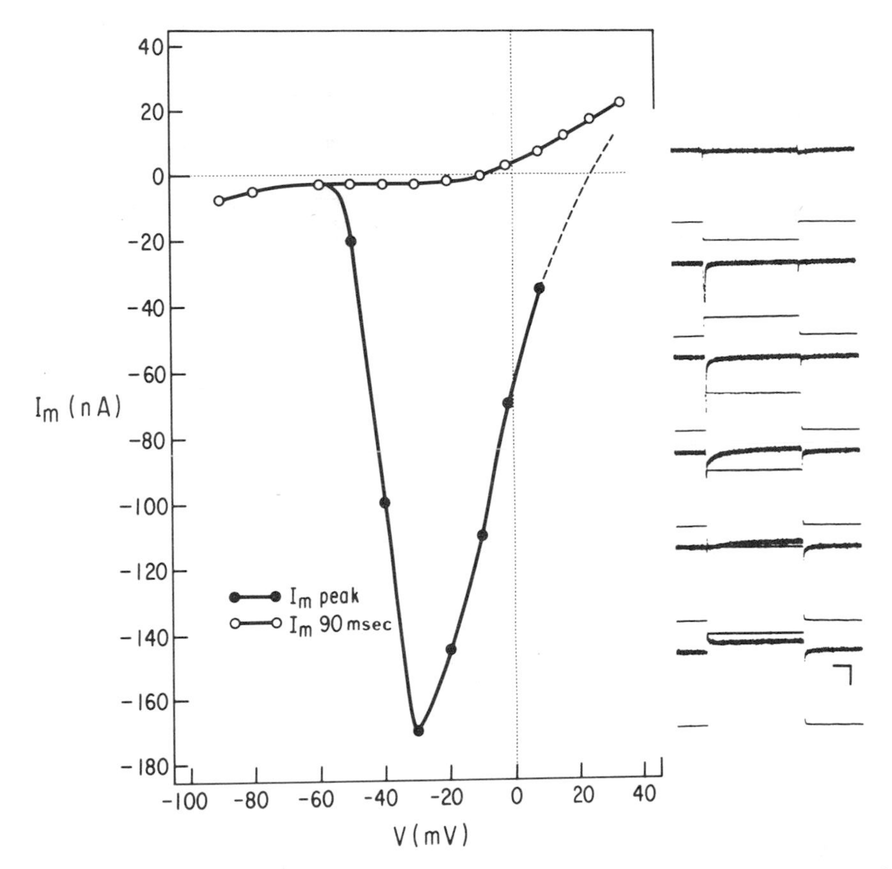

FIGURE 10. Current–voltage relationship of a cultured cluster. Inset shows the response of membrane current to steps of hyperpolarizing and depolarizing voltages. Calibration: vertical bar, 20 mV and 50 nA; horizontal bar, 20 msec.

4. POLYSTRAND

Ion transport studies of heart cells in tissue culture can provide quantitative steady-state flux data which may then be correlated with electrophysiological data obtained under similar conditions (Horres and Lieberman, 1977). As a consequence, such results provide a direct approach to the study of membrane permeability and ionic currents in cardiac muscle. In addition, it is feasible to pursue flux determinations in non-steady-state conditions to correlate events associated with active transport to electrogenic phenomena (Horres *et al.*, 1979).

Central to flux determinations under steady-state conditions are measurements of tracer kinetics. Although it is a relatively simple matter to obtain such measurements in biological preparations, the interpretation of these results must be approached with caution (Jones, 1975). The tendency is to consider tracer kinetics as a simple first-order process as can be theoretically predicted for tracer exchange from an idealized single compartment into a well-stirred bathing medium. This simplifying assumption more often is not the case because (1) diffusional delays are inherent in the morphology of the preparations, (2) cell populations are heterogeneous, (3) trauma induced by the dissection will compromise the functional integrity of a subpopulation of cells, and (4) steady-state conditions do not exist.

In considering the influence of diffusional delays on the interpretation of tracer kinetics in multicellular preparations, it must be recognized that measured exchange kinetics are determined by the rate of tracer exchange from the cells to the extracellular space, a process which will vary inversely with the volume to surface area (V/A) of the cells at a given transmembrane flux, and the rate of tracer exchange from the extracellular space to the surface of the preparation and/or to the vascular spaces (MacDonald *et al.*, 1974). In quiescent preparations, tracer exchange from the extracellular space is determined largely by (1) the diffusion coefficient in the extracellular matrix and (2) the distance and nature of the pathway that must be traversed by the ion(s) in question. At the surface of the vascular or extracellular space, additional barriers may be encountered, e.g., epithelial layers and/or unstirred fluid boundaries. Finally, the exchange characteristics of the experimental chamber or vessel must be considered. In a simplistic analogy one can envision a series system in which the largest resistance dominates exchange kinetics. Thus, the transmembrane exchange can be very rapid, as occurs in highly permeable cells characterized by a small V/A, but the experimentally observed tracer kinetics may be slow because of a large diffusional distance in extracellular space or because of any one or more of the factors mentioned above. The importance of diffusional delays should not be considered lightly in cardiac tissue since distances as small as 200–300 μm can have profound effects on the tracer kinetics of rapidly exchanging ions. Indeed, when the problem of diffusional delays is combined with those of cellular heterogeneity and loss of steady-state conditions, the interpretation of tracer results becomes a formidable task (Horres *et al.*, 1979).

In view of the above considerations, tissue culture stands out as an important method to be incorporated into studies of ion transport in excitable cells. Of paramount importance is the ability to grow viable preparations whose dimensions are sufficiently small to obviate most diffusional problems. In addition, a true steady state can be achieved, whereas dissected preparations rarely, if ever, can be maintained in a functional state for an equivalent period of time. Less well appreciated are the capabilities for evaluating contributions of varying cell populations to tracer kinetics and for empirically determining the effects of preparation geometry on tracer kinetics. The latter becomes important when trying to relate results obtained in cultured preparations to the

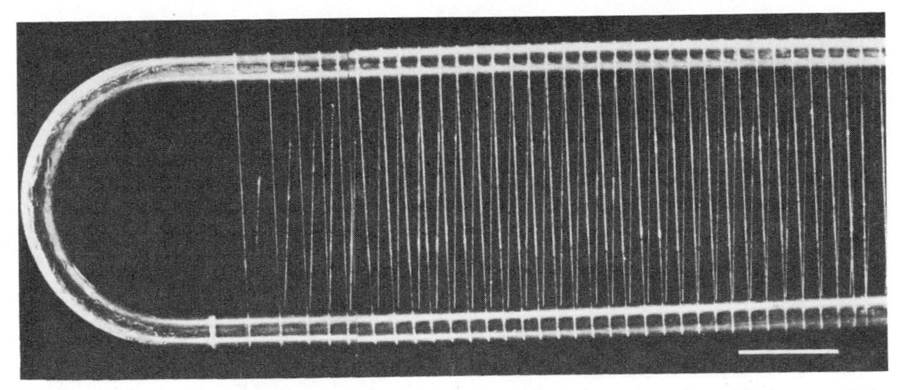

FIGURE 11. Photomicrograph of the polystrand preparation. Note that cellular growth is restricted to the central portion of the nylon substrate formed by a continuous winding around a silver wire. Scale: 3.8 mm.

often artifactual results obtained in large-diameter, naturally occurring preparations. In this regard, we have successfully utilized growth orientation in tissue culture to correlate tracer flux measurements with results obtained electrophysiologically (Horres *et al.*, 1979). Several of the techniques we have developed for transport studies will be described to illustrate the advantages of tissue culture preparations for ion transport studies.*

4.1. Morphological and Chemical Characterization

Measurement of morphological and chemical characteristics of the preparation are important in quantitating ion fluxes from tracer kinetics in steady state. Of primary concern are the freely exchangeable intracellular concentration of the ionic species being measured and the ratio of cell volume to cell surface area through which tracer exchange is occurring. Correlation of the ionic currents obtained by voltage-clamp techniques with ionic flux requires comparable chemical gradients and geometric characteristics of the preparation.

To minimize the dimensions of the preparations for ionic flux studies, we developed a nylon-supported growth-oriented preparation of heart cells (Horres *et al.*, 1977). A growth chamber was designed to foster the development of a longitudinally oriented preparation in which the cells are grown about a 20 μm nylon core as an annulus, the thickness of which is between 30 to 40 μm, giving an overall diameter of 80 to 100 μm. Tissue mass was increased by designing the preparation in the form of a parallel array of 70 segments, 5 mm long, supported by a continuous nylon monofilament wrapped about a U-shaped silver clip (Fig. 11). This configuration allows for convenient

*Preliminary results of sodium tracer kinetics have recently been reported by Wheeler *et al.* in *Fed Proc.* **39**(6):1841, 1980.

manipulation of the tissue through the various experimental solutions. The resolution of tracer and chemical measurements is also enhanced because tissue mass is increased without compromising the surface area. The monofilament can also be removed from the silver frame to provide a useful tare for tissue wet weight and dry mass determinations.

4.1.1. Wet Weight Determination

Preparations with large surface area/volume ratios tend to minimize diffusional limitations and are highly responsive to changes in the environment. Although this characteristic can be exploited experimentally, it must be carefully controlled to avoid untoward effects from variables other than those under investigation. For example, relatively brief exposure (less than 1 min) of the polystrand to room air produces a dramatic loss of water, temperature, and buffer capacity. To prevent these problems from occurring, we have modified a tissue culture incubator to permit the manipulation of preparations within an environment that is saturated with H_2O and buffered with 5% CO_2 air at 37°C (Fig. 12).

FIGURE 12. Modified tissue culture incubator for ion transport studies. Glove ports were installed to allow the manipulation of preparations at 37.5° C in a humidified atmosphere of 5% CO_2/air. Electrical strip heaters on the doors retard moisture condensation which otherwise would impede visibility. The incubator was illuminated internally and provisions were made for electrical access to the weighing chamber of the electrobalance.

As stated earlier, the polystrand preparation is of sufficient mass (0.6 mg dry weight) to allow conventional wet and dry mass determinations. The cells, together with their nylon support, can be slipped from the silver support in the form of a coil that can then be snared onto the wire hangdown of a Cahn electrobalance. Since the nylon support contributes only 0.3 mg to the preparation mass, it allows the use of the most sensitive range of the electrobalance. For wet weight determinations, the weighing chamber is placed inside the modified incubator to slow the rate of H_2O loss from the preparation. Removal of extraneous water from tissue cultured preparations by the traditional blotting technique used with intact muscle preparations is not recommended because adherent blotting material may not only damage the preparation but may add significantly to its weight, thus requiring unacceptably large correction factors. We have found that a liquid blotting technique using a fluorocarbon (FC-80, 3M Company) that is more dense than water can significantly reduce adherent water. The polystrand preparation is covered with about a centimeter of the fluorocarbon in a 13 × 100 mm test tube and gently tapped to promote the release of water to the surface of the fluorocarbon, where it is removed with a cotton-tipped applicator. In control tests, exposures to FC-80 of up to 60 min did not affect contractile activity of the preparations. Wet weights, obtained at 1–min intervals over a 10–min period, were determined by extrapolating back to the time the preparation was removed from the fluorocarbon (Fig. 13). To determine the volume of extracellular water, a gamma-emitting marker is incorporated into the bathing medium prior to blotting and weighing. The preparations are then digested with nitric acid and counted directly. Although

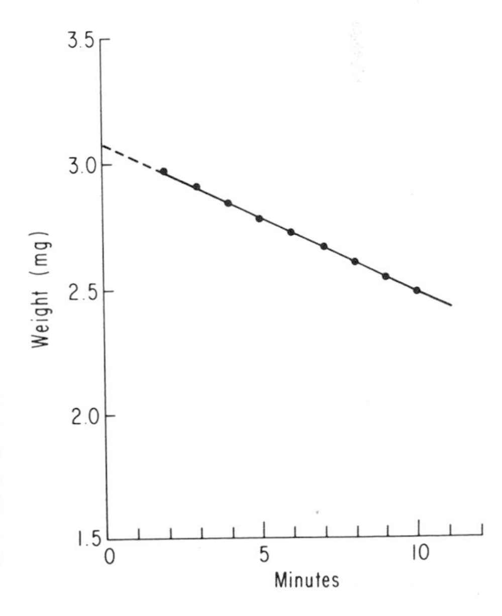

FIGURE 13. Tissue wet weight. The wet weight of the preparation was determined by extrapolating back to the time of removal from fluorocarbon. The slope of the curve indicates the high rate of moisture loss from these preparations. (From Horres *et al.,* 1977, with permission.)

each marker must be evaluated to determine its exclusion from intracellular space, the thinness of these preparations permits rapid penetration of markers throughout extracellular space. By exposing the preparation for a long period of time, it is therefore possible to determine whether the markers remain in extracellular space or are transported into the cells. A comparison of the results obtained with [^{125}I]-iothalamic acid and [^{51}Cr]-EDTA is shown in Fig. 14. The continual increase in tissue activity of [^{51}Cr]-EDTA indicates intracellular transport of this marker.

Intracellular sodium and potassium ion content can be determined by atomic absorption flame spectroscopy in the usual manner without pooling the preparations. However, unless the preparation is rinsed in a sodium-free medium prior to the measurement, a correction must be made for extracellular sodium. We have found that the extracellular compartment can be cleared of sodium without any significant loss of intracellular sodium by rinsing preparations in cold isotonic choline chloride for 15 sec (Lieberman *et al.*, 1978).

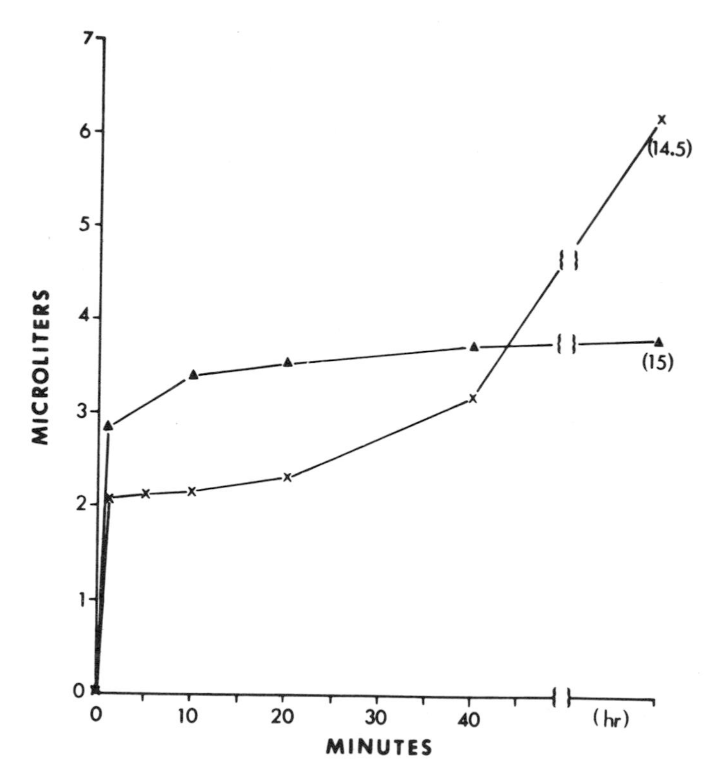

FIGURE 14. Evaluation of extracellular markers. Preparations were exposed to extracellular markers under culture conditions. [^{125}I]-Iothalamic acid (▲) quickly equilibrates and maintains a constant activity over many hours while the continued increase of [^{51}Cr]-EDTA (×) indicates entry of the marker into the cells.

4.1.2. Cell Geometry

Cell volume-to-surface area ratio (V/A) is a key factor in the determination of transmembrane fluxes from tracer kinetics. For a given value of transmembrane flux (J_K), the rate constant (k) for tracer exchange of the entire preparation will be proportional to the inverse of the cell V/A, in accord with the following equation: $k = J_K A/V [K]_i$ (Keynes, 1954). Thus, flattened cells with a small V/A will have a higher rate constant than will cylindrical or spherical cells of equal surface area for the same value of transmembrane flux. It is imperative that V/A ratios be determined in preparations analogous to those from which tracer kinetics are being obtained. We have used stereologic techniques (Weibel, 1973; Eisenberg *et al.*, 1974) to determine the V/A ratio in a variety of embryonic heart preparations (Table I).

Cell geometry will also have a major influence on diffusional errors in determining tracer kinetics. Increases in the rate of tracer movement across the cell membrane occurs in cells with small V/A and/or rapid flux. This provides the opportunity for tracer to accumulate in the extracellular space and thereby to reenter the cell when diffusion distance to the surface of the preparation is large. For a given preparation thickness and tissue transmembrane flux, diffusional errors will be greatest in a preparation consisting of tightly packed cells with a small V/A (MacDonald *et al.*, 1974).

4.2. Potassium Flux Determination

To resolve the contribution of potassium fluxes in steady state to the electrophysiological results under similar conditions, data must be obtained from cells with a constant intracellular ion content. Implicit in this condition is the requisite balance between inward and outward fluxes of the ionic species which requires the use of radioisotopic tracers to measure the unidirectional fluxes. Although numerous attempts, for technical reasons, have been made to substitute ionic species, such as ^{86}Rb for ^{42}K (Van Zweiten, 1968; deBarry *et al.*, 1977), it is preferable to use isotopes of the ionic species under consideration;

TABLE I. Volume to Surface Area Measurements of Embryonic Heart Cells

Preparation	$(V/A) \times 10^{-4}$ cm
6- to 8-day chick embryo ventricle[a]	1.24
Muscle cells of the cultured cluster[b]	1.28
Muscle cells of the synthetic strand[c]	1.14
Muscle cells of the polystrand preparation[d]	1.06
Nonmuscle cells of the polystrand preparation[d]	0.48

[a]Carmeliet *et al.* (1976).
[b]Ebihara *et al.* (1980).
[c]Lieberman *et al.* (1975); Sachs (1976).
[d]Horres *et al.* (1977).

otherwise, there will remain the possibility that the permeability of the membrane to the substitute tracer will differ from that of the ion of interest (Müller, 1965; Graves, 1979).

Having selected the appropriate tracer, it is then necessary to design an experimental approach that will allow accurate detection of the rate constants for tracer exchange. To increase resolution, sampling intervals should be short relative to the observed half-times of isotopic exchange. This can be conveniently accomplished by using a perfusion system in tracer efflux studies. For the polystrand preparation, we designed a perfusion chamber of minimal volume (1 ml) to allow control of temperature, pH, and stimulation rate during efflux experiments (Fig. 15). The perfusion rate represents a balance between a high rate which reduces the exchange constant of the perfusion system and a rate which optimizes counting statistics. In general, an acceptable ratio of time constants of the chamber to that of the preparation is 1/10.

Since the polystrand preparation is composed of fibroblasts and muscle cells (Horres *et al.*, 1977), it was necessary to evaluate the contributions of the two cell types to the overall measured flux. The efflux of ^{42}K under steady-state conditions is shown for a muscle-enriched polystrand preparation in Fig. 16. The presence of multiple compartments is indicated by the slowing of efflux kinetics at longer times. Further evidence is provided by varying the loading period (Fig. 17). In a single-compartment system, the rate constant of tracer efflux should be independent of loading time. However, in the presence of two cell types with different rate constants, the tracer kinetics represents a composite of the individual cells. By varying the loading time, the relative contributions can be changed such that short loading times will favor the more rapidly exchanging cells.

To determine whether two-compartment efflux kinetics could be related to cellular heterogeneity or to a diffusional problem, ^{42}K efflux kinetics of pure fibroblast preparations were evaluated under similar experimental conditions. The results in Fig. 16 show that tracer efflux from these cells can be described by a single exponential process with a rate constant of 0.015 min^{-1} (Horres and Lieberman, 1977). In analyzing the polystrand data, ^{42}K efflux kinetics of pure fibroblast preparations have been found to represent the slowly exchanging component. Incorporating this information into compartmental analyses (Horres and Lieberman, 1977) we were able to fit the results to a two-compartment model in which the fast compartment corresponds to muscle cells with a tracer exchange constant of 0.067 min^{-1}. From these data, a steady-state potassium efflux of 15.7 pmoles $\cdot$ cm^2 $\cdot$ sec^{-1} can be calculated for cardiac muscle cells. Combining these data with measurements of membrane potentials enables us to calculate the value of 3×10^{-7} cm $\cdot$ sec^{-1} for the membrane permeability of growth-oriented heart cells (Horres *et al.*, 1979), a value shown to be independent of the external potassium concentration in the range of 1.0–20 mM.

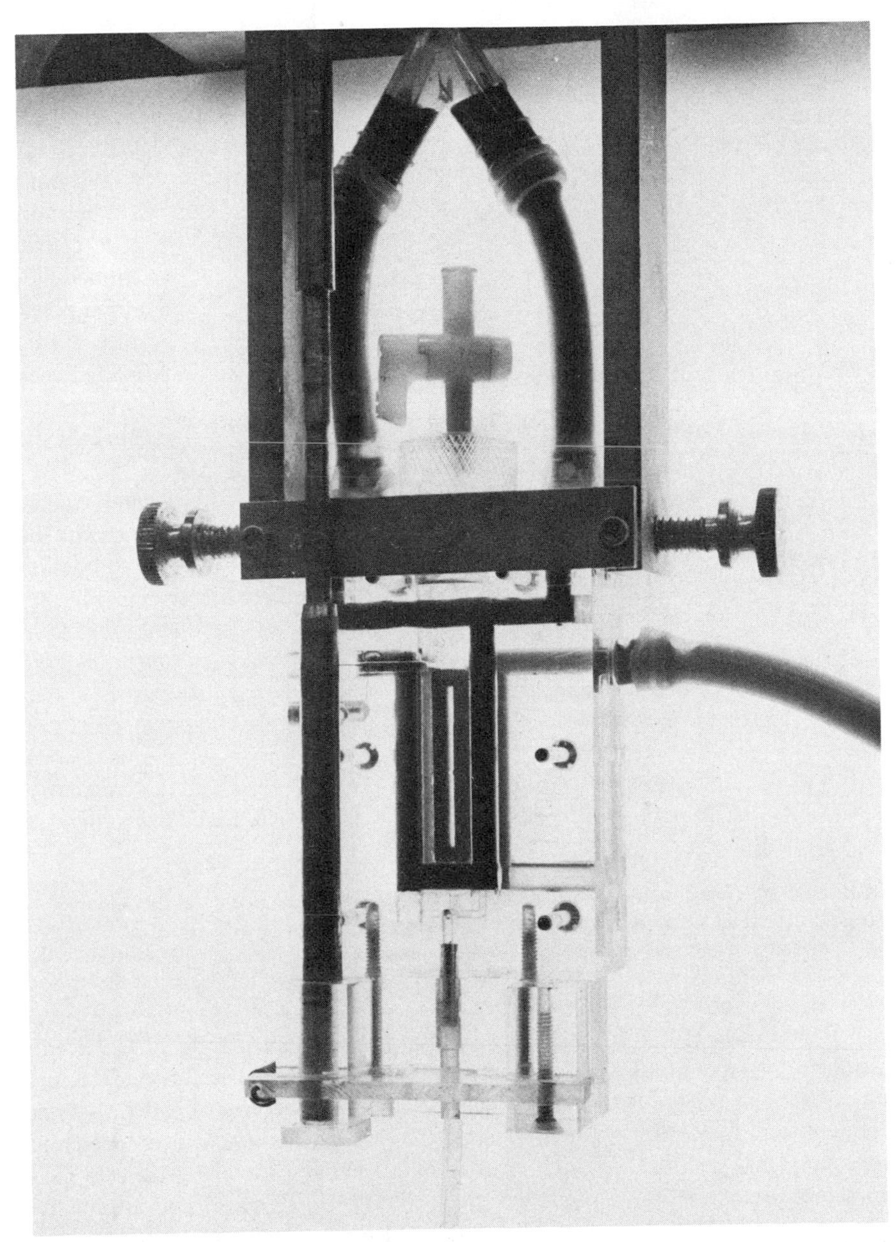

FIGURE 15. Photograph of flux chamber. The heating jacket of the lucite flux chamber is outlined by dye. Water enters through the port on the right, circulates through four vertical channels surrounding a central compartment in which the preparation is placed, and exits through the coaxial ports at the top. Perfusion fluid enters through the center of the coaxial tubing and exits through the bottom of the chamber. The preparation is aligned between two field stimulating electrodes with a central viewing window.

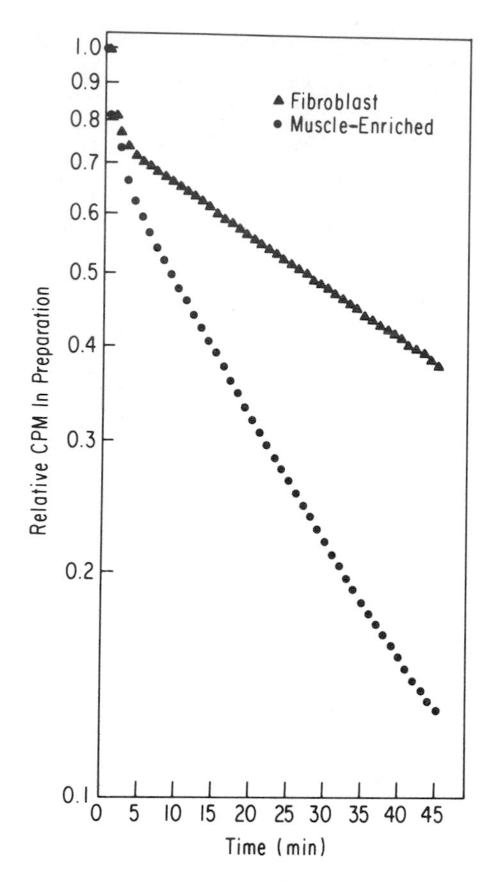

FIGURE 16. ^{42}K efflux kinetics of polystrands composed predominantly of cardiac muscle (●) and those of fibroblast-like cells (▲). The preparations were perfused at time zero with nonradioactive media containing 5.4 mM potassium at 37.5° C.

Since potassium efflux is associated with the repolarization phase of the cardiac action potential, there is considerable interest in the effects of stimulation rate on potassium efflux in cardiac muscle (Coraboeuf *et al.*, 1968; Polimeni and Vassalle, 1970; Juncker *et al.*, 1972). Tissue-cultured preparations are ideally suited for such experiments in that they are relatively free from mechanical artifacts produced by squeezing the extracellular space. An example of this type of experiment is shown in Fig. 18 in which increasing the rate of stimulation from rest to 250 min^{-1} produces an increase in efflux rate constant by a factor of 1.2 during the period of stimulation.

Growth orientation in tissue culture not only provides the investigator with the ability to develop preparations that are sufficiently thin to avoid the complications of tracer reflux (Keynes, 1954), but also allows one to evaluate, empirically, the effects of tracer reflux on tracer kinetics. By varying the initial

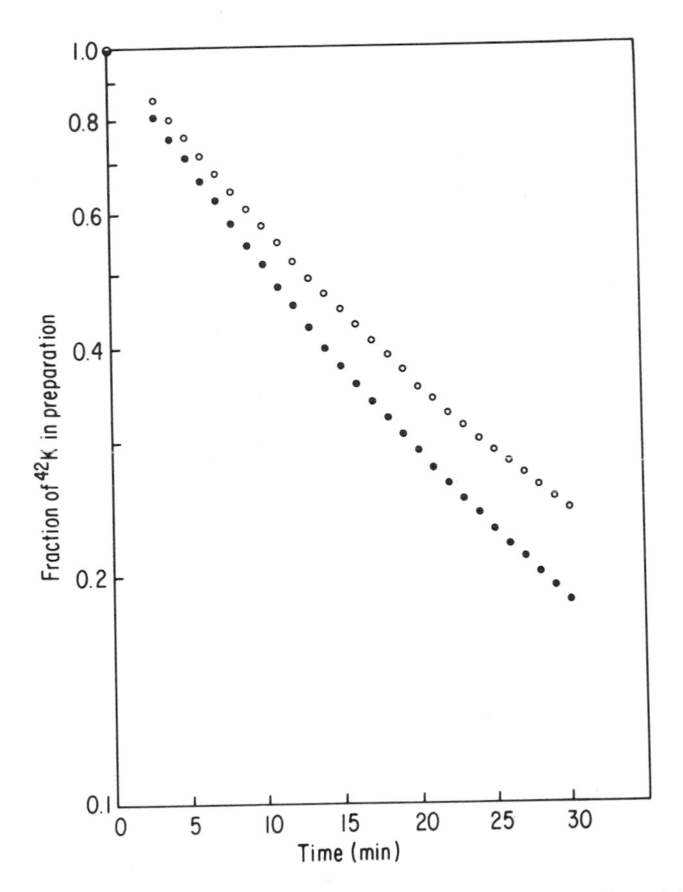

FIGURE 17. Effect of loading time on ^{42}K efflux kinetics of cardiac muscle enriched poly-strands in 5.4 mM potassium. Two preparations were loaded with ^{42}K for either 5 min (●) or 45 min (○). (From Horres *et al.*, 1979, with permission.)

cell density from a given suspension, preparations of different sizes can be grown in culture. An experiment designed to demonstrate the effects of diffusional limitations (tracer reflux) on tracer kinetics using nonmuscle cells is illustrated in Fig. 19. When ^{42}K-equilibrated preparations are perfused by K-free solution, tracer reflux is enhanced in thick preparations (Horres *et al.*, 1979) as is evident by the reduced slope in the efflux kinetics.

4.3. Active Transport

Active transport must play a major role in maintaining steady-state conditions in cardiac muscle because of the combined effects of small volume-to-surface area and high membrane permeability (comparable to that of large nerve and skeletal muscle cells). This point is well demonstrated in the poly-

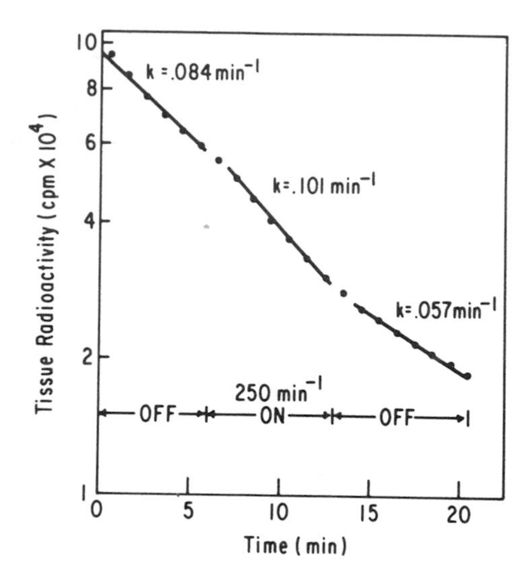

FIGURE 18. Effect of stimulation rate on ^{42}K efflux kinetics of cardiac muscle enriched polystrands in 5.4 mM potassium. Preparations were made quiescent by the addition of valinomycin (2.5 $\mu g \cdot ml^{-1}$).

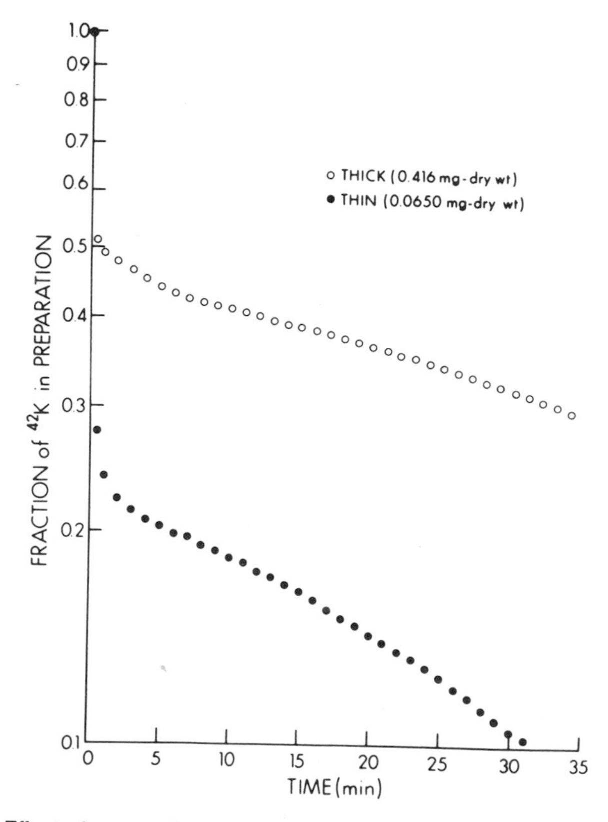

FIGURE 19. Effect of preparation size on ^{42}K efflux kinetics in K-free solution. The preparations were derived from nonmuscle cells of the same cell suspension.

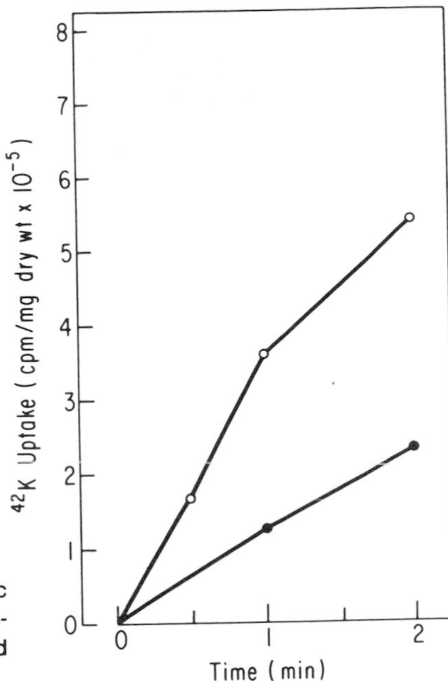

FIGURE 20. Rate of ^{42}K uptake for cardiac muscle enriched polystrands in 5.4 mM potassium under conditions of steady state (●) and during recovery from K-free perfusion (○).

strand preparation in which ouabain was shown to inhibit 79% of potassium influx (Lieberman *et al.*, 1978) and in which inhibition of active transport by K-free solutions reduces K content by 50% within 15 min (Horres *et al.*, 1979). Reintroduction of normal-K medium to study the recovery kinetics of potassium depletion results in a three-fold increase in ^{42}K influx (see Fig. 20).* These results, when correlated with the hyperpolarizing response seen in parallel electrical experiments (Fig. 21A), provide evidence for the existence of electrogenic transport (Horres *et al.*, 1979). Thick preparations (100 μm), on the other hand, do not readily depolarize during attempts to inhibit active transport and, upon reintroduction of K, generate a markedly attenuated hyperpolarizing response (Fig. 21B).

In light of the difficulties in quantitating radiotracer data from a heterogeneous population of cells, in nonsteady state, qualitative changes could be observed in muscle cells following active transport inhibition. Preparations exposed to K-free solution for 15 min were shown to contain electron-dense deposits in the mitochondria (Fig. 22). Although the composition of these deposits is unknown, similar results provide evidence that calcium phosphate could be their major component (Ashraf and Bloor, 1976; Trump *et al.*, 1976). To what extent this finding is related to stimulation of the Na–Ca exchange mechanism as a result of an increased Na_i (Reuter and Seitz, 1968) remains

*The actual stimulation of transport in muscle cells may be higher because of the contaminating effects of more slowly exchanging fibroblasts in these preparations.

FIGURE 21. Continuous recordings of transmembrane potentials from thin (A) and thick (B) spontaneously active polystrand preparations exposed to K-free perfusate for 8 min and then returned to 5.4 mM potassium.

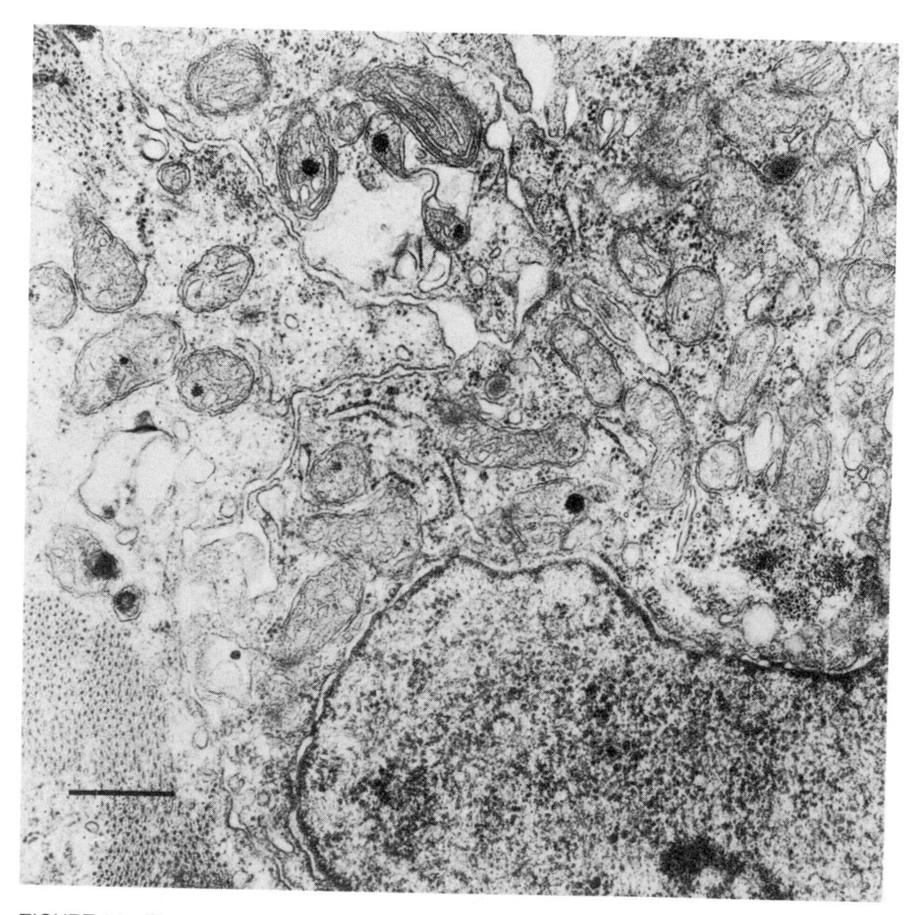

FIGURE 22. Electron micrograph of a polystrand preparation exposed to K-free solution for 15 min. Note the presence of electron-dense deposits in the mitochondria (× 77,000).

to be determined. The presence of mitochondrial deposits following active transport inhibition has potential clinical significance since comparable results have been reported in the ischemic myocardium (Shen and Jennings, 1972) and teleost nephron (Trump and Jones, 1977) as well as in preparations of cardiac and skeletal muscle subjected to several conditions associated with ischemia (Ganote *et al.,* 1975; Auclair *et al.,* 1976; Lipton, 1977).

5. CONCLUSION

Growth orientiation of cells in culture eliminates the need to make fundamental (and oftentimes erroneous) assumptions about the role of diffusional limitations on experimental results. Indeed, a unique feature of growth orientation in tissue culture is that variation in the size of the preparation can be achieved simply by altering the initial cell density. Thus, data of the kind described in this chapter can be evaluated empirically for the effects of tissue geometry on the measured parameters. In both electrical and transport studies, preparation geometry has been shown to have a profound influence on the experimental results. Such evidence offers a plausible explanation for the often observed differences in data obtained from naturally occurring preparations and those grown in tissue culture.

ACKNOWLEDGMENTS. We wish to acknowledge the technical assistance of Owen Oakeley, Phyllis Bullock, Pam Evans, and Dr. William Adam. We thank David Wheeler for his critical review of the manuscript and Pat Purcell for her secretarial assistance. This research was supported in part by grants from National Institutes of Health HL-12157, HL-23138, HL-07101, and GM-07184.

REFERENCES

Ashraf, M., and Bloor, C. M., 1976, X-ray microanalysis of mitochondrial deposits in ischemic myocardium. *Virchows Archiv* [*Cell Pathol.*] **22:**287.

Auclair, M. C., Adolphe, M., Moreno, G., and Salet, C., 1976, Comparison of the effects of potassium cyanide and hypoxia on ultrastructure and electrical activity of cultured rat myoblasts, *Toxicol. Appl. Pharmacol.* **37:**387.

Burrows, M. T., 1912, Rhythmische Kontraktionen der isolierten Herzmuskelzellen ausserhalb des Organismus, *Münch. Med. Wochenschr.* **59:**1473.

Burrows, R., and Lamb, J. F., 1962, Sodium and potassium fluxes in cells cultured from chick embryo heart cells, *J. Physiol. (Lond.)* **162:**510.

Carmeliet, E. E., Horres, C. R., Lieberman, M., and Vereecke, J. S., 1976, Developmental aspects of potassium flux and permeability of the embryonic chick heart, *J. Physiol. (Lond.)* **254:**673.

Chapman, J. B., and Johnson, E. A., 1976, Current–voltage relationships for theoretical electrogenic sodium pump models, *Proc. Aust. Physiol. Soc.* **7:**69.

Chapman, J. B., Kootsey, J. M., and Johnson, E. A., 1979, A kinetic model for determining the consequences of electrogenic active transport in cardiac muscle, *J. Theor. Biol.* **80:**405.

Coraboeuf, E., Delahayes, J., and Sjöstrand, 1969, A comparative study of K^{42} and Na^{24} movements during the cardiac cycle, *Acta Physiol. Scand.* **76:**40.

Crill, W. E., Rummery, R. E., and Woodbury, J. W., 1959, Effects of membrane current on transmembrane potentials of cultured chick embryo heart cells, *Am. J. Physiol.* **197:**733.

deBarry, J., Fosset, M., and Lazdunski, M., 1977, Molecular mechanism of the cardiotoxic action of a polypeptide neurotoxin from sea anemone on cultured embryonic cardiac cells, *Biochemistry* **16:**3850.

DeHaan, R. L., and Fozzard, J. A., 1975, Membrane response to current pulses in spheroidal aggregates of embryonic heart cells, *J. Gen. Physiol.* **65:**207.

Ebihara, L., Shigeto, N., Lieberman, M., and Johnson, E. A., 1980, The initial inward current in spherical clusters of chick embryonic heart cells, *J. Gen. Physiol.* **75:**437.

Eisenberg, B. R., Kuda, A. M., and Peter, J. B., 1974, Stereological analysis of mammalian skeletal muscle. I. Soleus muscle of the adult guinea pig, *J. Cell Biol.* **60:**732.

Eisenberg, R. S., and Engel, E., 1970, The spatial variation of membrane potential near a small source of current in a spherical cell, *J. Gen. Physiol.* **55:**736.

Fänge, R., Persson, H., and Thesleff, S., 1956, Electrophysiologic and pharmacological observations of trypsin-disintegrated embryonic chick hearts cultured in vitro, *Acta Physiol. Scand.* **38:**173.

Fischman, D. A., and Moscona, A. A., 1971, Reconstruction of heart tissue from suspensions of embryonic myocardial cells: Ultrastructural studies on dispersed and reaggregated cells, in: *Cardiac Hypertrophy* (N. Alpert, ed.), pp. 125–139, Academic Press, New York.

Galper, J. B., and Catterall, W. A., 1978, Developmental changes in the sensitivity of embryonic heart cells to tetrodotoxin and D600, *Dev. Biol.* **65:**216.

Ganote, C. E., Seabra-Gomes, R., Nayler, W. G., and Jennings, R. B., 1975, Irreversible myocardial injury in anoxic perfused rat hearts, *Am. J. Pathol.* **80:**419.

Goldman, Y., and Morad, M., 1977, Ionic membrane conductance during the time course of the cardiac action potential, *J. Physiol. (Lond.)* **268:**655.

Graves, J. S., 1979, Potassium transport in Chinese hamster ovary cells: Comparison of Rb-86 and K-42 as tracers, *J. Cell. Biol.* **83:**294.

Horres, C. R., Aiton, J. F., and Lieberman, M., 1979, Potassium permeability of embryonic avian heart cells in tissue culture, *Am. J. Physiol.* **236:**C163.

Horres, C. R., Aiton, J. F., Lieberman, M., and Johnson, E. A., 1979, Electrogenic transport in tissue cultured heart cells, *J. Mol. Cell Cardiol.* **11:**1201.

Horres, C. R., and Lieberman, M., 1977, Compartmental analysis of potassium efflux from growth-oriented heart cells, *J. Membr. Biol.* **34:**331.

Horres, C. R., Lieberman, M., and Purdy, J. E., 1977, Growth orientation of heart cells on nylon monofilament: Determination of the volume-to-surface area ratio and intracellular potassium concentration, *J. Membr. Biol.* **34:**313.

Hyde, A., Blondel, B., Matter, A., Cheneval, J. P., Filloux, B., and Girardier, L., 1969, Homo and heterocellular junctions in cell cultures: An electrophysiological and morphological study, in: *Progress in Brain Research, Vol. 31* (K. Akert and P. G. Waser, eds.), pp. 283–311, Elsevier, Amsterdam.

Johnson, E. A., and Lieberman, M., 1971, Heart: Excitation and contraction, *Annu. Rev. Physiol.* **33:**479.

Jones, A. W., 1975, Analysis of bulk-diffusion limited exchange of ions in smooth muscle preparations, in: *Methods in Pharmacology, Vol. 3* (E. E. Daniel and D. M. Paton, eds.), pp. 673–698, Plenum Press, New York.

Jongsma, H. J., and van Rijn, H. E., 1972, Electrotonic spread of current in monolayer cultures of neonatal rat heart cells, *J. Membr. Biol.* **9:**341.

Juncker, D. F., Greene, E. A., and Stish, R., 1972, Potassium efflux from amphibian atrium during the cardiac cycle, *Circ. Res.* **30**:350.

Kass, R. S., and Tsien, R. W., 1976, Control of action potential duration by calcium ions in cardiac Purkinje fibers, *J. Gen. Physiol.* **67**:599.

Katz, G. M., and Schwartz, T. S., 1974, Temporal control of voltage clamp membranes: An examination of principles, *J. Membr. Biol.* **17**:275.

Keynes, R. D., 1954, The ionic fluxes in muscle, *Proc. R. Soc. London* [*Biol.*] **142**:359.

Kobayashi, T., Ito, Y., and Rona, G., 1978, *Cardiac Adaptation, Recent Advances in Studies on Cardiac Structure and Metabolism, Vol. 12,* University Park Press, Baltimore.

Kootsey, J. M., 1975, Voltage clamp simulation, *Fed. Proc.* **34**:1343.

Kootsey, J. M., Vereecke, J., Shigeto, N., Lieberman, M., and Johnson, E. A., 1975, Deciphering nonlinear membranes: Distortion from cable length, *Biophys. J.* **15**:257a.

Lieberman, M., 1973, Electrophysiological studies of a synthetic strand of cardiac muscle, *Physiologist* **16**:551.

Lieberman, M., and Sano, T., 1976, *Developmental and Physiological Correlates of Cardiac Muscle,* Raven Press, New York.

Lieberman, M., Roggeveen, A. E., Purdy, J. E., and Johnson, E. A., 1972, Synthetic strands of cardiac muscle: Growth and physiological implication, *Science* **175**:909.

Lieberman, M., Horres, C. R., Purdy, J. E., and Halperin, L. R., 1978, Development of electrical activity in embryonic myocardial cells, in: *Fetal and Newborn Cardiovascular Physiology, Vol. 1, Developmental Aspects* (L. D. Longo and D. D. Reneau, eds.), pp. 237–255, Garland Press, New York.

Lieberman, M., Sawanobori, T., Kootsey, J. M., and Johnson, E. A., 1975, A synthetic strand of cardiac muscle: Its passive electrical properties, *J. Gen. Physiol.* **65**:527.

Lieberman, M., Sawanobori, T., Shigeto, N., and Johnson, E. A., 1976, Physiologic implications of heart muscle in tissue culture, in: *Developmental and Physiological Correlates of Cardiac Muscle* (M. Lieberman and T. Sano, eds.), pp. 139–154, Raven Press, New York.

Lieberman, M., Shigeto, N., Kootsey, J. M., and Johnson, E. A., 1975, Ionic currents in cardiac muscle, *Fed. Proc.* **34**:391.

Lipton, B. H., 1977, A fine-structural analysis of normal and modulated cells in myogenic cultures, *Dev. Biol.* **60**:26.

MacDonald, R. L., Mann, J. E., Jr., and Sperelakis, N., 1974, Derivation of general equations describing tracer diffusion in any two-compartment tissue with application to ionic diffusion in cylindrical muscle bundles, *J. Theor. Biol.* **45**:107.

McCall, D., 1979, Cation exchange and glycoside binding in cultured rat heart cells, *Am. J. Physiol.* **236**:C87.

McLean, M. J., and Sperelakis, N., 1976, Retention of fully differentiated electrophysiological properties of chick embryonic heart cells in culture, *Dev. Biol.* **50**:134.

McLean, M. J., and Sperelakis, N., 1974, Rapid loss of sensitivity to tetrodotoxin by chick ventricular myocardial cells after separation from the heart, *Exp. Cell Res.* **86**:351.

Mobley, B. A., and Page, E., 1972, The surface area of sheep cardiac Purkinje fibres, *J. Physiol. (Lond.)* **220**:547.

Moscona, A., 1952, Cell suspensions from organ rudiments of chick embryos, *Exp. Cell Res.* **3**:535.

Müller, P., 1965, Potassium and rubidium exchange across the surface membrane of cardiac Purkinje fibres, *J. Physiol. (Lond.)* **177**:453.

Noble, D., 1962, The voltage dependence of the cardiac membrane conductance, *Biophys. J.* **2**:381.

Noble, D., 1966, Applications of Hodgkin–Huxley equations to excitable tissues, *Physiol. Rev.* **46**:1.

Noble, D., and Stein, R. B., 1966, The threshold conditions for initiation of action potentials by excitable cells, *J. Physiol. (Lond.)* **187:**129.

Noble, D., and Tsien, R. W., 1972, The repolarization process of heart cells, in: *Electrical Phenomena in the Heart* (W. C. de Mello, ed.), pp. 133–161, Academic Press, New York.

Noma, A., and Irisawa, H., 1976, Membrane currents in the rabbit sinoatrial node cell as studied by the double microelectrode method, *Pflügers Arch.* **364:**45.

Polimeni, P. I., and Vassalle, M., 1970, Potassium fluxes in Purkinje and ventricular muscle fibers during rest and activity, *Am. J. Physiol.* **218:**1381.

Pomerat, C. M., 1954, Tissue culture technique in pharmacology, *Ann. N.Y. Acad. Sci.* **58:**971.

Purdy, J. E., Lieberman, M., Roggeveen, A. E., and Kirk R. G., 1972, Synthetic strands of cardiac muscle: Formation and ultrastructure, *J. Cell Biol.* **55:**563.

Reuter, H., and Seitz, N., 1968, The dependence of calcium efflux from cardiac muscle on temperature and external ion composition, *J. Physiol. (Lond.)* **195:**451.

Sachs, F., 1976, Electrophysiological properties of tissue cultured heart cells grown in a linear array, *J. Membr. Biol.* **28:**373.

Sachs, H. G., and DeHaan, R. L., 1973, Embryonic myocardial cell aggregates: Volume and pulsation rate, *Dev. Biol.* **30:**233.

Sperelakis, N., 1978, Cultured heart cell reaggregate model for studying cardiac toxicology, *Environ. Health Perspect.* **26:**243.

Trump, B. F., and Jones, R. T., 1977, Correlation of structure and active transport in the teleost nephron, *J. Exp. Zool.* **199:**365.

Trump, B. F., Mergner, W. J., Kahng, M. W., and Saladine, A. J., 1976, Studies on the subcellular pathophysiology of ischemia, *Circulation* **53:**I17.

van Zwieten, P. A., 1968, The use of ^{86}rubidium for the determination of changes in membrane permeability in the guinea pig atrail tissue, *Pflügers Arch.* **303:**81.

Weibel, E. R., 1973, Stereological techniques for electron microscopic morphometry, in: *Principles and Techniques of Electron Microscopy: Biological Applications* (M. A. Hayat, ed.), pp. 239–296, Van Nostrand Rheinhold Co., New York.

Index

Critical Essays on
RICHARD WRIGHT'S
Native Son

CRITICAL ESSAYS
ON
AMERICAN LITERATURE

James Nagel, General Editor
University of Georgia, Athens